DNA TOPOLOGY AND ITS BIOLOGICAL EFFECTS

Edited by

Nicholas R. Cozzarelli
University of California, Berkeley

James C. Wang
Harvard University

COLD SPRING HARBOR LABORATORY PRESS
1990

DNA TOPOLOGY AND ITS BIOLOGICAL EFFECTS

Monograph 20

Printed in the United States of America
Book design by Emily Harste

Library of Congress Cataloging-in-Publication Data

DNA topology and its biological effects / edited by
Nicholas R. Cozzarelli, James C. Wang.
p. cm. — (Cold Spring Harbor monograph series; 20)
Includes bibliographical references.
ISBN 0-87969-348-7
1. DNA—Conformation 2. DNA topoisomerase II. 3. DNA topoisomerase II.
I. Cozzarelli, Nicholas R. III. Wang, James C. III. Series.
[DNLM: 1. DNA Untwisting Proteins—physiology. QU 58 D6298]
QP624.D175 1990
574.87'3282—dc20
DLC 90-1540
for Library of Congress CIP

All Cold Spring Harbor Laboratory Press publications may be ordered directly from Cold Spring Harbor Laboratory, Box 100, Cold Spring Harbor, New York 11724. (Phone: 1-800-843-4388.) In New York State (516) 367-8325. FAX: (516) 367-8432.

DNA TOPOLOGY AND ITS BIOLOGICAL EFFECTS

COLD SPRING HARBOR MONOGRAPH SERIES

The Lactose Operon
The Bacteriophage Lambda
The Molecular Biology of Tumour Viruses
Ribosomes
RNA Phages
RNA Polymerase
The Operon
The Single-Stranded DNA Phages
Transfer RNA:
  Structure, Properties, and Recognition
  Biological Aspects
Molecular Biology of Tumor Viruses, Second Edition:
  DNA Tumor Viruses
  RNA Tumor Viruses
The Molecular Biology of the Yeast Saccharomyces:
  Life Cycle and Inheritance
  Metabolism and Gene Expression
Mitochondrial Genes
Lambda II
Nucleases
Gene Function in Prokaryotes
Microbial Development
The Nematode *Caenorhabditis elegans*
Oncogenes and the Molecular Origins of Cancer
Stress Proteins in Biology and Medicine
DNA Topology and its Biological Effects

Contents

Preface

Few subjects in biology can claim as broad an intellectual span as can DNA topology. The initial recognition of the formidable challenge that lay in the coiling and uncoiling of the two strands of DNA was made at the same time as the discovery of the double-helical structure of DNA. In the early 1960s, the discovery of double-stranded circular DNA and the finding that both strands of polyomavirus DNA and of SV40 DNA are continuous brought sharply into focus the unique topological problem of the separation of the two multiply-linked, single-stranded DNA rings during replication. The "swivel" requirement in the unlinking of the parental strands provided the impetus for the search for enzymes, now known as DNA topoisomerases, that transform the topology of DNA rings and loops.

Vinograd first had the insight that the three-dimensional shape of a duplex DNA ring is intimately related to the double-helical structure of DNA. He coined the terms "supertwisted," "superhelical," and "supercoiled" to describe the sinuous shape of a duplex DNA ring whose topology prevents it from assuming the more stable, relaxed form. Quantitative studies of supercoiled DNA received a boost when it became apparent that a theorem in differential geometry, derived in the late 1960s to describe the twist and writhe of a closed ribbon, was directly applicable to the description of supercoiled DNA.

The discovery of the DNA topoisomerases and their study in the last two decades have firmly established the importance of topology in all aspects of biological transactions involving DNA. The finding that topoisomerases are the targets of many antimicrobial and anticancer drugs has further expanded studies of DNA topology into clinical laboratories and hospital wards.

Despite the tendency of DNA topology to surface in all corners of the DNA world—an inevitable consequence of its deep roots in the double-helical structure of DNA—many students have only a vague notion of the topic, and many more view the discipline with awe and apprehension. Perhaps the reason lies in the term "topology"; the topological world has long been associated with fanciful forms, such as one-sided ribbons and bottles, unknotted knots, unlinked links, saucers that deform into tumblers, and teacups that deform into doughnuts. Few students in biology fully appreciate the connection between the abstraction in topology and the realism in biology.

At the 1986 Cold Spring Harbor meeting on DNA topology and its biological ramifications, the idea to compile a unique volume on this subject arose. Such a collection would help to introduce this ever-expanding field to students, as well as more established investigators, in disciplines ranging from mathematics and polymer chemistry to molecular biology and medicine. Four years later, this project has finally been realized.

The organization of this volume was designed to provide a structured introduction to the uninitiated, yet each chapter can be read as a separate unit for the more advanced. Authors were asked to write their chapters in a didactic style, operating on the assumption that the average reader would have little prior knowledge of the subject. The first chapter describes the structure of DNA and focuses on the major findings of the past decade. Chapters 2 and 3 introduce the topic of DNA bending. The importance of DNA bending first surfaced with the finding that double-stranded DNA is sharply curved around a histone core in the nucleosome, the basic structural unit of chromatin. DNA bending has since been found to influence transcription, replication, and site-specific recombination. The later chapters in this section discuss the basic topological and geometric parameters of DNA and the physiochemical properties of DNA rings. Particular emphasis is placed on supercoiled DNA and DNA structural transitions driven by supercoiling.

The first section of chapters focusing on the structural aspects of DNA topology is followed by a second group of chapters describing various topoisomerases. These enzymes are Nature's solutions to the topological problems of DNA. In the presence of topoisomerases, knotted DNA rings become unknotted, and linked DNA rings come apart in a display of biological sleight of hand. In the two decades since the discovery of these nimble enzymes, their ability to manipulate DNA strands has consistently fascinated biochemists and molecular biologists.

The final chapters describe the biological effects of DNA topology, presenting information gleaned primarily through studies on the biological roles of DNA topoisomerases, and closing with a discussion on DNA topoisomerases as targets of therapeutics, in particular antitumor agents. The available nucleotide sequences of DNA topoisomerases and their amino acid sequences are compiled in the Appendix.

This volume was not exempted from the usual perils associated with the publication of a book on a rapidly evolving subject. Several areas were omitted. Examples are the elegant biochemistry of site-specific recombination and the application of the mathematical theories of knots and tangles to recombinational topology, the higher order structure of chromosomes and the plausible roles of DNA topoisomerases in their organization, and the major role DNA topology has played in establishing that DNA is an interwound, right-handed double helix with a periodicity of 10.5 bp per turn. The effects of DNA topology on transcription, recombination, and repair were only touched on in passing and need fuller consideration.

The editors are particularly thankful to the authors for their scholarly contributions, for their patience during the process of putting the book together, and for their tolerance of our idiosyncrasies. We are also thankful to Jim Watson for his interest and support in the organization of the 1986 meeting and in the publication of this book; without his prodding, the editors could easily have abandoned both tasks. We are grateful to Nancy Ford, Managing Director of Publications, Ralph Battey, Technical Editor, and Joan Ebert, Editorial Assistant, at Cold Spring Harbor Laboratory Press, who expertly and painstakingly transformed a collection of chapters into a nicely printed volume. Finally, we thank all our past and present co-workers and all others in the field for making the study of DNA topology and its biological effects a wonderful and enjoyable pursuit.

J.C. Wang
N.R. Cozzarelli

DNA TOPOLOGY AND ITS BIOLOGICAL EFFECTS

1

New Approaches to DNA in the Crystal and in Solution

Horace R. Drew and Maxine J. McCall
CSIRO Division of Biotechnology
Laboratory for Molecular Biology
New South Wales, Australia

Chris R. Calladine
Department of Engineering
University of Cambridge
Cambridge CB2 1PZ, United Kingdom

I. INTRODUCTION

We review here the progress that has been made concerning the role of DNA in biology during the years 1979–1988. The first part of this chapter will deal with the use of X-ray crystallography to study the three-dimensional structure of DNA in single crystals at near-atomic resolution. This method led to the discovery of left-handed DNA (Pohl and Jovin 1972; Drew et al. 1978, 1980; Wang et al. 1979; Crawford et al. 1980) and also to the realization that the structure of normal, right-handed DNA depends on its nucleotide sequence (Dickerson and Drew 1981; Wang et al. 1982a; Shakked et al. 1983; McCall et al. 1985; Nelson et al. 1987).

 0-87969-348-7/90 $1.00 + .00

The second part of this chapter will describe how the results from X-ray crystallography made it possible to improve on the methods of studying DNA in solution using enzymes, chemicals, topological methods, "statistical sequencing," and gel electrophoresis. Such solution studies turned out to be essential in learning more about the properties of DNA at a length of 50–200 bp for which it is difficult if not impossible to form single crystals. In brief, it was found that (1) most of the X-ray structures of short pieces of DNA are relevant to long DNA in solution (Lomonossoff et al. 1981; Drew and Travers 1984, 1985a; Johnston and Rich 1985; Burkhoff and Tullius 1987), (2) the ability of DNA to fold into a tightly curved shape about proteins depends on its nucleotide sequence (Drew and Travers 1985b; Satchwell et al. 1986; Koudelka et al. 1987; Zahn and Blattner 1987), and (3) certain DNA molecules are stably curved even in the absence of protein owing to peculiarities in their base sequences (Marini et al. 1982; Diekmann and Wang 1985; Hagerman 1985, 1986; Griffith et al. 1986; Koo et al. 1986). Other methods of analysis of DNA in solution, for example by nuclear magnetic resonance (NMR) (Patel et al. 1982a, 1987; Lefevre et al. 1987) or by Raman spectroscopy (Nishimura 1986), are reviewed briefly.

The second section will also mention some early, promising work on several other subjects about which less is known: the tendency of thermal vibrations to be greater at some base sequences than at others (Gotoh and Tagashira 1981; Drew et al. 1985; Flick et al. 1986; McClellan et al. 1986), the evidence for Hoogsteen base-pairing in solution (Quigley et al. 1986; McLean and Waring 1988), and, finally, on the possibility for the design of an organic molecule that will recognize any particular sequence of bases depending on its exact construction (Kopka et al. 1985; Dervan 1986; Kissinger et al. 1987; Moser and Dervan 1987).

The third part of this chapter will summarize briefly some of the theories that have been put forward in recent years. These theories deal with (1) the general principles of DNA mechanics (Calladine 1982; Calladine and Drew 1984), (2) the sequence-dependent flexure of DNA around proteins (Calladine and Drew 1986), and (3) the stable curvature of DNA in solution (Calladine et al. 1988).

What was the spark that transformed the study of DNA and its physical properties from being an obscure subject just 10 years ago to being an area of active inquiry today? Perhaps, there will be no disagreement if we suggest that it was the discovery of left-handed DNA. This discovery came about in the following way. In 1972, F. Pohl and T. Jovin synthesized enzymatically a DNA polymer of sequence (dG-dC) and then characterized its spectral properties in solution. They found unexpectedly that the circular dichroism of poly(dG-dC) undergoes a reversal in sign as the salt concentration increases from 1 to 4 M. They postulated that the DNA molecule, in going from low to high salt, reverses its sense of coiling from right-handed to left-handed, since such a change would account for the circular dichroism results. (F. Pohl has informed us recently that he has been looking for left-handed DNA since 1967.)

When it became possible in the mid-1970s to synthesize DNA chemically in large amounts for the X-ray analysis of single crystals, the salt-induced transition of poly(dG-dC) was a problem of great interest. By 1978, crystals of both d(CGCG) and d(CGCGCG) had been grown (Drew et al. 1978; Wang et al. 1979). The X-ray patterns of these molecules in the crystal showed a helical form that had never been seen before: one with 12-bp per turn, a rise per base pair of 3.7 Å, and bases almost perpendicular to an imaginary helix axis. Also, the packing diameter of the helical DNA was an unusually small 18 Å, as compared with 20–23 Å for other forms of DNA.

II. DNA IN THE CRYSTAL

In 1979, A.H.-J. Wang (Wang et al. 1979) succeeded in preparing three heavy atom derivatives of d(CGCGCG) and solved its structure at 0.9 Å resolution. Aided by these results, Crawford et al. (1980) and Drew et al. (1980) solved two different crystalline forms of the tetramer d(CGCG) a few months later.

A. Left-handed DNA

These structures came as a shock to everyone, for all three turned out to be left-handed double helices. Figure 1 shows three d(CGCG) tetramers stacked end-on-end as a left-handed helix. One can see that the base pairs spiral in a counter-clockwise sense, completing a full 360° rotation after 12-bp steps (Fig. 1, bottom to top). The distance advanced along the helix axis after one turn is 12 x 3.7 Å = 44 Å. The helix is long and narrow, thereby accounting for its small packing diameter of 18 Å in crystals. The helical parameters for right-handed DNA are very different from those just mentioned. A typical right-handed helix (such as those described below) completes a full 360° turn after 10.5-bp steps and advances along the axis by 10.5 x 3.2 Å = 34 Å.

Are the left-handed structures of d(CGCG) and d(CGCGCG) equivalent to the low-salt or to the high-salt form of poly(dG-dC) studied by Pohl and Jovin (1972)? Studies of poly(dG-dC) by NMR (Patel et al. 1979) showed that the high-salt form of poly(dG-dC) has the symmetry of a dinucleotide, whereas the low-salt form has the symmetry of a mononucleotide. This was originally interpreted as evidence for a di-

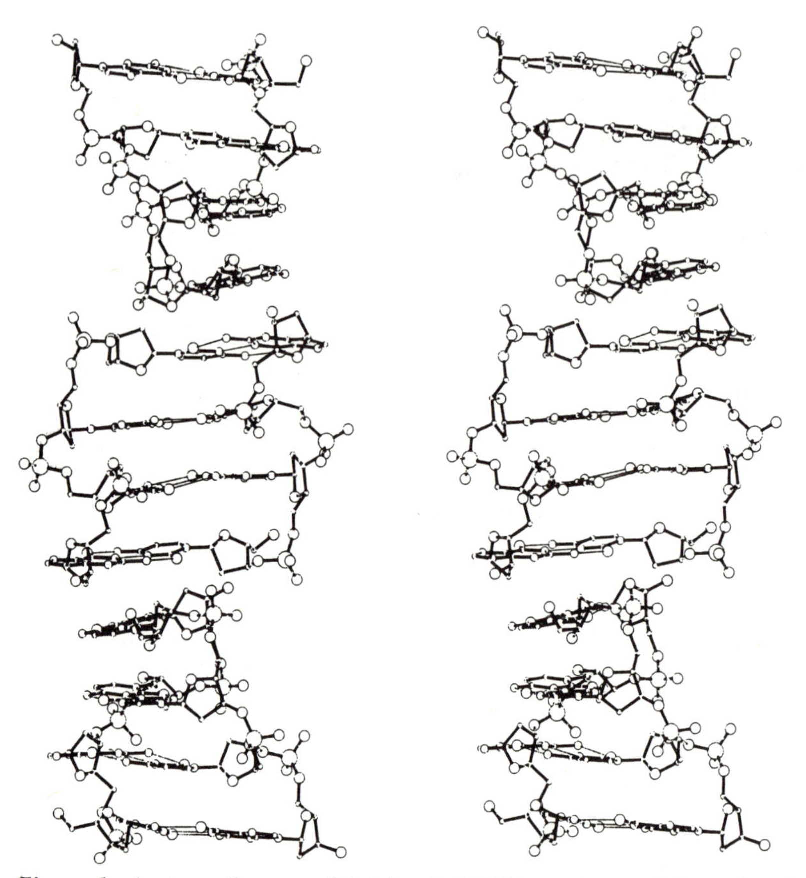

Figure 1 A stereodiagram of left-handed DNA, constructed from the single-crystal X-ray coordinates of three d(CGCG) tetramer double helices. This kind of left-handed helix is often called Z DNA because there is a strong alternation or "zig-zag" to the arrangement of successive base pairs.

nucleotide repeat at high salt in right-handed DNA; however, once the crystal structures were solved, it became clear that a left-handed helix would explain the experimental results nicely. A strong dinucleotide repeat arises from the structure shown in Figure 1 as a consequence of the striking alternation in helix twist angles, in going from one base-pair step to the next: The CpG step has a twist of just –15° as compared with –45° for the GpC step. Indeed, this left-handed helix is often called the "Z" form for its "zig-zag" nature. Subsequent NMR studies by Patel et al. (1982b) confirmed in detail that the low-salt form is right-handed and the high-salt form is left-handed. It is still not known how a change in salt concentration induces such a drastic change in helix structure.

Topological studies provide even more direct evidence that the high-salt form of poly(dG-dC) is left-handed. Normally, DNA in solution is a double helix with a structural periodicity of about 10 to 10.5 bases (Crick et al. 1979; Wang 1979). This is deduced through studying the relative mobilities of circular DNA molecules through agarose gels, where any addition of 10 bp to the length of the molecule leaves the overall DNA structure and hence the gel mobility nearly unaltered (Wang 1979). Also, it has been shown that the average DNA helix in solution is right-handed (Iwamoto and Hsu 1983). Finally, the oligo(dG-dC) helix at high salt or in negatively supercoiled plasmids adopts a helical sense opposite to that of normal right-handed DNA (e.g., Peck and Wang 1983). Other salt-induced transitions are known: for example, the transition of poly(dA-dT) that is induced by high concentrations of cesium fluoride (Vorlickova et al. 1980; Patel et al. 1981), but thus far the circular dichroism and NMR spectra have been unable to define clearly the structures that are involved, and no crystals of such sequences have yet been grown.

It is easy to show that salt per se is not necessary to get left-handed DNA. Thus, short segments of poly(dG-dC) can be induced to form left-handed DNA by the application of negative supercoiling in plasmids (Klysik et al. 1981; Nordheim et al. 1982; Peck and Wang 1983). The search for left-handed DNA in chromosomes has proved to be more difficult. Initial studies suggested that antibodies to the left-handed form of poly(dG-dC) would bind extensively to DNA within polytene chromosomes of the fruit fly. Later studies showed, however, that this was an artifact of treating the chromosomes with 45% acetic acid during their preparation and prior to adding the antibody (Hill and Stollar 1983). The strong acid not only denatures the DNA, but it extracts the histone proteins so as to leave the DNA under torsional and flexural stress. In fact, treatment of these chromosomes first with acid and then with a DNA-relaxing enzyme eliminates all traces of antibody binding (Hill and Stollar 1983).

Gross et al. (1985) have examined the conformation of long runs of oligo(dG-dT)$_n$ in total nuclear DNA. They find through the use of enzymes that these sequences are packaged quantitatively into nucleosomes in vivo and furthermore that the DNA within these nucleosomes is right-handed rather than left-handed. Thus, the most commonly occurring sequence, which has the potential to form left-handed DNA, does not in fact do so in vivo.

Peck and Wang (1985) have looked for left-handed DNA in bacteria, using as their assay the observation that a segment of left-handed poly(dG-dC) will cause termination of transcription by *Escherichia coli* RNA polymerase. They find that the polymerase stops in vitro when it reaches the (dG-dC) sequence in negatively supercoiled plasmids, but it does not stop in vivo when the same plasmid is introduced into a living cell. They interpret this to mean that the degree of DNA supercoiling in vivo is not sufficient to induce a conversion from right-handed to left-handed DNA. Other work has shown as well that the stress because of DNA supercoiling in living cells is relatively low (e.g., Bliska and Cozzarelli 1987) although it can be altered locally in the vicinity of genes that are being transcribed (Liu and Wang 1987).

Even if there is no strong evidence for left-handed DNA in vivo, could left-handed DNA arise transiently during various enzymatic processes? This is certainly a viable possibility, but so far the firm experimental evidence is lacking.

B. Right-handed DNA

Meanwhile, as the excitement over left-handed DNA subsided, the structure of right-handed DNA was being examined at higher resolution than in previous fiber X-ray studies. Inspired by the crystal structure of d(ATAT), which showed different sugar phosphate conformations at the adenine and thymine residues (Viswamitra et al. 1978), Klug and colleagues gathered evidence both from crystal and from solution to propose an alternating structure for poly(dA-dT). In the model of Klug et al. (1979), the ApT step has a twist angle of about 30° as compared with a twist of 40° for TpA.

This early suggestion of a sequence-dependent structure for DNA was substantiated shortly afterward. First, Wing et al. (1980) analyzed the crystal structure of a DNA dodecamer of sequence d(CGCGAA TTCGCG). In their preliminary report, they stated that the molecular structure of this DNA differed in two ways from that deduced by studying DNA in fibers: (1) The double helix was curved by 20° rather than straight, and (2) the base pairs were skewed by 10° to 20° in a propeller-like sense about their long axes rather than being perfectly flat. On refinement of the crystal structure at 1.9 Å resolution, Drew et al. (1981) showed that twist angles in different parts of the helix vary from 28° to 42°, about a mean of 36°. Hence, the variations of twist seen in this crystal were on the order of those proposed 2 years earlier in the model of Klug and colleagues for poly(dA-dT).

The culmination of this line of research came with the publication of two papers, by Dickerson and Drew (1981) and by Lomonossoff et al. (1981). Dickerson and Drew proved rigorously that the structure of d(CGCGAATTCGCG) in the crystal was influenced more strongly by its base sequence than by any other factor, such as its interaction with other molecules in the crystalline lattice. In simple terms, the structural variations seen in the crystal possess a strong, noncrystallographic, twofold symmetry that is due solely to the base sequence and is obscured only slightly by the curved helix axis. In the other paper, Lomonossoff et al. reported on the specific cleavage of three different molecules by the enzyme DNase I: poly(dA-dT), poly(dG-dC), and the dodecamer d(CGC GAATTCGCG). They first compared the cutting rate of the dodecamer in solution with its structure in the crystal and thereby found that DNase I cuts within this molecule most rapidly at steps of high helical twist. They then showed that DNase I cuts preferentially within the two polymers at steps ApT and GpC, respectively, rather than at steps TpA and CpG. They inferred from these data that both polymers have a sequence-dependent structure that is just as pronounced as in the dodecamer and furthermore that steps ApT and GpC were likely to have a higher helical twist (in the polymers) than do steps TpA and CpG. In their earlier model for poly(dA-dT), Klug et al. had tentatively assigned a low twist to ApT and a high twist to TpA for less direct reasons. Sequence-dependent effects were also observed in the crystal structures of DNA molecules of sequence d(GGCCGGCC) (Wang et al. 1982a), d(CCGG) (Conner et al. 1982), d(GGTATACC) (Shakked et al. 1983), and an RNA/DNA hybrid of sequence r(GCG)d(TATACGC) (Wang et al. 1982b).

McCall et al. (1985) determined the X-ray structure of d(GGGG CCCC) and, from the atomic coordinates of either d(GGGG) segment, derived a model for poly(dG)·poly(dC). Their model is shown in Figure 2A. One characteristic feature of the model is a lateral displacement of base pairs away from a central, imaginary helix axis. Whereas base pairs in classical right-handed DNA stack squarely on top of one another directly up the helix axis, base pairs in the model for poly(dG)·poly(dC) slide outward to take the shape of a broad coil. This creates in turn a deep central cavity within the confines of the helix, which can be seen clearly in the center of Figure 2A and also (from above) in Figure 2B. McCall et al. propose that it is the specific interaction of guanine rings, one above the other on the same sugar phosphate strand, that favors this kind of structure. In particular, they suggest that the five-membered ring of one guanine prefers to lie above the six-membered ring of its neighbor, as shown in Figure 2B. Thus, base pairs lie staggered sideways with respect to one another by 1–2 Å. Presumably, this reflects the preferred alignment of electrical charges within the rings as found, for example, in the calculations of Poltev and Teplukhin (1987). It is a simple matter to show that the staggered alignment of neighboring guanine rings, when repeated at each and every step of the helix, generates a structure of the kind shown in Figure 2A and B (Calladine and Drew 1984). Further evi-

A

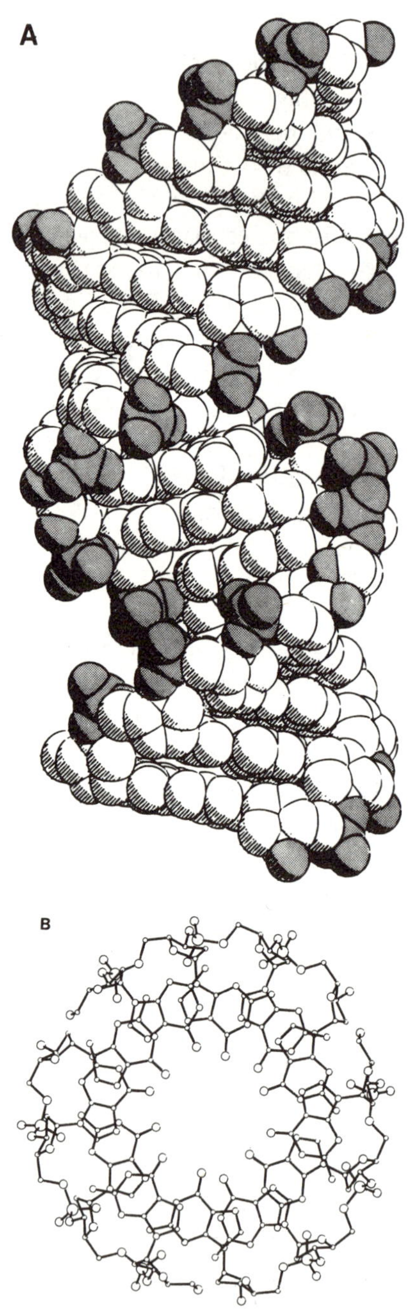

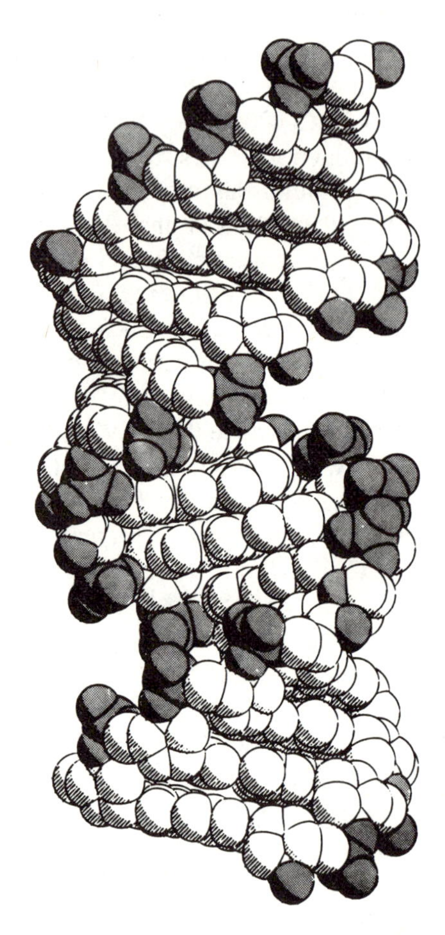

B

Figure 2 (See facing page for legend.)

dence for an outwardly coiled structure in sequences containing GpG has been presented by McCall et al. (1986) through determination of the crystal structure of d(GGATGGGAG) and its complementary strand d(CTCCCATCC).

Cruse et al. (1986) determined the X-ray structure of d(GCGCGC), the interpretation of which was complicated somewhat because every second phosphate contains a sulfur atom rather than an oxygen atom (i.e., a thiophosphate). Nevertheless, they observed a structure that alternates strongly in going from one base-pair step to the next. The alternation is not in the helix twist angle, as might be expected, but rather in the relative translation of base pairs along their short axes. Since the structural variations that Cruse et al. observe do not correlate either with the positions of sulfur atoms or with the identity of bases, it seems likely that this crystal structure represents just one of the many possible conformations that is available to d(GCGCGC) in solution.

C. Homopolymer (dA)·(dT)

One dramatic discovery about the structure of DNA was reported by Nelson et al. (1987). They analyzed the crystal structure of a DNA dodecamer of sequence d(CGCAAAAAAGCG) base paired to its complementary strand d(CGCTTTTTTGCG) (Fig. 3A,B). In Figure 3A, the six central A·T base pairs are shaded more darkly than the three G·C base pairs at either end. All of the A·T base pairs in the center are highly nonparallel about their long axes like the blades of a propeller. When we look at the central AT-rich region in more detail, as shown in Figure 3B, we can see clearly that the large adenine bases remain parallel to one another within their own strand. Similarly, the small thymine bases remain parallel to each other in their own strand. Yet, in going from one strand to its partner, the adenine and thymine bases within any base pair are angled with respect to one another by 20°. Studies of DNA in solution, using a variety of enzymes (Drew and Travers 1984, 1985a), have provided good evidence that the structure shown here is similar to that adopted in solution by both short and long runs of adenine bases. A care-

Figure 2 A stereodiagram of a model for poly(dG)·poly(dC), constructed from the single-crystal X-ray coordinates of 4 d(GGGG)·d(CCCC) tetramer double helices. (*A*) Side view, and the base pairs spiral outward from an imaginary central axis owing to their sideways stagger or "slide." (*B*) Top view, and the five-membered ring of each guanine base sits on the six-membered ring of a neighboring guanine base in what is called a "high-slide" arrangement. The cytosine bases, not shown in this figure, do not overlap one another at all.

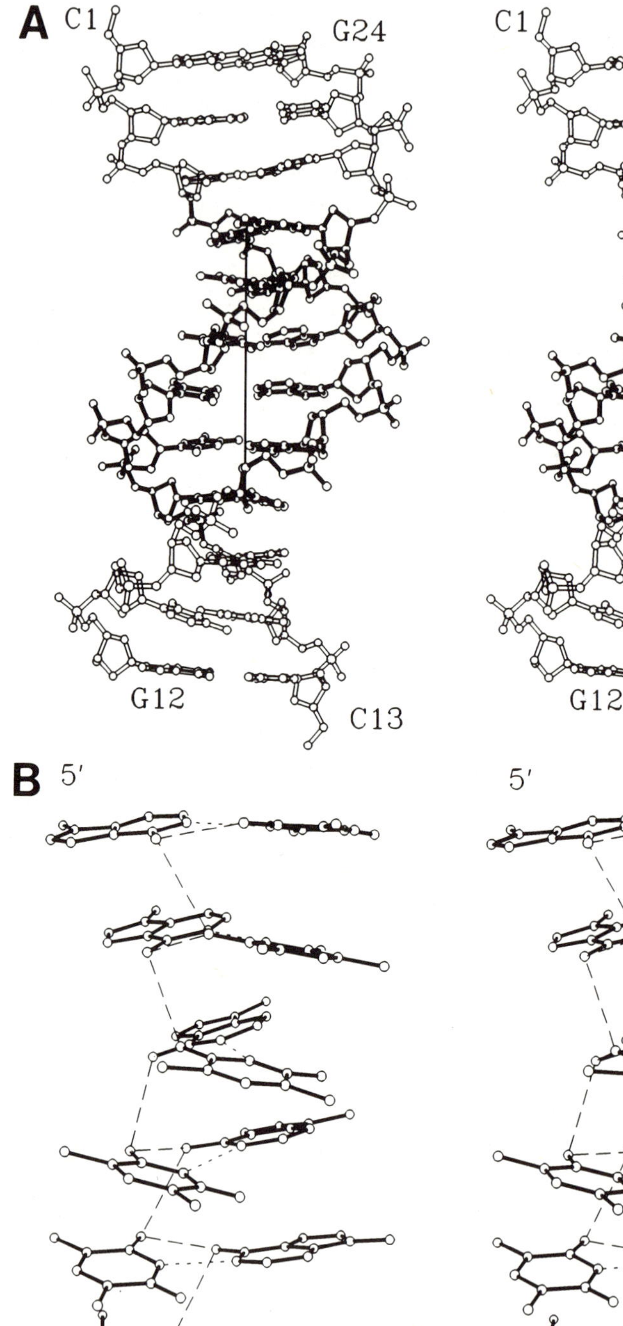

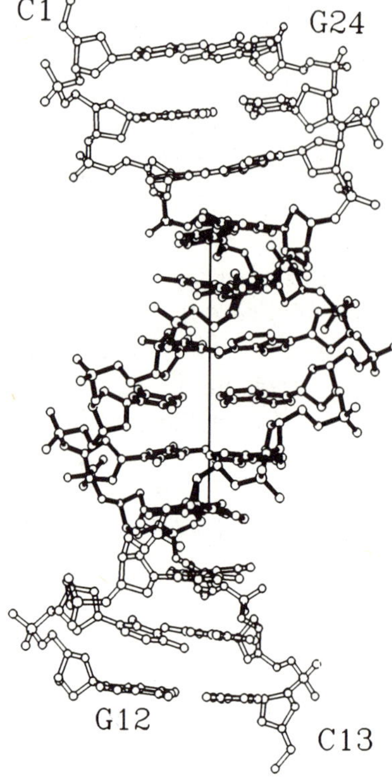

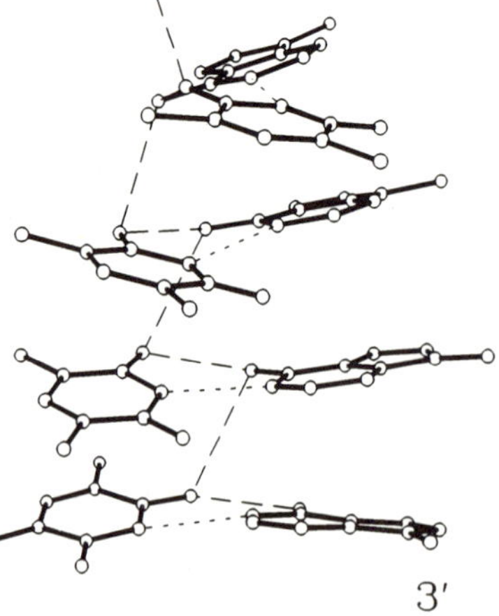

Figure 3 (*See facing page for legend.*)

ful study of the structure of poly(dA)·poly(dT) in solution by NMR methods also supports this conclusion (Lipanov and Chuprina 1987).

Several explanations may be offered for the high "propeller twist" at runs of adenine. Originally, it was thought that the tight coordination of several layers of ordered water within the minor groove of the AT-rich region could induce a large propeller twist and also cause the minor groove to become very narrow (Drew and Dickerson 1981; Fratini et al. 1982). A more important factor, however, might be the set of bifurcated hydrogen bonds that are located in the major groove of the AT-rich region and join adjacent bases (Fig. 3B). The base pairs are skewed so much that it is possible for the N_6 nitrogen of each adenine base to form hydrogen bonds with the O_4 oxygens of two different thymine bases. One hydrogen bond lies within the same base pair, and another connects to the base pair below. These bifurcated hydrogen bonds are represented as long, dashed lines in the figure to distinguish them from the usual Watson-Crick hydrogen bonds (drawn as short, dotted lines).

One reason why bifurcated hydrogen bonds are so attractive as an explanation for the unusual structure of poly(dA)·poly(dT) is that any sort of link between one base pair and the next should confer a certain rigidity on the double helix, and it is well known that homopolymer (dA)·(dT) is rigid. This was suspected first in the early studies of Rhodes (1979) and Simpson and Kunzler (1979), who found that neither homopolymer (dA)·(dT) nor homopolymer (dG)·(dC) could be wrapped about the usual set of eight histone proteins to form a "nucleosome core." More precise data concerning the relative flexural rigidities of (dA)·(dT) versus (dG)·(dC) were provided by Satchwell et al. (1986), who determined the base sequences of 177 different individual examples of nucleosome core DNA. They found that runs of adenine longer than about 10 bases tend to be located near either end of the nucleosomal DNA where the DNA is not very curved. Runs of (dG)·(dC) show no such distribution, but are located more generally throughout the curved regions of the core. Thus, although both homopolymers would appear to be rigid as compared with mixed-sequence DNA, the homopolymer (dA)·(dT) may be regarded as more flexurally rigid than the homopolymer (dG)·(dC), as evidenced by its inability to curve sharply around

Figure 3 A stereodiagram of the single-crystal X-ray structure of d(CGCA AAAAAGCG) and its complementary strand, courtesy of H.C.M. Nelson. (*A*) The central A·T base pairs are shaded more darkly than the terminal G·C base pairs. (*B*) One can see a close-up view of the d(AAAAAA)·d(TTTTTT) region and the set of bifurcated hydrogen bonds in the major groove connecting the N_6 nitrogen of each adenine to the O_4 oxygens of two different thymines.

the histone proteins. The special mechanical properties of poly(dA)·poly(dT) are likely to be relevant to the function of DNA in biological systems (Russell et al. 1983; Chen et al. 1987).

It is easy to verify that the special features seen in the crystal of Nelson et al. (1987) are possible only for a series of ApA or TpT steps and not for other sequences such as ApT or TpA. Presumably, it is the distinction between runs of adenine and other sequences that gives them their unusual helical repeat of 10.0 bp per turn versus 10.6 for other DNA (Peck and Wang 1981; Rhodes and Klug 1981), and it is this distinction that causes DNA containing many runs of adenine, located periodically once every helix repeat, to curve stably even in the absence of protein (Hagerman 1985; Koo et al. 1986; Calladine et al. 1988). Coll et al. (1987) have also observed bifurcated hydrogen bonds in a crystal of d(CGCAAATTTGCG) complexed with the antibiotic distamycin. Here again, the N_6 nitrogen of adenine is equidistant from two O_4 thymine oxygens, one within the same base pair and one on the base pair below.

D. DNA with Antibiotics

Before concluding, we should summarize some of the very informative work that has been done on the crystal structures of DNA bound to anticancer drugs and other antibiotics. Quigley et al. (1980) reported the first of these structures, which was of daunomycin bound to the sequence d(CGTACG): The drug intercalates twice within the sequence at both CpG steps (see also Wang et al. 1987). Wing et al. (1984) described the binding of cisplatin to d(CGCGAATTCGCG): Several different platinum atoms bind in the major groove in a monodentate fashion to the N_7 nitrogens of several different guanine rings. The investigators infer that the metal interacts also with the O_6 oxygen of guanine through the use of a bridging amino ligand or water molecule. Sherman et al. (1985) reported the structure of the same drug, cisplatin, bound in a bidentate fashion to the single-stranded dinucleotide d(GpG). It is not known which mode of binding of cisplatin to DNA is responsible for its ability to kill cancer cells. Viswamitra and Ramakrishnan (1987) determined the structure of mitoxanthrone bound to d(CpG) where it intercalates between adjoining dinucleotide double helices in the crystal. Kopka et al. (1985) studied the binding of netropsin to d(CGCGAATTCGCG) where it lies deep within the minor groove of the d(AATT) segment. Pjura et al. (1987) examined the binding of Hoechst 33258 to the same DNA and observed the drug in a similar location.

The most impressive of all these drug/DNA complexes has been the structure of triostin A bound to d(GCGTACGC) (Quigley et al. 1986).

The antibiotic triostin A contains two large aromatic rings so that it can intercalate twice per drug molecule and thereby form a kind of "sandwich" around the sequence CpG to which it binds strongly in solution (Waring and Wakelin 1974; Low et al. 1984). In the crystal, triostin A binds to the two CpG sequences near either end of the d(GCGTACGC) double helix in the expected way using both of its aromatic rings, but there are also some surprises.

A picture of the complex has been kindly provided to us by A.H-J. Wang and is shown in Figure 4. Each triostin A molecule unwinds the DNA by about 50°, leaving not so much a helix as a weakly twisted ladder. There are many specific contacts between the DNA and the drug, some of which are responsible for recognition of the sequence CpG. Also, the base pairs of the DNA on either side of the two darkly shaded aromatic rings of the antibiotic have forgone the usual Watson-Crick ge-

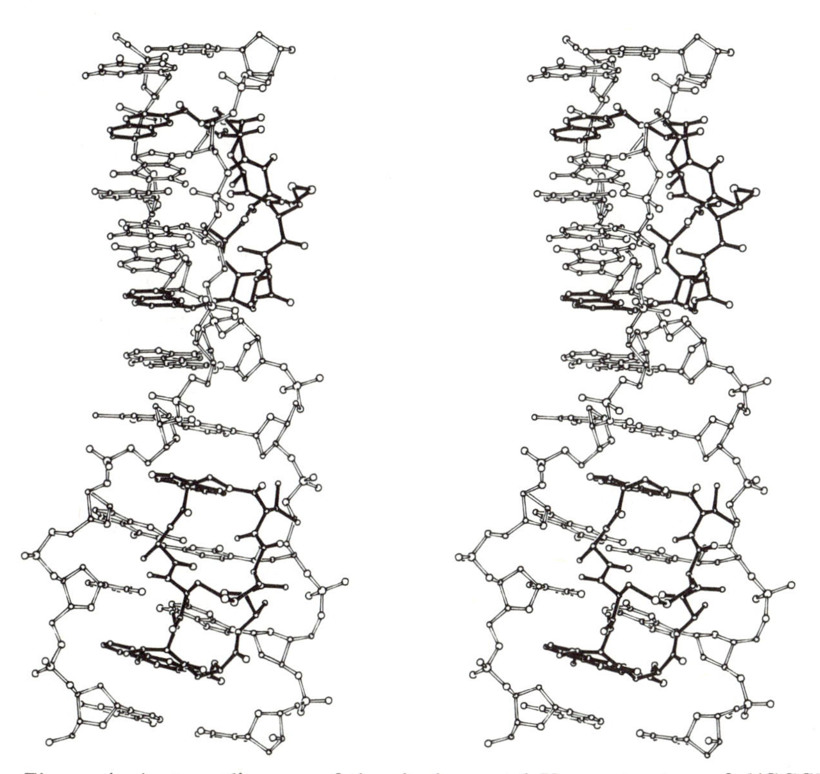

Figure 4 A stereodiagram of the single-crystal X-ray structure of d(GCGTA CGC) bound to two molecules of triostin A, which is a doubly intercalating antibiotic (courtesy of Dr. A.H.-J. Wang). The drug unwinds the helix by about 50° per drug molecule and also induces Hoogsteen base pairs in both G·C and A·T base pairs adjoining its binding site.

ometry in favor of a less-common Hoogsteen geometry. In a Hoogsteen base pair, it is the N_7 nitrogen of guanine or adenine that interacts with the N_3 nitrogen of cytosine or thymine. The advantage of a Hoogsteen pair in this particular situation would seem to be that it allows a close nonbonded fit between the sugar phosphate chains of the DNA and the peptide portion of the drug. There is some evidence that the binding geometry shown here in the crystal is relevant to the drug/DNA complex in solution (Mendel and Dervan 1987; Gao and Patel 1988).

E. Unusual Base Pairs

A number of studies have been made on short pieces of DNA that contain noncomplementary base pairs. For example, Brown et al. (1985) studied the pairing of guanine with thymine, Hunter et al. (1986) examined the pairing of adenine with cytosine, while two groups, Brown et al. (1986) and Privé et al. (1987), studied the pairing of guanine with adenine. In the last case, two different geometries for guanine with adenine were observed in two different crystals. The structure by Privé et al. (1987) shows an unusual hydrogen bond from the N_2 nitrogen of guanine in one base pair to the O_2 oxygen of thymine in an adjoining base pair. It also shows considerable variation in sugar phosphate torsion angles. This crystal diffracts to high resolution, and analysis of the native helix (without the mispairs) will prove informative. It is not clear at this time how any of these crystallographic studies relate to the "proofreading" of incorrect base pairs by cellular enzymes that are involved in DNA replication and repair.

III. DNA IN SOLUTION

Given that the methods for studying DNA in solution are generally qualitative and imprecise (with a few exceptions), why should anyone want to study DNA in solution at all? There are broadly two reasons for this: Most DNA molecules of biological interest, such as promoters and origins of replication, are difficult if not impossible to crystallize, and sometimes we are interested not in the three-dimensional structure of DNA per se but in the mechanical response of DNA to various torsional or flexural stresses that may be imposed, for example, by the binding of a protein. So, there are good reasons to develop methods for the study of DNA in solution. Nevertheless, since these methods are admittedly imprecise, it would be good to keep in mind the words of Huxley (1977), who said, "One of the great discoveries of modern times is the working hypothesis, which has replaced the idea of the dogma or the doctrine. We may form a hypothesis and be perfectly prepared to alter it as new facts appear; we do not have to stick to it through thick and thin and martyr other people because of it."

The experiments to be described below fall roughly into three categories: (1) those that probe the structure of DNA free in solution, (2) those that deal with the flexure of DNA bound to proteins, and (3) those that measure the mobility of DNA in gels as an indication of its intrinsic curvature. Other subjects, such as the thermal motion of DNA, alternative forms of base pairing, and the design of new DNA-binding ligands, are covered only briefly for reasons of limited space.

A. Methods for Probing DNA Structure in Solution

It is easy to measure gross changes in DNA structure, such as the "melting" of a double helix into its individual strands, by a variety of techniques. For example, one can measure how much ultraviolet light is absorbed by the DNA as a function of temperature in to determine the extent of base pairing. The problem has been to develop ways of studying DNA in solution that measure more subtle changes in structure, such as the conversion from a right-handed to a left-handed helix or variations within the structure of a right-handed helix because of its base sequence. Throughout the 1970s, scientists were largely limited to the methods of NMR and circular dichroism for studying DNA in solution. These techniques are useful in some applications, and indeed they were essential in the line of experimentation that led to the discovery of left-handed DNA (Pohl and Jovin 1972; Patel et al. 1979, 1982b), but they are not capable of determining the three-dimensional structure of DNA by themselves without reliance on some known structure as determined by X-ray crystallography.

By the mid-1970s, a few scientists were even questioning whether DNA was truly a helix. Some thought it was a kind of "flip-flop" structure, that turned right-handed for 5 bp, left-handed for another 5 bp, then right-handed again, and so on (Rodley et al. 1976). In other words, they thought that the two strands might not really be wound about one another at all. Few people took these ideas seriously, and yet they served a useful purpose in that they established how little we really knew about the structure of DNA in solution or in the fiber (Crick et al. 1979).

To provide stronger evidence concerning the helical nature of the molecule, two groups set out to measure what is now called the "helical periodicity" of DNA. Wang (1979) developed a method whereby one can insert short pieces of DNA into a long, circular plasmid and then tell from the relative mobilities of the original versus the modified plasmid in an agarose gel whether one has inserted an integral number of helix turns or some fraction thereof. By preparing inserts that varied in length from 1 to 60 bp, Wang was able to show that DNA in solution is indeed a helix and that it has a structural periodicity of about 10.4 bp. Rhodes and Klug (1980) took a different approach and laid long pieces of DNA down onto various inorganic surfaces such as mica or calcium phosphate (as in Liu and Wang 1978). They then treated the DNA with an enzyme that can cut at nearly every bond, if the DNA is free in solution. When the DNA is bound to a surface, however, the enzyme cuts much more slowly at bonds that lie close to the surface, as compared with bonds that lie far from the surface, and so the cutting pattern tells us the number of bonds between maximally exposed positions in the DNA. Rhodes and Klug performed this experiment for several different surfaces and for several different enzymes and in every case obtained a cutting periodicity of 10.6 bonds, which they took as the structural periodicity of the DNA. Peck and Wang (1981), after making some corrections to the gel mobility data, concurred with Rhodes and Klug that the helical repeat of DNA in solution is very close to 10.6 bp per turn. That is to say, the helix rotates about its central axis at a rate of $360^{\circ}/10.6 = 34.0^{\circ}$ on the average for each advance of 1-bp step (about 3.2 Å).

Early studies of DNA by fiber X-ray diffraction (Franklin and Gosling 1953) had led people to believe that DNA in solution would be the same as the "B" form, observed at high water content in fibers. The only other real choice seemed to be the "A" form, observed at low water content. Because of the close side-by-side packing of molecules in the fiber, the "B" form has an integral 10 bp per turn, whereas the "A" form has an integral 11. The measurements of Wang (1979) and Rhodes and Klug (1980) overturned these simple notions about the supposed equivalence of DNA in solution and in the fiber. They showed that the rate of twist of DNA in solution is 10.6 bp per turn and therefore not the same as in any of the previously observed fiber X-ray structures. The measurement of 10.6 as a helical repeat for DNA in solution also has strong implications for the interpretation of topological measurements concerning the path of DNA on the nucleosome (Klug and Lutter 1981).

The new methods led quickly to another discovery: The helical repeat of DNA is sequence-dependent. Peck and Wang (1981) and similarly Rhodes and Klug (1981) found that poly(dA)·poly(dT) has a helical repeat of just 10.0 bases, distinctly less than the values of 10.5 for poly(dA-dT), 10.6 for mixed-sequence DNA, and 10.7 for poly(dG)·poly(dC). Unfortunately, a measurement of helical repeat is of limited use in understanding the structure of DNA. It tells us about the mean twist of base-pair steps, but it does not tell us anything about other important aspects of base-pair geometry, such as roll, slide or propeller twist (Figs. 12 and 13). Nor does it tell us about the disposition of sugar phosphate chains relative to the bases, for example, whether the minor groove is wide or narrow. Therefore, other methods of probing the structure in solution had to be developed before our knowledge could progress further.

One key advance came with the investigation by Lomonossoff et al. (1981) of the specificity of the enzyme DNase I for different DNA sequences. As mentioned earlier, these investigators first compared the cutting specificity of DNase I within the sequence d(CGCGAATTCGCG) with the known structure of this molecule in the crystal (Dickerson and Drew 1981). They found that the enzyme cuts most rapidly at positions of high helical twist in this particular molecule. Then they measured patterns of DNase I cleavage for the DNA polymers poly(dA-dT) and poly(dG-dC), as well as for oligomers of sequence $(dG\text{-}dC)_n$, and found that DNase I cuts more rapidly at steps ApT and GpC than at steps TpA and CpG, respectively. They inferred from these data that both polymers possess a structure that is strongly sequence-dependent, just as in the dodecamer d(CGCGAATTCGCG) and furthermore that steps ApT and GpC in the polymers could possibly have a higher value of twist than do steps TpA and CpG. Thus, it seemed as if enzymes such as DNase I might provide a sensitive means of probing the structure of DNA in solution. The problems remaining were twofold: first, to investigate whether other enzymes in addition to DNase I would be useful as structural probes and, second, to calibrate the cutting specificities of these enzymes against the known structures of DNA as derived from X-ray crystallography.

With these ideas in mind, Drew (1984) investigated the structural specificities of five commonly used DNA nucleases. By measuring the rate at which a nuclease cuts the same base sequence when the DNA is folded into either a double helix or a hairpin loop, he was able to separate the influence of base sequence per se from that of helix structure and thereby show that many different proteins and chemicals can be used as probes of DNA structure in solution. From such measurements, he was able to show in detail that DNase I is sensitive to the width of the minor groove, DNase II to the conformation of an individual sugar phosphate strand, micrococcal and S1 nucleases to the exposure of a single strand, and copper phenanthroline to the availability of a base-pair step, perhaps for intercalation.

Drew and Travers (1984, 1985a) then extended these studies to long DNA molecules of biological origin, concentrating on the enzymes DNase I, copper phenanthroline, and DNase II. DNase I was an obvious choice because its cutting specificity had been shown earlier to depend on at least two different structural parameters: the local helix twist (Lomonossoff et al. 1981) and the width of the minor groove (Drew 1984). By measuring carefully patterns of cleavage from long double-helical DNA in solution, the investigators were able to show that DNase I cuts very poorly wherever it encounters a series of adenine bases on the same strand and also wherever it encounters a series of guanine bases on the same strand. The run of adenine remains resistant to DNase I if it includes an AT step, as in AAATTT, but not if it includes a TA step, as in TTTAAA. The run of guanine remains resistant so long as it remains homopolymeric, as in GGGGGG, and does not switch frequently between guanine and cytosine, as in GCGCGC. Some typical cutting data for DNase I are shown in Figure 5A to illustrate these principles.

Drew and Travers (1984) also correlated these observations with the structures of DNA as seen by X-ray crystallography. In the crystal, runs of adenine not containing the step TA are distinguished by their narrow minor grooves, that follow from the high propeller twist and bifurcated hydrogen bonds at a series of A·T base pairs (Fratini et al. 1982; Coll et al. 1987; Nelson et al. 1987). On the contrary, runs of guanine are distinguished by their wide minor grooves that follow from the low propeller twist and overlap of bases from opposite strands at a series of G·C base pairs (Wang et al. 1982a; McCall et al. 1985; Heinemann et al. 1987). If we suppose that DNase I prefers to bind a minor groove of intermediate width, or approximately 13 Å, as is typical of mixed-sequence DNA, one may suggest that the minor groove is too narrow (9 Å) at runs of adenine and too wide (16 Å) at runs of guanine to permit the optimal binding of DNase I. A model depicting these features is shown in Figure 6A. Changes in the width of the minor groove by external factors are known to cause changes in the rate of cutting by DNase I in accord with this model (Fox and Waring 1984; Low et al. 1984; Drew and Travers 1985a,b).

The crystal structure of DNase I bound to an octanucleotide of sequence d(GCGATCGC) shows that the model just described is correct, at least to a first approximation (Suck et al. 1988). In this crystal, the protein DNase I makes specific contacts along both strands of the DNA to phosphate groups and yet few if any contacts to base pairs. One can see therefore that the binding of DNase I to DNA will depend critically on where the phosphates are located in three-dimensional space. It is also observed, however, in this particular crystal that the minor groove of the DNA is slightly wider than expected, 15 Å rather than 13 Å. Does our model need to be changed so that the protein prefers to bind across a wide minor groove? Or is the minor groove wider than normal because

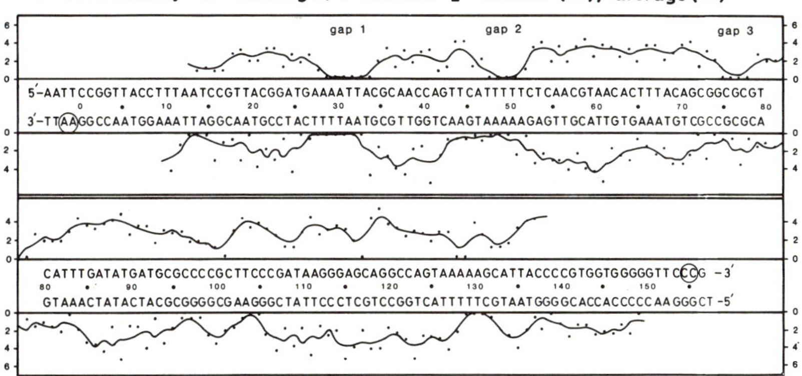

B ln(Probability of cleavage): DNAase II actual (.....), average (——)

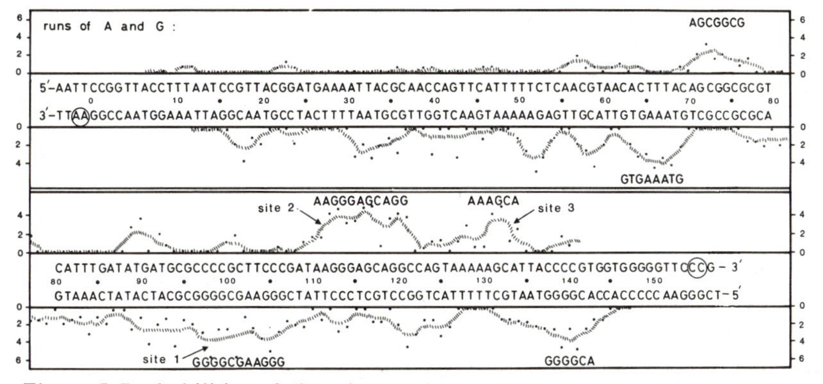

Figure 5 Probabilities of digestion by the enzymes DNase I and DNase II on the *tyrT* promoter from *E. coli*. All data are presented on a natural logarithmic scale so that the measured data span a range of cutting rates of exp(6.0) or about 400. (*A*) DNase I cuts poorly on both strands of the helix at long runs of AT or of GC. (*B*) DNase II cuts well on one strand of the helix at certain runs of purine bases.

of some other reason, perhaps because the DNA is cut by the enzyme before it forms a crystal? Further studies of DNase I with different DNA molecules of varied length and sequence should help to clarify this point.

The cutting pattern of copper phenanthroline is not shown, but it is much like that of DNase I. The width of helix grooves appears to influence the ability of this inorganic complex to intercalate between base pairs, if that is indeed its mode of binding.

Some typical cutting data for DNase II are shown in Figure 5B. DNase II cuts especially well within normal double-helical DNA at runs

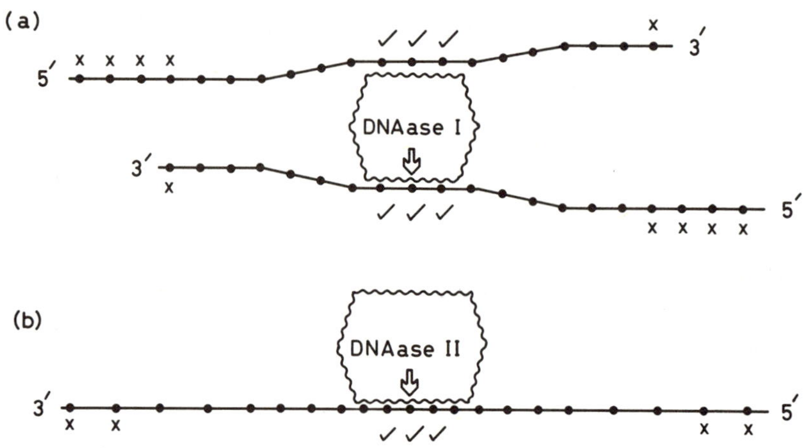

Figure 6 Models to explain the overall cutting specificities of DNase I and DNase II. (*a*) DNase I binds better to a minor groove of intermediate width rather than to a minor groove that is very narrow or very wide. (*b*) DNase II binds better to a sugar phosphate strand in which the phosphates are close together rather than far apart (as measured along the contour of the strand). Model *a* is supported by many sorts of experimental evidence, whereas model *b* has not yet been tested adequately.

of A and G (positions 100, 115, and 130) and yet especially poorly on the opposite strand at runs of T and C (in the same positions). In other experiments, where the structure of the helix has been altered by external factors, then the enzyme cuts also at mixtures of purine and pyrimidine bases and even at runs of pyrimidine (Drew and Travers 1984; Low et al. 1984). It seems therefore that the specificity of DNase II is for some particular conformation of the sugar phosphate strands and not for the identity of bases per se. What could DNase II see in a region of double helix containing a series of bases A and G? We do not yet have any single-crystal X-ray examples of such a helix for study, but one possible explanation derives from the common observation (e.g., McCall et al. 1985) that the purine bases A and G prefer to stack on one another more strongly than they stack on the pyrimidine bases T and C and much more strongly than the pyrimidine bases stack among themselves. Thus, the conformation of a double helix containing a series of bases A and G could be influenced by the strong stacking interactions of many successive purine bases, all on the same strand. In the model of Drew and Travers (1984), the purine strand becomes shorter than the pyrimidine strand as a consequence of these interactions, and it is this difference in conformation, shown in Figure 6B, that makes DNase II prefer to bind the purine strand rather than the pyrimidine.

How are we to account for another feature of the cutting patterns, which is the local variation in the rate of cutting when going from one base-pair step to the next? Indeed, it was this local variation in cutting that Lomonossoff et al. (1981) attributed to variation in local helix twist. The data of Figure 5, A and B, show these local variations clearly and provide enough new knowledge to demonstrate that the explanation of Lomonossoff et al. is correct but incomplete. For example, the step CG is known from X-ray studies always to have a low twist, and indeed it is usually cut by DNase I more slowly than its neighbors, as expected from the low twist. In a minority of cases, the step CG is cut more rapidly than its neighbors often when it is surrounded by other C and G bases, as in CCCGGG.

The explanation of Drew and Travers (1984) for these data is shown in Figure 7. There we can see that the orientation of the phosphate group relative to the rest of the helix is a variable quantity. In some places (Fig. 7, upper), the base pairs are arranged so that the surface of the phosphorus opposite O3′ faces outward into the solvent, whereas in other places (Figure 7, lower) the surface of the phosphorus opposite O5′ faces outward. Now DNase I and DNase II perform nucleophilic attack on the surface of the phosphorus opposite the bond they cut, which is O3′-P for DNase I or O5′-P for DNase II. It seems likely that such an attack will proceed most rapidly where the surface of the phosphorus required by the enzyme points out into the active site. In other words, with reference to Figure 7, DNase I should cut most rapidly at the upper phosphate and DNase II most rapidly at the lower. It has been experimentally verified that local rates of cleavage by DNase I and DNase II are anticorrelated with one another (Drew 1984; Drew and Travers 1984). The correlation between DNase I cutting rate and local helix twist, described previously by Lomonossoff et al. (1981), comes about because the favored phosphate geometry for DNase I often appears at sugar phosphate linkages of high twist in DNA crystals (Fratini et al. 1982; Cruse et al. 1986).

It is not possible to calibrate the cutting specificities of micrococcal nuclease and S1 nuclease in terms of the X-ray structures of double-helical DNA since these enzymes cut only where a single strand of the helix is temporarily unwound and exposed. Their use as probes of DNA unwinding will be discussed later.

In summary, we have explained how the cutting specificities of three different reagents, DNase I, copper phenanthroline, and DNase II, all provide a good means of measuring the structure of double-helical DNA in solution. As we study more carefully the cutting specificities of these reagents and calibrate their patterns of cleavage more precisely in terms

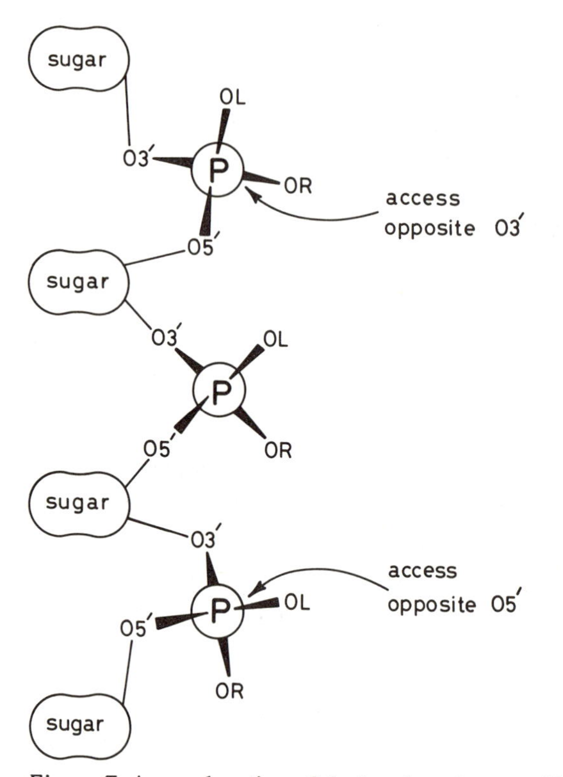

Figure 7 An explanation of the local cutting specificities of DNase I and DNase II. Depending on the base-pair geometries at the core of the helix, the phosphate group can rotate so as to leave either the surface of the phosphorus opposite O3′ accessible for cleavage, or else it can rotate so as to leave the surface opposite O5′ accessible for cleavage. DNase I cuts opposite O3′, whereas DNase II cuts opposite O5′.

of the structure of DNA seen in crystals, it may even become possible to calculate the observed patterns of cleavage from a knowledge of double-helical structure.

Now we have discussed three of the more important parameters that describe DNA structure in solution, namely the helical repeat (or twist), the width of the minor groove, and several different aspects of the sugar phosphate conformation. However, what about other parameters such as base-step roll, base-step slide, and propeller twist, which are necessary to describe completely the three-dimensional structure?

Calladine et al. (1988) have devised a way to obtain the roll angles of different base-pair steps for DNA in solution on the assumption that the conformation of DNA in solution is not greatly different from its conformation at low voltage within the pores of an acrylamide gel. From experimental data on the relative gel mobilities of curved versus straight DNA, they deduce that the roll angle differs by about 3° at steps AA/TT from that at other sequences. They then use the crystal structure of d(CGCAAAAAAGCG) by Nelson et al. (1987) to calibrate their model in absolute terms, and so they arrive at a roll angle of 0° for AA/TT, a roll of +3° for all other steps as a broad average, and a large roll of +7° for the step TA in the special context of a sequence such as TTTAAA. It is thought that the "clash" of adenine rings along their minor-groove edges causes the roll angle to be high in this instance (Calladine 1982).

No means are currently available for measuring either the slide or the propeller twist in solution with any accuracy. In the crystal, values of base-step slide have been observed to be larger for sequences containing GG/CC than for other DNA (McCall et al. 1985, 1986). Values of propeller twist in the crystal have been observed to be larger for homopolymer (dA)·(dT) than for other sequences (Nelson et al. 1987).

Many other studies of the cleavage of DNA by enzymes and chemicals have been carried out for the purpose of evaluating the helix structure in solution. We have space here to mention only a few of these. Burkhoff and Tullius (1987) have examined the cleavage of DNA by an inorganic complex of iron and EDTA. This complex cuts the phosphate indirectly by taking a proton from the deoxyribose ring of the sugar. The investigators find that the chemical cuts very slowly at runs of A in double-helical DNA, and they attribute this to the fact that the minor groove is very narrow at such sequences. They also suggest that the groove becomes wider at higher temperatures (Patel et al. 1982a; Drew and Travers 1984) and that the groove may be more narrow at the 3′ end of the run of A than at the 5′ end.

Johnston and Rich (1985) have studied the conformation of long tracts of oligo(dG-dC) and oligo(dG-dT) in supercoiled plasmids, using chemicals such as dimethylsulfate (DMS) or osmium tetroxide (OT). They find that both of these alternating purine/pyrimidine sequences adopt a left-handed form under strongly negative superhelical stress. Also, they describe many interesting structural effects that occur at the junctions of right-handed and left-handed DNA.

Enzymes and chemicals were also used to search for cruciforms in supercoiled DNA (Lilley 1984). It is found that the enzymes S1 nuclease and micrococcal nuclease cut near the tip of a cruciform, whereas T4 endonuclease IV and T7 endonuclease I cut at its base. Also, it has been shown that sequences near the very center of an inverted repeat have a greater influence on the rate at which a cruciform extrudes than do sequences elsewhere, suggesting that the cruciform starts as a small "bubble" at the center, which then grows larger (Murchie and Lilley 1987; Courey and Wang 1988).

Before concluding, we should mention briefly work that has been done using two other techniques, Raman spectroscopy and NMR. Nishimura (1986) has done the most thorough studies of DNA by Raman spectroscopy. He has calibrated his spectral lines in terms of the spectra of 30 different mononucleotides of known conformation in single crystals and then gone on to measure in solution the spectra from many different DNA polynucleotides and oligonucleotides of defined sequence. Thus, he has attempted to interpret the complicated spectra of long DNA molecules in terms of the simple spectra of their mononucleotide components. The technique has its limitations in that it can only look at vibrations in the sugar phosphate chain and cannot see the geometric overlap of the bases, but nevertheless, Nishimura has shown that these Raman spectra provide good evidence for variations in the structure of DNA because of its base sequence. He concludes that the structure of DNA in the crystal is very much like its structure in high-salt solution, owing to the fact that crystallization is a process of concentrating both DNA and its associated cations (Drew et al. 1980).

The literature on studies of DNA by NMR is far too extensive to be dealt with here. It has been reviewed elsewhere by Patel et al. (1982a, 1987). The technique of NMR is useful for studying in solution the close interactions between DNA and other kinds of molecules, such as antibiotics or proteins, but the method appears to be less useful for deducing the three-dimensional structure of DNA itself. The experimental data are limited to the measurement of very short (2–4 Å) proton-to-proton separations, and even these have not yet been interpreted reliably in terms of accurate proton-to-proton distances. The problems with this method have been discussed in detail by Lefevre et al. (1987).

Circular dichroism spectra of DNA have recently been calibrated in terms of the structures of DNA seen in single crystals (Fairall et al. 1989). These new measurements provide the first solid results obtained from such indirect spectroscopic studies.

B. The Sequence-dependent Flexure of DNA around Proteins

One of the most striking features of normal right-handed DNA is that its preferred curvature is determined largely by its base sequence. The main advance in this area came as a result of studying the coiling of DNA into nucleosomes. The nucleosome is a complex of 160–240 bp of DNA with nine polypeptides: two copies each of histones H2A, H2B, H3, and H4, and one copy of histone H1. It forms the simplest repeating unit of the chromosomes of higher organisms. Crystallographic studies of the protein/DNA complex (Finch et al. 1977; Richmond et al. 1984) have shown that the DNA coils into a flat, left-handed superhelix of almost two turns about eight of these protomers (all but H1). The DNA probably continues to coil about histone H1, as well, although this has not yet been demonstrated. In the late 1970s, there was a controversy over whether nucleosomes might occupy specific locations on the DNA. Some data argued in favor of this hypothesis, whereas other data argued against it (see Kornberg 1980). It was not until scientists began to reconstitute single nucleosomes onto DNA of defined sequence that the evidence swung in favor of the concept of "nucleosome positioning." Two studies in particular, by Simpson and Stafford (1983) and Ramsay et al. (1984), showed that the DNA would wrap specifically and reproducibly about the core octamer of histones (lacking H1), according to rules laid down in its nucleotide sequence. The problem was to determine what these rules might be.

We thought that certain aspects of the sequence-dependent structure of DNA, as seen in crystals, might be able to account for the results obtained with single nucleosomes. In particular, X-ray studies had shown that at runs of G and C the minor groove of the helix was exceptionally wide or could easily become wide (Wang et al. 1982a; McCall et al. 1985), whereas at certain runs of A and T the minor groove was narrow or could easily become narrow (Fratini et al. 1982; Nelson et al. 1987). In a curved DNA molecule, the grooves along the outside of the curve must be substantially wider on the average than those along the inside, and so one would expect that the wide minor grooves of the GC-rich regions should prefer to lie along the outside of any curved DNA, whereas the narrow minor grooves of the AT-rich regions should prefer to lie along the inside. The problem was to find a DNA molecule in which runs of G and C are typically half of a turn of helix away from runs of A and T, in order to test this hypothesis.

The *tyrT* (tyrosine tRNA) promoter from *E. coli* seemed to offer a natural example of that kind of DNA molecule (Lamond and Travers 1983). It contains runs of A and T at positions 10, 30, and 50, and also runs of G and C at positions 75 and 95, according to the numbering scheme shown in Figure 5A. Thus, several runs of A and T are located an integral number of helix turns away from one another and half of an integral number of turns away from runs of G and C. To test the hypothesis concerning the influence of base sequence on DNA curvature, we used the enzyme DNA ligase to close this 169-bp DNA into a covalently closed circle containing 16 double-helical turns, and then we used the enzyme DNase I to probe its structure (Drew and Travers 1985b). From the investigations of Rhodes and Klug (1980), concerning the cutting of

DNA on a surface, one expects that DNase I should cut bonds that lie along the outside of the circle faster than bonds that lie along the inside because the enzyme can gain access to the outer bonds more easily. Thus, prior to doing any experiment, we thought that DNase I would cut bonds in the small circle of *tyrT* DNA with a periodicity of 169/16 = 10.55 bonds.

The experimental data are shown in Figure 8. The experiment was performed by treating both the circular and the linear forms of *tyrT* DNA with DNase I in separate vials, measuring the pattern of digestion for each on a denaturing polyacrylamide gel and then quantifying these data by densitometry according to the method of Lutter (1978). Each individual point in Figure 8 represents the logarithm of cutting by DNase I in the circular form minus the logarithm of cutting by DNase I in the linear form, at each individual phosphate in the double helix. This procedure removes the component of sequence-specific cutting by DNase I to reveal the true accessibility of bonds in the circular DNA to the enzyme.

One can see immediately that the probabilities of cleavage in the figure are modulated with a wave-like periodicity of about 10.5 bonds. For example, the enzyme cuts well at positions 25, 36, 46, 57, and so forth, on the upper strand and at positions 22, 33, 43, 54, and so forth, on the lower strand. The uniform shift of three bonds between strands comes about because of the cutting characteristics of the enzyme. By taking the average of the data from the two strands, one can determine that the

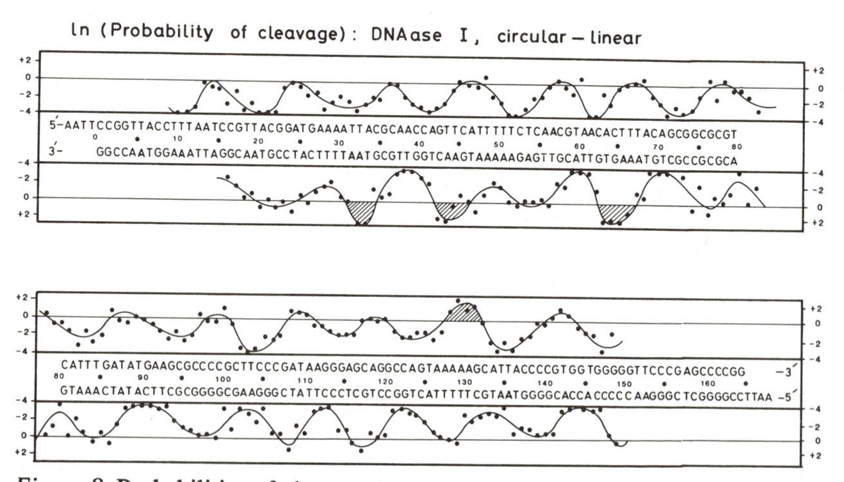

Figure 8 Probabilities of cleavage by DNase I for a 169-bp DNA molecule containing the *tyrT* promoter that has been closed covalently into a small circle. The measured data are plotted on a natural logarithmic scale and show the logarithm of cutting by DNase I in the circular form of this DNA minus the logarithm of cutting in the linear form.

minor groove of the DNA faces approximately outward in the circle at positions 23.5, 34.5, 44.5, 55.5, and so forth. The key point to note here is that all three of the AT-rich regions (at positions 10, 30, and 50) lie where the minor groove faces approximately inward, whereas both of the GC-rich regions (at positions 75 and 95) lie where the minor groove faces approximately outward. Not every run of G plus C or of A plus T can adhere to its own structural preferences owing to the topology of the circle and the difficulty of changing the twist of the DNA, and at position 130, we can find a run of A plus T where the minor groove points outward.

So, it seemed as if our hypothesis concerning the influence of base sequences on DNA curvature was confirmed. The next problem was to extend our results from a circle of free DNA to the same DNA curved around a nucleosome core. There the DNA wraps twice around a core of eight histone proteins into a very flat, highly curved, left-handed superhelix (Richmond et al. 1984). The number of base pairs in one turn of nucleosomal superhelix is about 80 or half that of the 169 bp in the *tyrT* circle. Thus, the extent of curvature within any double-helical turn is twice as great.

The results are shown in Figure 9 (Drew and Travers 1985b). Once again, we see that DNase I leaves a very periodic pattern of digestion with maxima in the difference plot occurring once every 10 to 10.5 bonds: Peaks in cutting can be seen at positions 25, 35, 46, 56, and so forth, on the upper strand, and at positions 22, 31, 42, and 53, and so forth, on the lower. The mean values for the two strands of 23.5, 33, 44, 54.5, and so forth, tell where the minor groove of the helix faces approximately outward. (The true situation is a bit more complicated than this because the protein does not cut exactly along the outermost surface of the DNA, but this simplified explanation is adequate for present purposes.) The key question was whether DNA bound to the histone octamer would adopt the same rotational setting as does free DNA in the *tyrT* circle. A comparison of Figures 8 and 9 shows that the rotational setting of the DNA is largely conserved in going from the circle to the nucleosome core. Thus, the minor grooves of the AT-rich regions at positions 10, 30, and 50 once again face approximately inward, whereas the minor grooves of the GC-rich regions at positions 75 and 95 face approximately outward. A few small differences in cutting can be detected on close comparison of these two plots, but most of these differences might be due to a change in helix twist from 10.56 to 10.2 bp per turn in going from the circle to the nucleosome (cf. Klug and Lutter 1981).

The tendency observed in the *tyrT* nucleosome, for AT-rich minor grooves to face inward and GC-rich minor grooves to face outward, was

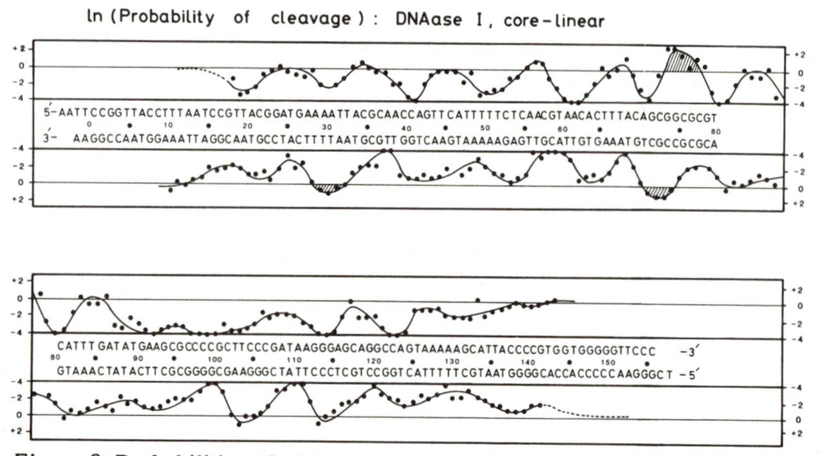

Figure 9 Probabilities of cleavage by DNase I for the same *tyrT* DNA wrapped about a histone octamer minus that of DNA free in solution. Compare with Fig. 8 to see that the maxima and minima of cutting remain in nearly the same positions, whether the DNA is curved into a small circle or else wrapped about a protein. The periodicity of cutting is slightly less here than in Fig. 8, 10.3 versus 10.56 bonds, owing to constraints imposed on the twist of the DNA by its wrapping about the protein.

seen independently by Rhodes (1985) in a reconstituted nucleosome of a frog gene for 5S RNA. It seemed important therefore to test whether the structural basis of sequence specificity might apply to all nucleosomes and perhaps in all instances where DNA wraps tightly about a protein. Our approach was to cut a mixture of billions of 145-bp molecules of nucleosomal DNA with DNase I obtained from digestion of chicken blood chromatin with micrococcal nuclease. DNA-binding drugs of known specificity are then used to protect the DNA from cleavage by DNase I wherever a binding site might be present (Drew and Travers 1985b). The results are shown in Figure 10. The lowest curve (Fig. 10a) shows the trace of core DNA without any enzyme. The next curve (Fig. 10b) shows the relatively featureless plot of the cutting of core DNA by DNase I in the absence of the drug. Curve c is of greater interest, and it shows the cutting of core DNA in the presence of an AT-specific drug called distamycin. Now the pattern of cleavage is modulated with a period of about 10 bonds with peaks appearing at positions 25, 36, 46, 56, and so forth, and valleys at positions 30, 41, 51, 61, and so forth. The valleys correspond to positions in the core DNA that contain a great many AT-rich regions that can bind distamycin and, hence, protect the DNA from cleavage. Finally, curve d shows cleavage by DNase I in the

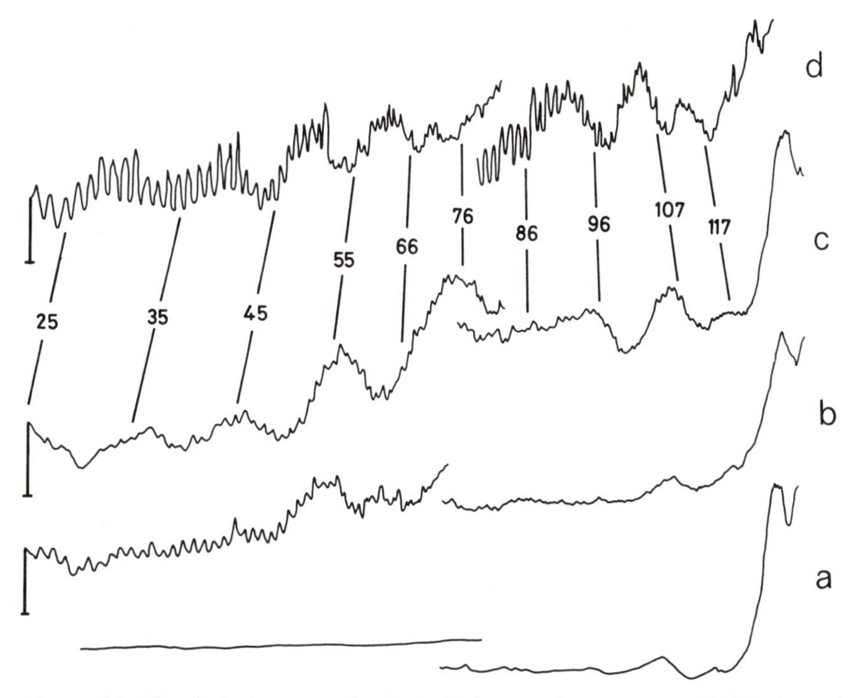

Figure 10 "Statistical sequencing" of chicken nucleosome core DNA through the use of DNase I and several small molecules of known DNA-binding specificity. (*a*) Core DNA with no DNase I; (*b*) the cutting of core DNA by DNase I in the absence of any ligand; (*c*) the cutting of core DNA by DNase I in the presence of distamycin; and (*d*) the cutting of core DNA by DNase I in the presence of chromomycin.

presence of a GC-specific ligand called chromomycin. Now, the pattern seen in curve c is inverted, with peaks at positions 30, 41, 51, 61, and so forth, and valleys at positions 25, 36, 46, 56, and so forth. The valleys correspond in this case to places in the core DNA that contain a great many GC-rich regions that can bind chromomycin.

By calibration of these data in terms of the known structure of DNA on the nucleosome core (Lutter 1978; Richmond et al. 1984), it is possible to demonstrate that the AT-rich regions indeed lie where the minor groove of the DNA points in toward the protein, whereas the GC-rich regions lie where the minor groove points outward from the protein. The next step was to determine the sequences of nucleosome core DNA in great detail. Results for 177 individual nucleosomes, containing a total of 25,000 bp, are summarized in Figure 11 (Satchwell et al. 1986). We constructed this figure first by separating all 25,000 bases of core DNA into their individual helix turns (each of 10.2 bases) and then by assigning a phase angle to each possible location within any helix turn. For example, a phase angle of 0° marks where the minor groove points outward; a phase of 180° marks where the minor groove points inward, whereas a phase of 270° marks where the minor groove lies along the top of the curved DNA. Finally, at each of 12 different phase angles, we counted how many times a given dinucleotide or trinucleotide appears and normalized the mean occurrence over all phases to 1.00 (Drew and Calladine 1987).

The 12 phase angles 0°, 30°, 60°, and so on, are listed along the top of the figure, whereas 10 dinucleotides and two special sequences, AGC and GGC, are listed along the right-hand side. It was necessary to split the dinucleotide GC into its trinucleotide components YGC (Y stands for pyrimidine), AGC, and GGC, because the 5′-flanking base seems to influence strongly the rotational preferences of this particular dinucleotide. One can see from the figure that GC-rich sequences such as GGC and CG (and also GG/CC to a lesser extent) are found most often at phase angles near 0°, where the minor groove points outward. In contrast, AT-rich sequences such as AA/TT (also AT and TA to a lesser extent) are found most often at phase angles near 180°, where the minor groove points inward. Thus, the extensive sequencing data provide a further confirmation of the previously mentioned results. In addition, they seem to provide a means of calculating the sequence-dependent flexure of DNA about any protein, at least with regard to the rotational setting of the DNA in a protein/DNA complex.

Before proceeding further, a few words of explanation are necessary. There are three different concepts to grasp in discussing the flexure of

0	30	60	90	120	150	180	210	240	270	300	330	360	
1.48	1.21	0.96	1.03	1.05	0.68	0.65	0.68	1.05	1.03	0.96	1.21	1.48	CG
1.23	1.06	0.95	1.07	1.02	0.84	0.88	0.84	1.02	1.07	0.95	1.06	1.23	TG/CA
0.84	0.89	0.83	0.96	1.14	1.23	1.05	1.23	1.14	0.96	0.83	0.89	0.84	TA
1.32	1.16	0.98	1.05	0.91	0.83	0.84	0.83	0.91	1.05	0.98	1.16	1.32	GG/CC
1.04	1.07	1.15	0.94	0.92	0.96	0.88	0.96	0.92	0.94	1.15	1.07	1.04	AG/CT
1.04	1.02	1.01	0.98	0.99	1.02	0.91	1.02	0.99	0.98	1.01	1.02	1.04	GA/TC
0.66	0.69	0.86	1.03	1.15	1.29	1.30	1.29	1.15	1.03	0.86	0.69	0.66	AA/TT
0.95	0.93	0.92	0.96	0.97	1.09	1.34	1.09	0.97	0.96	0.92	0.93	0.95	AT
0.91	1.10	1.14	0.96	0.96	0.90	0.98	0.90	0.96	0.96	1.14	1.10	0.91	GT/AC
1.07	1.07	1.22	0.75	0.96	1.08	0.76	1.08	0.96	0.75	1.22	1.07	1.07	(Y)GC
0.96	1.33	1.16	1.02	0.81	0.82	0.77	0.82	0.81	1.02	1.16	1.33	0.96	(A)GC
1.28	1.48	1.28	1.07	0.66	0.50	0.73	0.50	0.66	1.07	1.28	1.48	1.28	(G)GC

Figure 11 Matrix of rotational preferences based on aligned chicken sequences, with phase angle in degrees.

DNA. These are (1) isotropic flexibility, (2) anisotropic flexibility, and (3) intrinsic curvature. By the term isotropic flexibility, we mean that the DNA is equally flexible (either stiff or bendable) in all directions. For example, the sequence homopolymer (dA)·(dT) is stiffer than any other and generally avoids curved regions of the nucleosome entirely (Satchwell et al. 1986). By the term anisotropic flexibility, we mean that the DNA contains "hinges" (like a finger or a knee) that allow it to bend in one direction but not another. For example, the sequence GGC/GCC prefers to be bent in a direction that allows it to close the major groove and open the minor groove. Last, by the term intrinsic curvature, we mean that the DNA is curved stably in the absence of any protein owing to its sequence-dependent structure.

This discussion is relevant to use of the data listed in Figure 11. By averaging the results of Satchwell et al. (1986) into just a single helix turn, we have eliminated any information about isotropic flexibility, such as the known stiffness at runs of (dA)·(dT). Furthermore, since most molecules of nucleosome core DNA are not themselves bent in the absence of protein, it seems clear as well that the idea of intrinsic curvature does not apply to these data. What the figure does contain is precise, detailed information about the anisotropic flexibility of DNA bound to protein. This information has been presented in a form that is suitable for calculation of the sequence-dependent flexure (that is, anisotropic flexure) of DNA about repressors, activators, proteins involved in replication, and polymerases.

Two examples described below illustrate how the data in Figure 11 can be used to calculate protein/DNA specificity. Anderson et al. (1987) have determined the X-ray structure of a complex between the 434 repressor protein and a 14-bp DNA double helix of sequence d(AC AATATATATTGT). In this complex, the protein contacts the bases within both major grooves of the DNA at either end of the 14-mer but does not contact any bases within the minor groove at center. Still, the sequence of bases in the central region seems to contribute to the overall specificity of the protein/DNA interaction, as evidenced by the data of Koudelka et al. (1987) shown in Table 1. Replacement of the central d(ATAT) by d(GTAT) weakens the binding affinity by a factor of 2.5, replacement of d(ATAT) by d(GTAC) weakens it by a factor of 5, whereas replacement of d(ATAT) by d(ACGT) weakens it by a factor of 50. How is this accomplished? Koudelka et al. note that the DNA curves substantially about the 434 protein so as to close the minor groove at the sequence d(ATAT), and they point out that replacement of the AT-rich sequence by a sequence of mixed base composition should make it more difficult for the protein to induce curvature in the DNA. One can calcu-

Table 1 Relative binding affinities of 434 repressor for different DNA sequences

DNA sequence	Amount of protein required for 50% binding
ACAAT/AAAT/ATTGT	0.3
AATT	1.0
ATAT	1.0
TTAA	1.5
GTAT	2.5
GTAC	5.0
CTAG	5.0
AGAT	7.0
AGCT	50.0
ACGT	50.0

Data taken from Koudelka et al. (1987).

late that replacement of d(ATAT) by a GC-rich sequence such as d(GCGC) would weaken the binding by a factor of about 300.

In Figure 12A, we compare the likelihoods of flexure for all of the sequences listed in Table 1 (as calculated from the data shown in Fig. 11) with their relative binding affinities for 434 protein. The assay that Koudelka et al. (1987) use to measure binding is the concentration of protein necessary to obtain half-maximal protection from cleavage by DNase I; hence, the term $C_{1/2}$ on the left-hand side of the plot. Also, in this example the likelihood of flexure is calculated solely for the specific, curved geometry that is found in the X-ray structure of the 434/DNA complex; it is assumed that the structure of the complex will not change as the central sequence is altered. Subject to these assumptions, there is a good correlation between the likelihoods of flexure of many different DNA molecules and their ability to bind the protein. Data are shown also for the related Cro protein although the structure of its complex with DNA has not yet been determined. The shallower slope of the line for Cro protein than for 434 suggests that the DNA may not be bent so tightly in its complex with Cro as compared with its complex with 434. When trying to apply this approach to the naturally occurring sequences of DNA that are bound by 434 and Cro, one runs into trouble because base substitutions are not limited to the central four bases but extend further outward from the center into regions of DNA where the base pairs are contacted directly by the protein.

In a second example, Gartenberg and Crothers (1988) have studied the curvature of DNA when it is bound to the catabolite gene activator protein (CAP) of *E. coli.* Although they do not know the structure of the complex between DNA and CAP, they believe for various reasons that the DNA curves over 30 bp along one side of the protein. Furthermore, they think that some DNA molecules prefer to curve more tightly around this protein than do others, depending on the base sequence. To measure how tightly any given DNA molecule curves around the protein, they pass the complex of DNA and CAP through a gel of small pore size under the influence of an electrical field. Their hypothesis is that highly curved DNA molecules go more slowly in such a gel.

In Figure 12B, we have plotted the relative mobilities through a gel of complexes between the CAP protein and 12 related DNA molecules of the same length but slightly different base sequence. Then we compare these experimental values with the likelihood of flexure for each DNA as calculated from sequences of the nucleosome core (Fig. 11). The direction of bending in the calculation is that favored by Gartenberg and Crothers (1988), where the minor groove of the DNA points in toward the protein at the center of the binding site. There is a good correlation between the extent of curvature as measured by gel mobility and the ease of flexure of DNA about the protein. Only 1 of the 12 points does not lie along the smoothly fitted line: In this case, the base sequence of the DNA differs from the average only at the very edge of the proposed binding site for the CAP protein where it is not certain that the helix is curved. In summary, the sequence characteristics of nucleosome core DNA, as shown in Figure 11 account quantitatively for the flexure of DNA about two unrelated proteins, the 434 repressor and CAP.

How does this new information about DNA structure relate to the function of DNA in biological systems? One point that should be made is the following: It is now possible to derive information about the preferred curvature of DNA solely from its base sequence. For example, Amouyal and Buc (1987) have provided topological evidence to suggest that the DNA winds for one or more superhelical turns about *E. coli* RNA polymerase prior to strand separation and the initiation of transcription. If this is true, then alterations to the sequence of the DNA in regions flanking the promoter should have a measurable influence on the ability of the DNA to assume such a curved shape. In fact, two studies have shown that the curvature of DNA over a broad region of the promoter, extending upstream of and including the primary sites of contact for *E. coli* RNA polymerase, has a dramatic influence on the amount of RNA made by the polymerase enzyme in living cells (Bracco et al. 1989; Collis et al. 1989).

It should be mentioned that the matrix shown in Figure 11 accounts successfully for the rotational setting of DNA in a dozen different examples of single nucleosome cores, but problems are encountered when attempting to deal with long arrays of nucleosomes (Drew and Calladine 1987; Drew and McCall 1987). Nevertheless, certain long arrays of nucleosomes are known to be positioned specifically with respect to the

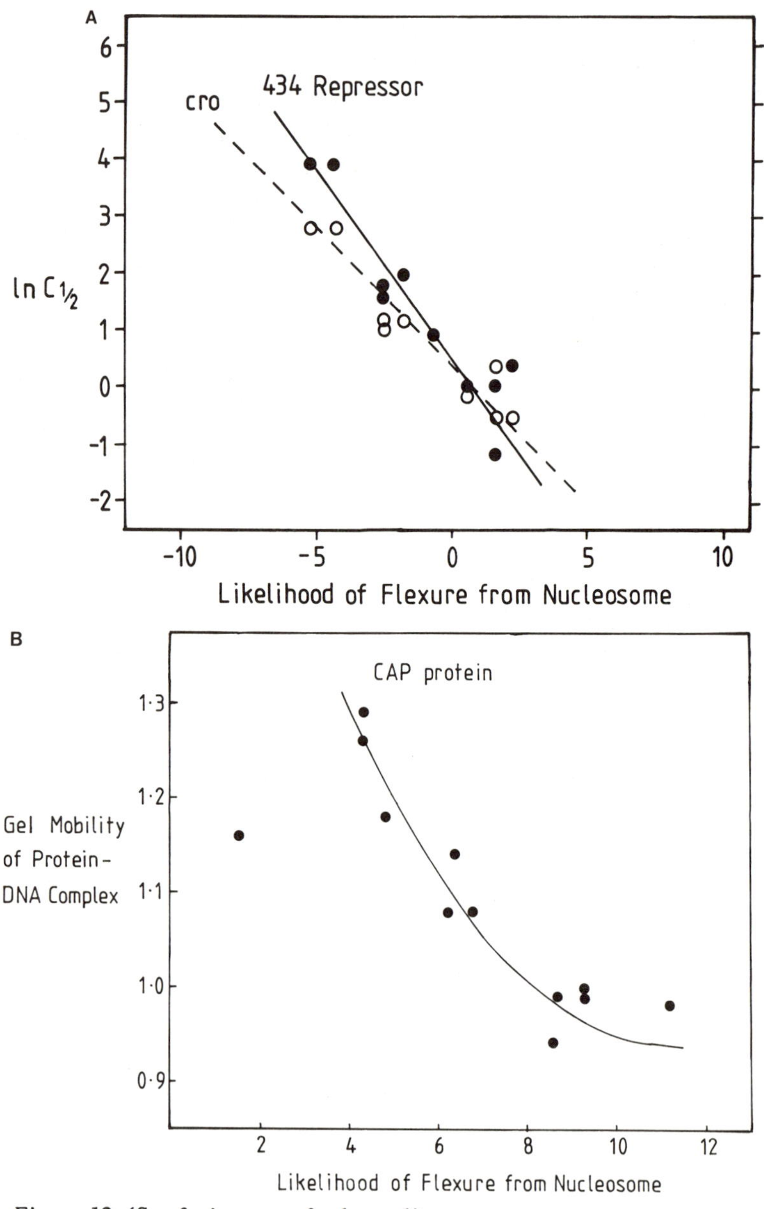

Figure 12 (*See facing page for legend.*)

base sequence (Cartwright and Elgin 1984; Benezra et al. 1986). In an extreme example, Hsieh and Griffith (1988) have shown that a single nucleosome will assemble preferentially over a segment of kinetoplast DNA (which is both intrinsically curved and anisotropically flexible) in the midst of a long DNA molecule.

C. The Intrinsic Curvature of DNA in Solution

This section will be kept brief since the topic of intrinsic curvature has been much discussed elsewhere. Basically, it should be clear by now that the structure of normal, right-handed DNA depends on its base sequence as evidenced by studies in the crystal and also studies in solution. The strongest sequence-dependent perturbation to the structure appears to come from runs of A or runs of T. Crystallographic analysis of a d(AA AAAA)·d(TTTTTT) segment of double helix shows that the base pairs are highly twisted in a propeller-like sense about their long axes, that the minor groove is very narrow, and that adjoining A·T base pairs form hydrogen bonds to one another via their major-groove edges (Nelson et al. 1987). When short runs of A or runs of T are repeated regularly, once every helix turn or thereabouts, then the DNA may be stably curved in the absence of any protein. This comes about as a consequence of the distinction in structure between runs of A and the other sequences.

The primary evidence for intrinsic curvature has been that such DNA molecules migrate more slowly than their mixed sequence counterparts through polyacrylamide gels in which the pores are very small. Presumably, the curved DNA molecules sweep out a larger volume, and consequently they can pass through fewer holes, than straight DNA of the same molecular weight. The altered mobility of intrinsically curved DNA through acrylamide gels has been studied by a large number of investigators, for example Marini et al. (1982), Diekmann and Wang (1985), Hagerman (1985, 1986), and Koo et al. (1986). The intrinsic cur-

Figure 12 (*A*) A plot of the logarithm of the concentration of 434 protein necessary to half-fill its binding site on DNA versus the likelihoods of flexure for many different DNA sequences. Experimental data were taken from Koudelka et al. (1987), and the calculation of flexure was from Figure 11. A plot is shown also for the Cro protein binding to DNA. (*B*) A plot of the mobility through a gel of a complex between the CAP protein and DNA versus the likelihoods of flexure for many different DNA sequences. The highly curved DNA molecules move more slowly through the gel than do the less curved molecules (Gartenberg and Crothers 1988).

vature of DNA has also been studied by electron microscopy (Griffith et al. 1986). A theory to explain all of the gel mobility data for repeating-sequence DNA in terms of the structures that have been observed by X-ray crystallography has been provided by Calladine et al. (1988) and is mentioned below. Currently there is no theory to predict the mobility of curved DNA of nonrepeating sequence because its shape is so complex.

It should be pointed out that the passage of DNA through a gel in the limit of low applied voltage is not expected to impose any significant forces on the intrinsic structure of the DNA, and it is by this method that the "intrinsic curvature" is commonly measured. In contrast, the wrapping of DNA around a protein imposes very significant forces on the intrinsic structure of the DNA, and it is the anisotropic flexibility of the structure that determines its response to these forces. Whereas DNA passing through a gel is generally curved by no more than 10° per helix turn, DNA bound to a protein is often curved by 20° to 50° per helix turn, so we see that intrinsic curvature per se cannot account for the specific wrapping of DNA around proteins unless the required curvature is very slight. Furthermore, the twin phenomena of intrinsic curvature and anisotropic flexibility follow rather different rules about the involvement of DNA base sequence, and they have had to be dealt with thus far in terms of separate theories (Calladine and Drew 1986; Calladine et al. 1988).

D. Other Subjects

We shall discuss here briefly three subjects for which no theoretical foundations are currently available. First, it is clear from many kinds of experiments that certain parts of a DNA double helix unwind more easily than do others. Studies of DNA by thermal denaturation, for example, have shown that the dinucleotide TpA unwinds from its double-helical form more easily than do the dinucleotides ApT, ApA, or TpT, even in the midst of mixed-sequence DNA (Gotoh and Tagashira 1981). Studies of DNA at room temperature but under the influence of negative superhelical stress reach basically the same conclusions through the use of S1 nuclease (Drew et al. 1985). Two different DNA-cutting reagents, copper phenanthroline and micrococcal nuclease, have both been found to cut preferentially at TpA sequences in linear DNA, and it is known that both reagents require helix unwinding for their activities (Flick et al. 1986). Studies of oligo(dA-dT) using chemicals imply that it is more torsionally deformable than other sequences (McClellan et al. 1986). In supercoiled plasmids at low salt and moderate temperature, the sequence oligo(dA-dT) adopts an apparently helical structure that is definitely not a cruciform but is somehow unwound relative to normal right-handed DNA (McClellan and Lilley 1987).

The widespread occurrence of TpA and TATA sequences at the sites of DNA unwinding in promoters, origins of replication, and cross-over points for recombination (Drew et al. 1985; Murchie and Lilley 1987) is therefore no accident. The sequence TATA can unwind more easily than others for structural reasons that are not yet fully understood. Indeed, Spassky et al. (1988) have drawn a good correlation between the abilities of bacterial promoters to unwind in this region (as measured by nuclease digestion) and their abilities to bind RNA polymerase.

A second subject concerns the question of whether DNA in solution contains only base pairs of the Watson-Crick geometry or whether there are Hoogsteen base pairs present at a low level. As one may recall from the first section of this chapter, certain antibiotics such as triostin A and echinomycin induce Hoogsteen base pairs in the immediate vicinity of their binding sites on DNA in a crystal (Quigley et al. 1986). Solution studies by Mendel and Dervan (1987), using the chemical diethylpyrocarbonate as a probe, have suggested that Hoogsteen pairs are present also in solution in the immediate vicinity of binding sites for the antibiotic. In addition, these investigators find that a long run of oligo(dA-dT) between two separate antibiotic binding sites becomes sensitive to diethylpyrocarbonate upon adding the antibiotic as if the run were a double helix containing only Hoogsteen base pairs. However, other investigators argue that diethylpyrocarbonate, the probe used in these experiments, does not detect Hoogsteen base pairs (McLean and Waring 1988).

A third subject concerns the attempts by many organic chemists to synthesize a molecule that will recognize any particular sequence of bases in the DNA depending on its exact construction. Past work has focused on synthetic analogs of the antibiotic distamycin, which binds like a "sausage" to DNA within the minor groove of AT-rich regions. For example, Dervan (1986) has made distamycin-like molecules of varying length in combination with other chemicals that intercalate between the base pairs or cut DNA. Kissinger et al. (1987) have made a more ambitious effort to alter the binding specificities of the various aromatic rings and charged groups that make up the distamycin molecule.

Recently, two groups have synthesized oligo(pyrimidine) chains that bind with some affinity within the major groove of DNA. These chains appear to recognize specific sequences in a long DNA molecule by means of specific hydrogen bonds of the Hoogsteen variety between pyrimidine bases in the synthesized strand and purine bases in the double helix (Le Doan et al. 1987; Moser and Dervan 1987). Furthermore, it has been reported that an oligo(purine) chain can also bind strongly within the major groove of DNA and can even repress the synthesis of RNA from a gene in a cell extract (Cooney et al. 1988).

This concludes our review of the progress that has been made by studying DNA in the crystal and in solution through various experimental approaches. Along with the experimental work has come a new theory or set of theories by which we now understand more clearly the physical reality of DNA. One might suppose that theories in biology are nonessential and in any case should take a back seat to experiment, but this is not true in the present circumstance. The experimentalists of the 1960s and 1970s were lost in a morass of incorrect assumptions about DNA that made it difficult, if not impossible, for them to progress further. They thought DNA could only be right-handed, they neglected the base pairs and worried about individual bonds in the sugar phosphate chain, and finally they thought the double helix was so stiff and rigid that it could only bend around proteins by "kinking" sharply at certain points. All of these notions look naive to us today, but we can see that they came about as a consequence of thinking about DNA in terms of an imaginary "helix axis" that ran through the center of the helix like a stiff rod. Thus, the problem facing the theorists was simply to learn to think about DNA without reference to any imaginary straight line or axis (Levitt 1983).

IV. THEORIES OF DNA STRUCTURE

In preceding sections there have been frequent references to the helical twist between successive base pairs, to the propeller twist between paired bases, and to the overall curvature and flexibility of DNA. Here, we shall attempt to bring those quantities and others within the scope of a single theoretical framework.

The classical Watson-Crick model for DNA shows that bases of arbitrary sequence can be accommodated in a simple double-helical structure that is uniform to first order and that is in strong contrast to the structure of proteins where higher-order configuration is determined by the sequence of amino acid residues. The more detailed structures now available from single-crystal X-ray studies make it clear that various perturbations are imposed on the well-known first-order structure. These perturbations are derived principally from the way in which successive bases stack upon each other.

The aims of a "mechanical" theory of DNA structure are twofold. The first is to explain the observed single-crystal X-ray structures in terms of base-stacking arrangements, and the second is to explain how various sequence-dependent mechanical properties of DNA such as intrinsic curvature and flexibility are involved in the recognition by proteins of specific DNA sequences.

A. Base-step Configurations

The first clue to the understanding of specific base-stacking arrangements is provided by the fact that in high-resolution X-ray studies of DNA the base pairs are not planar but have a significant amount of "propeller twist." This is permitted by the flexibility of the hydrogen bonds between Watson-Crick base pairs, and propeller twist tends to be higher in A·T than in G·C base pairs. Propeller twist probably comes about mainly because it affords better overlap of bases on the same strand in terms of van der Waals stacking energy.

There is obviously more kinematic or steric constraint in the stacking of propeller-twisted base pairs than with planar base pairs. At the crudest level, we may observe that purine bases R (= A or G) are larger than pyrimidine bases Y (= T or C), and so there is a possibility of cross-chain steric clash of purines on the minor-groove side in steps YR (Fig. 13). The potential clash on the minor-groove side in steps YR can be relieved in several different ways. These include (1) a low value of helical twist at the step, (2) a low value of slide at the step, or (3) a high value of roll at the step combined with a high value of slide (Calladine 1982).

These maneuvers or combinations of them involve three of the six possible degrees of freedom whereby the configuration of a base step may be described. These are shown schematically in Figure 14 as twist (τ), roll (ρ), and slide (σ). Altogether of course, there are six degrees of kinematic freedom at the step (three translations and three rotations), but it turns out that the other three are not so important being largely prevented by the van der Waals stacking effects (Calladine and Drew 1984).

A logical extension of this line of thought suggests that each type of step (GC, CG, AA, and so forth) has a particular, unique configuration that is conveniently described in terms of ρ, σ, and τ. This turns out to be an oversimplification since the available data, derived from atomic coordinates by a suite of computer programs devised by R.E. Dickerson, show that each type of step can exist in a range of configurations. These variations may be attributed to various factors including the obvious fact that base steps do not exist in isolation since they are connected to neighboring steps by the sugar phosphate chains that must be regarded as having some degree of elastic restraint.

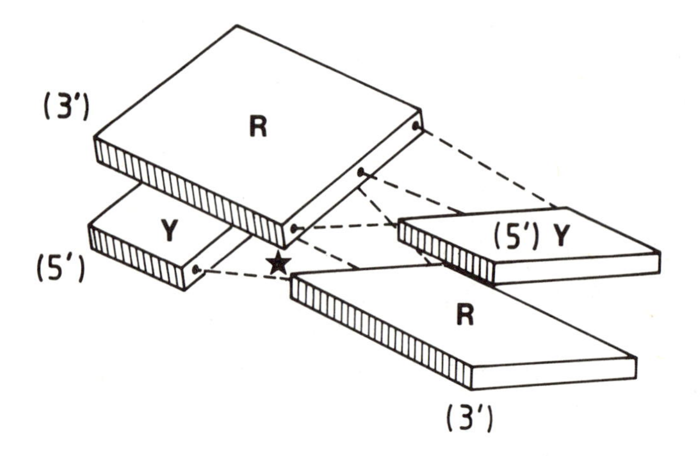

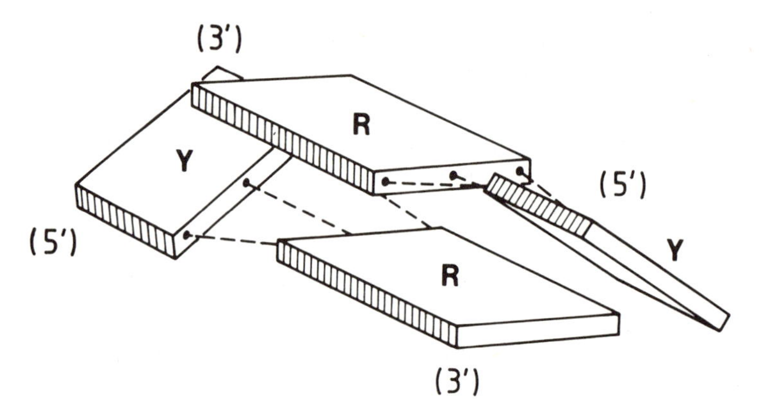

Figure 13 Perspective sketches to show stacking difficulties at a YR step in the presence of propeller twist. Pyrimidine Y and purine R bases are shown as small and large rectangles, respectively, linked by Watson-Crick hydrogen bonds, and the minor-groove edges of bases are shaded. (*Top*) There is good same-strand stacking so that roll ρ has a small value, and the "cross-chain purine clash" pushes slide σ to a low value. (*Bottom*) The purines adopt an overlapping configuration so that both σ and ρ adopt high values. These diagrams are highly schematic and in particular do not depict the helical twist of the step. Nevertheless, they describe well the observed "bistable" stacking arrangements of YR steps.

Nevertheless, the idea of considering the base step as the fundamental building block is important and useful. At a purely geometrical level, we can describe the path of the DNA as a sequence of clearly described steps; this will be useful later in describing the constraints imposed on successive base steps by a requirement, for example, that the DNA is

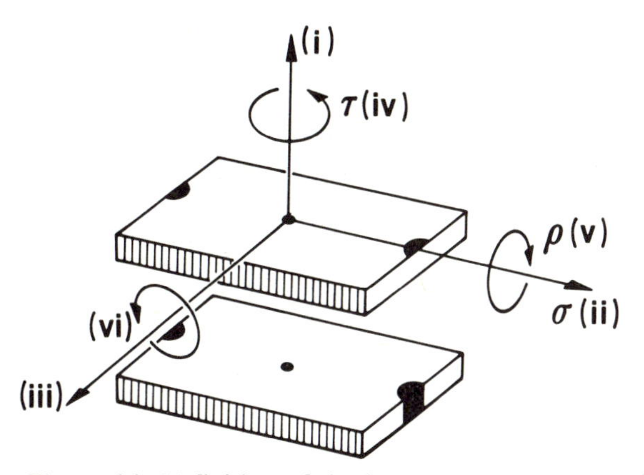

Figure 14 Definition of the base-step parameters twist (τ), roll (ρ), and slide (σ). Here the base pair is represented by a single rectangular block with C-6 (Y) and C-8 (R) atoms marked on the short edges and the minor-groove edges shaded. There are altogether six degrees of kinematic freedom at the step: three translations including σ and three rotations including ρ and τ. Translation i and rotation vi are inhibited by van der Waals interactions, whereas translation iii appears to be less significant than translation ii or σ.

wound onto a nucleosome core. At a more fundamental level, we can see that a quantitative description of DNA as a sequence of steps completely eliminates the need to consider a local helix axis in describing the configuration of DNA. Such an axis can be defined unambiguously only for regular DNA. For irregular DNA, its definition is at best ambiguous but in any case is not necessary.

As an illustration of the "base-step-centered" point of view, consider the set of DNA molecules depicted schematically in Figure 15. In Figure 15b, the base pairs have an incline of about 20° with respect to the axis of helical symmetry, but from a structural point of view it is better to describe the arrangement in terms of a 12° roll angle ρ at each step. The DNA of Figure 15a has both zero tilt and zero roll, and indeed it is straightforward to show that tilt as commonly described is a direct geometrical consequence of the roll over a series of base-pair steps in succession (Calladine and Drew 1984). Figure 15c and d show some nonuniform DNA. The trajectory of these molecules, each of which involves a bend, is best described in terms of a sequence of steps of specific geometry rather than in terms of a change in direction of a local helix axis because, although the concept of a helix axis can be used in this particular example, it would not be useful in any sequence of steps having irregular geometry.

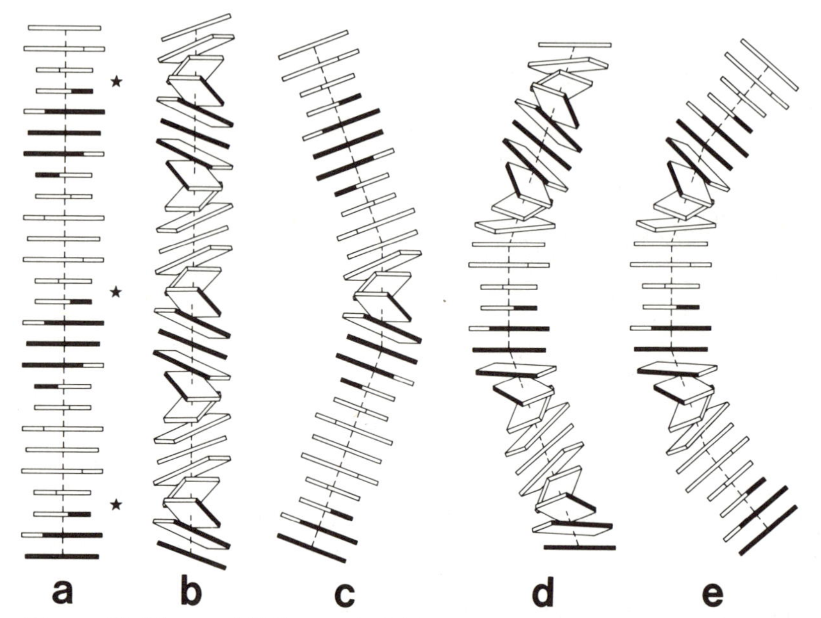

Figure 15 Pieces of DNA, each 25 base steps long, showing how the overall configuration of straight versus curved DNA depends on the arrangement of values of roll ρ. All steps have twist τ = 36°, so that the helical repeat is ten steps. Each base pair is represented by a single rectangular block with its minor-groove edge shown in black. (*a*) The roll angles ρ = 0°: All blocks are parallel, and the DNA is straight. (*b*) ρ = 12° at each step: Observe that the minor-groove edges are further apart than the major-groove edges. The DNA is also straight, but now the bases exhibit a 20° incline with respect to the axis of helical symmetry. (*c*) The first ten steps have ρ = 0°, then ρ = 12° for five steps, and the last ten steps have ρ = 0°. The central five steps have exactly the same geometry as in *b*, and the end parts are as in *a*, but now the DNA has a bend of 40° in the plane of the diagram. In *d*, the central five steps have ρ = 0°, whereas the outer parts have ρ = 12°. Again, a bend of 40° is formed in the plane of the diagram but of opposite sense from that of *c*. In *e*, successive batches of five steps have ρ = 0°, 12°, 0°, and so forth, and the DNA forms a curve in the plane of the diagram. Each five-step batch imparts a 20° turn, so the mean curvature is 4° per step.

When the configuration of discrete base steps in crystallized DNA oligomers is plotted in ρ, σ, and τ space, each kind of step is found to map to a range of points. For some steps, notably AA/TT, the range is small, whereas for others it is large. This suggests that the AA/TT step is stiff, whereas other steps are flexible to a greater degree. In relation to the task of bending around a protein, some flexibility (in the form of a wide range of allowable values of roll) is obviously useful. It is also found that steps YR in crystallized oligomers have bimodal distributions of roll; and indeed those examples with low roll also have low slide, whereas those with high roll also have high slide. This fits in with the idea that steps YR may be structurally bistable and that such bistability can be invoked to explain the transition between the classical B (σ = 0 Å, ρ = 0°) and A (σ = 1.5 Å, ρ = 12°) forms of DNA (Calladine and Drew 1984). This view of the dinucleotide steps of DNA as being relatively mobile in ρ, σ, and τ space, subject to elastic constraints imposed by the sugar phosphate chains and subject also to the stacking preferences of particular steps, is confirmed by the fact that the peculiarly rigid AA/TT steps have special features either of double-layer water bridging in the minor groove or of bifurcated cross-chain hydrogen bonds in the major groove (or possibly both) that by removing a degree of freedom confer a unique, rigid geometry to the step (Drew and Dickerson 1981; Nelson et al. 1987).

The above description of a base-step-centered view of DNA obviously omits much detail at both geometrical and mechanical levels. Thus, the diagrams of Figure 15 have been drawn with specially simple angles, and the description gives little clue that the observed configurations depend on extremely subtle interactions between the various parts of the molecule. Nevertheless, such a description is necessary for a proper understanding of the flexure of DNA.

B. Some Geometrical Aspects of DNA Curvature

Inspection of Figure 15c and d shows that the key feature of a bend in DNA is a short region having a different value of roll ρ from the remainder. Furthermore, the short region must be about five steps long, because it is clear that a region of different roll angle as long as ten steps would produce only a jog in the line of DNA and not a bend.

It is not difficult to extend these ideas to the situation where DNA adopts a curved form in order to wrap around a protein molecule (Calladine and Drew 1986). It is clear that a more or less continuous curve could be formed by having successive batches of five steps with alternating high- and low-roll angles as shown in Figure 15e. Such an arrangement, of course, is not the only way of making a curve for it is obvious that if any single base pair in the array were to be rotated about its long axis by a small amount, while all others were held fixed, then two local roll angles would be changed with no effect to the overall curve of the molecule.

At this point, we need a more systematic way of relating the sequence of roll angles along the DNA to the curvature of the molecule. In particular, we shall eventually need a way of accommodating a nonintegral helical repeat, i.e., the number of steps per turn of the sugar phosphate chains when the DNA is straightened out. Consider first the bending of a straight piece of DNA into a plane curve of radius R(Å). The most satisfactory way of describing curvature κ is as an angle of turning per unit of contour length. Thus, a complete circle turns through 2π radians over a contour length of $2\pi R$, so that

$$\kappa = 1/R$$

For example, to make a circle of radius 100 Å (which is about twice as large as for nucleosomal DNA) we require a curvature of

$$\kappa = 0.01 \text{ radians/Å}$$

or, in units of base-pair step of a nominal length 3.4 Å

$$\kappa = 57.3 \times 3.4/100 = 195/100 = 1.95^\circ/\text{step} \approx 2^\circ/\text{step}$$

So, to make a curve of radius 100 Å, we need an average rotation of about 2° per step about a moving axis that intersects the curve and is always perpendicular to the local plane defined by the curve.

Starting with straight DNA as in Figure 15a, the simplest plan would be to increase the angle of roll by 20° at the steps marked with an asterisk and separated by ten steps. This would clearly fold the DNA into a curve in the plane of the diagram with its center to the left. Alternatively, we could achieve a very similar configuration starting from the straight DNA by decreasing the angle of roll by 20° at the steps midway between those marked with an asterisk, or indeed we could achieve essentially the same trajectory by imposing a pattern of change of roll angles +10°, –10°, +10°, and so forth, at every fifth step.

Of course, none of these arrangements makes a truly smooth curve, but it is surely meaningless to argue too closely about the smoothness of bending an object as complex as DNA. These various alternative patterns of imposed roll angle for achieving a circle of radius 100 Å are set out below

```
           *                          *
... 0 0 0 20 0 0  0  0   0  0  0 0 0 20 0 0  0  0   0  0  0...
... 0 0 0  0 0 0  0  0 -20  0  0 0 0  0 0 0  0  0 -20  0  0...
... 0 0 0 10 0 0  0  0 -10  0  0 0 0 10 0 0  0  0 -10  0  0...
...-3 3 3  3 3 3 -3 -3  -3 -3 -3 3 3  3 3 3 -3 -3  -3 -3 -3...
           *                          *
```

The last line corresponds to the case based on Figure 15e although the curvature associated with a variation in roll from +3° to –3° is only half as large as that shown in the figure. Here the value of 3° has been calculated as described below to produce a curvature κ of 2° per step.

The common feature of all rows of the diagram is, of course, the periodic arrangement of numbers. The period is set equal to 10 in the present example because the helical repeat is 10, and so base steps separated by 10 in the straight DNA provide the parallel hinges that are particularly convenient for the formation of a plane figure. Any arbitrary sequence of imposed roll angles having a period of 10 would also produce a plane curve, by the same reasoning. So how can we calculate the magnitude of curvature in general?

Consider a piece of straight DNA that is to be converted into a circle by the imposition of roll at each step. Let the DNA lie on a horizontal reference plane so that the roll axis of one particular step, numbered n = 0, is vertical, and then let the orientation of the roll axis of step n as viewed along the DNA be θ_n measured clockwise in radians. For example, given regular DNA with a helix repeat of 10, we find that $\theta_1 = 36°$, $\theta_2 = 72°$, or in units of radians $\theta_n = 2\pi n/10$.

Now each imposed roll angle will, in general, be small, so it may be treated as a vector. Thus, its contribution to curvature in the reference plane is the component of the roll angle in the horizontal direction. Hence, successive imposed roll angles make contributions of

$$\rho_n \cos(\theta_n)$$

to the curvature. For the case where the helical repeat is 10 and the sequence of roll angles also has period 10, the angle turned through in 10 steps is given by

$$\phi_0 = \sum_{n=1}^{10} \rho_n \cos(2\pi n/10)$$

There is no guarantee, of course, that a given sequence of imposed roll angles will produce a curve in the horizontal plane, and indeed the sum

$$\phi_1 = \sum_{n=1}^{10} \rho_n \sin(2\pi n/10)$$

gives the contribution to curvature in the vertical plane. Finally, the magnitude of the curvature is given by

$$\kappa = (1/10)(\phi_0^2 + \phi_1^2)^{1/2}$$

and the plane of curvature makes an angle of arctan (ϕ_1/ϕ_0) with respect to the reference plane. Application of these formulas to the roll angles listed in the diagram above shows that they all give $\kappa = 2°$/step with the plane of bending so that a step marked with an asterisk has its minor groove facing outward from the center of curvature.

In the following sections, we shall describe the use of these formulas and generalizations of them to DNA with a nonintegral helical repeat, in several different situations. Here, we may note three important points that emerge from the analysis so far. First, the imposition of a constant roll at every step has no effect on curvature; second, curvature corresponds to the Fourier transform of the roll angles at a period equal to the helical repeat of the DNA; and third, it is important to consider the phasing of the curvature with respect to the sequence, which may be thought of most simply in terms of identifying those steps where the minor groove faces directly outward from the center of curvature.

C. The Sequence-dependent Flexure of DNA around Proteins

So far, our consideration of curvature has been purely geometrical. We now introduce some mechanics to the discussion by postulating some broad features of the behavior of particular types of base step. At present, we have insufficient information to be able to assign "stiffness constants" to individual steps, but we can make some simple statements about the allowable ranges of variation in roll ρ for particular types of step.

Ever since the detailed structure of the nucleosome core was first observed by X-ray crystallography (Finch et al. 1977), it has seemed possible that the "anisotropic flexural stiffness" of DNA might be partly responsible for the location of histone octamers on DNA in chromatin. In its simplest terms, anisotropic flexibility is the phenomenon that would occur if particular dinucleotide steps, which are prone to go into a high-roll configuration, were placed periodically along the DNA in phase with the helical repeat so that their roll axes were parallel. It would be easiest to bend such DNA around a protein core in such a way that these special steps have their minor grooves facing outward, and this would tend to locate the DNA on the core modulo the helical repeat. We must not expect this kind of effect to depend on a single type of step, of course, so we must be prepared to investigate the disposition of the various steps as a whole (i.e., all possible dinucleotide steps).

Drew and Travers (1985b) found that in small-circle DNA there was a tendency for GC- and AT-rich regions to have their minor grooves facing outward and inward, respectively; indeed, this correlates with the tendency observed in crystallized oligomers for steps in these groups to have high and low values of roll, respectively. Satchwell et al. (1986) investigated patterns in the location of all types of dinucleotide steps on a large number of samples of chicken nucleosome core DNA, and they found that AA and TT had a strong tendency to locate themselves so that their minor grooves faced inward, whereas GC steps had a strong tendency to locate themselves where their minor grooves faced outward. Indeed, all types of dinucleotide steps show a strong or weak periodic tendency of this sort. These statistical studies also revealed some special aspects of dinucleotide step location in the vicinity of the dyad of the nucleosome, that have not yet been explained satisfactorily although it seems likely that they are associated with an approximate "S-bend" of the DNA in this region. Furthermore, it is found that certain trinucleotide groups such as GGC/GCC have a strong preference to locate themselves where the minor grooves face outward and also that tracts of oligo(dA)·(dT) prefer to lie toward the ends of the 150-bp segment that is bound closely to the protein core.

Starting from a different point of view, Calladine and Drew (1986) tried to predict the phasing of the location of DNA on nucleosome cores (i.e., to find in which regions the minor grooves would face in or out) by assigning ranges of roll ρ to all types of steps in accordance with a simple scheme of base-stacking deduced from a study of crystallized oligomers. For example, steps YR and some others were regarded as being potentially bistable with two possible configurations $\rho = 0^\circ$ or 9°, whereas others (AA/TT and GA/TC) were regarded as being monostable at 0°, and a third group, steps RY, were taken as being bistable at $\rho = -9^\circ$ or 0° because of the possibility of cross-chain clash or else overlap on the major-groove side. For any postulated location of the DNA, a comparison was made between an "ideal" set of roll angles varying sinusoidally with a period of about 10.2 steps (which is the observed helical repeat) and those allowed by the base sequence as described above but with 2° of latitude. The sum of the squares of differences in roll angle was evaluated for all possible positions of the DNA on the core, and positions in which the sum was near a minimum did correctly locate the angular position of the DNA for about three-quarters of the samples of chicken core DNA.

The favored positions of a typical DNA molecule on the nucleosome core as indicated by this calculation are regularly spaced along the DNA at a period of about 10 bases (see Fig. 7 in Calladine and Drew 1986).

This indicates a preference of configuration in a hypothetical "smoke-ring" rolling motion, which is what we mean by a preferred angular configuration. At present, it is easier to determine the angular position in the wrapping of a piece of DNA around a nucleosome core than it is to decide which of several particular configurations, all having the same angular position, is the one that represents the actual location of the DNA on the core. This stage is described as translational positioning.

There are insufficient data from crystallized oligomers to construct a table of allowable roll values that will correctly locate all samples of chicken core DNA according to our algorithm, but a table constructed by combining data from crystallized oligomers with statistical studies (Satchwell et al. 1986) correctly locates not only virtually all of the chicken sequences, but also a sequence from a frog gene. The roll angles used in this algorithm are summarized in the central columns of Table 2. It should be noted that calculations using these roll angles are equivalent in their result to other calculations using the likelihoods of flexure given above in Figure 11; so, either means of calculation, by roll angles or by likelihoods, is currently acceptable. Further applications of this general theory are described in Drew and Calladine (1987) and also in Drew and McCall (1987), where the theory locates successfully the angular positions of DNA on at least 10 different reconstituted nucleosome cores. It still remains a problem to locate correctly the nucleosomes along the length of the DNA, in a translational sense. Although these studies are rather rudimentary, it is clear that these ideas provide a structural basis for the location of proteins on DNA by virtue of its sequence-dependent flexure, and indeed it seems likely that the redundancy in the three-letter nucleic acid code for amino acids can be deployed to this end in the form of a second level of coding for specific flexural properties within the DNA itself.

Because of limitations of space, we are not able to discuss here the stress-free curvature of certain DNA sequences in solution. Our theory concerning the curvature of these sequences is presented in Calladine et al. (1988) in terms of the different space curves adopted by long pieces of DNA when they are passing through gels of small pore size. The values of roll angle ρ, deduced from our theory of DNA motion through gels, are listed in the right-hand column of Table 2.

The sequence-dependent flexure of DNA is very important in biology, and many particular examples of its importance are beginning to emerge. For example, Otwinowski et al. (1988) report that the tryptophan (*trp*) repressor protein recognizes its DNA-binding site through the use of hydrogen bonds to the phosphates, rather than to the base pairs. Apparently, the sequence of bases in the DNA provides for a specific

Table 2 Roll angles for different sequences in DNA

Kind of step	Bending around nucleosome core		Stress-free in solution
CG	3 ±2;	9 ± 2	3.3
TG/CA	3 ±2;	9 ± 2	3.3
TA	–1 ±6		6.6
GG/CC	2 ±2;	9 ± 2	3.3
AG/CT	1 ±3;	8 ± 3	3.3
GA/TC	–4 ±3;	5 ± 3	3.3
AA/TT	0		0
AT	–5 ±2;	2 ± 2	3.3
GT/AC	2 ±2;	9 ± 2	3.3
(Y)GC	1 ±3;	8 ± 3	3.3
(A)GC	7 ±3		3.3
(G)GC	11 ±3		3.3

Values of roll ρ (degrees) at various dinucleotide steps as deduced from two different studies. Column 1 lists the dinucleotide steps. Columns 2 and 3 give allowable roll in relation to the wrapping of DNA around nucleosome cores. Steps such as CG have two possible ranges, as indicated ($1 \le \rho \le 5$ or $7 \le \rho \le 11$), corresponding to a bistable configuration; step TA has a single, wide range for ρ and is flexible, whereas step AA/TT has a unique value of ρ and is rigid. These values were obtained from statistical information about chicken erythrocyte core DNA ("model 2" of Calladine and Drew 1986) with reference also to oligonucleotide structures determined by X-ray methods. Note that the behavior of step GC is affected by its neighbor on the 5′ side, but the hypothesis that roll depends on the composition of the dinucleotide step alone is otherwise satisfactory to a first order. Column 4 gives unique roll angles at dinucleotide steps in unstressed DNA, as deduced from data on the gel-running of repeating-sequence DNA together with data from crystalized oligomers (Calladine et al. 1988). The gel-running data are well-fitted by a scheme in which all steps except TA (in the context TTTAAA) and AA/TT have the same value of roll.

phosphate conformation that can be bound by the *trp* protein about 10,000 times more tightly than that at DNA of mixed sequence. In another case, Summers and Sherratt (1988) have found that a segment of curved DNA contributes to the efficiency of recombination in bacteria although the details of this process remain to be established. Also, Schnos et al. (1988) have found that the binding of a protein to curved DNA at a viral origin of replication causes adjacent DNA sequences to unwind in a way that seems linked to the initiation of replication. Finally, Wahle and Kornberg (1988) have found that the tight binding of a gyrase protein to a segment of curved DNA provides for the nonrandom distribution of a plasmid between daughter cells following replication.

ACKNOWLEDGMENTS

We thank Drs. R. Cowan, A. Klug, H.C.M. Nelson, D. Rhodes, and A.A. Travers for help, and M. Dowell and A. McGill for preparing the manuscript.

REFERENCES

Amouyal, M. and H. Buc. 1987. Topological unwinding of strong and weak promoters by RNA polymerase: A comparison between the *lac* wild-type and the UV5 sites of *Escherichia coli*. *J. Mol. Biol.* **195:** 795–808.

Anderson, J.E., M. Ptashne, and S.C. Harrison. 1987. Structure of the repressor-operator complex of bacteriophage 434. *Nature* **326:** 846–852.

Benezra, R., C.R. Cantor, and R. Axel. 1986. Nucleosomes are phased along the mouse β-major globin gene in erythroid and nonerythroid cells. *Cell* **44:** 697–704.

Bliska, J.B. and N.R. Cozzarelli. 1987. Use of site-specific recombination as a probe of DNA structure and metabolism *in vivo*. *J. Mol. Biol.* **194:** 205–218.

Bracco, L., O. Kennard, A. Kolb, S. Diekmann, and H. Buc. 1989. Synthetic curved DNA sequences can act as transcriptional activators in *E. coli*. *EMBO J.* **8:** 4289–4296.

Brown, T., W.N. Hunter, G. Kneale, and O. Kennard. 1986. Molecular structure of the G•A base pair in DNA and its implications for the mechanism of transversion mutations. *Proc. Natl. Acad. Sci.* **83:** 2402–2406.

Brown, T., O. Kennard, G. Kneale, and D. Rabinovich. 1985. High-resolution structure of a DNA helix containing mismatched base pairs. *Nature* **315:** 604–606.

Burkhoff, A.M. and T.D. Tullius. 1987. The unusual conformation adopted by the adenine tracts in kinetoplast DNA. *Cell* **48:** 935–943.

Calladine, C.R. 1982. Mechanics of sequence-dependent stacking of bases in B-DNA. *J. Mol. Biol.* **161:** 343–352.

Calladine, C.R. and H.R. Drew. 1984. A base-centred explanation of the B-to-A transition in DNA. *J. Mol. Biol.* **178:** 773–782.

———. 1986. Principles of sequence-dependent flexure of DNA. *J. Mol. Biol.* **192:** 907–918.

Calladine, C.R., H.R. Drew, and M.J. McCall. 1988. The intrinsic curvature of DNA in solution. *J. Mol. Biol.* **201:** 127–137.

Cartwright, I.L. and S.C.R. Elgin. 1984. Chemical footprinting of 5S RNA chromatin in embryos of *Drosophila melanogaster*. *EMBO J.* **3:** 3101–3108.

Chen, W., S. Tabor, and K. Struhl. 1987. Distinguishing between mechanisms of eukaryotic transcriptional activation with bacteriophage T7 RNA polymerase. *Cell* **50:** 1047–1055.

Coll, M., C.A. Frederick, A.H.-J. Wang, and A. Rich. 1987. A bifurcated hydrogen-bonded conformation in the d(A)•d(T) base pairs of the DNA dodecamer d(CGCAAATTTGCG). *Proc. Natl. Acad. Sci.* **87:** 8385–8389.

Collis, C., P.L. Molloy, G. Both, and H.R. Drew. 1989. Influence of the sequence-dependent flexure of DNA on transcription in *E. coli*. *Nucleic Acids Res.* **17:** 9447–9468.

Conner, B.N., T. Takano, S. Tanaka, K. Itakura, and R.E. Dickerson. 1982. The molecular structure of d(^{I}CpCpGpG), a fragment of right-handed double helical A-DNA. *Nature* **295:** 294–299.

Cooney, M., G. Czernuszewicz, E.H. Postel, S.J. Flint, and M.E. Hogan. 1988. Site-specific oligonucleotide binding represses transcription of the human c-*myc* gene *in vitro*. *Science* **241:** 456–459.

Courey, A.J. and J.C. Wang. 1988. Influence of DNA sequence and supercoiling on the process of cruciform formation. *J. Mol. Biol.* **202:** 35–45.

Crawford, J.L., F.J. Kolpak, A.H.-J. Wang, G.J. Quigley, J.H. van Boom, G. van der Marel, and A. Rich. 1980. The tetramer d(CpGpCpG) crystallizes as a left-handed double helix. *Proc. Natl. Acad. Sci.* **77:** 4016–4020.

Crick, F.H.C., J.C. Wang, and W.R. Bauer. 1979. Is DNA really a double helix? *J. Mol. Biol.* **129:** 449–461.

Cruse, W.B.T., S.A. Salisbury, T. Brown, R. Cosstick, F. Eckstein, and O. Kennard. 1986. Chiral phosphorothioate analogues of B-DNA: The crystal structure of Rp-d[Gp(S)CpGp(S)CpGp(S)C]. *J. Mol. Biol.* **192:** 891–905.

Dervan, P.B. 1986. Design of sequence-specific DNA-binding molecules. *Science* **232:** 464–471.

Dickerson, R.E. and H.R. Drew. 1981. Structure of a B-DNA dodecamer: Influence of base sequence on helix structure. *J. Mol. Biol.* **149:** 761–786.

Diekmann, S. and J.C. Wang. 1985. On the sequence determinants and flexibility of the kinetoplast DNA fragment with abnormal gel electrophoretic mobilities. *J. Mol. Biol.* **186:** 1–11.

Drew, H.R. 1984. Structural specificities of five commonly-used DNA nucleases. *J. Mol. Biol.* **176:** 535–557.

Drew, H.R. and C.R. Calladine. 1987. Sequence-specific positioning of core histones on an 860 base-pair DNA: Experiment and theory. *J. Mol. Biol.* **195:** 143–173.

Drew, H.R. and R.E. Dickerson. 1981. Structure of a B-DNA dodecamer: Geometry of hydration. *J. Mol. Biol.* **151:** 535–556.

Drew, H.R. and M.J. McCall. 1987. Structural analysis of a reconstituted DNA containing three histone octamers and histone H5. *J. Mol. Biol.* **197:** 485–511.

Drew, H.R. and A.A. Travers. 1984. DNA structural variations in the *E. coli tyrT* promoter. *Cell* **37:** 491–502.

———. 1985a. Structural junctions in DNA: The influence of flanking sequence on nuclease digestion specificities. *Nucleic Acids Res.* **13:** 4445–4467.

———. 1985b. DNA bending and its relation to nucleosome positioning. *J. Mol. Biol.* **186:** 773–790.

Drew, H.R., R.E. Dickerson, and K. Itakura. 1978. A salt-induced conformational change in crystals of the synthetic DNA tetramer d(CpGpCpG). *J. Mol. Biol.* **125:** 535–543.

Drew, H.R., J.R. Weeks, and A.A. Travers. 1985. Negative supercoiling induces spontaneous unwinding of a bacterial promoter. *EMBO J.* **4:** 1025–1032.

Drew, H.R., T. Takano, S. Tanaka, K. Itakura, and R.E. Dickerson, R.E. 1980. High-salt d(CpGpCpG), a left-handed Z′ DNA double helix. *Nature* **286:** 567–573.

Drew, H.R., R.M. Wing, T. Takano, C. Broka, S. Tanaka, K. Itakura, and R.E. Dickerson. 1981. Structure of a B-DNA dodecamer: Conformation and dynamics. *Proc. Natl. Acad. Sci.* **78:** 2179–2183.

Fairall, L., S. Martin, and D. Rhodes. 1989. The DNA-binding site of *Xenopus* transcription factor IIIA has a non-B form structure. *EMBO J.* **8:** 1809–1817.

Finch, J.T., L.C. Lutter, D. Rhodes, R.S. Brown, B. Rushton, M. Levitt, and A. Klug. 1977. Structure of nucleosome core particles of chromatin. *Nature* **269:** 29–36.

Flick, J.T., J.C. Eissenberg, and S.C.R. Elgin. 1986. Micrococcal nuclease as a DNA structural probe: Its recognition sequences, their genomic distribution, and correlation with DNA structure determinants. *J. Mol. Biol.* **190:** 619–633.

Fox, K. and M. Waring. 1984. DNA structural variations produced by actinomycin and distamycin as revealed by DNase I footprinting. *Nucleic Acids Res.* **12:** 9271–9285.

Franklin, R.E. and R.G. Gosling. 1953. Structure of sodium thymonucleate fibres: Importance of water content. *Acta Crystallogr.* **6:** 673–677.

Fratini, A.V., M.L. Kopka, H.R. Drew, and R.E. Dickerson. 1982. Reversible bending and helix geometry in a B-DNA dodecamer: CGCGAATTBrCGCG. *J. Biol. Chem.* **257:** 14686–14707.

Gao, X. and D.J. Patel. 1988. NMR studies of echinomycin bisintercalation complexes with d(ACGT) and d(TCGA) duplexes in aqueous solution. *Biochemistry* **27:** 1744–1751.

Gartenberg, M.R. and D.M. Crothers. 1988. DNA sequence determinants of CAP-induced bending and protein binding affinity. *Nature* **333:** 824–829.

Gotoh, O. and Y. Tagashira. 1981. Stabilities of nearest-neighbor doublets in double-helical DNA determined by fitting calculated melting profiles to observed profiles. *Biopolymers* **20:** 1033–1042.

Griffith, J., M. Bleyman, C.A. Rauch, P.A. Kitchin, and P.T. Englund. 1986. Visualization of the bent helix in kinetoplast DNA by electron microscopy. *Cell* **46:** 717–724.

Gross, D.S., S.-Y. Huang, and W.T. Garrard. 1985. Chromatin structure of the potential Z-forming sequence $(dT\text{-}dG)_n \cdot (dC\text{-}dA)_n$: Evidence for an "alternating-B" conformation. *J. Mol. Biol.* **183:** 251–265.

Hagerman, P.J. 1985. Sequence dependence of the curvature of DNA: A test of the phasing hypothesis. *Biochemistry* **24:** 7033–7037.

———. 1986. Sequence-directed curvature of DNA. *Nature* **321:** 449–450.

Heinemann, U., H. Lauble, R. Frank, and H. Blocker. 1987. The crystal structure of d(GCCCGGGC). *Nucleic Acids Res.* **15:** 9531–9550.

Hill, R.J. and B.D. Stollar. 1983. Dependence of Z-DNA antibody binding to polytene chromosomes on acid fixation and DNA torsional strain. *Nature* **305:** 338–340.

Hsieh, C.-H. and J.D. Griffith. 1988. The terminus of SV40 DNA replication and transcription contains a sharp sequence-directed curve. *Cell* **52:** 535.

Hunter, W.N., T. Brown, N.N. Anand, and O. Kennard. 1986. Structure of an adenine•cytosine base pair in DNA and its implications for mismatch repair. *Nature* **320:** 552–555.

Huxley, A. 1977. Latent human potentialities. In *The Human Situation: Lectures at Santa Barbara, 1959*, p. 247. Triad-Granada, United Kingdom.

Iwamoto, S. and M.T. Hsu. 1983. Determination of twist and handedness of a 39-base-pair DNA in solution. *Nature* **305:** 70–72.

Johnston, B.H. and A. Rich. 1985. Chemical probes of DNA conformation: Detection of Z-DNA at nucleotide resolution. *Cell* **42:** 713–724.

Kissinger, K., K. Krowicki, J.C. Dabrowiak, and J.W. Lown. 1987. Molecular recognition between oligopeptides and nucleic acids: Monocationic imidazole lexitropsins that display enhanced GC sequence-dependent DNA binding. *Biochemistry* **26:** 5590–5595.

Klug, A. and L.C. Lutter. 1981. The helical periodicity of DNA on the nucleosome. *Nucleic Acids Res.* **9:** 4267–4283.

Klug, A., A. Jack, M.A. Viswamitra, O. Kennard, Z. Shakked, and T.A. Steitz. 1979. A hypothesis on a specific sequence-dependent conformation of DNA and its relation to the binding of the *lac*-repressor protein. *J. Mol. Biol.* **131:** 669–680.

Klysik, J., S.M. Stirdivant, J.E. Larson, P.A. Hart, and R.D. Wells. 1981. Left-handed DNA in restriction fragments and a recombinant plasmid. *Nature* **290:** 672–677.

Koo, H.-S., H.-M. Wu, and D.M. Crothers. 1986. DNA bending at adenine•thymine tracts. *Nature* **320:** 501–506.

Kopka, M.L., C. Yoon, D. Goodsell, P. Pjura, and R.E. Dickerson. 1985. Binding of an antitumour drug to DNA: Netropsin and CGCGAATT^{Br}CGCG. *J. Mol. Biol.* **183:** 553–563.

Kornberg, R. 1980. The location of nucleosomes in chromatin: Specific or statistical? *Nature* **292:** 579–580.

Koudelka, G.B., S.C. Harrison, and M. Ptashne. 1987. Effect of non-contacted bases on the affinity of *434* operator for *434* repressor and *Cro*. *Nature* **326:** 886–888.

Lamond, A.I. and A.A. Travers. 1983. Requirement for an upstream element for optimal transcription of a bacterial tRNA gene. *Nature* **305:** 248–250.

Le Doan, T., L. Perrouault, D. Praseuth, N. Habhoub, J.-L. Decout, N.T. Thuong, J. Lhomme, and C. Hélene. 1987. Sequence-specific recognition, photocrosslinking and cleavage of the DNA double helix by an oligo-[α]-thymidylate covalently linked to an azidoproflavine derivative. *Nucleic Acids Res.* **15:** 7749–7760.

Lefevre, J.F., A.N. Lane, and O. Jardetsky. 1987. Solution structure of the *trp* operator of *E. coli* determined by NMR. *Biochemistry* **26:** 5076–5090.

Levitt, M. 1983. Computer simulation of DNA double-helix dynamics. *Cold Spring Harbor Symp. Quant. Biol.* **47:** 251–262.

Lilley, D.M.J. 1984. DNA: Sequence, structure and supercoiling. *Trans. Biochem. Soc.* **12:** 127–140.

Lipanov, A.A. and V.P. Chuprina. 1987. The structure of poly(dA)•poly(dT) in a condensed state and in solution. *Nucleic Acids Res.* **15:** 5833–5844.

Liu, L.F. and J.C. Wang. 1978. DNA-DNA gyrase complex: The wrapping of DNA outside of the enzyme. *Cell* **15:** 979.

———. 1987. Supercoiling of the DNA template during transcription. *Proc. Natl. Acad. Sci.* **84:** 7024–7027.

Lomonossoff, G.P., P.J.G. Butler, and A. Klug. 1981. Sequence-dependent variation in the conformation of DNA. *J. Mol. Biol.* **149:** 745–760.

Low, L., H.R. Drew, and M. Waring. 1984. Sequence-specific binding of echinomycin to DNA: Evidence for conformational changes affecting flanking sequences. *Nucleic Acids Res.* **12:** 4865–4879.

Lutter, L.C. 1978. Kinetic analysis of deoxyribonuclease I cleavages in the nucleosome core: Evidence for a DNA superhelix. *J. Mol. Biol.* **124:** 391–420.

Marini, J.C., S.D. Levene, D.M. Crothers, and P.T. Englund. 1982. Bent helical structure in kinetoplast DNA. *Proc. Natl. Acad. Sci.* **79:** 7664–7668.

McCall, M.J., T. Brown, and O. Kennard. 1985. The crystal structure of d(GGGG CCCC): A model for poly(dG)•poly(dC). *J. Mol. Biol.* **183:** 385–396.

McCall, M.J., T. Brown, W.N. Hunter, and O. Kennard. 1986. The crystal structure of d(GGATGGGAG): An essential part of the binding site for transcription factor IIIA. *Nature* **322:** 661–664.

McLean, M.J. and M.J. Waring. 1988. Chemical probes reveal no evidence of Hoogsteen base pairing in complexes formed between echinomycin and DNA in solution. *J. Mol. Recog.* **1:** 138–151.

McClellan, J.A. and D.M.J. Lilley. 1987. A two-state conformational equilibrium for alternating $(A\text{-}T)_n$ sequences in negatively supercoiled DNA. *J. Mol. Biol.* **197:** 707–721.

McClellan, J.A., E. Palecek, and D.M.J. Lilley. 1986. $(A\text{-}T)_n$ tracts embedded in random sequence DNA: Formation of a structure which is chemically reactive and torsionally deformable. *Nucleic Acids Res.* **14:** 9291–9309.

Mendel, D. and P.B. Dervan. 1987. Hoogsteen base pairs proximal and distal to echinomycin binding sites on DNA. *Proc. Natl. Acad. Sci.* **84:** 910–914.

Moser, H.E. and P.B. Dervan. 1987. Sequence-specific cleavage of double helical DNA by triple helix formation. *Science* **238:** 645–650.

Murchie, A.I. and D.M.J. Lilley. 1987. The mechanism of cruciform formation in supercoiled DNA: Initial opening of central base pairs in salt-dependent extrusion. *Nucleic Acids Res.* **15:** 9641–9654.

Nelson, H.C.M., J.T. Finch, B.F. Luisi, and A. Klug. 1987. The structure of an oligo(dA)•oligo(dT) tract and its biological implications. *Nature* **330:** 221–226.

Nishimura, Y. 1986. A Raman spectroscopic study on the sequence-dependent conformations of DNA oligomers. *Nucleic Acids Symp. Ser.* **17:** 195–198.

Nordheim, A., E.M. Lafer, L.J. Peck, J.C. Wang, B.D. Stollar, and A. Rich. 1982. Negatively supercoiled plasmids contain left-handed Z-DNA segments as detected by specific antibody binding. *Cell* **31:** 309–318.

Otwinowski, Z., R.W. Schevitz, R.-G. Zhang, C.L. Lawson, A. Joachimiak, R.Q. Marmorstein, B.F. Luisi, and P.B. Sigler. 1988. Crystal structure of the *trp* repressor-operator complex at near-atomic resolution. *Nature* **335:** 321–329.

Patel, D.J., L.L. Canuel, and F.M. Pohl. 1979. Alternating B-DNA conformation for the oligo(dG-dC) duplex in high-salt solution. *Proc. Natl. Acad. Sci.* **76:** 2508–2511.

Patel, D.J., A. Pardi, and K. Itakura. 1982a. DNA conformation, dynamics and interactions in solution. *Science* **216:** 581–590.

Patel, D.J., L. Shapiro, and D. Hare. 1987. DNA and RNA: NMR studies of conformation and dynamics in solution. *Q. Rev. Biophys.* **20:** 35–112.

Patel, D.J., S.A. Kozlowski, A. Nordheim, and A. Rich. 1982b. Right-handed and left-handed DNA: Studies of B and Z-DNA by using proton nuclear Overhauser effect and P NMR. *Proc. Natl. Acad. Sci.* **79:** 1413–1417.

Patel, D.J., S.A. Kozlowski, J.W. Suggs, and S.D. Cox. 1981. Right-handed alternating DNA conformation: poly(dA-dT) adopts the same dinucleotide repeat with cesium, tetraalkylammonium, and steroid cations in aqueous solution. *Proc. Natl. Acad. Sci.* **78:** 4063–4067.

Peck, L.J. and J.C. Wang. 1981. Sequence-dependence of the helical repeat of DNA in solution. *Nature* **292:** 375–378.

———. 1983. Energetics of B-to-Z transition in DNA. *Proc. Natl. Acad. Sci.* **80:** 6206–6210.

———. 1985. Transcriptional block caused by a negative-supercoiling-induced structural change in an alternating CG sequence. *Cell* **40:** 129–137.

Pjura, P.E., K. Grzeskowiak, and R.E. Dickerson. 1987. Binding of Hoechst 33258 to the minor groove of B-DNA. *J. Mol. Biol.* **197:** 257–271.

Pohl, F. and T. Jovin. 1972. Salt-induced cooperative conformational change of a synthetic DNA: Equilibrium and kinetic studies with poly(dG-dC). *J. Mol. Biol.* **67:** 375–396.

Poltev, V.I. and A.V. Teplukhin. 1987. Interaction of bases and conformational manifestation of repetitive nucleotide sequences. *Mol. Biol.* (U.S.S.R.) **21:** 102–115.

Privé, G.G., U. Heinemann, S. Chandrasegaran, L.S. Kan, M.L. Kopka, and R.E. Dickerson. 1987. Helix geometry, hydration and G/A mismatch in a B-DNA decamer. *Science* **238:** 498–504.

Quigley, G.J., G. Ughetto, G.A. van der Marel, J.H. van Boom, A.H.-J. Wang, and A. Rich. 1986. Non-Watson-Crick G•C and A•T base pairs in a DNA-antibiotic complex. *Science* **232:** 1255–1258.

Quigley, G.J., A.H.-J. Wang, G. Ughetto, G. van der Marel, J.H. van Boom, and A. Rich. 1980. Molecular structure of an anticancer drug-DNA complex: Daunomycin plus d(CGTACG). *Proc. Natl. Acad. Sci.* **77:** 7204–7208.

Ramsay, N., G. Felsenfeld, B. Rushton, and J.D. McGhee. 1984. A 145 base-pair DNA sequence that positions itself precisely and asymmetrically on the nucleosome core. *EMBO J.* **3:** 2605–2611.

Rhodes, D. 1979. Nucleosome cores reconstituted from poly(dA-dT) and the octamer of histones. *Nucleic Acids Res.* **6:** 1805–1816.

———. 1985. Structural analysis of a triple complex between the histone octamer, a *Xenopus* gene for 5S RNA, and transcription factor IIIA. *EMBO J.* **4:** 3473–3482.

Rhodes, D. and A. Klug. 1980. Helical periodicity of DNA determined by enzyme digestion. *Nature* **286:** 573–578.

———. 1981. Sequence-dependent helical periodicity of DNA. *Nature* **292:** 378–380.

Richmond, T.J., J.T. Finch, B. Rushton, D. Rhodes, and A. Klug. 1984. Structure of the nucleosome core particle at 7 Å resolution. *Nature* **311:** 532–537.

Rodley, G.A., R.J. Scobie, R.H.T. Bates, and R.M. Lewitt. 1976. A possible conformation for double-stranded polynucleotides. *Proc. Natl. Acad. Sci.* **73:** 2959–2963.

Russell, D.W., M. Smith, D. Cox, V.M. Williamson, and E.T. Young. 1983. DNA sequences of two yeast promoter-up mutants. *Nature* **304:** 652–654.

Satchwell, S.C., H.R. Drew, and A.A. Travers. 1986. Sequence periodicities in nucleosome core DNA. *J. Mol. Biol.* **191:** 659–675.

Schnos, M., K. Zahn, R.D. Inman, and F.R. Blattner. 1988. Initiation protein-induced helix destabilization at the λ origin: A prepriming step in DNA replication. *Cell* **52:** 385–395.

Shakked, Z., D. Rabinovich, O. Kennard, W.B.T. Cruse, S.A. Salisbury, and M.A. Viswamitra. 1983. Sequence-dependent conformation of an A-DNA double helix: The crystal structure of the octamer d(GGTATACC). *J. Mol. Biol.* **166:** 183–201.

Sherman, S.E., D. Gibson, A.H.-J. Wang, and S.J. Lippard. 1985. X-ray structure of the major adduct of the anticancer drug cisplatin with DNA: *cis*-$[Pt(NH_3)_2(d(pGpG))]$. *Science* **230:** 412–417.

Simpson, R.T. and P. Kunzler. 1979. Chromatin and core particles formed from the inner histones and synthetic polydeoxyribonucleotides of defined sequence. *Nucleic Acids Res.* **6:** 1387–1415.

Simpson, R.T. and D.W. Stafford. 1983. Structural features of a phased nucleosome core particle. *Proc. Natl. Acad. Sci.* **80:** 51–55.

Spassky, A., S. Rimsky, H. Buc, and S. Busby. 1988. Correlation between the conformation of *E. coli* -10 hexamer sequences and promoter strength. *EMBO J.* **7:** 1871–1879.

Suck, D., A. Lahm, and C. Oefner. 1988. Structure refined to 2Å of a nicked DNA octanucleotide complex with DNase I. *Nature* **332:** 464–468.

Summers, D.K. and D.J. Sherratt. 1988. Resolution of Col E1 dimers requires a DNA sequence implicated in the three-dimensional organization of the *cer* site. *EMBO J.* **7:** 851–858.

Viswamitra, M.A. and B. Ramakrishnan. 1987. Crystal structure of deoxyCpG-mitoxanthrone complex. In *Collected Abstracts from the 14th International Congress of Crystallography,* Perth, Australia, August 1987. Abstract 04.X-6.

Viswamitra, M.A., O. Kennard, P.G. Jones, G.M. Sheldrick, S. Salisbury, L. Falvello, and Z. Shakked. 1978. DNA double-helical fragment at atomic resolution. *Nature* **273:** 687–688.

Vorlickova, M., J. Kypr, V. Kleinwachter, and E. Palecek. 1980. Salt-induced conformational changes of poly(dA-dT). *Nucleic Acids Res.* **8:** 3965–3973.

Wahle, E. and A. Kornberg. 1988. The partition locus of plasmid pSC101 is a specific binding site for DNA gyrase. *EMBO J.* **7:** 1889–1895.

Wang, A.H.-J., S. Fujii, J.H. van Boom, and A. Rich. 1982a. Molecular structure of the octamer d(GGCCGGCC): Modified A-DNA. *Proc. Natl. Acad. Sci.* **79:** 3968–3972.

Wang, A.H.-J., G. Ughetto, G.J. Quigley, and A. Rich. 1987. Interactions between an anthracycline antibiotic and DNA: Molecular structure of daunomycin complexed to d(CpGpTpApCpG) at 1.2 Å resolution. *Biochemistry* **26:** 1152–1163.

Wang, A.H.-J., S. Fujii, J.H. van Boom, G.A. van der Marel, S.A.A. van Boeckel, and A. Rich. 1982b. Molecular structure of r(GCG)d(TATACGC): A DNA-RNA hybrid helix joined to double-helical DNA. *Nature* **299:** 601–604.

Wang, A.H.-J., G.J. Quigley, F.J. Kolpak, J.L. Crawford, J.H. van Boom, G. van der Marel, and A. Rich. 1979. Molecular structure of a left-handed double-helical DNA fragment at atomic resolution. *Nature* **282:** 680–686.

Wang, J.C. 1979. Helical repeat of DNA in solution. *Proc. Natl. Acad. Sci.* **76:** 200–203.

Waring, M.J. and L.P.G. Wakelin. 1974. Echinomycin: A bifunctional intercalating antibiotic. *Nature* **252:** 653–657.

Wing, R.M., P. Pjura, H.R. Drew, and R.E. Dickerson. 1984. The primary mode of binding of cisplatin to a B-DNA dodecamer: CGCGAATTCGCG. *EMBO J.* **3:** 1201–1206.

Wing, R., H.R. Drew, T. Takano, C. Broka, S. Tanaka, K. Itakura, and R.E. Dickerson. 1980. Crystal structure analysis of a complete turn of B-DNA. *Nature* **287:** 755–758.

Zahn, K. and F.R. Blattner. 1987. Direct evidence for DNA bending at the lambda replication origin. *Science* **236:** 416–422.

2

Bending of DNA in Nucleoprotein Complexes

Andrew A. Travers and Aaron Klug
Medical Research Council
Laboratory of Molecular Biology
Cambridge CB2 2QH, United Kingdom

I. INTRODUCTION

When DNA is packaged, whether in a phage head or a eukaryotic chromosome, it is often tightly bent. In chromatin, the elementary unit of the structure is the nucleosome in which the DNA wraps approximately twice around an octamer of the histone proteins H3, H4, H2A, and H2B,

DNA Topology and Its Biological Effects

as a left-handed superhelix with a diameter of 86 Å (Finch et al. 1977; Richmond et al. 1984). The path of the DNA between histone octamers is not known, but it appears that variable lengths of "linker" DNA separate individual nucleosomes (Prunell and Kornberg 1982; Widom and Klug 1985). A single copy of the fifth histone, H1, is associated with the linker in a way that is not fully understood. When the linker DNA is digested away by nucleases, the H1 drops off leaving the so-called nucleosome core particle, consisting of about 145 bp wrapped in approximately 1.8 superhelical turns around the histone octamer. The two copies of each histone in the octamer are related by a unique axis of twofold symmetry, or *dyad,* that at one end passes through the midpoint of DNA. Although in the chromosome the histone octamers are associated with a great variety of DNA sequences, studies of nucleosome positioning in both reconstituted and naturally occurring systems have shown that these proteins can adopt well-defined, even precise locations with respect to the primary DNA sequence (Simpson and Stafford 1983; Edwards and Firtel 1984; Palen and Cech 1984; Ramsay et al. 1984; Drew and Travers 1985a; Rhodes 1985; Thoma and Simpson 1985; Ramsay 1986).

What are the molecular interactions that determine this precise positioning? The association of histone octamers with an immense variety of DNA sequences suggests that sequence-specific recognition, as classically exemplified by the interaction of the *lac* and λ repressors with their respective operators, is not a dominant determinant of nucleosome positioning. However, it has recently become apparent that the structural and mechanical properties of the DNA double helix vary in a sequence-dependent manner and could contribute significantly to the specific interaction of many proteins with DNA (Lomonossoff et al. 1981; Drew and Travers 1984). In this respect, DNA is not merely a passive ligand in binding to a protein, but its physical properties can actively influence the structure and stability of the complexes.

The property of DNA that is relevant to complexes in which it is tightly wrapped is the bendability of the molecule. DNA does not behave as an isotropic rod (Trifonov and Sussman 1980; Widom 1985; Calladine and Drew 1986); it may bend more easily in one plane rather than another, i.e., it has its anisotropic flexibility. In a long molecule, this will depend on the overall effect of the distribution of short sequences within it that are differentially flexible in different directions. In many large DNA/protein complexes, including those involved in the enzymatic manipulation of DNA in replication and recombination (Echols 1986), the DNA molecule is wrapped tightly in a nucleosome-like manner around a central core of protein molecules. In such structures, all the grooves (both major and minor) on the inside of the curve must narrow somewhat because of the compression associated with bending, whereas those on the outside of the curve become correspondingly wider. It is therefore clear that in such a DNA/protein complex the structure of the DNA must be able to accommodate this deformation. There are also sequences that, even in the absence of external forces imposed by interaction with proteins, impart a preferred direction of curvature on a DNA molecule, i.e., give an "intrinsic bend" (Marini et al. 1982; Marini and Englund 1983; Hagerman 1984; Wu and Crothers 1984). We shall discuss both kinds of bending and attempt to draw a relationship between them.

For the subsequent discussion, we need to state the terminology we use to describe the positioning of DNA on a nucleosome or similar complex. Two parameters must be considered: a translation, marking where the histone octamer is placed along the DNA, and a rotation, which defines the local orientation (azimuth) of the DNA relative to the direction of curvature. The DNA sequences that determine nucleosome positioning must also reflect these parameters. In a DNA molecule that is wrapped uniformly around a protein, any DNA sequences that favor particular rotational orientations should occur regularly with a periodicity equal to the local helical twist (Drew and Travers 1985a). In contrast, translational sequence markers would be expected to occur nonperiodically and to be found either at unique locations within the nucleosome, such as the dyad, or at positions symmetrically related about the dyad.

II. USE OF DNASE I AS A PROBE FOR DNA ORGANIZATION

When a DNA molecule is constrained either by binding to a surface or by circularization into a small circle, its three-dimensional organization may restrict the accessibility of a chemical reagent or enzyme relative to that in the unconstrained linear state. A carefully selected reagent can thus act as a probe for the configuration of the DNA molecule. The utility of such a probe depends on both its chemical specificity and the dependence of reactivity on the accessibility of the DNA substrate. One reagent that has been widely used in previous studies is the enzyme DNase I (Lutter 1978; Rhodes and Klug 1980). This enzyme has a broad chemical specificity for the cleavage of phosphodiester bonds in DNA (Drew 1984: Drew and Travers 1985a), and thus it can be used to investigate gross structural features without the complication of cleavage being restricted by high sequence specificity. In addition, the enzyme protein is large relative to the DNA double helix (Suck et al. 1984; Suck and Oefner 1986), and consequently, when a DNA molecule is lying on a surface, access of the enzyme to one side of the DNA helix is hindered. In this situation, only the most exposed phosphodiester bonds will be accessible to cleavage (Fig. 1). Because DNase I cleaves independently, the two phosphate backbones bordering the minor groove (Lutter 1978; Drew and Travers 1984; Suck and Oefner 1986), the location of cuts reveals the orientation of the DNA molecule relative to the surface on which it lies. A further characteristic of the complex of the nuclease with its DNA substrate is that the DNA is deformed, showing a 21.5° bend toward the major groove and away from the protein (Suck et al. 1988). This implies that any conformationally rigid sequence, for example, short oligo(dA-dT) or (dG-dC) tracts (see below), that cannot be readily deformed to the appropriate conformation would show a reduced rate of cleavage.

Although not directly relevant to the theme of this chapter, it should be noted that DNase I cleavage patterns can also be used to detect local variations in helical structure as well as the orientation of the double helix. This follows because in DNase I cleavage profiles there is a stagger between the cuts on the two strands. For classical B-form DNA, this stagger is on average about two to three nucleotides in the 3′ direction. However, the stagger is dependent on the distance between the two DNA strands and hence on the width of the minor groove. This means that A-form DNA, which has a wider minor groove, a high tilt, and a low average twist angle, would have a stagger of 0–1 bp (Rhodes and Klug

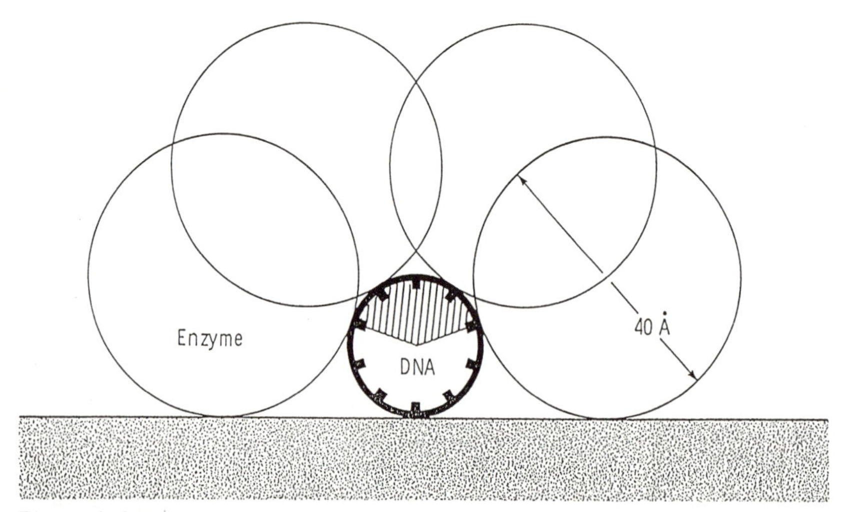

Figure 1 Schematic drawing of the accessibility of an enzyme to DNA lying on a flat surface. The surface restricts a spherical enzyme with a diameter of 40 Å to a maximum of four accessible phosphodiester bonds (redrawn from Rhodes and Klug 1980).

1986), whereas a DNA with a greater than average twist angle and hence a narrow minor groove (Drew and Travers 1984) would have a stagger of 3–4 bp.

III. THE BENDING PREFERENCE OF DNA IS A DETERMINANT OF NUCLEOSOME POSITIONING

The tight wrapping of DNA about the histone octamer strongly suggests that the anisotropic bendability of DNA may be a major determinant of nucleosome positioning. To test this hypothesis experimentally, we need to establish that the bending preference of a particular DNA molecule is the same when reconstituted into a nucleosome core particle as it is when constrained in the absence of any bound proteins. To approach this question, a 169-bp DNA fragment of bacterial origin (which had thus undergone no natural selection for binding to a histone octamer) was covalently closed into a small, relaxed circle (Drew and Travers 1985a). Such a molecule, if bent uniformly, would have an average diameter of about 170 Å, that is, about twice that of the superhelical turns on the nucleosome core. The problem is to determine by experiment whether the DNA in such a circle assumes a preferred direction of curvature and, if it does, to establish the rotational orientation of particular DNA sequences relative to the inside and outside of the circle. The probe used (see above) was the nuclease DNase I, which has an effective average diameter of 40 Å (Suck et al. 1984), chosen on the assumption that enzymatic access to the inside of the circle would be impeded relative to that on the outside. The experimental result, which shows this assumption to be correct and consistent with the structure of the DNase I/DNA complex (Suck et al. 1988), was that cuts on the circle follow a regular sinusoidal periodicity of 10.56 bp on both DNA strands in contrast with irregular cleavage pattern on the corresponding linear molecule. This shows immediately that in a small circle the DNA assumes a highly preferred configuration. The average cutting periodicity found has the value expected from the helical periodicity of mixed sequence DNA (Rhodes and Klug 1980) because the overall length of the molecule (169 bp) was carefully chosen so as to comprise a precisely integral number of double helical turns (16 x 10.56), thereby producing efficient closure during ligation (Shore and Baldwin 1983; Horowitz and Wang 1984).

Because DNase I cleaves DNA where the minor groove is accessible to the enzyme (Lutter 1978; Drew 1984; Drew and Travers 1985a; Suck and Oefner 1986), we can also deduce the angular orientation of the DNA sequence in the small circle. The general result is that AT-rich sequences are cleaved at a substantially lower rate in the circle than in the linear form, whereas cleavage at GC-rich sequences is not significantly reduced by circularization. This means that, on the average, AT-rich minor grooves face inward toward the center of the circle, whereas GC-rich minor grooves face outward. This experiment demonstrates that a short DNA molecule, when constrained in the absence of protein, adopts a highly preferred rotational orientation. Of course, in any given piece of mixed sequence DNA, it is unlikely that all the bending preferences in that sequence can be satisfied simultaneously by the configuration assumed by the whole molecule, and the overall setting will be determined by the balance of local preferences. Consequently, there will always be a few helix segments whose rotational position is imposed not by local constraints but by the preferred configuration of the whole molecule. At some sites, where the DNA is unfavorably positioned in this way, the rate of DNase I cleavage in the circle is greater than that on the linear molecule, suggesting a local deformation of the DNA structure. This effect is particularly apparent where AT-rich minor grooves face outward rather than inward.

Is the bending preference adopted in the absence of protein maintained when the same DNA fragment, now in a linear form, is placed on a histone octamer? Analysis of such a reconstituted nucleosome core particle showed that the angular setting of the DNA remained largely conserved in going from circle to nucleosome. The few slight differences that are observed may be attributed to a reduction in the average cutting periodicity from 10.56 bp in the circle to 10.32 bp upon nucleosome formation. As argued previously (Klug and Lutter 1981), this latter periodicity sets an upper bound to the average local twist of the DNA in the core particle.

IV. SEQUENCE DETERMINANTS OF DNA BENDING: STATISTICAL SEQUENCING OF NUCLEOSOME CORE DNA

If the rotational positioning of DNA on a nucleosome is influenced by particular sequences, we would expect that in a population of nucleosome core particles, the occurrence of such sequences would exhibit a periodic modulation that would reflect the structural periodicity of the DNA molecule lying on the surface of the histone octamer. To determine the general nature of such DNA sequences, a population of DNA molecules was extracted from a nucleosome core preparation isolated from chicken erythrocytes. This population had essentially the same dinucleotide composition as that determined previously for total chicken erythrocyte DNA (Swartz et al. 1962) and by this criterion was therefore representative of chicken DNA as a whole. The analysis of sequence content was done by a technique we have termed statistical sequencing (Drew and Travers 1985a). In this method (Fig. 2), the predominant locations of particular DNA sequences in a mixed population of DNA molecules are detected by binding antibiotic drugs of known sequence specificity to the DNA and then treating the drug/DNA complex with DNase I. At sites where the drug binds, the rate of DNase I cleavage is substantially reduced and can be measured. By this method of quantitative footprinting, any preferred location of such sites along the length of DNA can be determined.

Because it had been shown that the angular orientation of a particular DNA sequence (the *Escherichia coli tyrT* promoter) correlated with the disposition of AT-rich and GC-rich stretches, the antibiotic drugs chosen for the statistical sequencing of core DNA isolated from chicken erythrocytes were distamycin and chromomycin. The former binds selectively to runs of four or more A·T base pairs (Van Dyke et al. 1982; Fox and Waring 1985), whereas the latter has the reverse specificity, selecting runs of four or more G·C base pairs (Van Dyke and Dervan 1983; Fox and Howarth 1985). This analysis showed that on the core DNA the occurrence of short runs of both AT and GC are periodically modulated with an average period of 10.17 ± 0.05 bp. Also, these two modulations

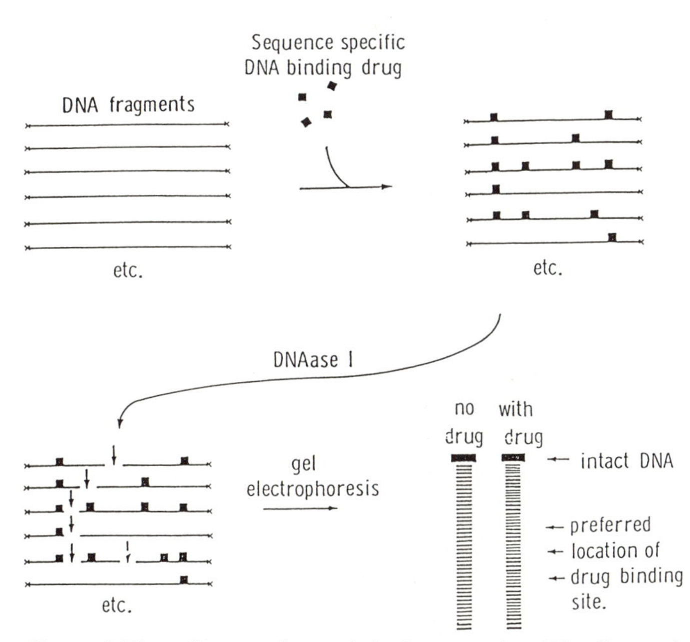

Figure 2 Flow diagram for statistical sequencing. The figure shows the method applied to an aligned population of short DNA molecules of uniform length.

are in opposite phases, a maximum in the occurrence of AT runs coinciding with a minimum in the occurrence of GC runs and vice versa. This periodic fluctuation in sequence content is thus not primarily in single-nucleotide composition but instead relates to the orientation of short runs of AT or of GC relative to the histone proteins. Since it had been shown that the minor groove points outward at the particle dyad (Lutter 1978; Richmond et al. 1984)—a position that corresponds to the midpoint of the population of DNA molecules analyzed—the average rotational orientation of such short runs could be deduced. The general conclusion from this experiment is that AT runs are preferentially placed where the minor groove faces approximately inward toward the histone octamer, whereas GC runs prefer to occupy positions where the minor groove points outward. This rule also holds for nucleosomes reconstituted from defined sequence DNA fragments and histone octamers (Simpson and Stafford 1983; Ramsay et al. 1984; Rhodes 1985). This kind of sequence variation argues strongly that the rotational orientation of DNA within the nucleosome is determined principally by certain directional bending preferences of the DNA rather than by any sequence-specified protein/DNA contacts.

V. SEQUENCE DETERMINANTS OF DNA BENDING: DIRECT SEQUENCING OF NUCLEOSOME CORE DNA

Statistical sequencing allows a description of the most dominant sequence features of nucleosome core DNA; however, it cannot describe the occurrences of all particular sequence combinations nor can it assess the detailed nature of helix curvature in regions such as those close to the protein dyad where the path of the DNA deviates markedly from a uniform superhelix (Richmond et al. 1984). To investigate these aspects of nucleosome core structure, 177 individual DNA molecules from the same DNA sample that was used for statistical sequencing were cloned and sequenced (Satchwell et al. 1986). These sequences were aligned at about their midpoint and analyzed for the occurrences of dinucleotides and trinucleotides. The most striking result from this analysis is a well-defined periodicity in the distribution of the dinucleotide ApA/TpT between base pair steps 1 and 56 where the dyad is defined as step 72.5 at the midpoint of the observed average DNA length of 145 bases (Fig. 3). These maxima of ApA/TpT occurrences are on average sited where the crystal structure shows the minor groove to point inward toward the histone octamer (Richmond et al. 1984). In this region, there are five marked peaks at an average spacing of 10.1 bases. This regular periodic

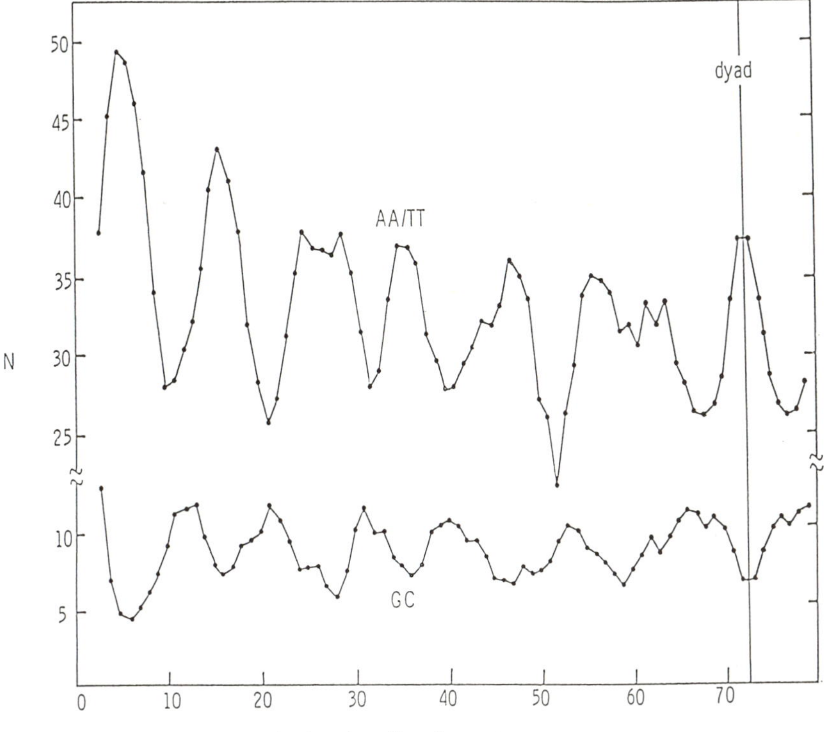

Figure 3 Variations in the occurrence of the dinucleotide AA/TT and GC versus position in the nucleosome core DNA sequence. N is the running 3-bond average of occurrence of a dinucleotide also averaged about the dyad at base-step 72.5. (Data are taken from Satchwell et al. 1986.)

pattern changes between position 56 and the dyad so that a maximum occurs at the dyad instead of the minimum that would have been expected had the previous pattern simply continued. This reversal of phase corresponds to a region of the nucleosome where the path of the DNA departs from a uniform superhelix with a "jog" (Fig. 4a) between the two adjacent turns of the supercoil (Richmond et al. 1984).

To obtain a quantitative measure of amplitude and phase for each of the ten possible dinucleotide steps, their distribution as a function of position were analyzed by Fourier transformation. This analysis confirmed the rotational orientation of the dinucleotide ApA/TpT and showed that its fractional variation in amplitude is ± 20% about the mean. Of the other dinucleotides with significant periodic modulations, GpC is the strongest and is found in the opposite phase to ApA/TpT with

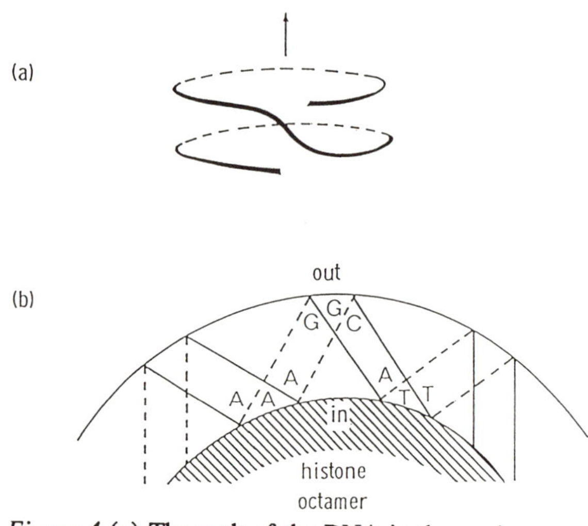

Figure 4 (*a*) The path of the DNA in the nucleosome core particle. The path is represented as a left-handed key ring; for clarity, the jog at the dyad is exaggerated in comparison with the actual structure (Richmond et al. 1984). (*b*) Preferred rotational orientation of trinucleotide sequences in the nucleosome core. The schematic drawing shows the relation of the sequences with respect to the minor groove of DNA.

a fractional variation in amplitude of ± 27%. Similarly the dinucleotides GpC/CpC and TpG/CpA are in phase with GpC, but their preferences for outward facing minor grooves are smaller, having variations in amplitude of ± 12% and ± 8%, respectively (Fig. 3).

It is clear from the crystal structures of short DNA fragments that the conformation at a particular base step depends not only on the base pairs on either side of it, but also on the sequence context in which it is located. We should thus ideally consider at least the flanking base steps on either side of a particular base step, that is, tetranucleotide sequences. However, the number of core DNA sequences available was insufficient to allow a reliable statistical evaluation of tetranucleotide occurrences, and therefore only the occurrences of the 32 possible trinucleotides were analyzed by Fourier transformation (assuming complementary trinucleotides to be equivalent). This analysis showed clearly that the positioning of some sequences on the nucleosome core was determined by a trinucleotide or longer sequence rather than by the dinucleotide components acting independently. Thus, of the seven trinucleotides containing the dinucleotide ApA/TpT, only three, ApApA/TpTpT, ApApT/ApTpT, and TpApA/ApTpT, show significant relative preferences for an inward facing minor groove with respective amplitudes of ± 36%, ± 30%, and ± 20% about their mean values. Indeed, the dinucleotide ApA/TpT in isolation (that is, the sequence ApA not flanked by A on either side and the sequence TpT not flanked by T) does not exhibit any significant periodic modulation. Again, the conclusion is that, for this class of sequences, rotational positioning is determined by at least a trinucleotide sequence. In contrast, the periodic modulations of the GpC-containing trinucleotides are all the same phase but differ substantially in amplitude.

VI. THE PERIODICITY OF NUCLEOSOME CORE DNA AND THE LINKING NUMBER PARADOX

When the results of the X-ray crystal analysis on the organization of DNA in the nucleosome core were first obtained and compared with physicochemical data, a paradox emerged that has come to be known as the linking number problem (Finch et al. 1977). The X-ray analysis showed that the double helix of DNA in a nucleosome is wound into about two turns of a shallow superhelix, whereas measurements on closed circular DNA extracted from SV40 chromatin gave a linking number reduction nearer to 1 per nucleosome (Germond et al. 1975) rather than the value of 2 that would be expected if the helical screw of the DNA remained the same. This subject is discussed by Finch et al. (1977) and Klug and Lutter (1981), who proposed that the explanation of the paradox lay in a difference in the screw of the DNA when in the nucleosome and free in solution.

The results on the periodic modulation of sequence in nucleosome core DNA have important implications with regard both to the change in linking number on nucleosome formation and to the packing of the two adjacent superhelical turns of DNA on the histone octamer. The main fact is that the periodicity in sequence of nucleosome DNA is different from the helical periodicity of DNA in solution (Drew and Travers 1985a; Satchwell et al. 1986). Using the method of statistical sequencing, Drew and Travers (1985a) obtained a value of 10.17 ± 0.05 bp for the modulation of sequence content within a mixed population of chicken core DNA molecules. A similar value of 10.21 bp was obtained by direct sequence analysis of a small sample of this same population (Satchwell et al. 1986). In both cases, the value for the periodicity is an average calculated across the 120–140 bp centered on the nucleosome dyad. To relate the sequence periodicity of core DNA to the problem of the linking number in the path of the DNA about the histone octamer, we must first consider how the sequence periodicity is related to the helical (i.e., structural) periodicity of the DNA molecule. The principal difficulty in discussions of the helical periodicity of the nucleosome is a failure to distinguish between frames of reference: namely, that which would be chosen by an observer outside the nucleosome (i.e., a "laboratory," also termed "absolute" or "intrinsic") frame and that which would be used by an observer following the path taken by the helical axis of the DNA in the supercoil (i.e., a "local" or "relative" frame). The local twist is strictly defined as the instantaneous torsion about a moving axis tangential to the space curve of the molecule (Crick 1953, 1976). In a supercoil, this is not the same function as the angle between two successive base pairs in the laboratory frame of reference.

It is the local frame that determines the periodic deformations of DNA structure consequent upon tight wrapping and is thus the frame sensed by nuclease probing. We note that the proposed solution of the linking number paradox explicitly referred to the local and not the laboratory frame (Klug and Lutter 1981). For a uniform superhelix wound on a cylindrical surface, the helical periodicities in the local and laboratory frames can be simply related (Crick 1976; Fuller 1978). In the local frame, the average twist angle per base step $\theta_{local} = 2\pi/n$, where n is the number of bases in one turn of the double helix. In the laboratory frame, there is an additional twist of $2\pi \sin \alpha$ associated with one turn of the superhelical path, where α is the pitch of the superhelix. This is distributed equally over the number of bases (n_b) in one turn of the superhelical path. Thus

$$\theta_{lab} = \theta_{local} + 2\pi/n_b$$

For the nucleosome core particle $\theta_{local} = 35.40^{o}$ (10.17 bp/turn, see below) $\alpha = -5.5^{o}$ and $n_b = 77.3$ bases (Richmond et al. 1984), and hence, in degrees

$$\begin{aligned}\theta_{lab} &= 35.4^{o} - 360\,(-0.096)/77.3 = (35.40 - 0.45)^{o} \\ &= 34.95^{o}\end{aligned}$$

which is equivalent to 360/34.95 = 10.31 bp/turn.

This relation between the local and absolute helical periodicities applies to DNA wrapped uniformly on a cylindrical surface. For the case of a general surface, as described by White and his colleagues (White and Bauer 1986; White et al. 1988; Cozzarelli et al., this volume), $Tw = STw + \phi$, where Tw and ϕ are the number of turns in the laboratory and local frames, respectively, whereas STw is the conversion factor from the local to laboratory frames.

We now turn to the four pieces of experimental evidence relevant to the linking number problem. These results taken together yield a coherent picture. The relation between these four quantities and the way they have been used are shown in Figure 5. First, the X-ray results show that the nucleosomal DNA is wrapped in 1.8 left-handed superhelical turns (Finch et al. 1977; Richmond et al. 1984). Second, the change in linking number associated with nucleosome formation is approximately −1 per nucleosome for nucleosome arrays (Germond et al. 1975; Simpson et al. 1985). Finally, nucleosomal DNA exhibits a sequence periodicity of 10.1 – 10.2 bp (Drew and Travers 1985a; Satchwell et al. 1986), which is clearly different from the fourth result, namely the helical periodicity of DNA in solution (10.6 bp/turn) (Wang 1979; Rhodes and Klug 1980). This recent result presents important implications about the change in linking number on nucleosome formation.

If a particular repeated sequence is to be used over and over again for preferred bending at equivalent positions on a uniform superhelical path, then the structural, i.e., the helical, periodicity in the local frame of reference must match the sequence periodicity. This conclusion remains true for a nonuniform superhelical path as on the nucleosome where the *average* sequence periodicity will correspond to the *average* local twist, provided that between positions of identical sequence phase at each extremity of the periodically modulated sequence there is no net change in the direction of curvature relative to the average superhelical axis. (In this context, the direction of maximum curvature is given by the normal in the osculating plane, defined as containing in the limit two successive

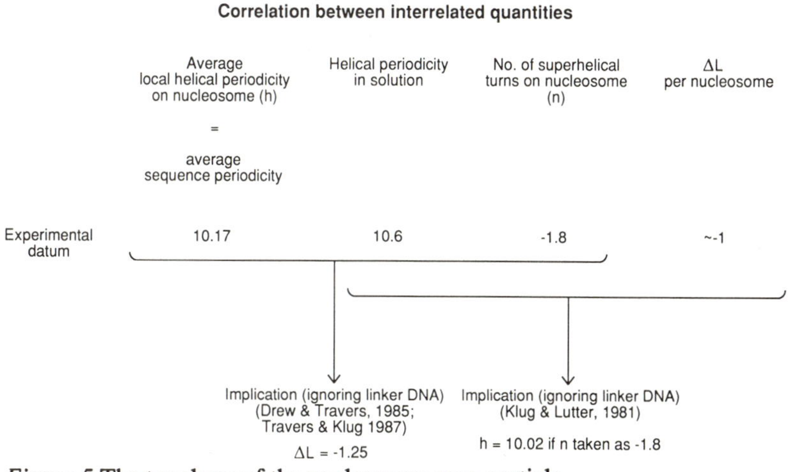

Figure 5 The topology of the nucleosome core particle.

tangents to the curve or three successive points [see, for example, Salmon and Rogers 1914].) If we assume that this condition is satisfied, then the average local twist on the nucleosome core would be 10.17 bp/turn.

The consequences for the change in linking number when DNA passes from solution, where its helical periodicity is 10.6 bp/turn (Wang 1979; Rhodes and Klug 1980; Peck and Wang 1981) are as follows. The change in linking number is equal to the number of superhelical turns if there is no change in the screw, as expressed in the local frame of reference (Finch et al. 1977). If the screw changes, this gives an extra contribution because the change of twist is not accounted for by the superhelical path. These contributions are additive provided the calculation is done in a local frame of reference (Fuller 1978). Then, in this case, for one superhelical turn, which contains 7.6 turns of the double helix (Richmond et al. 1984), and thus 77.3 bp (= 7.6 x 10.17)

$$\Delta Lk = -1 + ([77.3/10.17] - [77.3/10.6])$$

Hence

$$\Delta Lk = -1 + 0.31 = -0.69$$

where ΔLk is the change in linking number, –1 is due to the superhelical path, and 0.31 is due to the change in local twist.

For 1.8 superhelical turns on the nucleosome core

$$\Delta Lk = -0.69 - 1.8 = -1.24$$

This result for ΔLk can be derived in a different way by an argument made by J.T. Finch. When the superhelical path is pulled into an almost straight line, the writhe is reduced to almost zero, and all the linkage, which is conserved, is converted into twist. In this operation, the number of base pairs per turn in the local frame of reference is also conserved. Hence, for one superhelical turn on the nucleosome, the actual linking number is

$$Lk_{\text{nucleosome}} = -1 + (77.3\,/10.17)$$

Now, the linking number of DNA in solution is simply calculated from the twist. Hence for 77.3 bp

$$Lk_{\text{solution}} = 77.3\,/10.6$$

Hence

$$\Delta Lk = (-1 + [77.3/10.17]) - (77.3/10.6)$$

which is the same result as above. In this method, one is simply calculating the change in linkage between two, as it were, straight pieces of DNA so only changes in twist are involved.

Both methods of calculation demonstrate the essential point that the change in helical periodicity makes a substantial contribution to the change in linkage number as predicted previously (Finch et al. 1977; Klug and Lutter 1981). The value for ΔLk of −1.24 compares with the experimentally determined value of about −1 per nucleosome observed for nucleosome arrays containing four or more nucleosome core particles (Germond et al. 1975; Simpson et al. 1985).

The change in the linking number that is consequent upon formation of a single nucleosome is undetermined so we do not know whether this small discrepancy is due to a contribution from linker DNA or to a more complex topology for the path of nucleosome core DNA than we have assumed. However, recent measurements on reconstituted SV40 minichromosomes suggest that in this example the detectable torsional strain in linker DNA is no more than −0.02 to −0.05 turns/nucleosome (Ambrose et al. 1987). In addition, the reconstitution of a single histone octamer onto small, constrained DNA circles is associated with a linkage change of −1.1 ± 0.1 turns (Goulet et al. 1988; Zivanovic et al. 1988) although here it remains possible that the magnitude of the topological change is influenced by constraints imposed by the small size of the circles.

It should also be recalled that in the core particle there are deviations from a uniform superhelical path for the DNA (Richmond et al. 1984) and that they may contribute to the difference between the above calculation and the observed experimental value. If this were the case, the simple relationship between the twist in the local and laboratory frames of reference as described by Crick (1976) and by Fuller (1978) for a uniform superhelix may not be applicable.

It is probable that the average value of 10.17 bp/turn for the local twist of DNA in the nucleosome core particle obscures local variations in helical periodicity. Indeed, when the periodically modulated sequences in the region between 3 and 58 bp and 88 and 143 bp from the dyad are analyzed independently, the average periodicity of the sequence modulations of the trinucleotides with the greatest periodic modulation is 10.02 bp (B.F. Luisi et al., unpubl.). This means that to maintain an overall average periodicity of 10.17 bp, the average local twist of the central two double helical turns nearest the dyad must be close to 10.7 bp, i.e., to that of DNA in solution or even slightly underwound. Two lines of experimental evidence support this conclusion. First, the low cross-strand stagger of DNase I cleavage at and close to the dyad observed both for a population of chicken erythrocyte core particles (Lutter 1979) and for a nucleosome reconstituted with a defined, single DNA sequence (Rhodes 1985; Drew and Travers 1985a) is consistent with a low twist angle (Rhodes and Klug 1986). Second, the periodicity of cleavage by both DNase I and DNase II within the core increases in the vicinity of dyad (Lutter 1979, 1981; Cockell et al. 1983). Although the precise value of this cutting periodicity could be influenced by steric effects (for review, see Klug and Lutter 1981), the qualitative pattern of DNase I cleavage is closely paralleled by the sequence periodicities in core DNA (Fig. 6).

A local twist of 10.02 bp in the outer arms of nucleosome core DNA has a second important implication for nucleosome structure. As described previously (Finch et al. 1977; Klug and Lutter 1981), an integral number of base pairs per turn of the double helix would mean that the phosphate groups of the two adjacent superhelical turns would keep in phase. A periodicity of 10.0 bp/turn thus allows the same stabilizing interactions to occur repeatedly along the chain just as it does between adjacent molecules in fibers of B-DNA (Dover 1977). In addition, calculations using empirical energy functions suggest that smoothly bent DNA with a 45-Å radius of curvature is most stable with this same helical screw of 10.0 ± 0.1 bp/turn rather than 10.6 bp/turn (Levitt 1978). Hence, the observed sequence periodicity of core DNA strongly suggests that the structure of the nucleosome core particle optimizes the packing of adjacent superhelical turns of DNA, except in the neighborhood of the dyad. On the whole nucleosome, the exit and entry point of the DNA are also in the dyad region, which thus contains segments of three DNA helices whose precise relative configuration is as yet unknown.

We note finally that the above discussion of sequence periodicity relates to the "average" nucleosome core particle. However, it is clear that the average distance between DNase I cuts within reconstituted core particles is not constant but can vary within a narrow range from 10.2 to 10.5 bp (Drew and Travers 1985a; Rhodes 1985; Drew and Calladine 1987). One must not therefore equate the average helical periodicity with that of a particular core particle.

VII. SEQUENCE MARKERS FOR TRANSLATIONAL POSITIONING ON THE NUCLEOSOME

The periodically modulated sequences in nucleosome core DNA can be directly related to the periodic rotation of the DNA double helix as it wraps around the histone octamer and can thus define the angular orien-

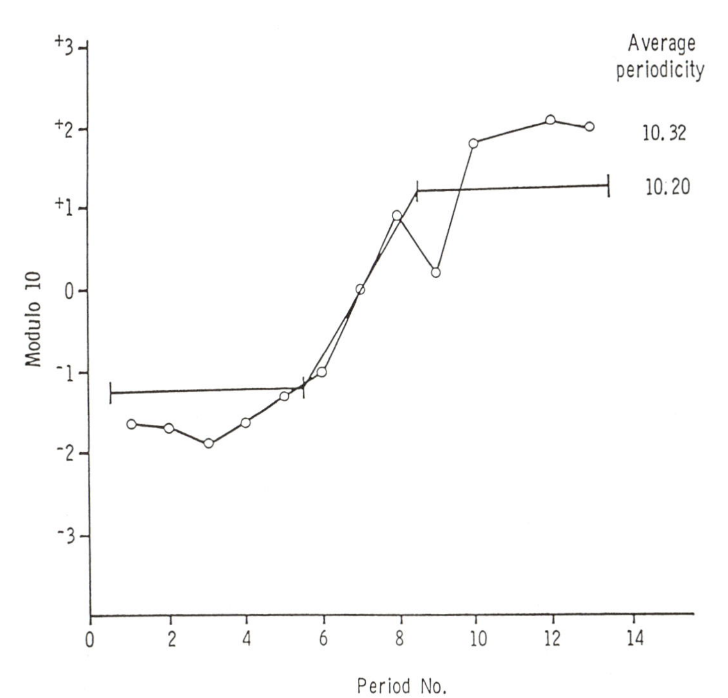

Figure 6 Comparison of average sequence periodicity with periodicity of cleavage by various nucleases in nucleosome core DNA. The average sequence periodicity (solid bars) for the outer arms of core DNA was calculated by Fourier analysis of the trinucleotide sequences with the strongest periodic modulation (GGC/GCC, AAA/TTT, AAT/ATT) separately over the two outer windows shown. These separate values were then averaged, and the periodicity in the region of the dyad was calculated from the difference of that number and the average sequence periodicity of 10.2 bp across the whole extent of nucleosome core DNA (Satchwell et al. 1986). The cleavage periodicities (open circles) are the average of the DNase I and DNase II cutting patterns of Lutter (1978, 1981). The dyad is taken as the origin, and thus any period in excess of 10 bp is scored as positive in the right-handed direction and negative in the left-handed direction. The difference between the two curves is a measure of the difference in angle of attack of the nucleases from perpendicularity to the superhelical axis produced by steric hindrance between neighboring superhelical turns (cf. Fig. 6 in Prunell et al. 1979; and Fig. 4 in Klug and Lutter 1981).

tation of the sequence relative to the direction of curvature of the helical axis. Note that the same average rotational setting can be achieved by simply translating the DNA along its axis relative to the histone octamer by a multiple of the helical periodicity. What determines this translational positioning? We might expect that local variations in both the magnitude and direction of bending, as well as variations in other physical parameters such as torsion, could select for the preferential location of sequences in a nonperiodic manner. The most striking example of such preferential location revealed by sequence analysis is the avoidance of the central superhelical turn of core DNA by runs of (dA-dT) greater than 8 bp in extent. Within core DNA, the occurrence of certain trinucleotides are also strongly favored or disfavored at particular positions. Such markers include a preference for TpTpG/CpApA at the dyad itself (Turnell et al. 1988) and for YpGpG/CpCpR in the regions immediately flanking the outer boundaries of the nucleosome (Drew and Calladine 1987).

With the available data, Drew and Calladine (1987) have constructed an algorithm based solely on the statistical occurrence of particular sequences in nucleosome core DNA. This algorithm assumes that the rotational sequence preferences for successive 10.2-bp periods are constant, except in the region of the dyad where a change in phase is necessary to accommodate changes in the direction of curvature. In addition, the translational position is determined by the relative frequency of occurrence of the trinucleotides ApApA/TpTpT and YpGpG/CpCpR over the same region. Even with this simplified picture of sequence distribution in core DNA, the algorithm accounts well for the rotational position of most reconstituted single nucleosomes as well as for that of three histone octamers reconstituted on a piece of DNA 860 bp long (Drew and Calladine 1987).

Three experimental studies have analyzed translational positioning by introducing local alterations in longer sequences that position precisely on a single histone octamer (Fitzgerald and Simpson 1985; Neubauer et al. 1986; Ramsay 1986). These alterations involve either successive deletions of the DNA sequence from the outer limits of the nucleosome core or, alternatively, internal insertions in the vicinity of the dyad. The latter alter the frames of part of the DNA sequence relative to its previous rotational orientation and to its contact points with specific histones. The general conclusion from all these studies is that the particular translational setting must arise through interaction with a relatively large segment of DNA. A secondary conclusion consistent with the sequence analysis and deduced largely from deletion studies is that those sequences that lie within two to three double helical turns of the dyad exert a stronger influence on translational positioning than sequences distant from the midpoint of the DNA. This conclusion is reinforced by an analysis of DNA sequences associated with nucleosomes containing histone H5 (Satchwell and Travers 1989). In these sequences, the translational sequence markers in the vicinity of the dyad are very similar to those in core DNA. In contrast, the rotational sequence signals are relatively weak in the H5-containing particles suggesting that the clamping of the structure of the nucleosome by this histone may be sufficient to override at least some of the sequence-dependent constraints on smooth DNA bending.

It also remains possible that sequences in the internucleosomal DNA or on the edge of the nucleosome core are equally or more important than the core DNA in determining translational positioning. Indeed, studies of the chromatin structure in the vicinity of the yeast *trp1* gene show that the neighboring origin of DNA replication is normally nucleosome-free in vivo (Thoma 1986). This region contains several short stretches of (dA-dT) and is intrinsically bent (Snyder et al. 1986). This raises the possibility that some structures of this type may be incompatible with the path of the DNA around the histone octamer and thus would exclude core nucleosome particles. We need also to ask to what extent do the sequence-dependent interactions of histones with DNA determine the positioning of nucleosomes in vivo. The preferred DNA sequence organization of nucleosome core particles (Satchwell et al. 1986) almost certainly defines high-affinity sites for histone octamers. Such sites, for example that on *Xenopus borealis* somatic 5S rDNA (Rhodes 1985) would be expected to be occupied efficiently in vivo. In vitro this particular nucleosome, in contrast with others, can survive repeated rounds of SP6 RNA polymerase transcription (Losa and Brown 1987; Lorch et al. 1988). Nevertheless, there are clear examples where the positions of nucleosome arrays in vivo do not depend solely on the DNA sequence to which the octamers are bound but instead can be shifted by alteration of the sequence context of the arrays (Fedor et al. 1988; Thoma and Zatchej 1989). In these situations, it has been proposed that the precise positions of individual nucleosomes may be dependent on interactions between histone octamers (Calladine and Drew 1986) or on the imposition of a boundary constraint by the presence of another protein (Fedor et al. 1988; Thoma and Zatchej 1988).

VIII. RELATION OF POSITION-DEPENDENT SEQUENCE PREFERENCES TO DNA STRUCTURE

To what extent are the rotational sequence preferences consistent with the known sequence-dependent polymorphism of DNA? The striking result of the sequencing of nucleosome DNA described above is that all such strongly periodically modulated sequences fall into two classes, whose settings on the nucleosome core differ in phase by approximately 180°. This means that the preferred base-pair steps of these two classes occupy positions separated by half a double helical turn. By reference to the crystal structure analysis (Richmond et al. 1984) and DNase I probing (Lutter 1978), we can deduce that these two classes correspond to positions where the minor groove points either inward or outward in relation to the histone octamer. In other words, the base pairs lie in an orientation that allows the major and minor grooves of the double helix to open and close smoothly as the DNA winds around the protein. This bending is consistent with the "roll" deformation in which the short axes of adjacent base-pair planes are inclined relative to each other (Dickerson and Drew 1981; Calladine 1982; Fratini et al. 1982; Dickerson 1983; Calladine and Drew 1984, 1986).

At present, there are insufficient data to relate the observed sequence periodicities to the many crystal structures of DNA that have been solved in the past few years. However, certain correlations are evident. For example, the sequence GpGpC in the crystal structure of d(GGGGCCCC) has a large total roll of +20° that opens the minor groove (McCall et al. 1985; see also Wang et al. 1982). The sequence ApApTpT in the crystal structure of d(CGCGAATTCGCG) has a slightly negative roll and a large "propeller twist" that closes the minor groove (Dickerson and Drew 1981). For the sequences ApApA and RpTpG, no detailed structural information is yet available.

Despite the paucity of X-ray data, Calladine and Drew (1986) have recently developed a second algorithm that accounts for the preferred positions of the histone octamer in terms of a set of preferred values of the roll angle at each of the ten types of dinucleotide steps. The algorithm measures the closeness of fit between an ideal distribution of roll angle values required to establish the given configuration of the DNA and the permissible values corresponding to a particular base sequence. The set of permissible values used is based partly on crystal structure data and partly chosen to fit the statistical preferences found in the observed data from nucleosome core DNA sequences. To obtain a good fit, most dinucleotide steps are allowed two possible values of roll angle in accordance with the bistability deduced from crystal data (Calladine and Drew 1984), and to introduce a degree of context dependence, the values for certain dinucleotide steps are chosen according to the preceding base pair.

This algorithm accounts successfully for the rotational setting observed for several DNA sequences that have been reconstituted with core histones and studied in solution (Simpson and Stafford 1983; Ramsay et al. 1984; Drew and Travers 1985a; Rhodes 1985; Drew and Calladine 1987). To account for the translational setting as well, Calladine and Drew (1986) found it necessary, as in the first algorithm based directly on statistical occurrences, to introduce a change of phase in the vicinity of the dyad as had been observed by Satchwell et al. (1986) for certain dinucleotide periodicities (e.g., Fig. 3, ApA) corresponding apparently to a jog in the path of the DNA, as described above.

This last point emphasizes the fact that sequence-dependent anisotropic flexibility is not the sole determinant of translational positioning on the nucleosome. Indeed, there are examples of strong sequence selection at particular positions that do not correlate with those for rotational orientation. One example already referred to is the nucleosome dyad, the point at which the two copies of histone H3 come into contact. Although the minor groove points outward at this position, the observed local sequence preferences show no correlation with the rotational preference expected for such an orientation on the nucleosome core (Satchwell et al. 1986; Turnell et al. 1988). Another possible example is the preferred occurrence of both TpGpG and CpGpG in locations other than in the region of the nucleosome dyad (Drew and Calladine 1987). In this case, it has been suggested that the torsional requirement for a high average twist angle in core DNA would result in the exclusion of these sequences from the central region because in the crystal structures of TpGpG and also of CpGpG the YpG steps are notable for an unusually low twist angle of 16° (Wang et al. 1982; McCall et al. 1986). Another possible determinant of nucleosome positioning would be specific chemical interactions between the DNA and the histone proteins at positions of close contact. At present, there is insufficient information to evaluate the contribution, if any, of such interactions.

IX. OTHER MODELS FOR DNA BENDABILITY

A number of largely theoretical, specific structural models have been proposed to account for the bendability of DNA, particularly in the context of the nucleosome core particle and also for the properties of intrinsically bent DNA. Trifonov and Sussman (1980) were the first to put forward the idea that there would be a relationship between a DNA sequence and its ability to be bent in a preferred direction. However, specific proposals on this relationship have not been sustained. In addition, many detailed predictions made by subsequent models are also not consistent with preferred positioning of particular sequences in nucleosome DNA.

The specific model proposed by Trifonov and his colleagues is the "wedge" structure for a DNA base pair step (Trifonov 1980, 1985; Mengeritsky and Trifonov 1983). They found by an iterative analysis of various eukaryotic DNA sequences that purine-purine steps in general, and ApA in particular, exhibit weak modulations with a 10.5-bp periodicity. Similarly, the occurrence of pyrimidine-pyrimidine steps appeared to be modulated with the same periodicity but in the opposite phase. They equated these weak periodic occurrences with the bending of DNA around the nucleosome core particle and suggested that the rotational orientation of these sequences was a consequence of a hypothetical wedge structure assumed by these dinucleotides (Trifonov and Sussman 1980). Neither the proposed periodicity nor the wedge structure is in accord with the experimental data. If, on the nucleosome core particle, ApA and TpT were to occupy positions about 5 bp apart, the occurrences of ApA and TpT combined should show a periodic fluctuation of 5 bp. Such a periodic modulation is notably absent (see Fig. 2 of Satchwell et al. [1986]). Instead, the occurrences of both ApA and TpT separately show an average modulation of 10.2 bp in identical phases. The wedge model thus does not account for the observed sequence modulations in nucleosome core DNA, and hence, such a proposed structure cannot be a major determinant of DNA bendability.

An alternative model was proposed by Zhurkin (1983, 1985) on the basis of the limited set of crystallographic data then available for DNA structures. He suggested that, when DNA bends, pyrimidine-purine steps prefer to adopt a position with the minor groove facing outward away from the direction of curvature, whereas purine-pyrimidine steps adopt the opposite orientation. Although such a relationship appears to hold for particular dinucleotide steps such as CpG, this suggestion is not generally true. Thus, both in crystals of G_4C_4 and in nucleosome core DNA, the purine-pyrimidine step GpC opens to the minor groove (McCall et al. 1985; Satchwell et al. 1986).

X. THE T·A BASE STEP

Not only can we correlate known DNA structures with certain observed sequence preferences in nucleosome core DNA, but also we can ask what these preferences tell us about the flexibility or rigidity of certain DNA sequences. For instance, simple alternating sequences that can wrap around the histone octamer must possess axial flexibility. There are two examples of such sequences, poly(dAT)·poly(dAT), which can be reconstituted into nucleosomes (Rhodes 1979), and sequences from the core DNA clones (Satchwell et al. 1986) that consist almost entirely of mixed purines on one strand and mixed pyrimidines on the opposite strand. These latter sequences thus contrast with the homopolymeric (dR-dY) sequences, poly(dA-dT) and poly(dG-dC), which cannot be reconstituted into nucleosomes (Rhodes 1979; Simpson and Kunzler 1979; see also, Kunkel and Martinson 1981; Prunell 1982) and are discussed in more detail below.

What are the distinguishing features of these axially flexible DNA sequences? Poly(dAT)·poly(dAT) contains two base steps that differ sub stantially in thermal stability with the dinucleotide step TpA melting at a temperature 20°C below the step ApT (Gotoh and Tagashira 1981). This difference in stability is presumably a consequence of a greater stacking overlap for the ApT step compared with the TpA step (Klug et al. 1979; Calladine and Drew 1984; Yoon et al. 1988). Several lines of evidence confirm the relative instability of the TpA step. First, in the crystal structure of pATAT, the helix is unstacked at its central TpA step, giving two separate ApT dinucleotide units (Viswamitra et al. 1978). Secondly, Patel et al. (1983), on determining the rate of exchange of the thymine imino proton by NMR spectroscopy as a measure of transient base-pair opening, found that such exchange was threefold faster for the sequence GTATAC than GAATTC and even faster for the sequence TATAAT. Thirdly, there are no kinetic barriers to the formation of cruciform structures by short poly(dAT)·poly(dAT) blocks in a negatively supercoiled plasmid, in contrast with other palindromic sequences that undergo this structural transition (Greaves et al. 1985; Haniford and Pulleyblank 1985). Finally, in both relaxed and supercoiled DNA, the TpA step is preferentially sensitive to cleavage by micrococcal nuclease and S1 nuclease (Dingwall et al. 1981; Hörz and Altenberger 1981; Drew et al. 1985; Flick et al. 1986). Because both these nucleases require an exposed single strand as a substrate (Drew 1984), these results suggest that some but not necessarily all sequences of the type NTAN are often, at least transiently, unwound (Drew et al. 1985). Taken together, these results show that the TpA dinucleotide step is less stable than other dinucleotide steps. This characteristic may account for the ubiquitous use of the TATA sequence in processes such as transcription, site-specific recombination, and the initiation of DNA replication, all of which involve DNA strand separation (Drew et al. 1985).

There is also reason to believe that the second example of simple sequence found in nucleosome core particles is easily unwound. The simplest asymmetric poly(dR-dY) sequence of this type is poly(dAG)·poly(dCT). Again, blocks of this alternating copolymer are preferentially cleaved by S1 nuclease both in relaxed and in negatively supercoiled DNA (Hentschel 1982; Mace et al. 1983; Htun et al. 1984; Pulleyblank et al. 1985; Evans and Efstratiadis 1986). It may therefore be more than a coincidence that both the simple repeating sequences of this type and also poly(dAT)·poly(dAT) can wrap around the histone octamer. Again, the deformability of the TpA step appears to be a major determinant of the selectivity of the interaction between the *trp* repressor and its operator site (see below; Otwinowski et al. 1988).

We emphasize that the crucial characteristic of simple DNA sequences that can be reconstituted into nucleosomes is that they possess conformational flexibility (see also Satchwell and Travers 1989). This property is not necessarily directly correlated with the melting temperatures of the sequence but, we suggest, may be related to the frequency of transient base-pair openings, as measured by the rate of exchange of the imino protons of guanine and thymine (Cheung et al. 1984; Leroy et al. 1988).

XI. INTRINSICALLY BENT DNA

We have so far been discussing the bendability of DNA with respect to its reaction to applied external forces imposed by interaction with proteins or by closure into a small circle. However, there are now clearly established cases of DNA existing in a permanently bent form without external constraints (Marini et al. 1982; Marini and Englund 1983; Hagerman 1984). This phenomenon was first recognized as a property of small circular DNA molecules found in the kinetoplast of certain flagellate protozoa (Simpson 1979; Marini et al. 1982; Marini and Englund 1983). It seems likely that the organized packing of these mini-circles into a catenated network is due to such an inherent structural feature of the DNA (Silver et al. 1986). The topic of intrinsic bending is now under intensive study by many groups, but some of the results on the bendability of DNA described above merge into the subject of permanently bent DNA. We can relate some of the features found in the disposition of sequences on the nucleosome directly to the structure of bent DNA.

It has been demonstrated, both for kinetoplast DNA and for artificially designed sequences, that any short runs of A or T nucleotides longer than 3 bp (such as AAAA or TTTTT), when periodically repeated at intervals of 10 to 11 bp, can confer a detectable amount of curvature on an isolated DNA molecule (Marini et al. 1982; Marini and Englund 1983; Wu and Crothers 1984; Hagerman 1985, 1986; Ulanovsky et al. 1986). This intrinsic bending is detectable by the anomalous migration in polyacrylamide gel electrophoresis (PAGE) of DNA fragments containing such sequences (Simpson 1979). From such measurements of mobility it remains unclear by what amount the DNA bends and in what direction (but see Calladine et al. 1988). However, direct measurements of birefringence decay show that a DNA molecule that is anomalously retarded on gel electrophoresis behaves in solution as though the average distance between the ends of the molecule is less than would be expected for a linear DNA molecule of average persistence length (Hagerman

1984; Levene et al. 1986; but see Diekmann and Pörschke 1987). These observations strongly suggest that DNA molecules with anomalous gel mobility possess a time-averaged net curvature. This conclusion is supported by two further observations. First, the proposed curvature can be visualized directly by electron microscopy (Griffith et al. 1986; Kitchin et al. 1986), and second, an otherwise normal DNA molecule may be induced to migrate abnormally by a rigid structure in the form of a DNA cruciform or a bound protein located near its center (Wu and Crothers 1984; Gough and Lilley 1985; Shuey and Parker 1986).

How do these results relate to the bending of DNA on the nucleosome? We know that on the nucleosome core there is a strong tendency for the centers of short runs of (dA-dT) such as AAA or AAAA to lie with their minor grooves along the inside of the DNA supercoil, whereas runs of somewhat longer length such as AAAAA and AAAAAA tend to be on the upper and lower surfaces of the supercoil (Fig. 7a). In other words, the junctions of these longer runs of (dA-dT) with their flanking sequences coincide (on average) on one side with a maximum in the distribution of the centers of the trinucleotide ApApA/TpTpT and on the other side with GpGpC/GpCpC (Fig. 4b). From the arguments advanced above, this means that these junctions are sited where the minor groove either closes or opens outward and the center of the homopolymer run lies in a region of minimum curvature. This is consistent with a structure in which the homopolymer run has little or no curvature, whereas at the junctions there is a significant "roll" angle responsible for a change in the direction of the helix axis. A structural discontinuity at certain such junctions is apparent from digestion studies with DNase I (Drew and Travers 1985a).

If we assume that the structure of longer (dA-dT) runs is the same in solution as it is on the nucleosome core particle, we can then attempt to predict the rotational orientation configuration of intrinsically bent DNA. The simplest sequence that forms such a structure is $(T_5A_5)_n$ (Hagerman 1985). This sequence has two junctions, an ApT step that can adopt a negative roll angle (Fratini et al. 1982) and a TpA step that in two crystals has a positive roll angle (Wang et al. 1982; Shakked et al. 1983). Because runs of $(dA)_n$, where $n \geq 4$, are sufficient for the formation of intrinsic bends (Koo et al. 1986), the sequence $(T_5A_5)_n$ should adopt a curved configuration with the ApT and TpA steps positioned where the minor groove is on the inside and outside of the curve, respectively (Fig. 7b). Similarly for the sequences $(GA_4T_4C)_n$ and $(CA_4T_4G)_n$, the ApT step would occupy an inward-facing minor groove, whereas the CpG and GpC steps would occupy a corresponding outward-facing position as they do on the nucleosome. These sequences indeed migrate unusually

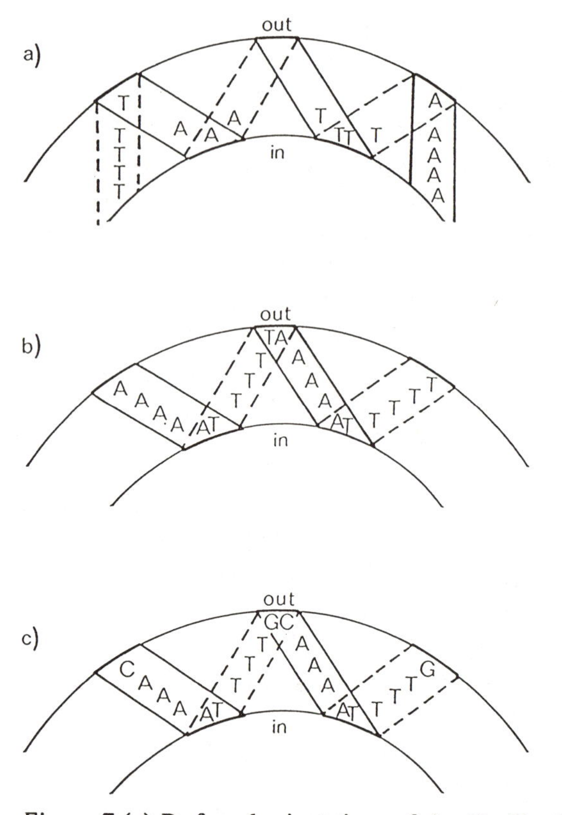

Figure 7 (*a*) Preferred orientations of A_3, T_4, T_5, A_5 sequences in nucleosome core DNA. Schematic details are the same as those in Fig. 4b. (*b*) Conjectured rotational orientation of the intrinsically bent DNA sequence $(A_5\ T_5)_n$ (Hagerman 1985). (*c*) Conjectured rotational orientation of the intrinsically bent DNA sequence $(CA_4T_4G)_n$ (Hagerman 1986).

slowly on PAGE. However, the related sequences $(GT_4A_4C)_n$ and $(CT_4A_4G)_n$, in which a TpA step replaces the ApT step, migrate normally (Hagerman 1986). This alteration could either be due to the instability of the TpA step or to a preference for this step with a positive roll angle to be positioned where the minor groove is on the outside of the curve. The former possibility is supported by two observations. First, that the (dA)-(dT) tracts adjacent to the TpA step are cleaved at a normal rate by a hydroxyl radical in sharp contrast with the reduced rate of cleavage of (dA)-(dT) tracts adjacent to an ApT step (Burkhoff and Tullius 1988). Second, the rate of thymine imino proton exchange is five times slower in the sequence $G_2A_3T_3C_2$ than in $C_2T_3A_3G_2$ (Leroy et al. 1988). This result shows that the two types of sequences are structurally distinct and thus implies that the particular sequences with a TpA step may not possess the structural characteristics necessary for the formation of intrinsic bends.

For the sequences discussed above, an alternation of positive and negative roll angles spaced on average at half-turn intervals could account for a curvature of the helical axis. However, such an alternation is not sufficient by itself (Diekmann 1986). The role of the oligo(dA-dT) stretch might therefore be, by virtue of its conformational rigidity, to stabilize the orientation of the base.

There are other indications that there is something special about oligo(dA)-(dT). It has long been apparent that poly(dA-dT) is structurally distinct from random sequence DNA, adopting a helical periodicity of 10 bp/turn in contrast with the average 10.5–10.6 bp/turn (Rhodes and Klug 1981; Peck and Wang 1981). Some of its other anomalous properties are discussed by Rhodes and Klug (1981). The reluctance of this homopolymer to reconstitute into nucleosome cores may arise because of a lack of conformational flexibility, a property that may be intrinsic to its unusual helical structure. In mixed-sequence DNA, runs of (dA-dT) 5 bp but not 3 bp in extent are relatively resistant to digestion by DNase I (Drew and Travers 1984). Thus, a critical length of 4–5 bp is required for runs of (dA-dT) to assume a DNase-I-resistant structure that we believe to be closely related or identical to poly(dA-dT). This is the same minimum length that is necessary for the formation of intrinsic bends (Koo et al. 1986; Diekmann 1986) and for the adoption of a structure with slow rate of exchange of thymine imino protons (Leroy et al. 1988).

The determination in this laboratory of the crystal structure of a DNA dodecamer containing a homopolymeric run of six A·T base pairs (Nelson et al. 1987) provides a structural explanation for the anomalous properties of oligo(dA-dT) tracts. In the crystal, the (dA-dT) tract is essentially straight, that is the average planes through the base pairs are parallel to each other and perpendicular to the helix axis, and thus they have zero roll. The base pairs do have an unusual structure because of the high propeller twist within each of them. This results in maximal overlap of the bases on each strand and the formation of a run of additional, non-Watson-Crick, cross-stranded hydrogen bonds. It is these latter features that could confer a conformational rigidity over and above that expected from base pairs with two hydrogen bonds. Similarly, we might expect these same features to limit the torsional flexibility of homopolymeric tracts of A·T base pairs as compared with tracts containing alternating (dAT-dAT) sequences. The distinctive physical features of oligo-(dA)·oligo(dT), the decreased helical repeat, and the reduced pitch and narrow minor groove in relation to B-type DNA are simply natural consequences of the compact base-stacking arrangement. We note that the similarity in crystal packing between the dodecamers d(CGCGAAT TCGCG) (Dickerson and Drew 1981) and d(CGCA$_6$GCG) (Nelson et al. 1987) implies that the overall bending of the molecules is very similar in both structures, possibly as a consequence of the packing interactions themselves. Whatever the cause, the details of the overall bending along the length of the molecules are different. In the d(CGCGAATTCGCG) structure, the bending is more continuously distributed, whereas in the d(CGCA$_6$GCG) structure the oligo(dA-dT) tract is straight with the bending concentrated at flanking base steps as large roll components coherent with each other. This has been attributed to the conformational rigidity of oligo(dA-dT) tracts.

XII. THE MOLECULAR BASIS FOR INTRINSIC BENDING

What are the implications of the structure of oligo(dA-dT) for the intrinsic bending of DNA? There must be something else about the poly(dA-dT) structure other than resistance to bending, or else permanent bending could be conferred by runs of oligo(dG-dC). It seems probable that both the perpendicularity of the mean plane of the A·T base pairs to the helix axis, i.e., zero roll, as well as the high degree of propeller twist within the base pairs must be significant. In the crystal of d(CGCA$_6$GCG), the most abrupt changes in the direction of the local helix axis occur at the CA step and the GC step 3′ to the (dA) tract (Nelson et al. 1987). At both these steps, the normal purine clash (Calladine 1982) is exaggerated by a difference in propeller twist resulting in a large positive roll angle for the pyrimidine-purine step and a corresponding negative roll angle for the purine-pyrimidine step.

Much of the contribution to overall bending could thus arise from large roll angles at the junctions of the A·T stretch with flanking base pairs. Because the angles at the 5′ and 3′ ends of the tract are opposite in sign, they would be additive when placed half a helical turn apart. However, it is not generally necessary to have all the change concentrated at the junctions; it could be more widely distributed over the region between the oligo(dA-dT) tracts. To give a permanent bend, there need only be a net roll accumulated in the intervening regions as discussed in general terms by Nelson et al. (1987). The mathematical basis for this is found in Calladine and Drew (1986), and its detailed application to bending was made by Calladine et al. (1988).

The failure of oligo(dG-dC) tracts (Koo et al. 1986) to confer intrinsic bending can also be explained by the above considerations. Homopolymeric runs of G·C base pairs would be expected to be conformationally rigid. However, in such runs, the GG-CC base steps are estimated to have an average positive roll angle of 4–5° (McCall et al. 1985). Because the average roll angle for all other base steps (with the exception of AA-TT in homopolymeric runs) is estimated to be approximately 3–4° (Calladine et al. 1988), this means that, for a DNA sequence containing phased runs of oligo(dG-dC), any net roll accumulated in the intervening regions will be compensated by that in the oligo(dG-dC) tracts, resulting in little or no net change in the direction of the helical axis.

Both the structure of oligo(dA-dT) tracts (Drew and Travers 1984) and the extent of bending as estimated from mobility in gels (Diekmann and Wang 1985; Koo et al. 1986; Diekmann 1987) are highly sensitive to temperature and to the ionic environment. In particular, the apparent intrinsic curvature is greatest at low temperatures and in the presence of a divalent cation (Koo et al. 1986; Diekmann 1987; Laundon and Griffith 1987). Because of this variation, we need to ask whether the crystal structure of oligo(dA-dT) is an appropriate model for the molecular basis of intrinsic curvature. It is clear that the conditions of crystallization (4°C, 10 mM Mg^{++}) fulfill the two principal requirements for maximum curvature, and the crystal structure of oligo(dA-dT) is therefore highly relevant. However, several molecular models for intrinsic curvature have so far been deduced from mobility measurements on different intrinsically bent sequences in gel buffers lacking a divalent cation. There is therefore no reason to assume that the *average* structure of an oligo(dA-dT) tract under these gel-running conditions is the same as that in the crystal structure (Nelson et al. 1987). Calladine et al. (1988) estimate from the mobility data that the intrinsic curvature conferred by a repeating $d(A_6)$ tract is 1.1°/base step, whereas the curvature estimated from circularization experiments (which necessarily include Mg^{++}) is approximately 2.3°/base step (Ulanovsky et al. 1986; Zahn and Blattner 1987).

XIII. OTHER MODELS FOR INTRINSIC BENDING

In the literature, there are several theories of bending that differ in their predictions as to where the actual changes in the direction of the helix axis occur. These essentially static models are based either on a semiqualitative analysis of the relative mobilities of defined sequence intrinsically bent species in gels (Koo et al. 1986; Ulanovsky and Trifonov 1987; Calladine et al. 1988; Koo and Crothers 1988) or on molecular mechanics calculations (von Kitzing and Diekmann 1987). Of those models that attempt to deduce the detailed structure of oligo(dA-dT) from mobilities in gels, that of Ulanovsky and Trifonov (1987) proposes that the AT tract is itself bent because of a combination of wedge roll and tilt at each and every AA-TT step. This large, noncoplanar wedge estimated to be around 9° would cause a continuous change in the helix direction within the oligo(dA-dT) tract. This proposal is clearly inconsistent with the structure of such a tract in the crystal of d(CGCA$_6$GCG) (Nelson et al. 1987).

In contrast, junction models (Koo et al. 1986; von Kitzing and Diekmann 1987; Koo and Crothers 1988) suggest that the bending occurs abruptly at or close to the ends of a straight AT stretch because of some specialized features of that region. In particular, Koo et al. (1986) and von Kitzing and Diekmann (1987) ascribe curvature to a change in the base-pair tilt at the junction while maintaining parallel stacking of the base pairs, as envisioned originally by Selsing et al. (1979) (see Fig. 8). The former model places the junction precisely at the base steps immediately flanking the oligo(dA-dT) run, whereas in the latter model the junctions occur at the first A-A step within a (dA-dT) tract and at a corresponding distance 3′ to the (dA) run. Both these models assume that the A·T base pairs assume a high negative tilt angle. Again, none of the essential features of these models are apparent in the crystal structure of d(CGCA$_6$GCG). A variant of the function model of Koo et al. (1986) invokes a combination of base step roll at the 5′ junction of (dA) tract with

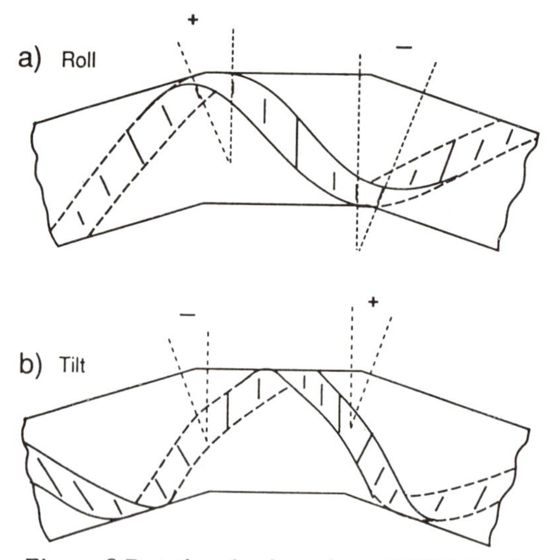

Figure 8 Rotational orientation of DNA bends produced by positive and negative roll angles separated by half a double helical turn (*a*) and positive and negative tilt angles separated by half a double helical turn (*b*). The schematics represent extreme models for bending. For clarity the angles are exaggerated.

a change in base pair tilt at both the 5′ and 3′ junctions (Koo and Crothers 1988). This model in common with the other models deduced from measurements of mobility in gels is subject to the caveats discussed above.

XIV. DNA BENDING, WRAPPING, AND LOOPING

In many biological interactions between DNA and proteins, the trajectory of the helical axis of the DNA is required to deviate from an average straight path, that is, the DNA is bent. We should, however, clearly distinguish between those situations, as on the nucleosome, where the DNA is tightly wrapped on a protein surface and those where DNA is looped between protein-binding sites that, although linearly distant from each other in the DNA sequence, are brought into close spatial proximity by cooperative interactions between the proteins bound at the separated sites (e.g., Krämer et al. 1987). In the case of DNA wrapping, both the configuration and the conformation of the DNA are highly constrained by contact with the protein surface, and in many cases, the magnitude of curvature can be large relative to the values normally assumed in solution. In contrast, when DNA make a long loop between proteins, the magnitude of curvature is generally much less, and the path of the DNA is not restricted to a precise configuration. Both these factors imply that the bendability of DNA is less significant when DNA loops than when it wraps. However, as we discuss below, although for looping the axial rigidity of the DNA generally is not an important consideration, its torsional rigidity limits the length of DNA in loops to certain discrete values.

The nucleosome core particle is an example of a nucleoprotein complex in which DNA is tightly wrapped and which does not make chemical interactions between histones and specific DNA sequences. In this case, bendability is a major determinant of specific positioning. In many other such complexes, however, DNA bending, throughout a major effect, seems to complement sequence-specific recognition. The relative contribution of bending to complex formation must depend both on the magnitude of curvature and on the length of the curved DNA segment. Because it seems that the energy required to deform DNA mildly in a nonpreferred direction is a good deal less than that which might be available for direct sequence recognition (by, for example, hydrogen bonding), the influence of bendability must generally be a second-order effect.

A crucial question is to what extent the sequence preferences observed on the nucleosome apply to other nucleoprotein complexes containing bent DNA. In two cases, that of the 434 repressor headpiece/operator complex (Anderson et al. 1987; Aggarwal et al. 1988) and of the catabolite-activator protein (CAP) interaction with the *lac* regulatory region (Gartenberg and Crothers 1988), there is sufficient information available to answer this question. Both these complexes are examples of interactions where it has been shown that conformational constraints on DNA complement identified or assumed direct base-specific recognition.

A. Catabolite-activator Protein

The complex of CAP with its binding site in the *lac* control region, a site for which the protein has the highest affinity of all characterized naturally occurring targets, is of particular interest. The existance of significant DNA bending has been inferred from the anomalous mobility of the complex in polyacrylamide gels (Wu and Crothers 1984), from the effect of CAP on the rate of ligation of small circular DNA molecules containing the *lac* control region (Kotlarz et al. 1986) and from electron micrographs of the complex (Gronenborn et al. 1984). In the complex, the strong binding interactions span 28–30 bp of DNA as indicated by ethylation interference experiments (Liu-Johnson et al. 1986). By relating the known crystal structure of free CAP protein complexed with cAMP at 2.5-Å resolution (McKay and Steitz 1981). Warwicker et al. (1987) have proposed models in which a ramp of positive electrostatic potential running around three sides of the protein provides a surface for interaction with 30–44 bp of DNA in a tightly bent configuration. These calculations suggest that the DNA is bent through approximately 150°, or on average, 35–50° per double helical turn. This approach does not yield a unique solution with the two most favored models differing principally in the planarity of bending: the first suggesting a planar bend, and the second suggesting a left-handed supercoil. The second is more consistent with the observed change of –0.18 in linking number on the binding of CAP to DNA (Buc et al. 1987).

Because sequence conservation in CAP binding sites is essentially confined to the central double helical turn, one might expect that bending of this magnitude would make use of the bending preferences of short DNA sequences within the CAP binding site (Travers and Klug 1987). This site contains an interrupted inverted repeat with potential major groove contacts through a helix-turn-helix motif on each side of the center of symmetry. Thus, the dyad at the center must have its minor groove pointing in toward the protein. At this position and at 10 bp on either side are short AT-rich sequences that on the nucleosome occur preferentially where the minor groove is narrowed (Fig. 9). Similarly, exactly out of

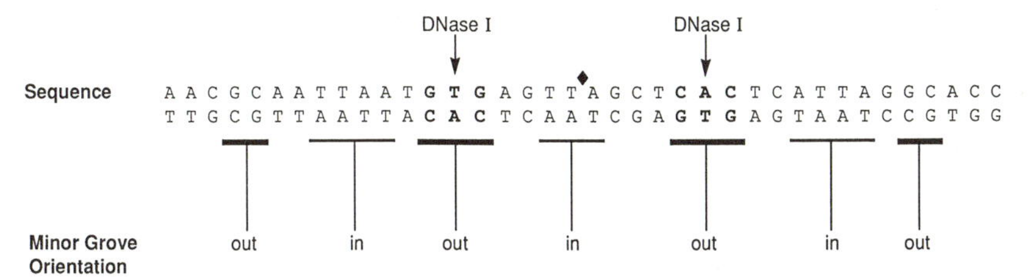

Figure 9 Sequence of the CAP binding site in the *E. coli lac* promoter region showing the relationship of the DNA sequence to the rotational orientation of the minor groove. A structural periodicity of 10 bp is assumed between the DNase I accessible regions (arrows), in accord with the observed cutting periodicity. (Data from Spassky et al. 1984.)

phase are short CC-rich sequences that preferentially occupy an outward-facing minor groove. This precise correspondence between the *lac* CAP binding site and nucleosomal DNA is consistent with both the known direction of bending in the CAP/DNA complex (Zinkel and Crothers 1987) and its inferred magnitude (Liu-Johnson et al. 1986; Warwicker et al. 1987). Of possible significance, this correspondence is not as apparent in other weaker naturally occurring CAP binding sites.

To test the inference (Travers and Klug 1987) that this pattern of sequences is related to the bending of DNA, Gartenberg and Crothers (1988) exhaustively mutagenized the CAP binding site, concentrating on two particular regions (–11 and –16) where the minor groove of DNA points respectively toward and away from the protein surface. They found that many of the mutations in these regions, but not mutations situated two double helical turns from the dyad, affected both the affinity of CAP for the binding site and also the mobility of CAP/DNA complexes in polyacrylamide gels. The results strongly suggest that the magnitude of DNA bending in the complex, as indicated by gel mobility, correlates with the binding affinity of CAP. From the effects of the sequence changes on gel mobility, Gartenberg and Crothers concluded that the inferred bending was more consistent with a dependence on the properties of dinucleotides than on those of mononucleotides, but there was insufficient data to test a dependence on trinucleotides. In general, the dinucleotide dependence of bending in the CAP/DNA complex correlates well with that observed on the nucleosome: AT-rich sequences being favored where the minor groove points inward and disfavored where the minor groove points outward with the opposite pattern for GC-rich sequences. Conversely, the sequence data from the nucleosomal DNA can be used to calculate the likelihood of bending of the mutated CAP binding sites. With one exception, there is an excellent correlation between the observed mobilities of the mutant complexes and the calculated flexural probability (Drew et al., this volume). This result establishes that the sequence-dependent bending preferences in the nucleosome are also applicable to other nucleoprotein complexes.

B. 434 Repressor and Cro Protein

A similar correspondence can also be shown to exist between the bending preferences in the nucleosome and the affinity of 434 repressor for its operator site. Recently, Aggarwal et al. (1988) have determined to 2.5-Å resolution the X-ray structure of a complex between the 434 repressor DNA-binding domain and a 20-bp operator fragment TATACAAGAAA GTTTGTACT. This structure is closely related to an earlier TATGTTCT TTCAAACATGAA crystal structure of the 434 DNA-binding domain with a 14-bp DNA fragment (Anderson et al. 1987) In both complexes, the DNA structure is distorted with an overall bend of an average radius of curvature of 65 Å. As in the nucleosome, bending is not smooth but is concentrated at two sites symmetrically disposed 2–3 base steps away from the central region, which is itself relatively straight. The most striking feature of the structure is a narrowing of the minor groove at the center. This narrowing is accompanied by an overwinding of the base steps in this region by several degrees. The crucial importance of the conformation of this central region for the interaction with the protein is demonstrated by the observation that the introduction of a nick on one strand at its central phosphodiester band increased the affinity of the operator for the repressor by approximately fivefold (Koudelka et al. 1988). In contrast, a similar nick 4 bp outside the operator had no effect on this affinity. In a complementary experiment, a mutation introduced into the repressor near the interface of the dimer resulted in a failure to discriminate between nicked and intact operators. It was assumed but not directly demonstrated that this mutation increases the flexibility of the dimer interaction.

The existence of constraints on the DNA conformation within the central region of the operator, where there are no direct amino acid base-pair interactions, is wholly consistent with the effect of mutations in this region on the affinity for the repressor. For example, replacement of the central d(ATAT) by d(ACGT) weakens the binding affinity by a factor of 50, whereas replacement by d(AAAA) increases it by a factor of 5. Significantly, the mutant repressor is less sensitive to these substitutions. Again, however, the clear pattern is that the required conformation is favored by AT-rich sequences and disfavored by GC-rich sequences. When the likelihood of flexure for the central region is calculated from the observed sequence-dependent bending preferences in the nucleosome, assuming that the structure of the complex does not change as the central sequence is altered, there is a strong correlation between the observed affinity of the operator for the repressor and the predicted likelihood of flexure (Drew et al., this volume). This result again points to the generality of the nucleosomal sequence preferences and indicates that the structural deformations at the center of the complexed 434 operator must be similar to those encountered at an inward-facing minor groove on the nucleosome.

A question at issue is whether these sequence preferences are related solely to bending or also to other properties of DNA, such as torsional flexibility or rigidity (Aggarwal et al. 1988). The structural feature common to inward-facing minor grooves on the nucleosome and the central region of the complexed 434 operator is an unusually narrow minor groove. The establishment of this structure is favored by a particular base-stacking conformation that concomitantly increases the twist. Any relaxation of the requirement for a narrow minor groove in the complexed operator would be expected to expand the minor groove at the center and at the same time slightly alter the angular register of the two bends. This prediction remains to be tested.

A second phage-coded protein that binds the 434 operator is the Cro repressor. The conformation of operator DNA complexed with this latter protein differs significantly from that of the same DNA complexed with the *cI* repressor (Wolberger et al. 1988). In the Cro/operator complex, the overall bend in the DNA is of comparable magnitude to that of <100-Å radius in the repressor/operator complex. Both complexes are similar in that there are no direct contacts between the base pairs and the protein in the vicinity of the operator dyads, yet the local conformation of the DNA in this region differs subtly with respect to the minor groove width in the two complexes. Significantly, the slightly wider minor groove in the Cro/operator complex, although still narrow relative to an average width, is reflected in a reduced response to sequence change when compared with the repressor/operator complex (Koudelka et al. 1987). In other words, the selection for a particular DNA sequence is less stringent because the conformational constraints are more relaxed in the complex with Cro than in that with the repressor.

C. *trp* Repressor

In the two examples discussed so far, the selectivity of the protein for a particular DNA sequence depends both on direct sequence recognition and also on local conformational constraints. Nevertheless, substantial selectivity can be established even in the absence of direct contacts between the amino acid side chains of a protein and the base pairs in the binding site. In the crystallized complex of the *E. coli trp* (tryptophan) repressor with its 18-bp operator, there are no direct hydrogen bonded or nonpolar contacts between the protein and the DNA that can account for the selectivity of the interaction (Otwinowski et al. 1988). Instead, direct hydrogen-bonded contacts are made principally to the phosphate groups in the operator although well-ordered water molecules can make polar contacts with the functional groups of mutationally sensitive base pairs (Bass et al. 1987). These detailed interactions suggest that the basis for specificity is the ability of the operator DNA to assume a particular overall conformation that allows for good interaction with a repressor dimer (Otwinowski et al. 1988). A necessary condition for the validity of this suggestion is that the energy required to deform other DNA sequences to the appropriate conformation must be too high to permit the formation of a stable complex.

The sequence of the *trp* operator in the crystallized complex is d(GTA CTAGTTAACTAGTAC), in which mutational studies show that the CTAG sequences centered 4 bp from the dyad are the most crucial for favorable binding (Bass et al. 1987). In the complex, this DNA structure is dominated by the flexibility of TpA steps. The three central TpA steps have a significant positive roll: The sequence dyad directs curvature away from the protein, whereas the two flanking TpA steps in the CTAG sequence direct curvature toward the protein. The essential element of the *trp* operator, CTAG, is unusual in that it is one of the rarest tetranucleotide sequences in *E. coli* and has the characteristic that it is preferentially cut by micrococcal nuclease (Flick et al. 1986), an enzyme that preferentially cleaves at unwound base steps (Drew 1984). The structure of CTAG in the bound operator may be related to that in the crystal structure of the self-complementary DNA oligomer d(CTCTAGAG) (Hunter et al. 1989). In this structure, the dominant feature is again at the TpA step where cross-strand purine-purine stacking forces the phosphate backbone into an unusual conformation. In this respect, the structures of the crystalline DNA and the operator site are comparable although the positive roll at TpA is greater in the latter case. Nevertheless, the thermal lability of the TpA step could in principle contribute to the formation of a stable DNA complex by providing a low-energy barrier to any necessary structural rearrangement on binding of the protein.

The *trp* operator/repressor complex when crystallized appears to be a good example of selectivity of protein binding that is determined in large part by the potential conformation of the DNA-binding site. Nevertheless, there are two caveats to this conclusion. First, although there are no direct contacts between the amino acid side chains and mutationally sensitive base pairs, three well-ordered water molecules mediate hydrogen-bonded interactions between each half repressor and base pairs in the half operator. This organization raises the possibility that recognition of base pairs need not require precise contacts between the protein and the individual base but instead could be mediated by proxy through a precisely positioned third molecule. Second, in the crystal structure the amino-terminal arms of the repressor do not form defined contacts with operator DNA although removal of these arms reduces the in vitro affinity of the repressor for the operator by 30- to 100-fold (Carey, quoted in Otwinowski et al. 1988).

D. Large Nucleoprotein Complexes

The influence of DNA bendability on nucleosome positioning also has other implications for protein/DNA interactions in general. Although the precise path of the DNA is not known, electron microscopy of three other large DNA/protein complexes involving the *E. coli* DNA gyrase and *dnaA* protein and bacteriophage λ integrase (Better et al. 1982; Fuller et al. 1984; Kirchausen et al. 1985) shows that a long stretch of DNA is tightly wrapped around an aggregate of protein molecules just as in the nucleosome. One consequence of this arrangement is the bringing together of parts of the DNA molecule that are far apart in the linear sequence and that are involved in a DNA transaction (Echols 1986). By inspection of the patterns of DNase I digestion of such complexes, it is possible to deduce the general direction of curvature of the DNA with respect to the protein. In all cases so far examined,[1] it appears the sequence-dependent preferences that are characteristic of the nucleosome core are conserved in the interactions of other proteins with DNA (A.A. Travers, unpubl.). The sequence conservation implies that the DNA curvature in protein/DNA complexes is positive; that is, the curvature is directed toward the protein. If this were not the case, no correlation would have been observed between minor grooves directed outward from the protein (and therefore accessible to DNase I) and sequences in corresponding positions on the core associated with a known direction of curvature.

[1]The particular proteins surveyed were *E. coli* DNA gyrase (Morrison and Cozzarelli 1981; Kirkegaard and Wang 1981), *dnaA* protein (Fuller et al. 1984), Tn*3* resolvase (Grindley et al. 1982; Sherratt et al. 1984), λ integrase (Ross et al. 1979), λ O protein (Zahn and Blattner 1985), and *E. coli* RNA polymerase holoenzyme (Siebenlist et al. 1980; Spassky et al. 1985).

The conformational constraints imposed by DNA wrapping should be contrasted with those for DNA looping. Here, the formation of looped complexes requires that the interacting proteins be in the correct angular orientation relative to each other. This means that for loops of less than about 500 bp the distance between double helical binding sites can only vary by integral multiples of double helical turns. This is because in this range of DNA length the torsional flexibility of DNA is insufficient to stabilize bending between sites that are not in angular register (Shore and Baldwin 1983). These restrictions on the distance between protein binding sites has been observed experimentally in several cases. In the regulatory region of the *E. coli araBAD* operon, repression depends on two binding sites for the *araC* protein normally separated by approximately 225 bp. This separation can be increased up to about 400 bp (R.F. Schleif, pers. comm.) but only in increments corresponding to double helical periods (Dunn et al. 1984). The requirement for quantized intervening lengths of this type implies that the optimal distance between binding sites will also be dependent on the average twist of the DNA. In the case of the tetrameric *lac* repressor, which can simultaneously bind two separated *lac* operators on the same DNA fragment (Mossing and Record 1986) and thus form a loop (Krämer et al. 1987), underwinding of DNA in the loop as a consequence of negative superhelical strain alters the optimal lengths for loop formation between the two *lac* operators (Krämer et al. 1988). In yet another striking but perhaps artificial example, the λ*cI* repressor can bind cooperatively to two operator sites separated by five or six double helical turns but not by five and one-half double helical turns (Hochschild and Ptashne 1986). In this example, the magnitude of curvature in the loop is sufficient to deform the local structure of the DNA, and hence, the binding energy stabilizing protein-protein contacts must be sufficient to overcome any adverse bending preferences. Although most well-characterized cases of DNA looping occur in eubacteria, it seems likely that this phenomenon is of general occurrence. Indeed, a requirement for quantized lengths of DNA separating the binding sites for different proteins have been reported in the region regulating the expression of early SV40 RNA (Takahashi et al. 1986).

XV. BIOLOGICAL IMPLICATIONS OF DNA BENDABILITY

We have argued that DNA bendability is a major determinant of the specific positioning in the interaction between a long piece of DNA and the histone octamer. The precise placement of core nucleosome particles in chromatin can thus be directly related to a particular structural property of DNA. Positioning of this nature has important implications for the regulation of eukaryotic genes transcribed by both RNA polymerase II and RNA polymerase III. Several cases have now been described where nucleosomes are positioned to overlap or occlude transcriptional control regions (Rhodes 1985; Almer and Hörz 1986; Richard-Foy and Hager 1987; Kefalas et al. 1988). Specific activation of such genes can require removal of these positioned nucleosomes. One such example is the yeast PHO5 gene where activation results in the removal of four nucleosomes upstream of the transcription starting point with the consequent unmasking of an additional regulatory sequence (Almer et al. 1986). A similar situation occurs in the mouse mammary tumor virus long terminal repeat where hormone-mediated promoter activation requires the interaction of the glucocorticoid receptor with DNA exposed on the surface of a positioned nucleosome (Richard-Foy and Hager 1987). This interaction alters the chromatin structure in this region to expose a binding site for a transcriptional activator, nuclear factor I.

In addition to DNA bendability, the intrinsic curvature of DNA may also influence DNA/protein interactions. Thus, in some situations, it may be functionally advantageous for a particular DNA sequence to remain nucleosome free. Such a role has been suggested for runs of (dA-dT) acting as constitutive upstream activating elements in yeast (Struhl 1985) and also for the presence of intrinsically bent DNA in a yeast replication origin (Snyder et al. 1986). In contrast, the interaction of SV40 T antigen with its binding site is dependent on the altered DNA configuration imposed by a short oligo(dA-dT) tract (Ryder et al. 1986). Similarly, it has been suggested that the condensation of heterochromatin in mouse nuclei may be dependent on short stretches of intrinsically bent DNA in a specialized satellite DNA (Radic et al. 1987). Another example is the facilitation of the initial location of a promoter site by *E. coli* RNA polymerase by the presence of intrinsic bends upstream of the conserved hexameric –35 and –10 sequences (Travers 1988). In this case, the facilitation of wrapping by intrinsic bending implies that one of the first steps in polymerase/promoter complex formation is recognition of a particular structure of DNA rather than a precise sequence (cf. Drew and Travers 1985b).

There is now substantial evidence that the tight wrapping of DNA is a necessary prelude to certain protein-mediated DNA transactions, in particular those involving the unstacking or unwinding of the DNA double helix in the initiation of both transcription and DNA replication and, additionally, in site-specific recombination. The origins of replication (*ori*) on both bacteriophage λ and *E. coli* DNA contain tandemly repeated sequences that bind an oligomeric complex of the λ O protein and the *E.*

coli dnaA protein (Fuller et al. 1984; Zahn and Blattner 1987). The initial recognition of these replication origins by their respective binding proteins involves cooperative binding to form a complex in which the DNA is wrapped in a negative supercoil around a protein core (Fuller et al. 1984). In the case of the *E. coli oriC* complex, *dnaA* protein in vitro can by itself induce strand separation within a 13-bp repeated sequence (Bramhill and Kornberg 1988), whereas for the λ complex additional proteins appear to be necessary for localized unwinding (Dodson et al. 1986). In both cases, it has been suggested that the wrapping of DNA in the initial complex drives the destabilization of the AT-rich regions at which unstacking occurs. This process would then be analogous to RNA chain initiation by *E. coli* RNA polymerase in which the unstacking of the base pairs in the –10 region is preceded at many promoters by DNA wrapping (for review, see Travers 1988).

The many examples of DNA bending in protein/DNA complexes show that the sequence-dependent preferences for bending a DNA double helix are of wide biological occurrence and utility. These sequence-dependent features constitute another physical level of information present in DNA sequences, secondary to the chemical code.

XVI. SUMMARY

The DNA in the nucleosome core particle constitutes the first well-studied example of protein-induced DNA bending. In this structure, the DNA is wrapped tightly around a histone octamer with approximately 80 bp per superhelical turn. Studies of both naturally occurring and reconstituted systems have shown that DNA sequences very often adopt well-defined locations with respect to the octamer. This precise positioning correlates with the preferred occurrence of certain short sequences in positions where the minor groove of DNA points in toward or away from the histone octamer and can therefore be explained in terms of the differential flexibility of different sequences and of departments from smooth bending. The simple geometry of the nucleosome allows one to draw conclusions that have more general applicability to other nucleoprotein complexes in which DNA is bent.

Certain DNA sequences are also intrinsically bent or curved. The relationship between this phenomenon and the sequence dependence of DNA bendability is related to the conformational flexibility or rigidity of short-sequence elements. Together, these sequence-dependent features constitute another physical level of information present in DNA sequences, secondary to the chemical code.

ACKNOWLEDGMENTS

We thank Drs. John Finch, Michael Levitt, Hillary Nelson, and Daniela Rhodes for stimulating discussions and helpful comments on the manuscript.

XVII. APPENDIX

A. Helix Geometry

In this section, we summarize the definitions for the helical parameters describing the structure of double-stranded DNA nucleic acids. These definitions have been successively discussed and expanded by Arnott et al. (1969), Fratini et al. (1982), Dickerson (1983), and von Kitzing and Diekmann (1987). It is important to note that until 1988 no one set of definitions was universally used for discussions of DNA curvature and therefore that the quantitative parameters deduced by different investigators may not necessarily be directly comparable. However, at a European Molecular Biology Organization workshop on DNA bending and curvature held in Cambridge in September 1988, a set of definitions for the structural parameters of DNA was agreed upon and is fully described by Dickerson et al. (1989).

1. Reference Frames

The geometry of a base pair or a base step may generally be described with reference to either a global or to a local helix axis. A global helix axis is the best helix axis for a single DNA molecule considered as a whole. This definition is useful for structural analysis of DNA fibers or crystals of short DNA oligomers but is clearly inappropriate for curved DNA.

A local helix axis can be defined with respect to a single base pair or to a base step. In the former case, the axis is perpendicular to the mean plane of the base pair. In the latter case, two types of definition are in common use. In the first, the helical twist axis is normal to the plane that bisects the solid angle between two adjacent base-pair planes. The parameters obtained relative to this axis are called wedge parameters. The second definition uses the approximate local cylinder symmetry of the double helix. The *cylinder twist axis* is defined such that base-pair plane 1 is transformed into base-pair plane 2 only by rotation around and translation along this axis (von Kitzing and Diekmann 1987).

2. Roll and Tilt Angles

Absolute roll and tilt angles describe the angular orientation of a base-pair plane with respect to a defined helical axis. Absolute roll, symbolized as ϕ_R (Fratini et al. 1982), is the inclination of the short axis of the mean base-pair plane to a plane normal to the *global* helix axis. Absolute tilt, ϕ_T, is the inclination of the long axis of the mean base-pair plane to a plane normal to the global helix axis. In these definitions, ϕ_R is positive if the base plane rotates counterclockwise when viewed along the base-pair axis from the strand 1 side, and ϕ_T is positive if the base plane rotates clockwise when viewed directly into the minor groove.

Relative roll and tilt angles, θ_R and θ_T, respectively, measure the *changes* in roll and tilt orientation from one base pair to the next; that is, they refer to the geometry of a base step. The angles θ_R and θ_T are independent of the coordinate frame in which the helix positions are expressed. θ_R is the rotation around the long axis of a base pair relative to that of its neighbor in the direction of the major or minor groove (by convention, it is positive when rotating away from the minor groove). θ_T is the corresponding relative rotation around the short axis toward either super phosphate backbone (Fratini et al. 1982). Relative roll and tilt angles described with respect to the wedge helical axis differ slightly from the values derived from the definitions of Fratini et al. (1982) because the former but not the latter are independent of the twist angle (Prunell et al. 1984; von Kitzing and Diekmann 1987).

3. *The Relation between Relative Roll and Absolute Tilt*

In a DNA structure in which each base step has a uniform roll angle ρ, the local helix axis itself assumes an overall helical path. This means that with respect to the global helix axis the base pairs will tilt from the horizontal by arctan (ρ/τ) where τ is the twist between adjacent base pairs (Calladine and Drew 1984). A similar relationship can also be derived for the wedge parameters (von Kitzing and Diekmann 1987).

REFERENCES

Aggarwal, A.K., D.W. Rodgers, M. Drottar, M. Ptashne, and S.C. Harrison. 1988. Recognition of a DNA operator by the repressor of phage 434: a view at high resolution. *Science* **242:** 899–907.

Almer, A. and W. Hörz, W. 1986. Nuclease hypersensitive regions with adjacent positioned nucleosomes mark the gene boundaries of the PHO5/PHO3 locus in yeast. *EMBO J.* **5:** 2681–2687.

Almer, A., H. Rudolph, A. Hinnen, and W. Hörz. 1986. Removal of positioned nucleosomes from the yeast PHO5 promoter upon PHO5 induction releases additional upstream activating DNA elements. *EMBO J.* **5:** 2689–2696.

Ambrose, C., R. McLaughlin, and M. Bina. 1987. The flexibility and topology of simian virus 40 DNA in minichromosomes. *Nucleic Acids Res.* **15:** 3703–3721.

Anderson, J.E., M. Ptashne, and S.C. Harrison. 1987. Structure of the repressor-operator complex of bacteriophage 434. *Nature* **326:** 846–852.
Arnott, S., S.D. Dover, and A.J. Wonacott. 1969. Least squares refinement of the crystal and molecular structures of DNA and RNA from X-ray data and bond lengths and angles. *Acta Crystallogr.* **B25:** 2192–2206.
Bass, S., P. Sugiono, D.N. Arvidson, R.P. Gunsalus, and P. Youderian. 1987. DNA specificity determinants of *Escherichia coli* tryptophan repressor binding. *Genes Dev.* **1:** 565–572.
Better, M., C. Lu, R.C. Williams, and H. Echols. 1982. Site-specific DNA condensation and pairing mediated by the int protein of bacteriophage lambda. *Proc. Natl. Acad. Sci.* **79:** 5837–5841.
Bramhill, D. and A. Kornberg. 1988. Duplex opening by dnaA protein at novel sequences in initiation of replication at the origin of the *E. coli* chromosome. *Cell* **52:** 743.
Buc, H., M. Amouyal, M. Buckle, M. Herbert, A. Kolb, D. Kotlarz, M. Menendez, S. Rimsky, A. Spassky, and E. Yeramian. 1987. Activation of transcription by the cyclic AMP receptor protein. In *RNA polymerase and the regulation of transcription* (ed. W.S. Reznikoff et al.), p. 115–125. Elsevier, New York.
Burkhoff, A.M. and T.D. Tullius. 1988. Structural details of an adenine tract that does not cause DNA to bend. *Nature* **331:** 455–457.
Calladine, C.R. 1982. Mechanics of the sequence-dependent stacking of bases in B-DNA. *J. Mol. Biol.* **161:** 343–352.
Calladine, C.R. and H.R. Drew. 1984. A base-centred explanation of the B-to-A transition in DNA. *J. Mol. Biol.* **178:** 773–782.
———. 1986. The principles of sequence-dependent flexure of DNA. *J. Mol. Biol.* **192:** 907–918.
Calladine, C.R., H.R. Drew, and M.J. McCall. 1988. The intrinsic curvature of DNA in solution. *J. Mol. Biol.* **201:** 127–137.
Cheung, A., K. Arndt, and P. Lu. 1984. Correlation of *lac* operator DNA iminoproton exchange kinetics with its function. *Proc. Natl. Acad. Sci.* **81:** 3665–3669.
Cockell, M., D. Rhodes, and A. Klug. 1983. Location of the primary sites of micrococcal nuclease cleavage on the nucleosome core. *J. Mol. Biol.* **170:** 423–446.
Crick, F.H.C. 1953. The Fourier transform of a coiled-coil. *Acta Crystallogr.* **6:** 685–689.
———. 1976. Linking numbers and nucleosomes. *Proc. Natl. Acad. Sci.* **73:** 2639–2643.
Dickerson, R.E. 1983. Base sequence and helix structure variation in B- and A-DNA. *J. Mol. Biol.* **166:** 419–441.
Dickerson, R.E. and H.R. Drew. 1981. Structure of a A-DNA dodecamer: Influence of base sequence on helix structure. *J. Mol. Biol.* **149:** 761–786.
Dickerson, R.E., M. Bansal, C.R. Calladine, S. Diekmann, W.N. Hunter, O. Kennard, R. Lavery, H.C.M. Nelson, W.K. Olson, W. Saenger, Z. Shakked, H. Sklenar, D.M. Soumpasis, C.-S. Tung, E. von Kitzing, A.H.-J. Wang, and V.B. Zhurkin. 1989. Definitions and nomenclature of nucleic acid structure parameters. *EMBO J.* **8:** 1–4.
Diekmann, S. 1986. Sequence specificity of curved DNA. *FEBS Lett.* **195:** 53–56.
———. 1987. Temperature and salt dependence of the gel migration anomaly of curved DNA fragments. *Nucleic Acids Res.* **15:** 247–-265.
Diekmann, S. and D. Pörschke. 1987. Electro-optical analysis of curved DNA fragments. *Biophys. Chem.* **26:** 207–216.
Diekmann, S. and J.C. Wang. 1985. On the sequence determinants and flexibility of the kinetoplast DNA fragment with abnormal gel electrophoretic mobilities. *J. Mol. Biol.* **186:** 1–11.
Dingwall, C., G.P. Lomonossoff, and R.A. Laskey. 1981. High sequence specificity of

micrococcal nuclease. *Nucleic Acids Res.* **9:** 2659–2673.

Dodson, M., H. Echols, S. Wickner, C. Alfano, K. Mensa-Wilmot, B. Gomes, J. Lebowitz, J.D. Roberts, and R. McMacken. 1986. Specialized nucleoprotein structures at the origin of replication of bacteriophage λ: Localized unwinding of duplex DNA by a six-protein reaction. *Proc. Natl. Acad. Sci.* **83:** 7638–7642.

Dover, S.D. 1977. Symmetry and packing in B-DNA. *J. Mol. Biol.* **110:** 699–700.

Drew, H.R. 1984. Structural specificities of five commonly used DNA nucleases. *J. Mol. Biol.* **176:** 535–557.

Drew, H.R. and C.R. Calladine. 1987. Sequence specific positioning of core histones on an 860 bp DNA: Experiment and theory. *J. Mol. Biol.* **195:** 143–173.

Drew, H.R. and Travers, A.A. 1984. DNA structural variations in the *E. coli tyrT* promoter. *Cell* 37: 491-502.

———. 1985a. DNA bending and its relation to nucleosome positioning. *J. Mol. Biol.* **186:** 773–790.

———. 1985b. Structural junctions in DNA: The influence of flanking sequence on nuclease digestion specificities. *Nucleic Acids Res.* **13:** 4445–4467.

Drew, H.R., J.R. Weeks, and A.A. Travers. 1985. Negative supercoiling induces spontaneous unwinding of a bacterial promoter. *EMBO J.* **4:** 1025–1032.

Dunn, T.M., S. Hahn, S. Ogden, and R.F. Schleif. 1984. An operator at −280 base pairs that is required for repression of *ara*BAD operon promoter: Addition of DNA helical turn between the operator and promoter cyclically hinders repression. *Proc. Natl. Acad. Sci.* **81:** 5017–5020.

Echols, H. 1986. Multiple DNA-protein interactions governing high precision DNA transactions. *Science* **233:** 1050–1056.

Edwards, C.A. and R.A. Firtel. 1984. Site-specific phasing in the chromatin of the rDNA in *Dictyostelium discoideum*. *J. Mol. Biol.* **180:** 73–90.

Evans, T. and A. Efstratiadis. 1986. Sequence dependent S1 nuclease hypersensitivity of a heteronomous DNA duplex. *J. Biol. Chem.* **261:** 14771–14780.

Fedor, M.J., N.F. Lue, and R.D. Kornberg. 1988. Statistical positioning of nucleosomes by specific protein-binding to an upstream activating sequence in yeast. *J. Mol. Biol.* **204:** 109–127.

Finch, J.T., L.C. Lutter, D. Rhodes, R.S. Brown, B. Rushton, M. Levitt, and A. Klug. 1977. Structure of nucleosome core particles of chromatin. *Nature* **269:** 29–36.

Fitzgerald, P.C. and R.T. Simpson. 1985. Effects of sequence alterations in a DNA segment containing the 5S RNA gene from *Lytechinus variegatus* on positioning of a nucleosome core particle *in vitro*. *J. Biol. Chem.* **260:** 15318–15324.

Flick, J.T., J.C. Eissenberg, and S.C.R. Elgin. 1986. Micrococcal nuclease as a DNA structural probe: its recognition sequences, their genomic distribution and correlation with DNA structure determinants. *J. Mol. Biol.* **190:** 619–633.

Fox, K.R. and N.R. Howarth 1985. Investigations into the sequence selective binding of mithramycin and related ligands to DNA. *Nucleic Acids Res.* **13:** 8695–8714.

Fox, K.R. and M.J. Waring 1984. DNA structural variations produced by actinomycin and distamycin as revealed by DNAse I footprinting. *Nucleic Acids Res.* **12:** 9271–9285.

Fratini, A.V., M.L. Kopka, H.R. Drew, and R.E. Dickerson. 1982. Reversible bending and helix geometry in a B-DNA dodecamer CGCGAATTBrCGCG. *J. Biol. Chem.* **257:** 14686–14707.

Fuller, F.B. 1978. Decomposition of the linking number of a closed ribbon: a problem from molecular biology. *Proc. Natl. Acad. Sci.* **75:** 3557–3561.

Fuller, R.S., B.E. Funnell, and A. Kornberg. 1984. The dnaA protein complex with the *E.*

coli chromosomal origin (*ori*C) and other DNA sites. *Cell* **38:** 889–900.

Gartenberg, M.R. and D.M. Crothers. 1988. DNA sequence determinants of CAP-induced bending and protein binding affinity. *Nature* **333:** 824-829.

Germond, J.E., B. Hirt, P. Oudet, M. Gross-Bellard, and P. Chambon. 1975. Folding of the DNA double helix in chromatin-like structures from simian virus 40. *Proc. Natl. Acad. Sci.* **72:** 1843–1847.

Gotoh, O. and Y. Tagashira. 1981. Stabilities of nearest-neighbour doublets in double-helical DNA determined by fitting calculated melting profiles to observed profiles. *Biopolymers* **20:** 1033–1042.

Gough, G.W. and D.M.J. Lilley. 1985. DNA bending induced by cruciform formation. *Nature* **313:** 154–156.

Goulet, I., Y. Zivanovic, A. Prunell, and B. Revet. 1988. Chromatin reconstitution on small DNA rings. I. *J. Mol. Biol.* **200:** 253–266.

Greaves, D.R., R.K. Patient, and D.M.J. Lilley. 1985. Facile cruciform formation by an $(A\text{-}T)_{34}$ sequence from a *Xenopus* globin gene. *J. Mol. Biol.* **185:** 461–478.

Griffith, J., M. Bleyman, C.A. Rauch, P.A. Kitchin, and P.T. Englund. 1986. Visualisation of the bent helix in kinetoplast DNA by electron microscopy. *Cell* **46:** 717–724.

Grindley, N.D.F., M.R. Lassick, R.C. Wells, R.J. Wityk, J.J. Salvo, and R.R. Reed. 1982. Transposon-mediated site-specific recombination: Identification of three binding sites for resolvase at the *res* I sites of γ δ and Tn3. *Cell* **30:** 19–27.

Gronenborn, A.M., M.V. Nermut, P. Eason, and G.M. Clore. 1984. Visualization of cAMP receptor protein-induced DNA kinking by electron microscopy. *J. Mol. Biol.* **179:** 751–757.

Hagerman, P.J. 1984. Evidence for the existence of stable curvature of DNA in solution. *Proc. Natl. Acad. Sci.* **81:** 4632–4636.

———.1985. Sequence dependence of the curvature of DNA: A test of the phasing hypothesis. *Biochemistry* **24:** 7033–7037.

———.1986. Sequence-directed curvature of DNA. *Nature* **321:** 449–450.

Haniford, D.B. and D.E. Pulleyblank. 1985. Transition of a cloned $d(AT)_n$-$d(AT)_n$ tract to a cruciform *in vivo*. *Nucleic Acids Res.* **13:** 4343–4363.

Hentschel, C.C. 1982. Homocopolymer sequences in the spacer of a sea urchin histone gene repeat are sensitive to S1 nuclease. *Nature* **295:** 714–716.

Hochschild, A. and M. Ptashne. 1986. Cooperative binding of λ repressors to sites separated by integral turns of the DNA helix. *Cell* **44:** 681–687.

Horowitz, D.J. and J.C. Wang. 1984. Torsional rigidity of DNA and length dependence of the free energy of supercoiling. *J. Mol. Biol.* **173:** 75–91.

Hörz, W. and W. Altenburger. 1981. Sequence specific cleavage of DNA by micrococcal nuclease. *Nucleic Acids Res.* **9:** 2643–2658.

Htun, H., E. Lund, and J.E. Dahlberg. 1984. Human U1 RNA genes contain an unusual nuclease S1 cleavage site within the conserved 3′ flanking region. *Proc. Natl. Acad. Sci.* **81:** 7288–7292.

Hunter, W.N., B.L. D'Estaintot, and O. Kennard. 1989. Structural variation in a deoxyoctanucleotide d(CTCTAGAG), implications for protein-DNA interactions. *Biochemistry* **28:** 2444–2451.

Kefalas, P., F.C. Gray, and J. Allan. 1988. Precise nucleosome positioning in the promoter of the chicken β^A globin gene. *Nucleic Acids Res.* **16:** 501–517.

Kirchausen, T., J.C. Wang, and S.C. Harrison. 1985. DNA gyrase and its complexes with DNA: Direct observation by electron microscopy. *Cell* **41:** 933–943.

Kirkegaard, K. and J.C. Wang. 1981. Mapping the topography of DNA wrapped around gyrase by nucleolytic and chemical probing of complexes of unique DNA sequences.

Cell **23:** 721–729.

Kitchin, P.A., V.A. Klein, K.A. Ryan, K.L. Gann, C.A. Rauch, D.S. Kang, R.D. Wells, and P.T. Englund. 1986. A highly bent fragment of *Crithidia fasciculata* kinetoplast DNA. *J. Biol. Chem.* **261:** 11302–11309.

Klug, A. and L.C. Lutter. 1981. The helical periodicity of DNA on the nucleosome. *Nucleic Acids Res.* **9:** 4267–4283.

Klug, A., A. Jack, M.A. Viswamitra, O. Kennard, Z. Shakked, and T.A. Steitz. 1979. A hypothesis on a specific sequence-dependent conformation of DNA and its relation to the binding of the *lac*-repressor protein. *J. Mol. Biol.* **131:** 669–680.

Koo, H.S. and D.M. Crothers. 1988. Calibration of DNA curvature and a unified description of sequence-directed bending. *Proc. Natl. Acad. Sci.* **85:** 1763–1767.

Koo, H.S., H.-M. Wu, and D.M. Crothers. 1986. DNA bending at adenine-thymine tracts. *Nature* **320:** 501–506.

Kotlarz, D., A. Fritsch, and H. Buc. 1986. Variation of intramolecular ligation rates allow the detection of protein-induced bends in DNA. *EMBO J.* **5:** 799–803.

Koudelka, G.B., S.C. Harrison, and M. Ptashne. 1987. Effect of non-contacted bases on the affinity of 434 operator for 434 repressor and Cro. *Nature* **326:** 886–888.

Koudelka, G.B., P. Harbury, S.C. Harrison, and M. Ptashne. 1988. DNA twisting and the affinity of bacteriophage 434 operator for bacteriophage 434 repressor. *Proc. Natl. Acad. Sci.* **85:** 4633–4637.

Krämer, H., M. Amouyal, A. Nordheim, and B. Müller-Hill. 1988. DNA supercoiling changes the spacing requirement of two *lac* operators for DNA loop formation with *lac* repressor. *EMBO J.* **7:** 547–556.

Krämer, H., M. Niemöller, M. Amouyal, B. Revet, B. von Wilcken-Bergmann, and B. Müller-Hill. 1987. *lac* repressor forms loops with linear DNA carrying two suitable spaced *lac* operators. *EMBO J.* **6:** 1481–1491.

Kunkel, G.R. and H.G. Martinson. 1981. Nucleosome will not form on double-stranded RNA or over poly(dA)•poly(dT) tracks in recombinant DNA. *Nucleic Acids Res.* **9:** 6869–6888.

Laundon, C.H. and J.D. Griffith. 1987. Cationic metals promote sequence-directed DNA bending. *Biochemistry* **26:** 3759–3762.

Leroy, J.L., E. Charretier, M. Kochoyan, and M. Guéron. 1988. Evidence from base-pair kinetics for two types of adenine tract structures in solution: Their relation to DNA curvature. *Biochemistry* **27:** 8894–8898.

Levene, S.D., H.-M. Wu, and D.M. Crothers. 1986. Bending and flexibility of kinetoplast DNA. *Biochemistry* **25:** 3988–3995.

Levitt, M. 1978. How many base-pairs per turn does DNA have in solution and in chromatin? Some theoretical calculations. *Proc. Natl. Acad. Sci.* **75:** 640–644.

Liu-Johnson, H.-N., M.R. Gartenberg, and D.M. Crothers. 1986. The DNA binding domain and bending angle of *E. coli* CAP protein. *Cell* **47:** 995–1005.

Lomonossoff, G.P., P.J.G. Butler, and A. Klug. 1981. Sequence-dependent variation in the conformation of DNA. *J. Mol. Biol.* **149:** 745–760.

Lorch, Y., J.W. La Pointe, and R.D. Kornberg. 1988. On the displacement of histones from DNA by transcription. *Cell* **55:** 743–744.

Losa, R. and D.D. Brown. 1987. A bacteriophage RNA polymerase transcribes in vitro through a nucleosome core without displacing it. *Cell* **50:** 801–808.

Lutter, L.C. 1978. Kinetic analysis of deoxyribonuclease cleavages in the nucleosome core: Evidence for a DNA superhelix. *J. Mol. Biol.* **124:** 391–420.

———. 1979. Precise location of DNase I cutting sites in the nucleosome core determined by high resolution gel electrophoresis. *Nucleic Acids Res.* **6:** 41–56.

———. 1981. DNase II digestion of the nucleosome core: precise locations and relative exposure of sites. *Nucleic Acids Res.* **9:** 4251–4265.

Mace, H.A.F., H.R.B. Pelham, and A.A. Travers. 1983. Association of an S1-nuclease sensitive structure with short direct repeats 5′ of *Drosophila* heat shock genes. *Nature* **304:** 555–557.

Marini, J.C. and P.T. Englund. 1983. Correction and retraction. *Proc. Natl. Acad. Sci.* **80:** 7678.

Marini, J.C., S.D. Levene, D.M. Crothers, and P.T. Englund. 1982. Bent helical structure in kinetoplast DNA. *Proc. Natl. Acad. Sci.* **79:** 7664–7668.

McCall, M., T. Brown, and O. Kennard. 1985. The crystal structure of d(GGGGCCCC). A model for poly(dG)•poly(dC). *J. Mol. Biol.* **183:** 385–396.

McCall, M., T. Brown, W.N. Hunter, and O. Kennard. 1986. The crystal structure of d(GGATGGGAG) forms an essential part of the binding site for transcription factor IIIA. *Nature* **322:** 661–664.

McKay, D.B. and T.A. Steitz. 1981. Structure of catabolite gene activator protein at 2.9 Å resolution suggests binding to left handed DNA. *Nature* **290:** 744–749.

Mengeritsky, G. and E.N. Trifonov. 1983. Nucleotide sequence-directed mapping of the nucleosomes. *Nucleic Acids Res.* **11:** 3833–3851.

Morrison, A. and N.R. Cozzarelli. 1981. Contacts between DNA gyrase and its binding site of DNA: Features of symmetry and asymmetry revealed by protection from nucleases. *Proc. Natl. Acad. Sci.* **78:** 1416–1420.

Mossing, M.C. and M.T. Record, Jr. 1986. Upstream operators enhance repression of the *lac* promoter. *Science* **233:** 899–892.

Nelson, H.C.M., J.T. Finch, B.F. Luisi, and A. Klug. 1987. The structure of an oligo(dA)-oligo(dT) tract and its biological implications. *Nature* **330:** 221–226.

Neubauer, R., W. Linxweiler, and W. Hörz. 1986. DNA engineering shows that nucleosome phasing on African green monkey α-satellite is the result of multiple additive histone-DNA interactions. *J. Mol. Biol.* **190:** 639–645.

Otwinowski, J., R.H. Schveitz, R.-G. Zhang, C.L. Lawson, A. Joachimiak, R.O. Marmorstein, B.F. Luisi, and P.B. Sigler. 1988. Crystal structure of *trp* repressor/operator complex at atomic resolution. *Nature* **335:** 321–329.

Palen, T.E. and T.R. Cech. 1984. Chromatin structure at the replication origins and transcription-initiation regions of the ribosomal RNA genes of Tetrahymena. *Cell* **36:** 933–942.

Patel, D.J., S.A. Kozlowski, S. Ikuta, K. Itakura, R. Bhatt, and D.R. Hare. 1983. NMR studies of DNA conformation and dynamics in solution. *Cold Spring Harbor Symp. Quant. Biol.* **47:** 197–206.

Peck, L. and J.C. Wang. 1981. Sequence dependence of the helical repeat of DNA in solution. *Nature* **292:** 375–378.

Prunell, A. 1982. Nucleosome reconstitution on plasmid-inserted poly(dA)•poly(dT). *EMBO J.* **1:** 173–179.

Prunell, A. and R.D. Kornberg. 1982. Variable center to center distance of nucleosomes in chromatin. *J. Mol. Biol.* **154:** 515–523.

Prunell, A., I. Goulet, Y. Jacob, and J. Goutorbe. 1984. The smaller helical repeat of poly(dA)•poly(dT) relative to DNA may affect the wedge property of the dA•dT base pair. *Eur. J. Biochem.* **138:** 253–257.

Prunell, A., R.D. Kornberg, L. Lutter, A. Klug, M. Levitt, and F.H.C. Crick. 1979. Periodicity of deoxyribonuclease I digestion of chromatin. *Science* **204:** 855–858.

Pulleyblank, D.E., D.B. Haniford, and A.R. Morgan. 1985. A structural basis for S1 nuclease sensitivity of double stranded DNA. *Cell* **42:** 271–280.

Radic, M.Z., K. Lundgren, and R.A. Hamkalo. 1987. Curvature of mouse satellite DNA and condensation of heterochromatin. *Cell* **50:** 1101–1108.

Ramsay, N. 1986. Deletion analysis of a DNA sequence that positions itself precisely on the nucleosome core. *J. Mol. Biol.* **189:** 179–188.

Ramsay, N., G. Felsenfeld, B.M. Rushton, and J.D. McGhee. 1984. A 145-base pair DNA sequence that positions itself precisely and asymmetrically on the nucleosome core. *EMBO J.* **3:** 2605–2611.

Rhodes, D. 1979. Nucleosome cores reconstituted from poly(dA-dT) and the octamer of histones. *Nucleic Acids Res.* **6:** 1805–1816.

———.1985. Structural analysis of a triple complex between the histone octamer, a *Xenopus* gene for 5S RNA and transcription factor IIIA. *EMBO J.* **4:** 3473–3482.

Rhodes, D. and A. Klug. 1980. Helical periodicity of DNA determined by enzyme digestion. *Nature* **286:** 573–578.

———. 1981. Sequence-dependent helical periodicity of DNA. *Nature* **292:** 378–380.

———. 1986. An underlying repeat in some transcriptional control sequences corresponding to half a double helical turn of DNA. *Cell* **46:** 123–132.

Richard-Foy, H. and G.L. Hager. 1987. Sequence-specific positioning of nucleosomes over the steroid-inducible MMTV promoter. *EMBO J.* **6:** 2321–2328.

Richmond, T.J., J.T. Finch, B. Rushton, D. Rhodes, and A. Klug. 1984. Structure of the nucleosome core particle at 7 Å resolution. *Nature* **311:** 532–537.

Ross, W., A. Landy, Y. Kukuchi and H. Nash. 1979. Interaction of int protein with specific sites on λ *att* DNA. *Cell* **18:** 297–307.

Ryder, K., S. Silver, A.L. DeLucia, E. Fanning, and P. Tegtmeyer. 1986. An altered DNA conformation in origin region I is a determinant for the binding of SV40 T antigen. *Cell* **44:** 719–725.

Salmon, G. and R.A.P. Rogers. 1914. Invariants and covariants of systems of quadrics. In *A treatise on the analytic geometry of three dimensions,* p. 228. Longmans, Green, London.

Satchwell, S.C. and A.A. Travers. 1989. Asymmetry and polarity of nucleosomes in chicken erythrocyte chromatin. *EMBO J.* **8:** 229–238.

Satchwell, S.C., H.R. Drew, and A.A. Travers. 1986. Sequence periodicities in chicken nucleosome core DNA. *J. Mol. Biol.* **191:** 659–675.

Selsing, E., R.D. Wells, C.J. Alden, and S. Arnott. 1979. Bent DNA: Visualisation of a base-paired and stacked A-B conformational junction. *J. Biol. Chem.* **254:** 5417–5422.

Shakked, Z., D. Rabinovich, O. Kennard, W.B.T. Cruse, S.A. Salisbury, and M.A. Visamitra. 1983. Sequence-dependent conformation of an A-DNA double helix: The crystal structure of the octamer d(GGTATACC). *J. Mol. Biol.* **166:** 183–201.

Sherratt, D., P. Dyson, M. Boocock, L. Brown, D. Summers, G. Stewart, and P. Chan. 1984. Site-specific recombination in transposition and plasmid stability. *Cold Spring Harbor Symp. Quant. Biol.* **49:** 227–233.

Shore, D. and R.L. Baldwin. 1983. Energetics of DNA twisting. II. Topoisomer analysis. *J. Mol. Biol.* **170:** 983–1007.

Shuey, D.J. and C.S. Parker. 1986. Bending of promoter DNA on binding of heat shock transcription factor. *Nature* **373:** 459–461.

Siebenlist, U., R.B. Simpson, and W. Gilbert. 1980. *E. coli* RNA polymerase interacts homologously with two different promoters. *Cell* **20:** 269–281.

Silver, L.E., A.F. Torri, and S.L. Hajduk. 1986. Organized packaging of kinetoplast DNA networks. *Cell* **47:** 537–543.

Simpson, L. 1979. Isolation of maxicircle component of kinetoplast DNA from hemoflagellate protozoa. *Proc. Natl. Acad. Sci.* **76:** 1585–1588.

Simpson, R.T. and P. Kunzler. 1979. Chromatin and core particles formed from the inner histones and synthetic polydeoxyribonucleotides of defined sequence. *Nucleic Acids Res.* **6:** 1387–1485.

Simpson, R.T. and D.W. Stafford. 1983. Structural features of a phased nucleosome core particle. *Proc. Natl. Acad. Sci.* **80:** 51–55.

Simpson, R.T., F. Thoma, and J.M. Brubaker. 1985. Chromatin reconstituted from tandemly repeated cloned DNA fragments and core histones: a model system for the study of higher order structure. *Cell* **42:** 799–808.

Snyder, M. A.R. Buchman, and R.W. Davis. 1986. Bent DNA at a yeast autonomously replicating sequence. *Nature* **324:** 87–89.

Spassky, A., S. Busby, and H. Buc. 1984. On the action of the cyclic AMP-cyclic AMP receptor protein complex at the *Escherichia coli* lactose and galactose promoter regions. *EMBO J.* **3:** 43–50.

Spassky, A., K. Kirkegaard, and H. Buc. 1985. Changes in the DNA structure of the *lac*UV5 promoter during formation of an open complex with RNA polymerase. *Biochemistry* **24:** 2723–2731.

Struhl, K. 1985. Naturally occurring (dA-dT) sequences are upstream promoter elements for constitutive transcription in yeast. *Proc. Natl. Acad. Sci.* **82:** 8419–8423.

Suck, D. and C. Oefner. 1986. Structure of DNase I at 2.0 Å resolution suggests a mechanism for binding to and cutting DNA. *Nature* **321:** 620–625.

Suck, D., A. Lahm, and C. Oefner. 1988. Structure refined to 2 Å of a nicked DNA octanucleotide complex with DNase I. *Nature* **332:** 464–468.

Suck, D., C. Oefner, and W. Kabsch. 1984. Three-dimensional structure of bovine pancreatic DNase I at 2.5 Å resolution. *EMBO J.* **3:** 2423–2430.

Swartz, M.N., T.A. Trautner, and A. Kornberg. 1962. Enzymatic synthesis of deoxyribonucleic acid. XI. Further studies on nearest neighbour base sequences in deoxyribonucleic acids. *J. Biol. Chem.* **237:** 1961–1967.

Takahashi, K., M. Vigneron, H. Matthes, A. Wildeman, M. Zenke, and P. Chambon. 1986. Requirement of stereospecific alignments for initiation from the simian virus 40 early promoter. *Nature* **319:** 121–126.

Thoma, F. 1986. Protein-DNA interactions and nuclease-sensitive regions determine nucleosome positions on yeast plasmid chromatin. *J. Mol. Biol.* **190:** 177–190.

Thoma, F. and R.T. Simpson. 1985. Local protein-DNA interactions may determine nucleosome positions on yeast plasmids. *Nature* **315:** 250–252.

Thoma, F. and M. Zatchej. 1988. Chromatin folding modulates nucleosome positioning in yeast minichromosomes. *Cell* **55:** 945–953.

Travers, A.A. 1988. Protein induced DNA bending. *Nucleic Acids Mol. Biol.* **2:** 136–148.

Travers, A.A. and A. Klug. 1987. DNA wrapping and writhing. *Nature* **327:** 280–281.

Trifonov, E.N. 1980. Sequence-dependent deformational anisotropy of chromatin DNA. *Nucleic Acids Res.* **8:** 4041–4053.

Trifonov, E.N. 1985. Curved DNA. *CRC Crit. Rev. Biochem.* **19:** 89–106.

Trifonov, E.N. and J.L. Sussman. 1980. The pitch of chromatin DNA is reflected in its nucleotide sequence. *Proc. Natl. Acad. Sci.* **77:** 3816–3820.

Turnell, W.R., Satchwell, S.C. and Travers, A.A. 1988. A decapeptide motif for binding to the minor groove of DNA. A proposal. *FEBS Lett.* **232:** 263–268.

Ulanovsky, L.E. and E.N. Trifonov. 1987. Estimation of wedge components in curved DNA. *Nature* **326:** 720–722.

Ulanovsky, L., M. Bodner, E.N. Trifonov, and M. Choder. 1986. Curved DNA: Design, synthesis and circularization. *Proc. Natl. Acad. Sci.* **83:** 862–866.

Van Dyke, M.W. and P.B. Dervan. 1983. Chromomycin, mithramycin and olivomycin binding sites on heterogeneous deoxyribonucleic acid. Footprinting with (Methidium-propyl-EDTA) iron (II). *Biochemistry* **22:** 2373–2377.

Van Dyke, M.W., R.P. Hertzberg, and P.B. Dervan. 1982. Map of distamycin, netropsin and actinomycin binding sites on heterogeneous DNA: DNA cleavage-inhibition patterns with methidiumpropyl-EDTA.Fe (II). *Proc. Natl. Acad. Sci.* **79:** 5470–5474.

Viswamitra, M.A., O. Kennard, P.C. Jones, G.M. Sheldrick, S. Salisbury, L. Falvello, and Z. Shakked. 1978. DNA double helical fragment at atomic resolution. *Nature* **273:** 687–688.

von Kitzing, E. and S. Diekmann. 1987. Molecular mechanics calculations of dA_{12}•dT_{12} and of the curved molecule $d(GCTCGAAAA)_4$•$d(TTTTCGAGC)_4$. *Eur. Biophys. J.* **15:** 13–26.

Wang, A. H.-J., S. Fujii, J.H. van Boom, and A. Rich. 1982. Molecular structure of the octamer d(GGCCGGCC): modified A-DNA. *Proc. Natl. Acad. Sci.* **79:** 3968–3972.

Wang, J.C. 1979. Helical repeat of DNA in solution. *Proc. Natl. Acad. Sci.* **76:** 200–203.

Warwicker, J., B.P. Engelman, and T.A. Steitz. 1987. Electrostatic calculations and model-building suggest that DNA bound to CAP is tightly bent. *Proteins* **2:** 283–289.

White, J.H. and W.R. Bauer. 1986. Calculation of the twist and the writhe for representative models of DNA. *J. Mol. Biol.* **189:** 329–341.

White, J.H., N.R. Cozzarelli, and W.R. Bauer. 1988. Helical repeat and linking number of surface wrapped DNA. *Science* **241:** 323–327.

Widom, J. 1985. Bent DNA for gene regulation and DNA packaging. *Bioessays* **2:** 11–14.

Widom, J. and A. Klug. 1985. Structure of the 300 Å chromatin filament: X-ray diffraction from oriented samples. *Cell* **43:** 207–213.

Wolberger, C., Dong, Y., Ptashne, M. and Harrison, S.C. 1988. Structure of a phage 434 Cro/DNA complex. *Nature* **335:** 789–795.

Wu, H.-M. and D.M. Crothers. 1984. The locus of sequence-directed and protein-induced DNA bending. *Nature* **308:** 509–513.

Yoon, C., C.C. Privé, D.S. Goodsell, and R.E. Dickerson. 1988. Structure of an alternating-B DNA helix and its relationship to A-tract DNA. *Proc. Natl. Acad. Sci.* **85:** 6332–6336.

Zahn, K. and F.R. Blattner. 1985. Binding and bending of the λ replication origin by the phage O protein. *EMBO J.* **4:** 3605–3616.

———. 1987. Evidence for DNA bending at the λ replication origin. *Science* **236:** 416–422.

Zhurkin, V.B. 1983. Specific alignment of nucleosomes on DNA correlates with periodic distribution of purine-pyrimidine and pyrimidine-purine dimers. *FEBS Lett.* **158:** 293–297.

———. 1985. Sequence-dependent bending of DNA and phasing of nucleosomes. *J. Biomol. Struct. Dynamics* **2:** 785–804.

Zinkel, S.S. and D.M. Crothers. 1987. DNA bend direction by phase sensitive detection. *Nature* **328:** 178–181.

Zivanovic, Y., I. Goulet, B. Revet, M. Le Bret, and A. Prunell. 1988. Chromatin reconstitution on small DNA rings. II. DNA supercoiling on the nucleosome. *J. Mol. Biol.* **200:** 267–290.

3

Protein-Protein Interactions and DNA Loop Formation

Ann Hochschild
Department of Microbiology and Molecular Genetics
Harvard Medical School
Boston, Massachusetts 02115

I. INTRODUCTION

Recent studies of a number of prokaryotic operons have revealed that regulatory proteins bound at widely separated sites on the DNA can interact, thereby inducing the formation of a DNA loop. Such loops appear to play a critical role in the transcriptional regulation of several of these operons. DNA loop formation was first implicated in the regulation of the galactose and arabinose operons of *Escherichia coli*. In both cases, transcriptional repression was found to depend on operator sites located at considerable distances from the promoter regions, leading to the suggestion that repression was mediated by an interaction between repressors bound at widely separated sites.

This chapter will begin with a description of a model system for studying DNA loop formation, the binding of λ repressor to artificially separated operator sites. This will be followed by a qualitative discussion of the energetics of loop formation. The last section will contain a review of several prokaryotic systems in which loop formation has been studied.

II. DNA LOOP FORMATION BY λ REPRESSOR

λ repressor binds cooperatively to adjacent operators on the phage chromosome (Johnson et al. 1979). That is, the presence of a strong repressor binding site adjacent to a weaker site increases the affinity of the weaker site for repressor. Consider, for example, the interaction of λ repressor with the right operator (O_R) in a λ lysogen (see Fig. 1a) (for review, see Ptashne 1986a). O_R1 binds repressor relatively tightly, whereas O_R2 and O_R3 each have an intrinsically weaker affinity for repressor. (The intrinsic affinity of a single operator site for repressor is determined after inactivating the other two sites by mutation.) In a lysogen, a repressor bound at O_R1 stabilizes the association of a second with O_R2 so that its apparent affinity for repressor is almost as high as that of O_R1. O_R1 and O_R2 are thus fully occupied; O_R3 is not because

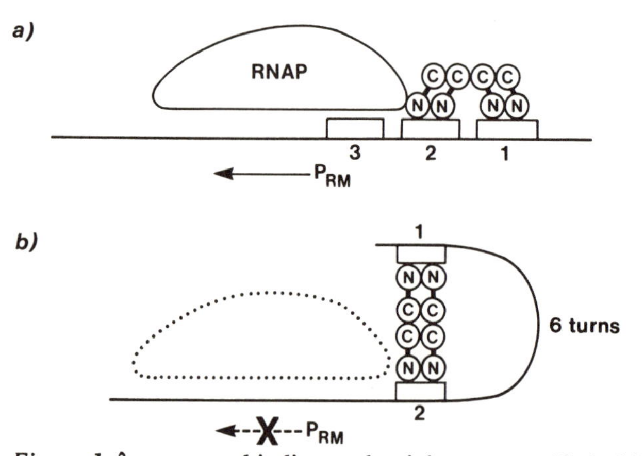

Figure 1 λ repressor binding to the right operator (O_R). (*a*) A normal λ lysogen. Repressor dimers bind cooperatively to O_R1 and O_R2, the dimer at O_R1 helping a second to bind to O_R2. The dimer at O_R2 contacts RNA polymerase to activate transcription from P_{RM}. The repressor monomer consists of two domains; the amino domain is the DNA-binding domain, and the carboxyl domain mediates dimer formation and the cooperative interaction between adjacently bound dimers. (*b*) An artificial template that has been constructed by inserting DNA between O_R1 and O_R2 to give a center-to-center spacing of six turns of the DNA helix. (Also, O_R3 has been inactivated by multiple mutations.) Repressor dimers again bind cooperatively to O_R1 and O_R2, but the dimer at O_R2 is unable to stimulate transcription from P_{RM} efficiently. Note that this figure suggests that the protein-protein interaction is stronger for nonadjacently bound repressors because both monomers of each dimer participate in the interaction. Actually, we do not know how many monomers are involved in either case, and thus the strength of the interaction may be the same for adjacently and nonadjacently bound repressors.

only two repressors cooperate at one time, and the O_R1-bound repressor preferentially associates with the O_R2-bound repressor.

Repressor is a two-domain protein (Pabo et al. 1979) that binds each operator site as a dimer. Its amino domain contacts the DNA (Sauer et al. 1979; Pabo and Lewis 1982), whereas its carboxyl domain, which contains the most important dimer contacts, mediates the cooperative interaction between adjacently bound dimers (Johnson et al. 1979). Repressor's amino domain, when separated from the carboxyl domain, binds noncooperatively according to the intrinsic affinity of each site for repressor.

The unexpected observation that λ repressor also binds cooperatively to nonadjacent operator sites indicated that the repressor might be able to induce the formation of DNA loops (Hochschild and Ptashne 1986). Cooperative binding was demonstrated using a DNase footprinting assay to measure the affinity of a weak site for repressor in the absence or presence of a strong site located at a nonadjacent position on the same DNA fragment. The extent to which the presence of the strong site increases the affinity of the weak site is a measure of the cooperative effect. This cooperativity was found to be a periodic function of the spacing between the two operator sites. Specifically, the binding is cooperative when the center-to-center distance between the two operators is an integral number n of turns of the DNA helix (5, 6, or 7), assuming 10.5 bp/turn. On the other hand, the binding is noncooperative when the distance between the two operators is close to $n + 1/2$ turns of the DNA helix (4.6, 5.5, or 6.4).

The periodic dependence of cooperativity on the spacing between the two operator sites suggested that cooperative binding was mediated by a protein-protein contact, necessitating the formation of a smooth bend in the DNA spanning the two sites. Dunn et al. (1984) first observed a periodic effect of spacing on repression of the arabinose (*ara*) operon and proposed the formation of a DNA loop to explain this effect. Since repressor binds primarily to one side of the DNA helix, the bound repressor dimers are located on the same side of the helix when the two sites are separated by an integral number of turns and on opposite sides of the helix when the sites are separated by $n + 1/2$ turns. To bring together two repressor dimers that are bound on opposite sides of the DNA helix, bending of the DNA is insufficient; twisting and/or writhing is also required.

Cooperative binding is observed in the case of two operator sites separated by a nonintegral number of turns if a 4 nucleotide gap is introduced into one strand of the DNA between the two operator sites. Under these circumstances, the DNA is expected to turn freely around the single-stranded region, and therefore the coming together of the bound repressor dimers is not much more difficult energetically than the coming together of dimers bound on the same side of the DNA helix.

The distortion of the DNA backbone that is induced by the cooperative binding of λ repressor to separated operator sites results in a change in the sensitivity of the DNA between the two sites to DNase I. When a DNase footprint assay is used to measure the binding of repressor to separated operator sites, cooperative binding is always accompanied by the appearance of an alternating set of enhanced and diminished cleavages affecting the DNA between the two integrally spaced operators. The enhanced and diminished cleavages are separated by roughly 5 bp. Assuming that the interaction between the bound repressors induces a smooth bend in the DNA, the enhanced cleavages map to the outside of the protein/DNA complex, whereas the diminished cleavages map to the inside of this complex. A similar pattern of enhanced and diminished cleavages was previously observed with a small circular DNA molecule (Drew and Travers 1985); the phenomenon is thus diagnostic of loop formation, and it has also been observed in other experimental systems (Kramer et al. 1987; Salvo and Grindley 1988; Valenzuela and Ptashne 1989; P. Beachy, unpubl.).

Cooperative binding to separated operator sites, like cooperative binding to adjacent operator sites, depends on repressor's carboxyl domain; the purified amino domain binds noncooperatively to operators separated both by integral and nonintegral numbers of turns (Hochschild and Ptashne 1986). Furthermore, a single amino acid substitution in repressor's carboxyl domain eliminates cooperative binding to both nonadjacent and adjacent operator sites (Hochschild and Ptashne 1988). The strategy used to isolate this mutant repressor will be outlined below. The behavior of the mutant protein suggests that the same protein-protein contact mediates the cooperative interaction between adjacently and nonadjacently bound repressors.

A direct visualization of the DNA loops induced by the cooperative binding of λ repressor was accomplished using electron microscopy (Griffith et al. 1986). Repressor was incubated separately with four different DNA fragments, two of which carried integrally spaced operator sites and two of which carried nonintegrally spaced operator sites. Bent DNA molecules of the expected configuration were observed only when the fragments carried integrally spaced operators. Protein was visible on the inside of the bend, and the arms of the DNA fragments were of the predicted lengths based on the location of the operator sites.

Cooperative binding of λ repressor to separated operator sites becomes progressively weaker as the distance between the two operator sites is increased beyond about eight turns of the DNA helix (J. Douhan III, unpubl.); at a spacing of 20 turns the cooperative effect as measured by DNase footprinting is barely detectable. In addition, the periodicity of the effect damps out as the spacing between the two operator sites is increased from 8 to 20 turns. (As discussed below, the cost of twisting the DNA by half a turn decreases as the length of the DNA fragment over which the twist is distributed increases.)

λ repressors can also interact when bound to separated operators in vivo (Hochschild and Ptashne 1988). The periodicity of the effect is the same as in vitro. This interaction between nonadjacently bound repressors in vivo has an unexpected effect on transcriptional regulation. Normally, repressor bound at O_R2 (see Fig. 1a) stimulates transcription from the adjacent promoter P_{RM}, using its amino domain to contact RNA polymerase (for review, see Ptashne 1986a). Activation depends only on the occupancy of O_R2, and in a λ lysogen, cooperative binding to O_R1 and O_R2 ensures its occupancy. However, when O_R1 is moved an integral number of turns away from O_R2 (see Fig. 1b), the repressor bound at O_R2 interacting with a repressor bound at a distance is impaired for transcriptional activation. As occurs when the sites are adjacent, the interaction between repressors bound at O_R1 and O_R2 when these two sites are separated increases the affinity of O_R2 for repressor. However, in this case, the interaction prevents normal activation of P_{RM}, perhaps because the repressor at O_R2 is constrained so that it cannot effectively contact RNA polymerase or perhaps because of some effect of the loop itself.

This form of negative regulation was exploited to isolate the mutant repressor mentioned above that is unable to bind cooperatively. Repressor was produced at a level such that O_R2 would be occupied whether or not the binding was cooperative, and a mutant was sought that would stimulate transcription efficiently on a template that carried O_R1 an integral number of turns upstream of O_R2. A mutant unable to bind cooperatively and therefore unable to mediate loop formation is predicted to restore efficient transcription from P_{RM} under these circumstances.

Cooperative binding to separated operator sites has also been observed using the repressor proteins of the related phages 434 and P22 (Valenzuela and Ptashne 1989). Like λ repressor, these repressors bind cooperatively to adjacent operator sites on their respective phage chromosomes (Poteete and Ptashne 1982; Wharton et al. 1984). Cooperative binding is in each case mediated by the carboxyl domain, and the two proteins share considerable sequence homology with λ repressor in this region (Sauer et al. 1982). Both 434 and P22 repressors bind cooperatively to separated operator sites in vitro as measured by DNase footprinting. In the case of P22 repressor, cooperative binding to separated operator sites has also been measured in vivo using a repression assay. The same assay was used to isolate mutants defective in cooperative binding. Six different mutants were obtained, each bearing a single amino acid change within the carboxyl domain. Two of the mutants have been purified and shown to bind noncooperatively to adjacent as well as separated operator sites in vitro (Valenzuela and Ptashne 1989).

The experiments performed in vivo using λ and P22 repressors illustrate two different forms of "negative regulation at a distance." P22 repressor was used to show that repression can be accomplished when a protein bound well upstream of a promoter facilitates the binding of a second molecule to a site that overlaps the promoter. However, the artificial λ system described above shows that repression can also be accomplished when a protein bound well upstream of a promoter interacts with a second molecule bound adjacent to the promoter at a site from which it would normally activate transcription. The physiological systems that will be discussed in the last section include examples of each of these forms of negative regulation.

III. ENERGETICS OF DNA LOOP FORMATION

How can we estimate the probability that two DNA-bound proteins will interact and induce the formation of a DNA loop? It is useful to distinguish between two extreme cases: first, the case of two interacting proteins that are bound relatively close together on the DNA so that the size of the protein is significant compared with the distance separating the two binding sites (i.e., up to at least 20 turns of the DNA helix), and second, the case of proteins that are bound sufficiently far apart on the DNA so that the size of the protein is negligible compared with the size of the loop. In the first case, the energy required to bend the DNA is a critical parameter, whereas in the second case the loop is sufficiently large so that DNA flexibility is no longer a limiting factor, but the configurational entropy of the DNA chain is. We will consider each case in turn.

A. Formation of Relatively Small Loops

In the λ example discussed above, the protein-binding sites are separated by 5–7 turns of the DNA helix. In this and similar cases, we can model the looped complex as roughly a semicircle bridged by the interacting protein molecules, as is illustrated schematically in Figure 2. We know

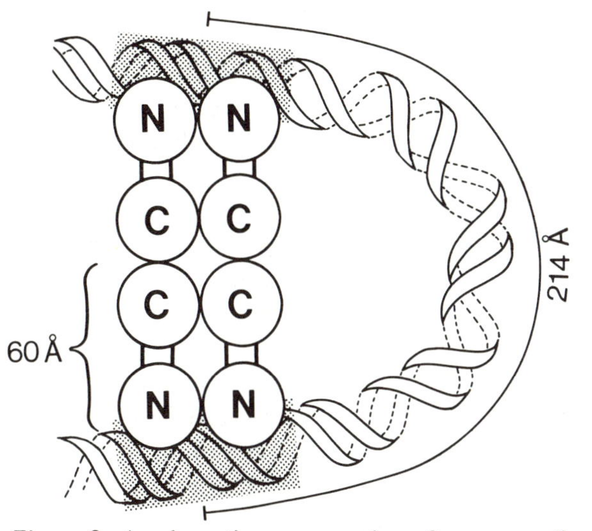

Figure 2 A schematic representation of repressor dimers binding to operators separated by six turns of the DNA helix. The DNA spanning the two operators is bent along the arc of a circle with a radius of 60 Å, the approximate height of a repressor molecule. The center-to-center distance between operators is 63 bp = 214 Å.

that the dissociation constant, K_d, for the dissociation of λ repressor tetramers into dimers in solution is 10^{-6} M (Chadwick et al. 1971; R.T. Sauer, unpubl.). Assuming that the same protein-protein interaction is involved when the DNA-bound dimers come together, loop formation will be favored provided that the effective concentration of one DNA-bound dimer in the vicinity of the other exceeds 10^{-6} M.

Although we will use the concept of effective concentration in discussing the formation of large loops, we do not know how to estimate the effective concentration of proteins that are bound relatively close to one another along the DNA. We will therefore formulate the problem in an alternative way by asking whether the interaction energy of the DNA-bound proteins exceeds the energetic cost of deforming the DNA. That is, our strategy is to obtain separate estimates for the strength of the interaction between the bound repressor dimers on the one hand and the cost of deforming the DNA the requisite amount on the other.

First, we consider the interaction of the DNA-bound proteins, treating the DNA as a completely flexible tether. As mentioned above, the K_d for the tetramer-dimer equilibrium in solution is 10^{-6} M, corresponding to a ΔG of −8.5 kcal/mole for tetramer formation. However, still assuming that the association of the DNA-bound dimers is chemically identical with the association of dimers free in solution, the two reactions are not necessarily equivalent entropically. One way in which they might differ is that, whereas the intermolecular association of two dimers in solution involves the loss of one set of translational and rotational entropies, the intramolecular association of the DNA-bound dimers, which are largely immobilized on the DNA, might not have as large an entropic loss. According to this line of argument, the ΔG for the association of the DNA-bound dimers could be significantly more favorable than –8.5 kcal/mole (still ignoring the cost of DNA bending). This argument is analogous to one that has been made to help explain enzyme catalysis (see Fersht 1977 and references therein).

In discussing the interaction energy for the association of DNA bound dimers, we have assumed that there is no allosteric effect of DNA-binding on the tetramerization reaction. On the basis of an estimate for the cost of bending the DNA (see below), it is not necessary to invoke such an effect for explaining the formation of the loop in the λ system.

We now consider the deformation of the DNA. An estimate for the energetic cost can be based on a formula for calculating the free energy of bending smoothly an isotropic rod of length L: $\Delta G_b = RT/2\ (pL/r^2)$, where p is the persistence length of the DNA (a parameter that describes its intrinsic stiffness) and r is the radius of curvature (Landau and Lifshitz 1958). Using the example of λ repressor binding to operators separated by six helical turns (see Hochschild and Ptashne 1986), the radius is defined by the height of a repressor dimer (~60 Å), and L is 156 Å because the two operator sites are separated by 46 bp (from inside edge to inside edge). Assuming a persistence length of 500 Å, $\Delta G_b = (0.3)(500)(156)/60^2 = 6.5$ kcal/mole. Thus, assuming a ΔG of ≤ -8.5 kcal/mole for the association of the DNA-bound dimers, the cost of DNA bending should be more than compensated by the anticipated interaction energy (see Fig. 3).

As discussed above, the formation of relatively small DNA loops depends on the exact spacing between operator sites; the loop forms only when the sites are separated by an integral number of turns of the DNA helix. The amount of energy required to twist the DNA and bring sites that are on opposite sides of the helix into alignment for loop formation can be derived from the free energy of supercoiling, ΔG_τ, which has been measured as a function of DNA length by Horowitz and Wang (1984). According to their data, the cost of misalignment by half a turn is less than 0.1 kcal/mole for separations greater than approximately 2500 bp and approximately 3 kcal/mole for separations approximately 200 bp in length. With separations that are less than 100 bp, the cost of misalignment is predicted to be very high: approximately 7 kcal for 100 bp and

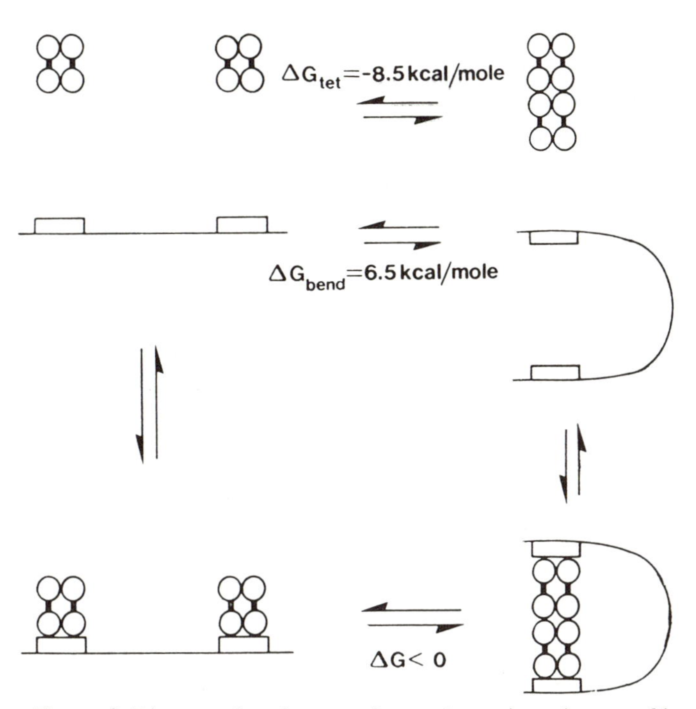

Figure 3 Diagram showing two thermodynamic pathways of loop formation. Indicated are an estimate for the cost of bending the DNA as well as the experimentally determined ΔG for the tetramerization of free dimers. As discussed in the text, the tetramerization of the DNA-bound dimers may be more favorable entropically.

approximately 13 kcal for 50 bp (see Wang and Giaever 1988 and references therein). Thus, when binding sites are separated by around six helical turns, the cost of twisting by half a turn is expected to be considerably more than the cost of bending the DNA.

The estimate for the cost of bending relatively short pieces of DNA even in the absence of twisting is subject to various uncertainties deriving from properties of both the DNA and the interacting proteins. The formula for calculating the energy of bending was derived for an isotropic rod that is weakly bent, whereas the looped protein/DNA complexes involve curvatures that are quite large. Furthermore, the ease or difficulty of bending the DNA will presumably depend to some extent on its sequence. Finally, flexibility in the protein may also reduce the strain on the DNA; thus, the DNA may not have to bend as much as would be expected assuming rigid proteins. (Protein flexibility could also reduce the amount of twisting that would be required when the sites are not optimally aligned.)

In the case of λ repressor, there are several reasons for believing that the carboxyl domain is flexibly tethered to the amino domain, thus allowing some deformation of the protein. First, repressor binds cooperatively to operator sites on the λ chromosome that have center-to-center spacings ranging between 1.9 and 2.3 turns of the DNA helix, and the magnitude of the cooperative effect is the same despite these differences in angular alignment (Johnson 1980; Ackers et al. 1982), suggesting that the protein accommodates. (Note that since the λ operator is 17 bp long, 3 bp between operators gives a center-to-center spacing of 1.9 turns.) Second, it appears that the cooperative interaction between adjacently and nonadjacently bound repressors involves the same portion of the carboxyl domain; a single amino acid substitution near the carboxyl terminus of the protein eliminates cooperative binding to both adjacent operators and operators separated by six and seven turns of the helix (Hochschild and Ptashne 1988). If the protein were a rigid structure, it is hard to imagine how the same surface patch of repressor could mediate the contact between adjacently and nonadjacently bound molecules.

B. Formation of Large Loops and the Jacobson-Stockmayer Factor

When two DNA-bound proteins are far enough apart on the DNA, the looped complex will resemble a DNA ring. In this case, we can estimate the probability of close approach of the DNA-bound proteins by considering an analogous reaction: the cyclization of DNA fragments with complementary single-stranded ends. The cyclization probabilities, called Jacobson-Stockmayer (J-S or *j*) factors have been measured previously as a function of fragment length (Wang and Davidson 1966b; Mertz and Davis 1972; Shore et al. 1981). As explained below, the use of the empirically derived *j* values greatly simplifies the problem of predicting whether or not a loop will form in a given case. However, the analogy with DNA ring formation is not valid at short separations (up to at least 20 helical turns) because the size of the protein is significant compared with the size of the loop.

The *j* value can be understood as the effective concentration of one end of the molecule in the vicinity of the other. It can be evaluated (Jacobson and Stockmayer 1950) as a ratio of two equilibrium constants, K_c/K_a, where K_c is the cyclization constant and K_a is the bimolecular equilibrium constant for joining two linear molecules. K_c can be measured at very low concentrations of DNA fragment, conditions under which dimer formation is negligible; K_a (which is independent of fragment length) can be measured using sheared half-molecules that cannot circularize (Wang and Davidson 1966a,b). To understand the interpretation of the j value as an effective concentration consider the dissociation of a dimeric DNA molecule (A_2) into one linear molecule (A) and one circularized molecule (cA) (see Levene and Crothers 1986): $A_2 \rightleftharpoons cA + A$ (see Fig. 4). The equilibrium constant for this reaction $(cA)(A)/(A_2)$ is equal to K_c/K_a (i.e., j) because $K_c/K_a = (cA)(A)^2/(A)(A_2)$. Thus, the j value is the concentration of free A at which the concentration of circles equals the concentration of dimers, or the probability of an intramolecular collision equals the probability of an intermolecular collision. So, the higher the j value, the more effectively circularization competes with dimer formation, i.e., the higher is the effective concentration of one end of the molecule in the vicinity of the other.

Both theoretical and experimental analyses indicate that the j value reaches a maximum of approximately 10^{-7} M for DNA fragments of approximately 700 bp (Shore et al. 1981; Shore and Baldwin 1983; Shimada and Yamakawa 1984). With decreasing fragment length, the j values fall off sharply, and with increasing fragment length, the j values diminish more gradually. The position of the maximum value depends on the intrinsic stiffness of DNA, which is described by the persistence length. (Under the conditions used by Shore et al. [1981] the persistence length is about 150 bp.) For short DNA fragments close to the persistence length or less, the DNA can be modeled as a relatively stiff rod, and the j value is determined primarily by an enthalpic term that describes the cost of deforming and bending the DNA. For relatively long DNA fragments, the DNA can be modeled as a flexible coil, and the j value is determined primarily by an entropic term that describes the loss in configurational entropy of the DNA molecule when its ends are held in proximity to one another. This loss increases with fragment length; that is, the probability of one end finding the other decreases with increasing fragment length.

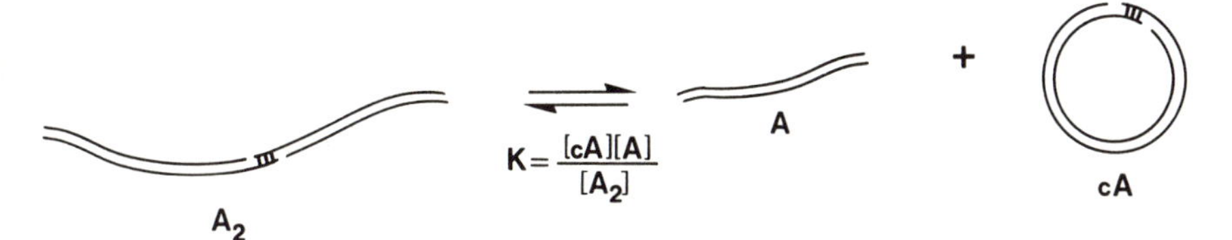

Figure 4 The dissociation of a dimeric DNA molecule (A_2) to give one linear molecule (A) and one circular molecule (cA). The equilibrium constant for this reaction equals j (see text), and therefore the j value can be thought of as the concentration of free A at which the concentration of dimers equals the concentration of circles, a value that increases with the probability of circle formation.

For DNA fragments less than 500-bp long, the cyclization probabilities depend not only on fragment length, but also on the helical alignment of the ends of the DNA (Shore and Baldwin 1983). The *j* values thus show a periodic fluctuation with local maxima occurring approximately every 10 bp. When the number of base pairs in the DNA fragment is not an integral multiple of the helix repeat, the free energy required to twist the ends of the DNA and bring them into alignment was found to lower the *j* value by more than a factor of 10 for DNA fragments of about 250 bp.

On the basis of the measured *j* values, would we expect λ repressor to induce the formation of DNA loops at very large separations? Consider the intermolecular reaction of two half-molecules, each of which carries one of the repressor binding sites to which a repressor dimer is prebound (see Fig. 5a). The equilibrium constant for this reaction (K_a) can be approximated by the equilibrium constant for the association of dimers to form tetramers in solution (10^6 liter moles^{-1}). Now consider the cyclization of a molecule that carries both binding sites to which repressor dimers are prebound (see Fig. 5b). The equilibrium constant for this reaction (K_c) is equal to $j(K_a)$ because $j = K_c/K_a$. Since the *j* value reaches a

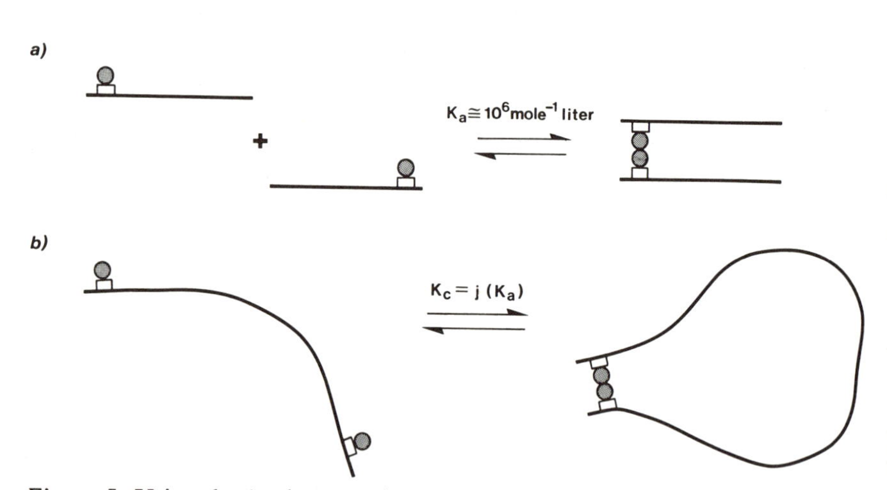

Figure 5 Using the *j* value to estimate the probability of loop formation by λ repressor. (*a*) The association of two half-molecules. Each half-molecule carries one operator to which a repressor dimer is prebound. K_a approximately equals the equilibrium constant for the association of free dimers to form tetramers (10^6 liter moles^{-1}). (*b*) The circularization of a full-length molecule. The molecule carries two operators to which repressor dimers are prebound. K_c equals $j(K_a)$ by definition and is therefore less than or equal to 0.1 since the *j* value reaches a maximum of 10^{-7} M.

maximum at 10^{-7} M (see above) and since $K_a = 10^6$ liter moles^{-1}, $K_c \leq 0.1$. Thus, we would not expect to observe loops at large separations, an expectation consistent with the experimental data (see Section II).

C. Cooperative Binding

If two DNA-bound proteins interact, then this interaction energy will be reflected in an increased affinity of the protein for its binding sites (cooperative binding). That is, the sum of the binding energies will be greater when both sites are present on the same piece of DNA (and so aligned as to allow the interaction) than would be measured for each site independently. This is a consequence of the decreased free energy (increased stability) of the final bound complex relative to the unlooped complex. This extra binding energy will be reflected primarily in an apparent increase in the affinity of the protein for the weaker binding site if the two sites differ in strength (see Table 1 and legend for some sample values and derivation). This is because among the singly bound DNA molecules the strong site is much more likely to be occupied than the weak.

D. Loop Formation by Preformed Protein Multimers

DNA looping occurs not only when two DNA-bound proteins interact, but also when a protein multimer (or a multifunctional protein, such as the λ *int* protein, see Section IV) interacts simultaneously with two nonadjacent binding sites. *lac* repressor, for example, forms stable tetramers that are capable of binding two *lac* operator sites (Kania and Müller-Hill 1977). How does the case of *lac* repressor differ from that of λ repressor? In the case of the λ repressor, the protein-protein interaction is much weaker than the protein-DNA interaction, whereas in the *lac* case the protein-protein interaction is much stronger. Thus, in the case of the λ repressor, when the concentration of DNA is low, the DNA-binding sites will be saturated at protein concentrations well below those required to drive the formation of tetramers in solution. For this reason, the probability of loop formation can be described by a simple equilibrium involving the association and dissociation of the DNA-bound dimers under conditions where there are essentially no unbound DNA molecules and no free tetramers.

In the case of *lac*, however, since the DNA-binding species is the preformed tetramer, the tendency for loops to form does not depend on the strength of the contact between the dimers in the tetramer. To define a simple equilibrium that describes the probability of loop formation in

Table 1 The effect of cooperativity on site occupancy

K_1 (ΔG_1)	K_2 (ΔG_2)	K_{coop} (ΔG_{coop})	K_2^* (ΔG_2^*)	K_2/K_2^* ($\Delta G_2^* - G_2$)
3.3×10^{-9} M (−11.7 kcal/mole)	3.3×10^{-9} M (−11.7 kcal/mole)	3.6×10^{-2} M (−2.0 kcal/mole)	6.3×10^{-10} M (−12.7 kcal/mole)	5 (−1.0 kcal/mole)
3.3×10^{-9} M (−11.7 kcal/mole)	5.0×10^{-8} M (−10.1 kcal/mole)	3.6×10^{-2} M (−2.0 kcal/mole)	3.5×10^{-9} M (−11.7 kcal/mole)	14 (−1.6 kcal/mole)
3.3×10^{-9} M (−11.7 kcal/mole)	5.0×10^{-7} M (−8.7 kcal/mole)	3.6×10^{-2} M (−2.0 kcal/mole)	2.1×10^{-8} M (−10.6 kcal/mole)	24 (−1.9 kcal/mole)
3.3×10^{-9} M (−11.7 kcal/mole)	5.0×10^{-6} M (−7.3 kcal/mole)	3.6×10^{-2} M (−2.0 kcal/mole)	1.8×10^{-7} M (−9.3 kcal/mole)	28 (−2.0 kcal/mole)

Consider two operators, O_1 and O_2. K_1 and K_2 are dissociation constants that describe the intrinsic affinities of O_1 and O_2 for repressor. K_{coop} is a constant that describes the cooperative term; if ΔG_{coop} is a value less than 0 that denotes the decrease in $|\Delta G_1 + \Delta G_2|$ under conditions where the binding is cooperative, K_{coop} is related to ΔG_{coop} by the familiar relation $1/K_d = e^{-\Delta G/RT}$. K_{coop} can be determined experimentally by comparing the intrinsic affinity of each operator site for repressor with the affinity of that site when the binding is cooperative. The value chosen (3.6×10^{-2}) is based on the λ system (see Ackers et al. 1982). Likewise, the values chosen for K_1 (all lines) and for K_2 (line 2) are based on the λ system; 3.3×10^{-9} M and 5×10^{-8} M are the intrinsic dissociation constants for the binding of repressor to O_R1 and O_R2, respectively. K_2^* is a dissociation constant for the binding of repressor to O_2 when the binding is cooperative, and the ratio K_2/K_2^* is thus the factor by which binding to O_2 increases because of cooperativity. Table 1 shows that for fixed values of K_1 and K_{coop}, the ratio K_2/K_2^* increases as O_2 is weakened. That is, the weaker O_2 is compared with O_1, the more the energy of cooperativity is reflected in an increase in the affinity of O_2 for repressor. For the given value of K_{coop}, the ratio K_2/K_2^* approaches a limiting value of approximately 28 when essentially all of the energy of cooperativity contributes to an increase in the affinity of O_2 for repressor.

The following calculation was performed to compute K_2/K_2^* for each value of K_2 (for a detailed discussion of the λ system, see Ackers et al. 1982). Let fO_2 stand for the fractional occupancy of O_2. We can write

$$fO_2 = \frac{(RO_2) + (R_2O)}{(O) + (RO_1) + (RO_2) + (R_2O)}$$

where (RO_1) is the concentration of DNA molecules on which O_1 alone is occupied, (RO_2) is the concentration of DNA molecules on which O_2 alone is occupied, (R_2O) is the concentration of DNA molecules on which both O_1 and O_2 are occupied, and (O) is the concentration of free DNA molecules. Dividing top and bottom by (O) and using the equilibrium relations, $K_1 = (R)(O)/(RO_1)$, $K_2 = (R)(O)/(RO_2)$, and $K_1K_2K_{coop} = (R)^2(O)/(R_2O)$, we can rewrite the equation

$$fO_2 = \frac{(R)/K_2 + (R)^2/K_1K_2K_{coop}}{1 + (R)/K_1 + (R)/K_2 + (R)^2/K_1K_2K_{coop}}$$

We can now set $fO_2 = 0.5$ and solve for (R), which equals K_2^*. Performing the appropriate algebra gives the simplified equation

$$(R)^2/K_1K_2K_{coop} + ([K_1 - K_2]/K_1K_2)(R) - 1 = 0$$

or

$$(R)^2 + (K_1 - K_2)\,K_{coop}(R) - K_1K_2K_{coop} = 0$$

Using the quadradic formula we can solve for (R)

$$(R) = \frac{(K_2 - K_1)\,K_{coop} + ([K_1 - K_2]^2\,K_{coop}^{\;2} + 4K_1\,K_2\,K_{coop})^{1/2}}{2}$$

Note that when $K_2 = K_1$, $(R) = K_1(K_{coop})^{1/2}$, and when $K_2 >> K_1$ (O_2 significantly weaker than O_1)

$$(R) = \frac{K_2K_{coop} + (K_2^{\;2}\,K_{coop}^{\;2} + 4\,K_1K_2K_{coop})^{1/2}}{2}$$

$$= \frac{K_2K_{coop} + (K_2K_{coop}\,[K_2K_{coop} + 4K_1])^{1/2}}{2}$$

Finally, when K_2 is several orders of magnitude greater than K_1, $K_2K_{coop} >> 4K_1$, (R) approaches the limiting value K_2K_{coop}, and the ratio K_2/K_2^* approaches $1/K_{coop}$.

this case, we consider a DNA molecule containing one strong binding site and one weak binding site. We find a concentration of *lac* repressor that is sufficiently high so that the strong site is fully bound but not high enough so that the weak site would be occupied were it present on a separate DNA fragment. The relevant equilibrium involves the association and dissociation of the DNA-bound tetramer and the weak site (see Fig. 6). Specifically, loop formation will be favored provided that the effective concentration of the DNA-bound tetramer in the neighborhood of the weak site exceeds the K_d for the binding of *lac* repressor to that site.

When the distance between the two sites is relatively short, we can attempt to compare the affinity of the DNA-bound tetramer for the weak site with the cost of bending the DNA (taking into account the possible entropic advantage of the intramolecular reaction as compared with the corresponding intermolecular reaction). When the distance between sites is sufficiently long, we can again use the *j* value to estimate the probability of loop formation, as outlined in Figure 7.

E. Eukaryotic Activators and Synergism

The idea that transcriptional activation in eukaryotes involves protein-protein interactions with the formation of DNA loops has been widely discussed (see Ptashne 1986b). Many eukaryotic activators appear to function synergistically, i.e., the amount of transcription measured in the presence of two activator-binding sites is considerably more than twice that measured in the presence of a single site. What follows is a brief discussion of some possible sources of this synergism.

In the simplest case, consider two transcriptional activators that interact with one another directly. Because these activators bind cooperatively, they will function synergistically provided that the concentration of

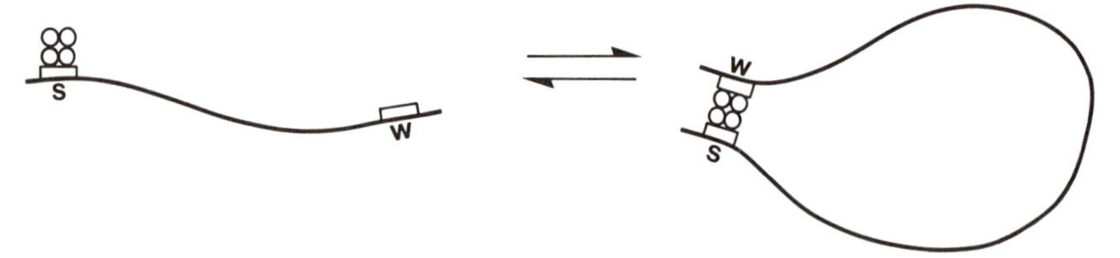

Figure 6 Loop formation by a preformed tetramer. The tetramer is prebound to a strong binding site on a DNA molecule that also carries a weak binding site. The probability of loop formation depends on the cost of looping the DNA and on the interaction energy for the binding of the DNA-bound tetramer to the weak site.

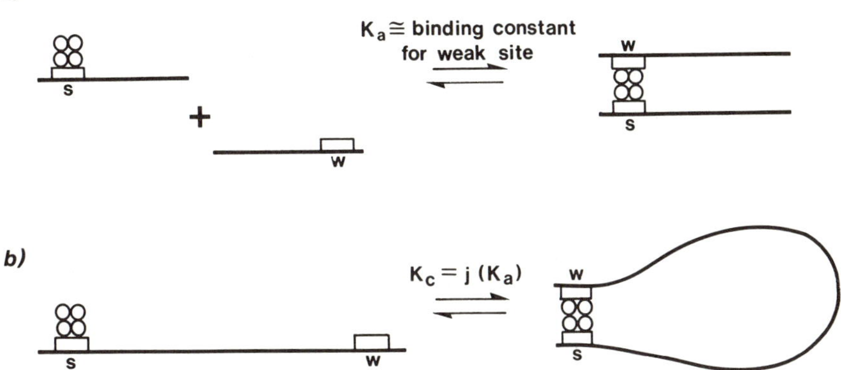

Figure 7 Using the j value to estimate the probability of loop formation by a preformed tetramer. (*a*) The association of two half-molecules. One of the half-molecules carries a strong binding site to which the tetramer is prebound and the other carries an unoccupied weak site. K_a approximately equals the equilibrium constant for the binding of a free tetramer to the weak site. (*b*) The circularization of a full-length molecule. K_c equals $j(K_a)$ by definition, and since the j value reaches a maximum of 10^{-7} M, loop formation will be favored provided K_a is greater than 10^7 liter moles^{-1}.

activator is not so high that either one of the single sites alone would be saturated. If all of the synergism is due to cooperative binding, then the effect will weaken as the concentration of the activator is increased or the strength of the binding sites is increased.

Consider, however, the case of two activators (A and B) that do not themselves interact but instead interact with a common target protein (T) bound near the startpoint of transcription (see Ptashne 1988). Assume that transcriptional activity reflects the extent to which the available sites on the target protein are occupied by activator protein. A and B will function synergistically provided that they can interact simultaneously with T (see Fig. 8a). The synergism derives in part from cooperative DNA binding; A and B will bind cooperatively to DNA sites a and b, just as they would if they interacted directly because the interaction of one DNA-bound activator (A for example) with the target brings the target into the vicinity of the second activator binding site (b) where it can stabilize the association of the second activator (B) with that site. However, synergism in the case of activators that interact with a common target derives not only from cooperative DNA binding, but also from the cooperative binding of A and B to the target. Unlike the case where A and B interact directly, the two activators will continue to function synergistically even under conditions where sites a and b are saturated.

a)

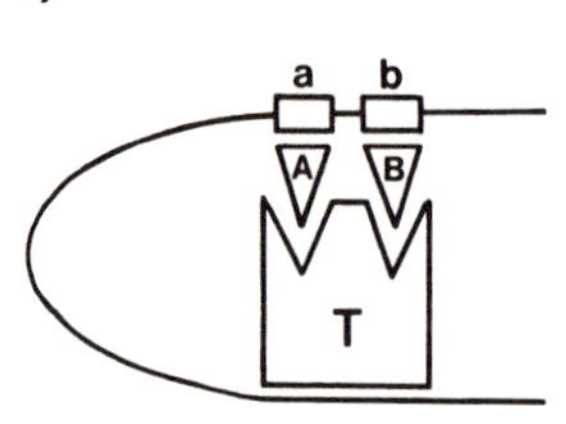

b)

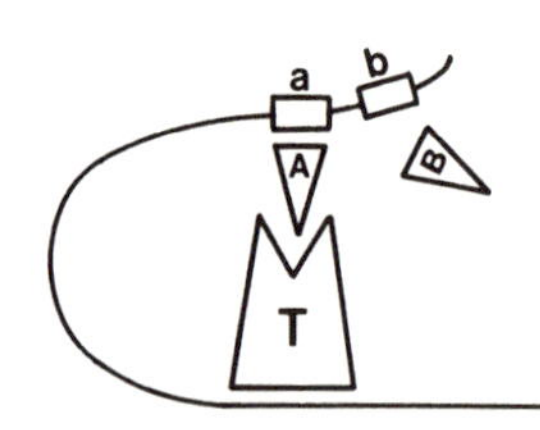

Figure 8 Interaction of activators A and B with target protein T. (*a*) Activators A and B function synergistically by virtue of their interaction with a target that contains at least two activator-binding sites. (*b*) Activators A and B do not function synergistically because the interaction of DNA-bound A with T can neither increase the probability that B will bind the DNA nor that B will bind the target.

This is because the interaction of either DNA-bound activator with T increases the probability that the other DNA-bound activator will also interact with T since the price of forming the looped complex is paid only once when the first activator binds T. Of course, the synergism will be masked if the activator/target interaction is very strong and a single DNA-bound activator suffices to saturate one of the sites on T. Assuming that complete occupancy of one of the two target sites results in 50% of the maximal stimulation, the addition of the second activator can boost stimulation by no more than a factor of two under conditions where a single activator suffices to saturate both the DNA site and the target site. (Note that under conditions where DNA sites a and b are not saturated, the weaker the activator/target interaction, the less cooperative the binding of A and B to the DNA. This is true for the same reason that the strength of the activator/activator contact determines the magnitude of the cooperativity in the case where the two proteins interact directly.)

In contrast, consider the nonsynergistic case in which the target can only interact with one activator at a time (see Fig. 8b). In this case, the presence of both sites a and b will actually have a less than additive effect on transcription because A and B compete for a single site on the target. We will calculate the amount of activity expected using an arbitrary example. For simplicity, assume sites a and b are fully occupied in all cases and that T is prebound to the DNA. Let us say that when a single site is present (either a or b), 20% maximal stimulation is observed, corresponding to 20% occupancy of T by the activator. The fractional occupancy of T by A (f_A) equals (A·T)/([A·T]+[T]), where (A·T) is the concentration of target complexed with activator and (T) is the concentration of uncomplexed target. Defining an equilibrium constant K_A = (A·T)/(T), we can write (A·T)/([A·T]+[T]) = $K_A/(K_A+1)$, and in our example $K_A = 0.25$. Similarly, $K_B = 0.25$. What is the predicted fractional occupancy of T when both sites (a and b) are present? f = ([A·T]+[B·T])/ ([A·T]+[B·T]+[T]), which equals $(K_A+K_B)/(K_A+K_B+1)$. Thus, in our example, f = 0.33, and we expect to observe 33% maximal stimulation.

IV. DNA LOOP FORMATION AND TRANSCRIPTIONAL REGULATION IN PROKARYOTES

A. *araC* Protein

A role for DNA loop formation in the regulation of transcription was first established for the case of the arabinose operon of *E. coli*. Transcription of the *araBAD* operon is both positively and negatively regulated by the *araC* protein (Sheppard and Englesberg 1967; Englesberg et al. 1969). Arabinose induces transcription by a factor of approximately 100, and this induction depends absolutely on *araC* protein, which binds to a site (*araI*) located just upstream of the promoter (Ogden et al. 1980; Lee et al. 1981). (The operon is also positively regulated by CAP protein.) Under noninducing conditions, *araC* protein negatively regulates *araBAD* transcription in the following special sense. It binds to a site located upstream of the I site and depresses the basal level of transcription.

The site *araO*$_2$ to which *araC* protein binds to repress basal transcription under noninducing conditions is located at approximately –280 relative to the startpoint of transcription (Dunn et al. 1984). Deletion of this site results in a tenfold increase in the amount of transcription observed in the absence of arabinose. Full induction, however, still requires the addition of arabinose; in its presence, transcription increases by another factor of ten. An indication that the mechanism of repression from this distant site involves an interaction between proteins bound at widely separated sites came from analysis of several spacing mutants containing small insertions and deletions between *araO*$_2$ and the promoter region (Dunn et al. 1984). Repression was impaired with deletions of 8 and 16 bp and with insertions of 5, 15, and 24 bp. However, repression was normal with insertions of 11 and 31 bp, insertions that roughly preserve the angular alignment between the promoter region and *araO*$_2$, assuming a helical repeat of 10.5 bp.

Using a constitutive promoter mutant that results in a high basal level of transcription, Hahn et al. (1984) demonstrated that *araC*-mediated repression under noninducing conditions was eliminated either by deletion of *araO*$_2$ or by mutation of *araI*. This finding suggests that repression of the constitutive promoter and, by extension, the wild-type promoter, involves an interaction between *araC* molecules bound at *araO*$_2$ and at *araI*. Subsequently, Martin et al. (1986) showed by footprinting in vivo that the induction site, *araI*, is occupied by *araC* protein under both inducing and noninducing conditions, as required by the model. Furthermore, they demonstrated cooperative binding to *araO*$_2$ by showing that the deletion of downstream sequences (including *araI*) eliminated binding to that site. This demonstration is consistent with the idea that *araC* binds cooperatively to *araI* and *araO*$_2$ under noninducing conditions, inducing the formation of a DNA loop.

More recent experiments have established that *araC* protein can also mediate the formation of an alternative loop (Huo et al. 1988). Another *araC*-binding site (*araO*$_1$) is located between *araO*$_2$ and *araI*, overlapping the promoter for the *araC* gene (P_C). (This promoter points in the opposite direction from P_{BAD}.) Occupancy of *araO*$_2$, as measured by footprinting in vivo, depends on the presence of either *araI* or *araO*$_1$ on the same DNA template. In the absence of arabinose, *araO*$_1$ is largely unoccupied, and after the addition of arabinose the occupancy of *araO*$_1$ gradually increases. The investigators propose a model according to which the *araO*$_2$-*araI* loop predominates under noninducing conditions and the *araO*$_2$-*araO*$_1$ loop predominates under inducing conditions.

Several interesting mechanistic questions remain to be elucidated in this system. How does the addition of arabinose result in the destabilization of the *araO*$_2$-*araI* loop and the stabilization of the *araO*$_2$-*araO*$_1$ loop? Also, how does the interaction between *araC* bound at *araO*$_2$ and *araI* inhibit P_{BAD} transcription under noninducing conditions?

Two models have been proposed to explain the mechanism of repression of basal transcription. Schleif and colleagues (see, for example, Martin et al. 1986) have suggested that the *araC* molecule bound at *araI* is held in an inactive conformation by its interaction with the molecule bound at *araO*$_2$. Another mechanism has been proposed recently by Lee et al. (1987). They have defined a second binding site for *araC* protein, required for activation, that is located between the previously defined I site and the promoter. They suggest that cooperative binding to the two adjacent I sites ensures the occupancy of the new site in the presence of arabinose. In the absence of arabinose, a preferential interaction between *araC* molecules bound at *araO*$_2$ and the upstream I site helps ensure that the promoter-proximal I site remains unoccupied. According to both models, the addition of arabinose promotes a shift in the pattern of cooperative interactions.

B. *gal* Repressor

DNA loop formation has also been implicated in the repression of the galactose (*gal*) operon of *E. coli*. This operon is transcribed from two overlapping promoters, P_1 and P_2. The CAP protein stimulates transcription from P_1 and represses transcription from P_2 (Musso et al. 1977), and both P_1 and P_2 are negatively regulated by the *gal* repressor (Adhya and Miller 1979; Kuhnke et al. 1986).

The first *gal* operator mutations to be mapped and sequenced were located upstream of the CAP site, at approximately –60 from the startpoint of P_1 transcription (Adhya and Miller 1979; diLauro et al. 1979). Subsequently, a second set of operator constitutive mutants was isolated that defined a new *gal*-repressor-binding site located within the first structural gene, approximately 40-bp downstream from the startpoint of transcription (Fritz et al. 1983; Irani et al. 1983). The existence of two operator sites, one (O_E) located upstream of and the other (O_I) located downstream from the promoter region, led to the suggestion that repression might involve the formation of a DNA loop held together either by one tetrameric repressor molecule or by two interacting dimers (Fritz et al. 1983; Majumdar and Adhya 1984).

What is the mechanism of repression of the *gal* operon? One possibility is that the occupancy of each site (O_E and O_I) contributes independently to repression and that cooperative binding to the two sites ensures the stable occupancy of both. Another possibility is that repression depends on the looped protein/DNA structure per se, for example, because of a change in the conformation of the DNA in the promoter region or because RNA polymerase is trapped within the looped domain. An elegant set of experiments has been done by Haber and Adhya (1988) to test the idea that efficient repression depends on an interaction between repressors bound at O_E and O_I. They converted one or the other site into a *lac* operator and assayed repression in the presence of high levels of both *gal* repressor and *lac* repressor. If repression depends on site occupancy alone, a heterologous pair of operators should be able to mediate full repression, assuming both operators are fully occupied. They found, however, that each heterologous pair mediated only partial repression, and a homologous combination of operators (either *gal* or *lac*) was required for full repression, implying that the looped structure contributes directly to repression.

C. *lac* Repressor

The *lac* operon contains a primary *lac*-repressor-binding site (O_1) that overlaps the promoter as well as two so-called pseudo-operators, one (O_3) located 93-bp upstream of O_1 and the other (O_2) located 401-bp downstream from O_1 in the structural gene for β-galactosidase (Reznikoff et al. 1974; Gilbert et al. 1976). It had been assumed previously that these weak auxiliary sites have no physiological significance, but

Eismann et al. (1987) recently showed that the destruction of O_2 reduces fivefold the repression of the *lac* operon in vivo. Since *lac* repressor is a stable tetramer with the potential for binding two operator sites (Kania and Müller-Hill 1977), a tetramer bound at O_1 could increase the occupancy of either pseudo-operator by the formation of a DNA loop (see discussion in section III).

Two studies performed in vivo using artificial constructs first suggested that *lac* repressor may bind cooperatively to separated operator sites, 231-bp apart in one case (Besse et al. 1986) and 283- or 185-bp apart in the other case (Mossing and Record 1986). Mossing and Record placed a wild-type *lac* operator at various distances upstream of a *lac* promoter/operator region bearing a mutation within the operator. In their experiments, the *lac* promoter was removed from its natural context so that neither pseudo-operator was present. They found that repression in vivo was significantly enhanced by operators positioned 185- and 283-bp upstream of the mutant operator but only slightly enhanced by an operator positioned 118-bp upstream, implying that the binding of repressor to a wild-type site suitably positioned upstream can increase the occupancy of the mutant site within the promoter. They argued that the weakness of the effect at 118 bp reflects the relative difficulty of bending short DNA fragments. It is also possible that a "side of the helix" effect is responsible for the failure of the 118-bp construct to show enhanced repression, especially in light of a subsequent demonstration in vitro of loop formation with a separation between operators of only 63 bp (see below).

An extensive analysis of the binding of *lac* repressor to separated operator sites in vitro has established that *lac* repressor can induce the formation of DNA loops by binding simultaneously to operator sites separated by 63–535 bp (Kramer et al. 1987). The formation of loops was assayed using three different techniques, a gel-binding assay (in which three protein/DNA complexes could be distinguished based on their mobilities: a singly bound DNA molecule, a molecule that is bound by two noninteracting repressors, and the looped complex), the DNase footprinting assay, and electron microscopy. The larger loops appeared to be more stable than the smaller loops, and loop formation was a periodic function of the spacing between the operator sites in the range examined, from 14.6 to 16 turns of the DNA helix. The investigators also demonstrated that, at high concentrations of *lac* repressor, unlooped structures in which both operators were occupied by *lac* repressor tetramers replaced the looped structures.

Several groups have investigated the effect of supercoiling on loop formation in the *lac* system. Whitson et al. (1987) examined the binding of *lac* repressor to linear and negatively supercoiled operator-containing plasmids using a filter binding assay. They found that the dissociation rate constants for plasmids containing a single operator decreased as a function of the negative supercoil density but that the effect was much more dramatic for plasmids that carried the pseudo-operators in addition to the primary operator, suggesting that supercoiling stabilizes a looped complex in which *lac* repressor is bound to both the primary operator and one of the pseudo-operators.

Borowiec et al. (1987) examined the binding of repressor to a plasmid carrying just O_1 and O_3 both in vivo and in vitro using a dimethylsulfate (DMS) footprinting assay. In vivo, they observed an unexpectedly high occupancy of O_3 (the repressor binding affinity of which is at least 100-fold lower than that of O_1 as measured in vitro) under conditions where there was insufficient repressor to bind all the copies of O_1, suggesting that a significant fraction of the repressor was binding simultaneously to both O_1 and O_3. To test this idea, they performed a titration experiment in vitro under conditions where repressor was limiting and the binding was stochiometric. Using a linear template carrying both operators, they saw no occupancy of O_3, but using a supercoiled template they detected increasing occupancy of O_3 with the addition of increasing amounts of repressor. Meanwhile, the occupancy of O_1 increased just as on the linear template with the titration endpoint occurring at the same concentration of repressor on both templates. There being no free repressor available to bind O_3 below the titration endpoint, the experiment thus demonstrates that supercoiling promotes the joint occupancy of O_1 and O_3 by a single tetramer. The DMS footprinting experiment also revealed one hyperreactive residue located between the two operators that was observed in the presence of repressor using supercoiled DNA. Using an additional modification reagent (potassium permanganate), the investigators obtained evidence for a local disruption of the helix spanning approximately 5 bp at the site of the hyperreactive residue. They suggest that supercoiling might facilitate the melting of this 5-bp sequence to allow for a sharp bend in the DNA at that position.

A more recent study supports the interpretation of the above study: Sasse-Dwight and Gralla (1988) showed by footprinting in vivo that *lac* repressor binds cooperatively to O_1 and O_3 on a plasmid carrying these two sites. The binding of repressor to O_3 depends on the presence of O_1 on the same plasmid. Furthermore, the use of the drug coumermycin, which is an inhibitor of supercoiling, had a similar effect on the occupancy of O_3, as did the deletion of O_1, implying that cooperative binding is dependent on DNA supercoiling in vivo as well as in vitro.

Finally, Kramer et al. (1988) examined the effect of supercoiling on loop formation in vitro using a set of DNA minicircles carrying two synthetic *lac* operators separated by different distances. Various topoisomers of the minicircles were separately incubated with *lac* repressor. Loop formation was examined by means of the gel-binding assay. The experiments showed, first, that negative supercoiling increases the stability of the looped complexes; second, that the optimal interoperator spacing for loop formation changes with linking number; and third, that increasing negative supercoiling relaxes the requirement for an optimal spacing between the two operators, presumably because the supercoiled molecules constitute a mixed population in which the change in linking number is variably partitioned into changes in twist and writhe so that the helical repeat differs from molecule to molecule.

Why does negative supercoiling increase the stability of DNA loops? In part, writhe provides an intrinsic bend to the DNA. However, as pointed out both by Borowiec et al. (1987) and by Kramer et al. (1988) negative supercoiling may also promote local structural alterations in the DNA that in turn facilitate bending or kinking.

D. *deoR* Repressor

DNA loop formation has also been implicated in the regulation of the *deo* operon, which encodes four enzymes involved in nucleoside catabolism. This operon is transcribed from two promoters, P1 and P2, which are separated by some 600 bp (Albrechtsen et al. 1976; Valentin-Hansen et al. 1982). The operon is negatively regulated by the product of the *deoR* gene, which controls transcription from both P1 and P2, each of which contains a binding site for the repressor. Dandanell and Hammer (1985) demonstrated that ordinarily both operators are required to achieve full repression in vivo. That is, separate fusions of each promoter to an assayable gene were inefficiently repressed. Full repression was restored by introducing a second repressor-binding site upstream of each promoter or by overproduction of the repressor. On the basis of these observations, the investigators proposed that the *deoR* repressor might interact with both sites simultaneously and cause the formation of a DNA loop. The fact that overproduction of *deoR* repressor can mediate full repression when only one repressor-binding site is present implies that repression does not depend on the loop per se but on the increased site occupancy that cooperative binding to separated operator sites accomplishes.

Subsequently, Valentin-Hansen et al. (1986) discovered a third *deoR*-repressor-binding site located 270-bp upstream of P1. This operator also enhances repression, especially when one of the other operators is deleted. Using artificial constructs, Dandanell et al. (1987) have demonstrated that repression of the isolated *deoP2* promoter can be increased by inserting a second repressor-binding site 1 to 5 kb downstream with the amount of repression declining with increasing distance between the two operators.

E. NR_1

The transcriptional activation of nitrogen-regulated (NR) genes in *E. coli* and *Klebsiella pneumoniae* also appears to involve DNA loop formation (for reviews, see Gussin et al. 1986; Magasanik and Neidhardt 1987). It has been suggested that the relevant activators may interact directly with RNA polymerase while bound to sites a considerable distance upstream. The *glnA* operon of *E. coli* is positively controlled by NR_1, the product of the *glnG* gene, during nitrogen-limited growth (Hirschman et al. 1985; Hunt and Magasanik 1985). The activation of this operon also depends on a specialized σ factor called σ^{54}. NR_1 binds to multiple sites upstream of the promoter and has been shown to activate transcription both in vivo (Reitzer and Magasanik 1986) and in vitro (Ninfa et al. 1987) when these sites are moved to positions as far as 1000-bp upstream or downstream. In the presence of very high concentrations of NR_1 (in vivo or in vitro), transcriptional activation occurs in the absence of NR_1-binding sites, per haps because NR_1 binds nonspecifically to the DNA when present at high concentrations. Ninfa et al. (1987) have postulated that at lower concentrations, the binding of NR_1 to specific sites on the DNA increases the probability of its interaction with RNA polymerase through the looping of the intervening DNA.

F. Recombination Proteins and Replication Proteins

This chapter has focused on transcriptional regulators that induce the formation of DNA loops. However, proteins involved in other processes, such as site-specific recombination, also induce the formation of DNA loops (for reviews, see Echols 1986; Gellert and Nash 1986; Craig 1988). For example, long range protein-protein interactions are known to be involved in the integration of bacteriophage λ into the bacterial chromosome. The phage attachment site, *attP*, which spans 240 bp and contains ten binding sites for the recombination proteins, *int* and the integration host factor (IHF) (for review, see Weisberg and Landy 1983), condenses into a structure known as the intasome that is required for the integration reaction (Better et al. 1982). More specifically, *attP* consists of a 15-bp core sequence, containing the crossover site, and two flanking arms. *int* binds to a pair of sites within the core and also recognizes a distinct set of sites located 50- and 150-bp away on the flanking arms (see Weisberg and Landy 1983). The formation of the intasome is mediated by cooperative interactions between nonadjacently bound proteins and depends on DNA supercoiling (Richet et al. 1986; Thompson et al. 1987). Furthermore, the demonstration that *int* protein contains two distinct DNA-binding domains that can bind simultaneously raises the possibility that a single *int* protomer may bridge a core site and an arm site on the same DNA molecule (Moitoso de Vargas et al. 1988). Finally, a recent observation suggests that the role of IHF may simply be to facilitate DNA bending. Another DNA-binding protein (*E. coli* CAP) that also bends DNA can substitute for IHF in the integration reaction when an IHF site is replaced with a CAP site (Goodman and Nash 1989).

A second example of a recombinational system in which protein-protein interactions mediate loop formation is the site-specific inversion reaction that controls expression of the flagellin genes in *Salmonella.* The Hin protein, the site-specific recombinase, interacts specifically with the two 26-bp recombination sites (Johnson and Simon 1985). Efficient inversion depends not only on the Hin protein and the recombination sites, but also on a *cis*-acting enhancer-like sequence that is normally located within the invertible segment 103-bp away from the closest recombination site (Johnson and Simon 1985).

This recombinational enhancer, which consists of two binding sites for a host protein called Fis (Johnson et al. 1986; Koch and Kahmann 1986), can function at various positions within the invertible segment and also when placed in either orientation outside of the invertible segment as far away as 4 kb from the closest recombination site (Johnson and Simon 1985; Johnson et al. 1986). The only restriction appears to be that the enhancer sequence cannot be placed too close (within 48 bp) to either of the recombination sites. Johnson et al. (1987) suggest that the enhancer serves as a scaffold for the assembly of a synaptic complex held together by interactions between the Hin and Fis proteins. The formation of this structure would require the looping out of the DNA between each recombination site and the enhancer. This model is consistent both with the observation that the enhancer can be flexibly positioned relative to the recombination sites and also with the observation that there must be some minimal distance between enhancer and recombination site (so as to allow the DNA to loop). Similar enhancer sequences have been identified in the related DNA inversion systems of phage Mu (Kahmann et al. 1985) and phage P1 (Huber et al. 1985).

Another process in which protein-mediated DNA looping has been demonstrated is replication. Mukherjee et al. (1988) have described an enhancer of replication on the plasmid R6K. *ori*γ (which is infrequently used as a replication origin) functions as a *cis*-acting enhancer of replication from *ori*α located 2000-bp away and from *ori*β located 1200-bp away. Replication depends on the replication initiator protein that binds tightly to seven 22-bp repeats within *ori*γ. Mukherjee et al. (1988) suggest that the initiation of replication involves the transfer of initiator protein to *ori*α or *ori*β by DNA looping. They demonstrate DNA looping and cooperative binding using several different techniques. First, they show that the addition of initiator protein to a linear DNA fragment containing a copy of *ori*γ close to one end and a copy of *ori*β close to the other end increases the rate of T4 ligase-induced circularization. Second, they use an exonuclease III protection assay and a DMS protection assay to demonstrate cooperative binding; protection of *ori*β sequences depends on the presence of *ori*γ sequences further than 2000-bp away. Finally, they use electron microscopy to visualize the looped complexes.

V. CONCLUSION

Transcriptional regulation often depends on interactions between DNA-bound proteins, and when these interactions involve proteins that are bound at a distance from one another, the formation of DNA loops is induced. The physiological significance of these loops may differ from case to case. If cooperative binding with enhanced site occupancy is the physiologically relevant factor, then loop formation contributes only passively to the process. The same is true if the critical event is the association of two proteins. In some cases (such as the *gal* system), however, the loop itself may be important because of the structural constraint it imposes on the DNA. As new examples of DNA looping emerge, it will be important to examine the effect of site occupancy independent of looping as a first step in exploring the mechanistic significance of loop formation. As exemplified by the arabinose system, loop formation may be a dynamic process that is itself subject to regulatory controls. Systems in which different protein-protein interactions prevail under different circumstances may serve as models for the more complex regulation characteristic of eukaryotes.

ACKNOWLEDGMENTS

I thank J.C. Wang, M. Ptashne, H.A. Nash, E. Giniger, Jerry Orloff, and members of the Ptashne lab for many helpful discussions, and J.O. Schwartz, G. Gill, and G. Hochschild for comments on the manuscript.

REFERENCES

Ackers, G.K., A.D. Johnson, and M. Shea. 1982. Quantitative model for gene regulation by lambda phage repressor. *Proc. Natl. Acad. Sci.* **79:** 1129–1133.

Adhya, S. and W. Miller. 1979. Modulation of the two promoters of the galactose operon of *Escherichia coli*. *Nature* **279:** 492–494.

Albrechtsen, H., K. Hammer-Jespersen, A. Munch-Petersen, and N. Fiil. 1976. Multiple regulation of nucleoside catabolizing enzymes: Effects of a polar *dra* mutation on the deo enzymes. *Mol. Gen. Genet.* **146:** 139–145.

Besse, M., B. von Wilcken-Bergmann, and B. Müller-Hill. 1986. Synthetic *lac* operator mediates repression through *lac* repressor when introduced upstream and downstream from *lac* promoter. *EMBO J.* **5:** 1377–1381.

Better, M., C. Lu, R. C. Williams, and H. Echols. 1982. Site-specific DNA condensation and pairing mediated by the int protein of bacteriophage λ. *Proc. Natl. Acad. Sci.* **79:** 5837–5841.

Borowiec, J.A., L. Zhang, S. Sasse-Dwight, and J.D. Gralla. 1987. DNA supercoiling promotes formation of a bent repression loop in *lac* DNA. *J. Mol. Biol.* **196:** 101–111.

Chadwick, P., V. Pirotta, R. Steinberg, N. Hopkins, and M. Ptashne. 1971. The λ and 434 phage repressors. *Cold Spring Harbor Symp. Quant. Biol.* **35:** 283–294.

Craig, N.L. 1988. The mechanism of conservative site-specific recombination. *Annu. Rev. Genet.* **22:** 77–105.

Dandanell, G. and K. Hammer. 1985. Two operator sites separated by 599 base pairs are required for *deoR* repression of the deo operon of *E. coli. EMBO J.* **4:** 3333–3338.

Dandanell, G., P. Valentin-Hansen, J.E. Love Larsen, and K. Hammer. 1987. Long-range cooperativity between gene regulatory sequences in a prokaryote. *Nature* **325:** 823–826.

diLauro, R., T. Taniguchi, R. Musso, and B. de Crombrugghe. 1979. Unusual location and function of the operator in the *Escherichia coli* galactose operon. *Nature* **279:** 494–500.

Drew, H.R. and A.A. Travers. 1985. DNA bending and its relation to nucleosome positioning. *J. Mol. Biol.* **186:** 773–790.

Dunn, T.M., S. Hahn, S. Ogden, and R.F. Schleif. 1984. An operator at –280 base pairs that is required for repression of araBAD operon promoter: Addition of DNA helical turns between the operator and promoter cyclically hinders repression. *Proc. Natl. Acad. Sci.* **81:** 5017–5020.

Echols, H. 1986. Multiple DNA-protein interactions governing high-precision DNA transactions. *Science* **233:** 1050–1056.

Eismann, E., B. von Wilcken-Bergmann, and B. Müller-Hill. 1987. Specific destruction of the second *lac* operator decreases repression of the *lac* operon in *Escherichia coli* fivefold. *J. Mol. Biol.* **195:** 949–952.

Englesberg, E., C. Squires, and F. Meronk. 1969. The L-arabinose operon in *Escherichia coli B/r:* A genetic demonstration of two functional states of the product of a regulator gene. *Proc. Natl. Acad. Sci.* **62:** 1100–1107.

Fersht, A. 1977. In *Enzyme structure and mechanism*, p. 44–48. W.H. Freeman, San Francisco.

Fritz, H.-J., H. Bicknase, B. Gleumes, C. Heibach, S. Rosahl, and R. Ehring. 1983. Characterization of two mutations in the *Escherichia coli galE* gene inactivating the second galactose operator and comparative studies of repressor binding. *EMBO J.* **2:** 2129–2135.

Gellert, M. and H. Nash. 1986. Communication between segments of DNA during site-specific recombination. *Nature* **325:** 401–404.

Gilbert, W., A. Majors, and A. Maxam. 1976. How proteins recognize DNA sequences. *Life Sci. Res. Rep.* **4:** 167–178.

Goodman, S.D. and H.A. Nash. 1989. Functional replacement of a protein-induced bend in a DNA recombination site. *Nature* **341:** 251–254.

Griffith, J., A. Hochschild, and M. Ptashne. 1986. DNA loops induced by cooperative binding of λ repressor. *Nature* **322:** 750–752.

Gussin, G.N., C.W. Ronson, and F.M. Ausubel. 1986. Regulation of nitrogen fixation genes. *Annu. Rev. Genet.* **20:** 567–591.

Haber, R. and S. Adhya. 1988. Interaction of spatially separated protein-DNA complexes for control of gene expression: Operator conversions. *Proc. Natl. Acad. Sci.* **85:** 9683–9687.

Hahn, S., T.M. Dunn, and R.F. Schleif. 1984. Upstream repression and CRP stimulation of the *Escherichia coli* L-arabinose operon. *J. Mol. Biol.* **180:** 61–72.

Hirschman, J., P.-K. Wong, K. Sei, J. Keener, and S. Kustu. 1985. Products of nitrogen regulatory genes *ntrA* and *ntrC* of enteric bacteria activate *glnA* transcription *in vitro* evidence that the *ntrA* product is a σ factor. *Proc. Natl. Acad. Sci.* **82:** 7525–7529.

Hochschild, A. and M. Ptashne. 1986. Cooperative binding of λ repressors to sites separated by integral turns of the DNA helix. *Cell* **44:** 681–687.

———. 1988. Interaction at a distance between λ repressors disrupts gene activation. *Nature* **336:** 353–357.

Horowitz, D. and J.C. Wang. 1984. Torsional rigidity of DNA and length dependence of the free energy of DNA supercoiling. *J. Mol. Biol.* **173:** 75–91.

Huber, H.E., S. Iida, W. Arber, and T.A. Bickle. 1985. Site-specific DNA inversion is enhanced by a DNA sequence element in *cis*. *Proc. Natl. Acad. Sci.* **82:** 3776–3780.

Hunt, T.P. and B. Magasanik. 1985. Transcription of *glnA* by purified *Escherichia coli* components: Core RNA polymerase and the products of *glnF, glnG,* and *glnL*. *Proc. Natl. Acad. Sci.* **82:** 8453–8457.

Huo, L., K.J. Martin, and R. Schleif. 1988. Alternative DNA loops regulate the arabinose operon in *Escherichia coli*. *Proc. Natl. Acad. Sci.* **85:** 5444–5448.

Irani, M.H., L. Orosz, and S. Adhya. 1983. A control element within a structural gene: The gal operon of *Escherichia coli*. *Cell* **32:** 783–788.

Jacobson, H. and W.H. Stockmayer. 1950. Intramolecular reaction in polycondensations. I. The theory of linear systems. *J. Chem. Phys.* **18:** 1600–1606.

Johnson, A.D. 1980. "Mechanism of action of the lambda cro protein." Ph.D. thesis, Harvard University, Cambridge, Massachusetts.

Johnson, A.D., B.J. Meyer, and M. Ptashne. 1979. Interactions between DNA-bound repressors govern regulation by the lambda phage repressor. *Proc. Natl. Acad. Sci.* **76:** 5061–5065.

Johnson, R.C. and M.I. Simon. 1985. Hin-mediated site-specific recombination requires two 26 bp recombination sites and a 60 bp recombinational enhancer. *Cell* **41:** 781–791.

Johnson, R.C., M.F. Bruist, and M.I. Simon. 1986. Host protein requirements for in vitro site-specific DNA inversion. *Cell* **46:** 531–539.

Johnson, R.C., A.C. Glasgow, and M.I. Simon. 1987. Spatial relationships of the Fis binding sites for Hin recombinational enhancer activity. *Nature* **329:** 462–465.

Kahmann, R., F. Rudt, C. Koch, and G. Mertens. 1985. G inversion in bacteriophage Mu DNA is stimulated by a site within the invertase gene and a host factor. *Cell* **41:** 771–780.

Kania, J. and B. Müller-Hill. 1977. Construction, isolation and implications of repressor-galactosidase•β-galactosidase hybrid molecules. *Eur. J. Biochem.* **79:** 381–386.

Koch, C. and R. Kahmann. 1986. Purification and properties of the *Escherichia coli* host factor required for inversion of the G segment in bacteriophage Mu. *J. Biol. Chem.* **261:** 15673–15678.

Kramer, H., M. Amouyal, A. Nordheim, and B. Müller-Hill. 1988. DNA supercoiling changes the spacing requirement of two *lac* operators for DNA loop formation with *lac* repressor. *EMBO J.* **7:** 547–556.

Kramer, H., M. Niemoller, M. Amouyal, B. Revet, B. von Wilcken-Bergmann, and B. Müller-Hill. 1987. *lac* repressor forms loops with linear DNA carrying two suitably spaced *lac* operators. *EMBO J.* **6:** 1481–1491.

Kuhnke, G., A. Krause, C. Heibach, U. Gieske, H.-J. Fritz, and R. Ehring. 1986. The upstream operator of the *Escherichia coli* galactose operon is sufficient for repression of transcription initiated at the cyclic AMP-stimulated promoter. *EMBO J.* **5:** 167–173.

Landau, L. and E. Lifshitz. 1958. In *Statistical physics*, p. 478–482. Pergamon Press, London.

Lee, N., C. Francklyn, and E.P. Hamilton. 1987. Arabinose-induced binding of AraC protein to *$araI_2$* activates the *araBAD* operon promoter. *Proc. Natl. Acad. Sci.* **84:** 8814–8818.

Lee, N.L., W.O. Gielow, and R.G. Wallace. 1981. Mechanism of *araC* autoregulation and the domains of two overlapping promoters, P_C and P_{BAD}, in the L-arabinose regulatory region of *Escherichia coli. Proc. Natl. Acad. Sci.* **78:** 752–756.

Levene, S.D. and D.M. Crothers. 1986. Ring closure probabilities for DNA fragments by Monte Carlo simulation. *J. Mol. Biol.* **189:** 61–72.

Magasanik, B. and F. C. Neidhardt. 1987. Regulation of carbon and nitrogen utilization. In Escherichia coli *and* Salmonella typhimurium: *Cellular and Molecular Biology* (ed. F.C. Neidhardt et al.), p. 1318–1325. American Society for Microbiology, Washington, D.C.

Majumdar, A. and S. Adhya. 1984. Demonstration of two operator elements in gal: *In vitro* repressor binding studies. *Proc. Natl. Acad. Sci.* **81:** 6100–6104.

Martin, K., L. Huo, and R.F. Schleif. 1986. The DNA loop model for *ara* repression: AraC protein occupies the proposed loop sites *in vivo* and repression negative mutations lie in these same sites. *Proc. Natl. Acad. Sci.* **83:** 3654–3658.

Mertz, J.E. and R.W. Davis. 1972. Cleavage of DNA by R_I restriction endonuclease generates cohesive ends. *Proc. Natl. Acad. Sci.* **69:** 3370–3374.

Moitoso de Vargas, L., C.A. Pargellis, N.M. Hasan, E.W. Bushman, and A. Landy. 1988. Autonomous DNA binding domains of λ integrase recognize two different sequence families. *Cell* **54:** 923–929.

Mossing, M.C. and M.T. Record. 1986. Upstream operators enhance repression of the *lac* promoter. *Science* **233:** 889–892.

Mukherjee, S., H. Erickson, and D. Bastia. 1988. Enhancer-origin interaction in plasmid R6K involves a DNA loop mediated by initiator protein. *Cell* **52:** 375–383.

Musso, R.E., R. diLauro, S. Adhya, and B. de Crombrugghe. 1977. Dual control for transcription of the galactose operon by cyclic AMP and its receptor protein at two interspersed promoters. *Cell* **12:** 847–854.

Ninfa, A.J., L.J. Reitzer, and B. Magasanik. 1987. Initiation of transcription at the bacterial glnAp2 promoter by purified *E. coli* components is facilitated by enhancers. *Cell* **50:** 1039–1046.

Ogden, S., D. Haggerty, C.M. Stoner, D. Kolodrubetz, and R. Schleif. 1980. The *Escherichia coli* L-arabinose operon: Binding sites of the regulatory proteins and a mechanism of positive and negative regulation. *Proc. Natl. Acad. Sci.* **77:** 3346–3350.

Pabo, C.O. and M. Lewis. 1982. The operator-binding domain of λ repressor: Structure and DNA recognition. *Nature* **298:** 443–447.

Pabo, C.O., R.T. Sauer, J.M. Sturtevant, and M. Ptashne. 1979. The λ repressor contains two domains. *Proc. Natl. Acad. Sci.* **76:** 1608–1612.

Poteete, A.R. and M. Ptashne. 1982. Control of transcription by the bacteriophage P22 repressor. *J. Mol. Biol.* **157:** 21–48.

Ptashne, M. 1986a. *A genetic switch.* Cell Press, Palo Alto, California and Blackwell Scientific, Oxford, England.

———. 1986b. Gene regulation by proteins acting nearby and at a distance. *Nature* **322:** 697–701.

———. 1988. How eukaryotic transcriptional activators work. *Nature* **335:** 683–689.

Reitzer, L.J. and B. Magasanik. 1986. Transcription of *glnA* in *E. coli* is stimulated by activator bound to sites far from the promoter. *Cell* **45:** 785–792.

Reznikoff, W.S., R.F. Winter, and C.K. Hurley. 1974. The location of the repressor binding sites in the *lac* operon. *Proc. Natl. Acad. Sci.* **71:** 2314–2318.

Richet, E., P. Abcarion, and H.A. Nash. 1986. The interaction of recombination proteins with supercoiled DNA: Defining the role of supercoiling in λ integrative recombination. *Cell* **46:** 1011–1021.

Salvo, J.J. and N.D.F. Grindley. 1988. The γδ resolvase bends the *res* site into recombinogenic complex. *EMBO J.* **7:** 3609–3616.

Sasse-Dwight, S. and J.D. Gralla. 1988. Probing co-operative DNA-binding *in vivo* — The *lac* O_1:O_3 interaction. *J. Mol. Biol.* **202:** 107–119.

Sauer, R.T., C.O. Pabo, B.J. Meyer, M. Ptashne, and K.C. Backman. 1979. Regulatory functions of the λ repressor reside in the amino-terminal domain. *Nature* **279:** 396–400.

Sauer, R.T., R.R. Yocum, R.F. Doolittle, M. Lewis, and C.O. Pabo. 1982. Homology among DNA-binding proteins suggests use of a conserved super-secondary structure. *Nature* **298:** 447–451.

Sheppard, D. and E. Englesberg. 1967. Further evidence for positive control of the L-arabinose system by gene *araC*. *J. Mol. Biol.* **25:** 443–454.

Shimada, J. and H. Yamakawa. 1984. Ring-closure probabilities for twisted wormlike chains: Application to DNA. *Macromolecules* **17:** 689–698.

Shore, D. and R.L. Baldwin. 1983. Energetics of DNA twisting. I. Relation between twist and cyclization probability. *J. Mol. Biol.* **170:** 957–981.

Shore, D., J. Langowski, and R.L. Baldwin. 1981. DNA flexibility studied by covalent closure of short fragments into circles. *Proc. Natl. Acad. Sci.* **78:** 4833–4837.

Thompson, J.F., L. Moitoso de Vargas, S.E. Skinner, and A. Landy. 1987. Protein-protein interactions in a higher order structure direct lambda site-specific recombination. *J. Mol. Biol.* **195:** 481–493.

Valentin-Hansen, P., H. Aiba, and D. Schumperli. 1982. The structure of tandem regulatory regions in the *deo* operon of *Escherichia coli* K12. *EMBO J.* **1:** 317–322.

Valentin-Hansen, P., B. Albrechtsen, and J.E. Love Larsen. 1986. DNA-protein recognition: Demonstration of three genetically separated operator elements that are required for repression of the *Escherichia coli deoCABD* promoters by the DeoR repressor. *EMBO J.* **5:** 2015–2021.

Valenzuela, D. and M. Ptashne. 1989. P22 repressor mutants deficient in co-operative binding and DNA loop formation. *EMBO J.* **8:** 4345–4350.

Wang, J.C. and N.R. Davidson. 1966a. Thermodynamic and kinetic studies on the interconversion between the linear and circular forms of phage lambda DNA. *J. Mol. Biol.* **15:** 111–123.

———. 1966b. On the probability of ring closure of lambda DNA. *J. Mol. Biol.* **19:** 469–482.

Wang, J.C. and G.N. Giaever. 1988. Action at a distance along a DNA. *Science* **240:** 300–304.

Weisberg, R.A. and A. Landy. 1983. Site-directed recombination in phage lambda. In *Lambda II* (ed. R.W. Hendrix et al.), p. 211. Cold Spring Harbor Laboratory, Cold Spring Harbor, New York.

Wharton, R.P., E.L. Brown, and M. Ptashne. 1984. Substituting an α-helix switches the sequence-specific DNA interactions of a repressor. *Cell* **38:** 361–369.

Whitson, P.A., W.-T. Hsieh, R.D. Wells, and K.S. Matthews. 1987. Supercoiling facilitates *lac* operator-repressor-pseudooperator interactions. *J. Biol. Chem.* **262:** 4943–4946.

4

Primer on the Topology and Geometry of DNA Supercoiling

Nicholas R. Cozzarelli and T. Christian Boles[1]
Department of Molecular and Cell Biology
University of California, Berkeley
Berkeley, California 94720

James H. White
Department of Mathematics
University of California, Los Angeles
Los Angeles, California 90024

I. HISTORICAL INTRODUCTION

In 1953, Watson and Crick proposed their model for the structure of DNA in which two polydeoxyribonucleotides are wound as right-handed helices around each other and around a common axis (Watson and Crick 1953). It was not until 12 years later that Vinograd and his colleagues discovered that the helix axis can also be coiled: This higher-order structure was named supercoiling (Vinograd et al. 1965, 1968; Vinograd and Lebowitz 1966; Bauer and Vinograd 1968). The key initial observation

[1]Present address: Graduate Department of Biochemistry, Brandeis University, Waltham, Massachusetts 02254.

was that the DNA of the polyoma virus had a higher sedimentation coefficient than an equal length of linear DNA. They concluded that this was due to compaction both by cyclization and by supercoiling and that supercoiling required both DNA strands be free of nicks or gaps. It soon became clear that circular DNA from numerous sources was supercoiled. An important further generalization from the work of Pettijohn, Laemmli, and Worcel, was that even linear DNA can be supercoiled because inside the cell it is constrained into topologically distinct domains (Stonington and Pettijohn 1971; Worcel and Burgi 1972; Pettijohn and Hecht 1974; Drlica and Worcel 1975; Benyajati and Worcel 1976; Paulson and Laemmli 1977). Indeed, virtually all DNA in vivo is supercoiled to approximately the same degree (for review, see Bauer 1978).

From the outset, it was clear that understanding DNA supercoiling required a combination of mathematics and experimentation. A quantitative relationship between supercoiling and the coiling of the strands of the DNA double helix was initially developed by the Vinograd group in an intuitive way (Vinograd and Lebowitz 1966; Bauer and Vinograd 1968), and Glaubiger and Hearst in 1967 introduced the mathematical concept of a linking number for the description of supercoiled DNA (Glaubiger and Hearst 1967). A rigorous mathematical treatment of the analogous problem for space curves was first formulated by White (1969), but the mathematicians and molecular biologists were unaware of each other's efforts. Fuller (1971) was the first to apply the mathematical concepts of writhe and twist to DNA. Although Crick (1976) modestly characterizes his 1976 contribution as "an expansion and clarification of part of [Fuller's treatment]," his paper delineates clearly many of the features of supercoiling that would occupy future investigators. In recent years, White and colleagues have significantly amplified the mathematical analysis of DNA supercoiling (White and Bauer 1986, 1987, 1988; White et al. 1988; White 1989).

The enzymology of supercoiling was initially studied by Wang, who discovered the first member of the class of enzymes, called topoisomerases, that introduce and remove supercoils (Wang 1971). These enzymes, like supercoils, are now known to be ubiquitous (Cozzarelli 1980; Wang 1985; Maxwell and Gellert 1986).

Supercoiling plays many important physiological roles in the cell. The linking number of DNA equals writhe, a quantity related to supercoiling, plus double helical twist (White 1969). Because the linking number is a constant for each topologically closed DNA domain, writhe must change in an equal and compensating fashion to all changes in the twist of the double helix. This accommodation occurs frequently because fundamental processes such as DNA replication and transcription unwind the double helix. This removal of supercoils is energetically favorable, and indeed many biological processes are parasites of supercoiling that are unable to proceed in its absence. Supercoiling is also the chief means by which DNA is compacted in an orderly manner into the narrow confines of the cell (Olins and Olins 1974). Supercoiling not only has essential physiological roles, but has been an invaluable experimental tool for studying DNA structure and the interaction of DNA with proteins (Wang 1986).

II. SCOPE OF THIS CHAPTER

DNA supercoiling, despite its importance, is not widely understood by bioscientists. Our major goal is to define and illustrate each of the basic geometric and topological parameters of supercoiled DNA in a fashion that is rigorous but readily accessible to molecular biologists. The term supercoiling has been used in several different ways, and this led to some confusion. We will therefore delineate clearly the different usages and present three different quantitative descriptors of supercoiling and how they are interrelated. We have recently shown that the division of the linking number into three rather than two components clarifies the relationship between topological and geometric properties of DNA and experimentally measurable quantities (White et al. 1988). Because this subdivision is new and has been presented only in technical form, we will develop it more broadly and intuitively in this chapter. We shall then summarize recent work on the size and shape of two important forms of supercoiled DNA to give a concrete description of supercoiling. A Glossary of Terms concludes this chapter.

In the quantitative description of DNA structure, complex entities such as twist, writhe, supercoiling, and helical repeat are represented by single numbers. Clearly, in this reduction, some structural information is lost. However, good quantitative indices preserve critical information, are insensitive to irrelevant features, can be measured accurately, and are related mathematically to other well-defined parameters. An example from physics is that, in calculating gravitational forces, the entire Earth, despite its complex structure, can be represented by a single number: its mass. Similarly, for many purposes, the linking number is a good DNA index: It reduces the complex three-dimensional winding of the strands of the double helix about each other to a single number that aids in understanding how DNA is manipulated in the cell. However, it lacks information on the shape of DNA and therefore is not useful for some purposes.

III. BASIC DEFINITIONS

Supercoiling means literally the coiling of a coil. It is a general term that applies not only to DNA, but also to proteins, ropes, and abstract curves. In considering DNA structure, it is convenient to define an axis for the ordinary B-type double helix as an imaginary line down the center of the molecule between the helical curves W and C, which represent the strands. Thus, for DNA, supercoiling refers to the coiling of the axis of the double helix in space.

Bauer and Vinograd (1968) pointed out that there are two basic forms of DNA supercoiling, solenoidal (or toroidal) and plectonemic (or interwound). Solenoidal supercoiling (Fig. 1A) is exemplified by the wrapping of DNA around histones in nucleosomes (Richmond et al. 1984). Plectonemic supercoiling (Fig. 1B) is the form for underwound DNA in solution (Vinograd et al. 1965; Spengler et al. 1985). In either case, the supercoiling arises from constraints imposed on the DNA. For the nucleosome model, the DNA wraps in a left-handed solenoidal manner about the cylindrical surface created by the histone octamer, a geometric constraint. With underwound DNA in solution, the duplex axis winds plectonemically because of a change in linking number from the relaxed state, a topological constraint. It will be clear by the end of this chapter that the two forms of supercoiling differ in many fundamental ways.

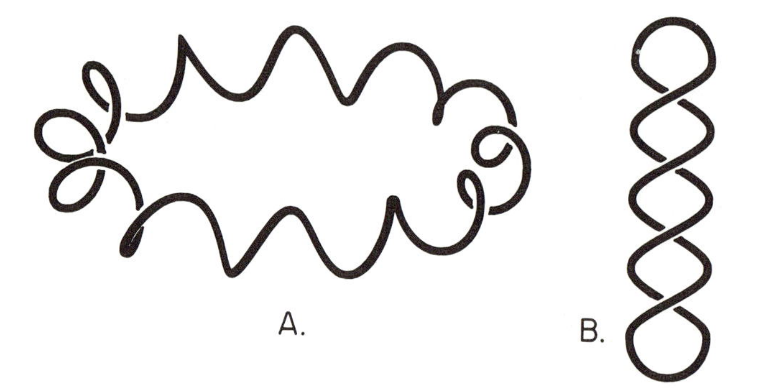

Figure 1 Two forms of supercoiled DNA. The double helix is depicted as a solid tube, and both DNAs are negatively supercoiled. (*A*) Solenoidal, or toroidal, supercoiling in which the DNA wraps helically about a circular superhelix axis. The term toroidal comes from the fact that the DNA can be thought of as winding about a torus (a doughnut-shaped surface). (*B*) In plectonemically supercoiled DNA, the DNA winds helically up and back down about a superhelix axis. In the plectonemic form, regions of the DNA that are distant in primary sequence wind around each other. Hence, plectonemic DNA is often called interwound.

Because topological and geometric changes in DNA can give rise to and alter supercoiling, it is essential to distinguish carefully topological and geometric properties. A topological property is defined as one that remains unchanged under continuous deformations; i.e., those that maintain continuity of the backbone. Continuous deformations include conformational changes because of thermal motion or the binding of proteins, RNA, or drugs; discontinuous deformations involve strand breakage by nucleases, topoisomerases, or physical forces. The simplest example of a topological property is the linking number. As long as a circular DNA molecule remains intact, its linking number is fixed although its shape may change dramatically during isolation and analysis. Topological properties are also usually quantized, often to integral values. A geometric property, on the other hand, may change under deformations and can be described by quantities that specify size and shape. Curvature, bending, and indeed supercoiling itself are all geometric properties. We shall present simple mathematical equations that relate the topological and geometric properties of DNA.

For a segment of DNA to have a topological property such as a linking number, the beginning and end must be fixed relative to each other. This constraint obviously applies to closed circular DNA because beginnings and ends are exactly the same points. Inside the cell, a number of factors conspire to impede the relative movements of the ends of segments of DNA to give topologically closed domains within linear and interrupted circular DNA (Stonington and Pettijohn 1971; Worcel and Burgi 1972; Pettijohn and Hecht 1974; Drlica and Worcel 1975; Benyajati and Worcel 1976; Paulson and Laemmli 1977; Wu et al. 1988).

IV. THE LINKING NUMBER

We begin with a review of linking number *Lk*. This critical property of closed DNA describes the intertwining of the *W* and *C* strands. Mathematically, it is usually more convenient instead to use the linking number of *C* (or *W*) with the helix axis *A* because writhe and twist are defined in terms of the axis. Fortunately, the linking number of *C* with *W* equals that of *C* (or *W*) with *A*. This is because either *W* or *C* can be continuously deformed into the axis without changing the linking number (White 1989). We will usually illustrate our relationships with the *C* strand, but the *W* strand could be used equivalently.

The linking number of two curves depends on their orientation. We establish the convention that *W* and *C* are oriented in the same direction, i.e., in a parallel sense, and that the DNA axis also points in this direction. This topological definition of strand orientation clearly disregards phosphodiester bond polarity but facilitates the mathematics.

There are two simple ways to define *Lk*. The first uses the concept of a spanning surface that has the DNA axis as a boundary curve or edge, much as a soap film extends across a wire frame (Figs. 2 and 3). The orientation of the spanning surface is defined by choosing a vector perpendicular to the surface using the orientation of the boundary curve *A* and the standard right-hand rule (Fig. 2). Thus, in Figure 2A, this surface normal vector points upward, whereas in Figure 2B it points downward. The direction of the normal vector distinguishes the two sides of the spanning surface and allows computation of *Lk* as follows. Suppose that *C* intersects the spanning surface some number of times. To each such intersection, a number is associated, +1 or −1, depending on whether *C* punctures the surface in the same direction as the surface normal or opposite to it, respectively. *Lk* is defined as the sum of all these index numbers. Use of this definition to obtain the *Lk* of left-handed and right-handed double helices is shown in Figure 3. Two important consequences immediately follow from the definition. First, *Lk* is necessarily an integer because it is just a count (signed) of the number of punctures. Second, if the orientations of both *C* and *A* are reversed, *Lk* is unchanged. Topological analysis proves that this definition of *Lk* is independent of the choice of spanning surface. Thus, the simplest such surface may be used to compute *Lk*.

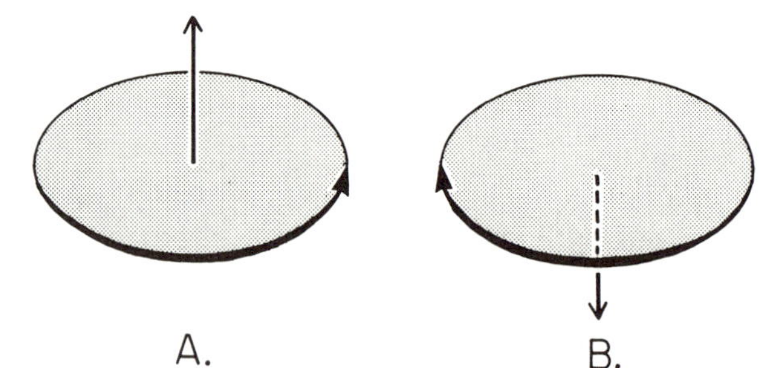

Figure 2 Definition of orientation for spanning surfaces. The spanning surfaces across the circular boundary curves are shown as shaded regions. The orientation of the curves is indicated by the arrow heads. The orientation of the surface is defined by the standard right-hand rule. Place your right hand along the boundary curve with your index finger pointing in the same direction as the curve. The direction indicated by your extended thumb defines a unique sidedness for the surface and the direction of the reference vector perpendicular to the surface. (*A*) This surface normally points up. (*B*) The orientation of the curve is reversed, and therefore the orientation of the surface is also flipped.

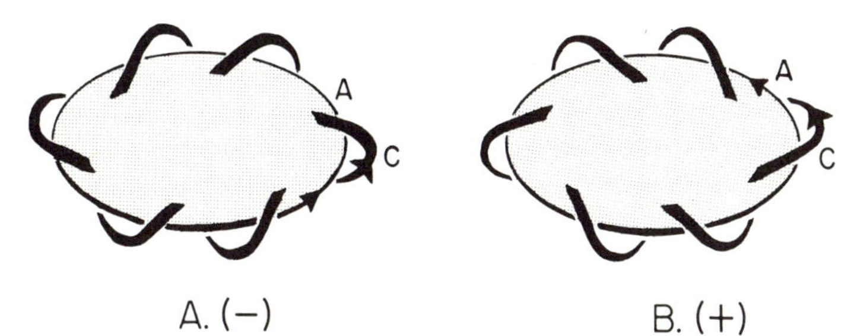

Figure 3 The spanning surface definition applied to left- and right-handed DNA duplexes. The curves *A* and *C* represent the duplex axes and one of the phosphodiester backbone strands of DNA, respectively. Because the axis has the same direction in both parts *A* and *B*, the surface normal points up for both molecules. The spanning surface (stippled) is bounded by *A*. *C* by convention has the same orientation as the axis. Punctures by *C* of the spanning surface in the direction opposite to the surface normal, as in the left-handed DNA in part *A*, are given a negative sign. Punctures in the same direction as the surface normal, as for the right-handed DNA in part *B*, are positive. The sum of all such signed punctures is the *Lk* for the molecule. Thus, the *Lk* of the left-handed DNA is −6 and that of the right-handed DNA is +6.

Another definition of *Lk* uses a type of plane projection of two curves. A simple plane projection can be considered to be the shadow of a curve on a flat screen. Such a projection is perfectly two-dimensional because all extension along the direction of illumination is lost. To define *Lk*, we need to modify the projection so that a critical piece of three-dimensional information, the relative overlay of crossing segments of the curves, is retained. This is usually done graphically by interrupting the underpassing segments but leaving the crossing overpass intact. In any such embellished projection, one of the curves may cross over or under the other at a number of points. To each of these crossings or nodes is associated a number, +1/2 or −1/2; the sign is fixed according to the rule explained in Figure 4. *Lk* for DNA is defined as the sum of all the numbers associated with the crossings of *C* with *A*. Use of this definition for calculating the *Lk* of left-handed and right-handed DNA is shown in Figure 5. Comparison of Figures 3 and 5 shows that there are two nodes for each spanning surface puncture (that is why each node has half the value of a puncture) and that the sign conventions in the two definitions are equivalent. There must be an even number of nodes in closed DNA because whenever one curve passes over the other it must also pass under it to end up at the starting position. Therefore, *Lk* is integral. The value of *Lk* is independent of projection. Thus, the simplest projection, usually the one with the smallest number of nodes, can be used to calculate *Lk*.

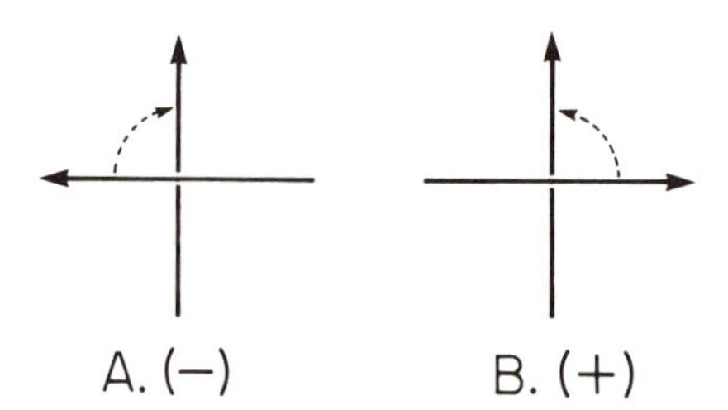

Figure 4 Sign convention for crossings of curves. The arrows indicate the orientation of two crossing curves. To determine the sign of the crossing, also called a node, the upper segment is rotated by an angle less than 180° so that it points in the same direction as the lower segment. If this requires a clockwise rotation, as in *A*, then the node is given a (–) sign. If the rotation is counterclockwise, then the node is (+), as in *B*.

A third definition of *Lk* is the one originally devised by Gauss over one and one half centuries ago (Gauss 1833) although not published until 1867. This is now called the Gauss integral and is useful in computations and in the rigorous proof that the first two definitions of *Lk* are equivalent (White 1989).

Preparations of closed DNA generally have a narrow distribution in *Lk* rather than a unique value. Molecules that differ only in a topological property, such as *Lk*, are called topoisomers. In describing such preparations, it is convenient to refer to the average *Lk* of the population, which will usually not be an integer.

To separate the *W* and *C* strands from each other during semiconservative DNA replication, *Lk* must be reduced to zero. This task is

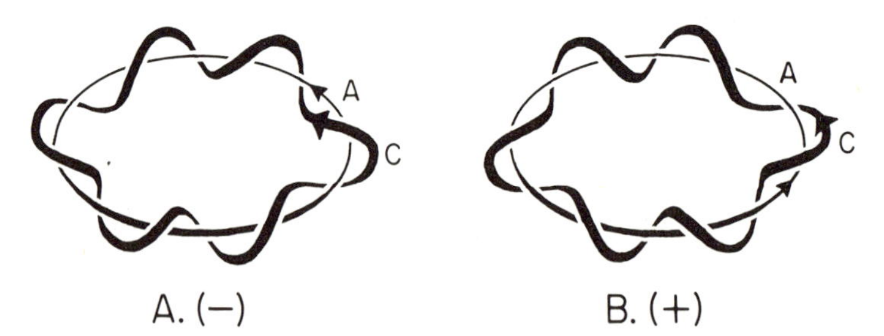

Figure 5 Plane projection definition of *Lk* for left- and right-handed duplex DNA. The DNA axis *A* and one of the DNA strands *C* are depicted. Each crossing of *C* with *A* is assigned a value of ±1/2, with the sign given by the convention illustrated in Fig. 4. The sum of these values over the whole molecule is *Lk*. In part *A*, there are 12 (–) nodes for this left-handed molecule, and therefore the *Lk* equals –6. Changing the handedness of winding reverses the sign of the nodes, and therefore the right-handed molecule in part *B* has an *Lk* of +6.

accomplished by topoisomerases that reduce *Lk* in steps of one or two from values for mammals that are over 10^8 per genome.

V. WRITHE

The linking number is a topological property that is the sum of two geometric properties, twist (*Tw*) and writhe (*Wr*), that measure important structural features of DNA. This is expressed by the well-known equation (White 1969):

$$Lk = Tw + Wr \tag{1}$$

An intuitive feeling can be given for this relationship. *Lk* is the sum of the signed indices associated with the crossings between *C* and the DNA axis. These nodes can arise in either of two ways: the local winding of the strands of the double helix, which is related to *Tw*, and the crossings of the helix axis itself, which is measured by *Wr*. We next define *Wr* and *Tw* more precisely.

Writhe is defined in terms of a single curve, the DNA axis *A*. It may seem paradoxical that self-crossings of *A* contribute to *Lk*, which is a measure of the crossing of *C* with *A*, but crossings of the axis necessarily result in strand-axis crossings. In any plane projection, the DNA axis *A* may intersect itself a number of times. To each such crossing, we assign a number +1 or –1, whose sign is fixed according to the convention presented above (Fig. 4). Adding all these numbers for a given projection yields a sum, Wr_p, or the projected writhing number. Unlike *Lk*, the value of Wr_p depends on the projection chosen, as illustrated by the curve shown in Figure 6. In some projections, a figure eight with a single (–) node is seen (Fig. 6A); in other projections, no nodes are seen (Fig. 6B). The writhe of the axis A is defined as the average of Wr_p over all possible projections. Thus, for the curve in Figure 6, *Wr* is between 0 and –1.

Because of this averaging, *Wr* will generally not be an integer although Wr_p is integral for each projection. However, should a curve pass very close to itself, the crossing so induced will give a contribution of +1 or –1 in almost all views. This is why the DNA model that is coiled as a left-handed helix in Figure 7A has a *Wr* nearly equal to –3. If the superhelix is progressively extended as shown in Figure 7, there will be an increasing number of views that show no crossings of the axis. In the limit where the DNA segment is pulled perfectly straight, the supercoils are gone (Fig. 7C), and *Wr* is zero in accord with the common-language meaning of the word writhe. Because *Wr* does change with deformation,

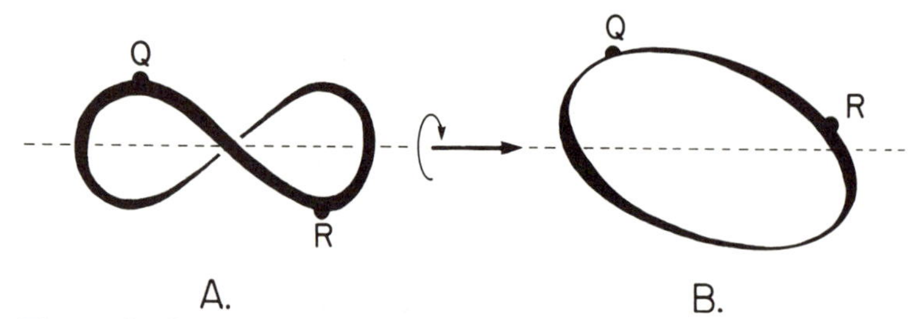

Figure 6 Illustration of projected *Wr*. The duplex axis of the same nonplanar closed DNA molecule is shown in two different projections obtained by rotating the molecule about the dashed line. The points *Q* and *R* on the DNA help illustrate the rotation. The segment *QR* crosses in front in part *A* but is in the upper rear in part *B*. The projected writhing number Wr_p is the sum of all crossings in a given projection with the sign given by the convention in Fig. 4. Wr_p is −1 in *A* and 0 in *B*. *Wr* is the average of Wr_p for all possible projections.

it is clearly not a topological invariant, but the compensating advantage is that it is a sensitive, continuously varying measure of the overall shape of the DNA.

Writhe is exactly zero not only for the straight piece of DNA in Figure 7C, but for any DNA that is planar, no matter how curved its path. This is because for writhe to be non-zero, there must be a projection in which a node is seen, but this requires at least one crossing DNA segment to be out of the plane.

It can be difficult to calculate the *Wr* of supercoiled DNA by adding up the values of Wr_p for all possible projections. *Wr* can, however, be calculated if the geometry of the supercoils is relatively simple as we illustrate below. For more complex geometric cases, a Gauss integral definition of *Wr* analogous to the one for *Lk* can be very useful (White 1989). *Wr* is calculated frequently from the difference between other parameters of DNA shape that can be measured experimentally.

There is a common misconception about writhe. As a DNA is negatively supercoiled by a reduction in *Lk*, say by the action of DNA gyrase, it acquires progressively more negative writhe. A consequence of this is that the DNA becomes more compact, and up to a limit its electrophoretic mobility through agarose gels increases (Depew and Wang 1975; Pulleyblank et al. 1975). Because a population of DNA topoisomers forms discrete electrophoretic bands rather than a smear, it is often thought that during supercoiling *Wr* changes in integral amounts to match the changes in *Lk* or at the least that the value of *Wr* is constant for a given topoisomer. Actually, both *Wr* and *Tw* change with *Lk*, and therefore both change in nonintegral steps. Because the interconversion be

A.

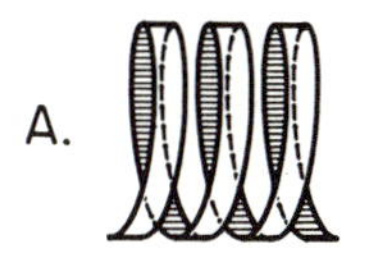

B.

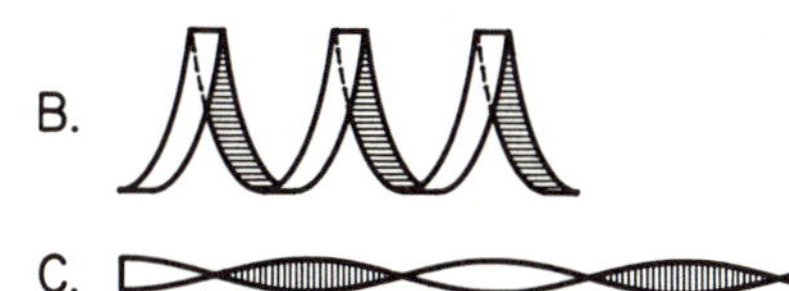

C.

Figure 7 The relation between *Wr* and superhelical pitch. The duplex DNA is schematized as a ribbon whose edges are the *W* and *C* strands. In part *A*, the DNA contains three nearly flat (–) supercoils. Wr_p for nearly all views is –3, making $Wr \approx -3$ as well. The only views for which no nodes are seen are those perpendicular or nearly so to the superhelix axis. As the ends of the DNA are pulled apart in parts *B* and *C*, and the superhelical pitch increases, the number of views where there are no crossings increases, and $|Wr|$ decreases. Already in part *B*, Wr_p is zero. In the limit, part *C*, where the DNA axis is a straight line, no crossings of the axis are possible, and *Wr* is exactly zero. Since rotation of the ribbon ends was prevented during this process of pulling out the superhelix, the edges of the ribbon cross progressively more as the pitch increases; this indicates that (–) *Tw* is incorporated into the DNA. The change in *Tw* exactly compensates for the change in *Wr* so that for the linear form $Tw = -6$. *Tw* in part *C* is exactly –3 but *Wr* in part *A* is a little greater than –3 because in part *A* there is a little (–) *Tw*. Note that in the progression from part *A* to *C*, as the superhelical winding angle increases, the radius of the superhelix decreases.

tween *Wr* and *Tw* is rapid on the time-scale of electrophoresis, the average electrophoretic mobilities of topoisomers differ by discrete values, giving rise to the familiar sharp ladder of bands on agarose gels.

VI. TWIST

Twist is a geometric entity that is conceptually difficult to understand. This is because *Tw* is not just a measure of how many times a backbone strand winds about the axis *A*, but how the strand winds. To define precisely the twist of *C* about *A*, we need to establish a unique point on *C* for each point on *A*. The easiest way to do this is to take a cross-section of the DNA perpendicular to the axis. Such a section will contain unique points, *a* on *A* and *c* on *C* (Fig. 8A). Let v_{ac} be a vector of length one directed along the line from *a* to *c*, starting at *a*. As the intersecting plane is moved along the DNA, the vector v_{ac} will turn about the axis *A*. Twist is a particular measure of this turning of v_{ac}. If *A* is a straight line or a planar curve, twist is simply the number of times v_{ac} spins about *A*. In more general cases, twist must also take into account the spatial configuration of A as well as the winding of C. To illustrate, at each point a on A, we place the origin of the x, y, and z axes of a Cartesian coordinate system (Fig. 8B). We choose the x direction to be along the vector v_{ac} and the z direction to be along the tangent vector to A at point a; thus, the y direction is mutually perpendicular to these two vectors. As point a moves along A, the direction of v_{ac} will change, thereby changing that of the x, y, z frame. The change of v_{ac} will have a component in the xy plane and a component along the z axis; Tw measures only the former component as the point a is moved by infinitesimal amounts along A. It may seem arbitrary to consider only the xy component of the change in v_{ac}. Tw, however, is a measure of how the C strand coils about the helix axis, and any change in v_{ac} along the z axis, i.e., along A, cannot contribute to the winding of C about A. When A is straight or planar, all change of v_{ac} must be in the xy plane because the change in v_{ac} is always normal to z. As noted above, under these conditions Tw reduces to just the number of times C winds about A.

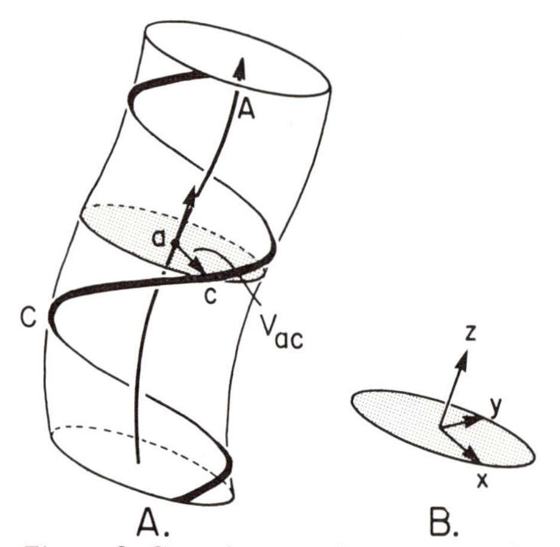

Figure 8 Cartesian coordinate system for defining Tw. In part A is a model for DNA, showing the duplex axis A and one of the DNA strands C. C is shown winding around a cylindrical surface. We construct the origin of a Cartesian coordinate system at a given point a on A. The plane perpendicular to A at a intersects the curve C at c, and the unit vector along the line ac, v_{ac}, defines the x axis. The tangent vector to A is along the z axis, and the y axis is mutually perpendicular to the z and x axes, as shown in part B. The x and y axes are in the plane of intersection. Tw is the component of the rotation of v_{ac}, that occurs in the xy plane as a is moved in infinitesimal steps along the entire DNA molecule. Notice that the coordinate system is changing constantly as a is moved.

Because it is important to understand the precise component of v_{ac} motion that makes up *Tw*, we give a second definition of *Tw*. We start as above with the vector v_{ac} embedded in a Cartesian coordinate system. We designate this vector $v_{ac}{}^0$ at some particular place and $v_{ac}{}^1$ after it has moved an infinitesimal distance along *A*. We now move $v_{ac}{}^1$ in a parallel fashion so that its origin coincides with that of $v_{ac}{}^0$, but it retains its original direction. The *projection* of the translated $v_{ac}{}^1$ in the *xy* plane containing $v_{ac}{}^0$ subtends an angle with $v_{ac}{}^0$. This angle, integrated over the length of *A* and expressed as turns, is *Tw*. The two definitions of *Tw* are equivalent because the projection into the *xy* plane loses all change of v_{ac} along the *z* axis.

We showed that the *Wr* of the DNA in Figure 7 goes from about –3 to 0 as the left-handed superhelix is pulled straight. *Tw* behaves in an exactly compensatory way. To demonstrate this, we fixed the ends of the ribbon model for the DNA segment shown in Figure 7 so that all changes are limited to this region. When the *Wr* is removed completely in Figure 7C, no self-crossings of *A* remain, but the curves formed by the two edges of the ribbon now show six (–) nodes. Because these edges represent the *W* and *C* strands of DNA, each of these crossings contribute –1/2 to *Lk*, giving a *Tw* of –3 for the segment. If the supercoil had been right-handed, then *Tw* for the straight segment would have been +3.

For any closed DNA, the *Lk* of *C* with *A* equals that of *A* with *C* because interchanging the names of the two curves leaves the nodes unchanged. Because *C* winds about *A*, the *Wr* of *C* will not equal the *Wr* of *A*. From these two facts and Equation 1, it follows that the twist of *C* about *A* will not equal the twist of *A* about *C*. In summation, interconversion of *A* and *C* changes the geometric properties of *Wr* and *Tw*, but not the topological property of *Lk*.

VII. WINDING NUMBER, SURFACE TWIST, AND SURFACE LINKING NUMBER

In this section, we describe a new way of looking at DNA structure that is both valuable in its own right and provides an illuminating conceptual framework for considering the classical concepts of linking number, twist, and writhe (White et al. 1988). We pointed out in the last section that *Tw* is a measure of not only how many times *C* winds about *A*, but also of the configuration of *A*. We now derive explicit measures of these two aspects of twist. To do so, we consider DNA to lie on a surface *M* (Fig. 9A). The best-known example of DNA on a surface is the nucleosome core particle (Richmond et al. 1984). For this structure, the DNA axis *A* can be depicted as wrapping nearly twice as a left-handed

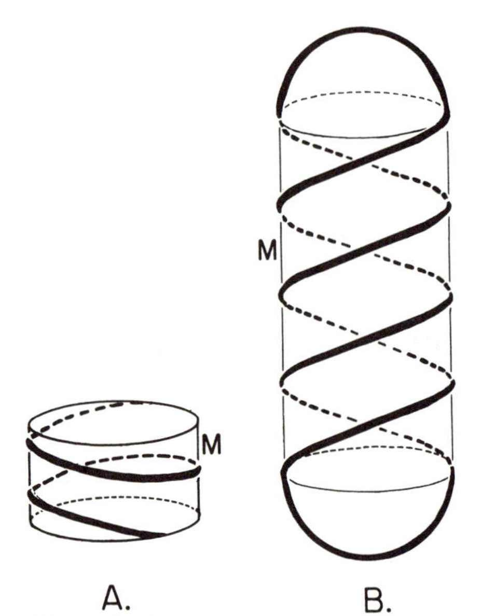

Figure 9 Examples of solenoidal and plectonemic surface winding. The bold curves represent the axis of a DNA double helix. To model solenoidal supercoiling in part *A*, we imagine the DNA winding helically on the surface of a cylinder. For the plectonemic case in part *B*, the DNA winds helically up and back down on the surface of a cylinder with hemispherical caps on either end. In both cases shown, the molecules are negatively supercoiled.

helix on the surface of a cylinder (the surface *M*). Even in the absence of a real physical surface, DNA can be considered to lie on an appropriate virtual surface. For example, supercoiled DNA in solution can be considered as wrapped on a capped cylinder, as shown in Figure 9B. It is convenient mathematically to consider the axis *A* rather than the backbone to lie on the surface.

As we have seen in the last section, the vector ν_{ac} turns about *A* as the DNA is traversed. The surface *M* now provides a reference frame in which to measure this rotation. We choose another vector ν of unit length starting from *A* but extending along the normal to the surface as the reference vector (Fig. 10); both ν and ν_{ac} lie in a plane intersecting the axis. We define the winding number, Φ, to be the number of times ν_{ac} rotates past ν as *A* is traversed. For a closed DNA, Φ must necessarily be an integer because ν_{ac} and ν are always in the same plane and must return to their original positions after one transit. The integral nature of Φ can be seen in another way that is analogous to the spanning-surface definition of *Lk*. Because the axis *A* lies on the surface *M*, *C* punctures M as the DNA double helix winds right-handedly about *A*. Because there are

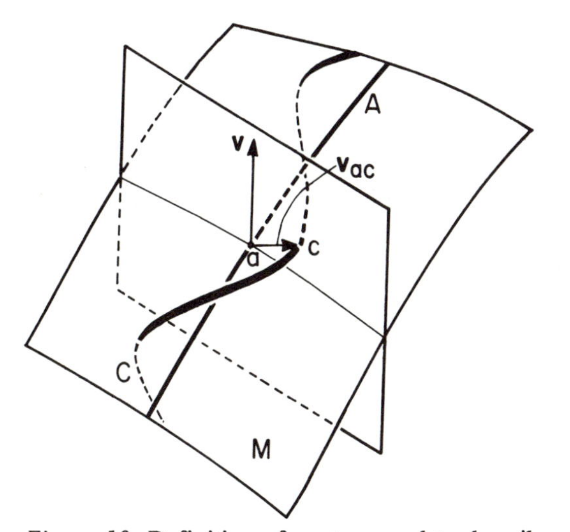

Figure 10 Definition of vectors used to describe surface winding. The duplex axis A lies on a virtual surface M. Thus, the phosphodiester backbone C will pass above and below M as it winds about A. To describe surface winding, we define two vectors originating from A at an arbitrarily chosen point a. One of these vectors is the unit vector normal to surface M at a, denoted ν. Because the surface is curved, the direction of ν changes as a is moved along A. The other vector is the strand-axis vector ν_{ac}, which is along the line that connects a and its corresponding point c on C. Both ν and ν_{ac} lie in the plane which is perpendicular to A at a. STw describes how ν winds about the duplex axis A as a is moved along A. Φ is the number of times that ν_{ac} rotates through ν as the axis is traversed.

two such punctures for each complete revolution of ν_{ac} about ν, we assign an index of +1/2 to each such puncture; Φ is just the sum of these indices. For closed DNA, the number of punctures must be even to return to the starting place (C must begin and end on the same side of the surface M); therefore, Φ must be an integer.

Φ is directly related to the repeat of the double helix h defined with reference to a surface. If N is the total number of base pairs in a DNA, then the average value of h is N/Φ. The helical repeat, and therefore Φ, can often be measured experimentally. For example, if DNA wrapped around a protein or lying on a surface is treated with a nuclease or a chemical probe, there will be a periodic maximal exposure to the agent on each strand when it is farthest away from the surface (Lutter 1978; Rhodes and Klug 1980; Tullius and Dombroski 1985). Barring some additional systematic protection of the DNA from the probe, the distance in base pairs between the positions of maximal exposure is equal to h, and the sum total of exposures is Φ. This definition of helical repeat is identical to what has been called the helical repeat in the local frame (Finch et al. 1977; Travers and Klug 1987). Another measure of local helical structure that has been called the helical repeat in the laboratory frame (Finch et al. 1977; Travers and Klug 1987) is equal to N/Tw. We shall not consider this parameter further, so that in this paper helical repeat will always refer to N/Φ.

Φ is a quantitative measure of the winding of C about A. It is just one of the two components of Tw because it gives no information about the shape of A. The second component of twist is called surface twist (STw) (White and Bauer 1988). One may therefore express Tw by the formula

$$Tw = \Phi + STw \tag{2}$$

Just as Tw is a measure of how v_{ac} changes as the axis is traversed, STw is a measure of how the reference vector v changes and therefore of how the surface near the DNA is altered. Surface twist can be interpreted in terms of a mathematical artifice called a displacement curve. This is the curve obtained by moving a small distance ε away from A along the surface normal v and is designated A_ε. The choice of ε is unimportant as long as A_ε does not cross A in the displacement. Examples are shown in Figure 11. As with any two closed curves, A and A_ε have a twist relating them. Surface twist is just the twist of A_ε about A.

An intuitive feeling can be given for Φ and STw by considering the reference vector v to be an observer. This observer walks on the reference surface and along the axis so that his body is always locally upright, i.e., perpendicular to the surface, no matter how convoluted is his path. By analogy, someone walking around the world is always perpendicular to the surface of the Earth. The observer can count how many times the vector v_{ac} moves past his line of vision during one transit of the DNA; this number is Φ. Equivalently, the observer can count the number of times a feature of the helix such as the minor groove appears, and this is also just Φ. The observer cannot tell, however, if he has himself revolved around A any more than the globetrotter, halfway around the world, perceives that he is upside-down compared to his starting position. This determination requires a second observer outside the reference frame. We can also give a concrete interpretation of STw. The head of the observer traces out a displacement curve A_ε. During the walk along A, the coiling of A may force the observer to go head-over-heels (i.e., A_ε winds around A) in a forward or backward motion or by a lateral cartwheeling or by combinations of these. STw measures only the lateral motions (changes in observer v perpendicular to A) and ignores forward and backward movements (changes in *v* along *A*) because it is only the former that measures the observer's rotation about the axis. It is immediately clear why surface twist is zero for planar curves or equatorial curves on a sphere. For planar cases, our observer always remains upright, whereas for an equator there will be no sideward motion although our observer will go head-over-heels in the forward direction while circumnavigating a globe.

STw need not be integral. Although our first observer, the vector *v*, must go head-over-heels an integral number of times in one pass around closed DNA to get back exactly to his original position, it is only the lateral component of the motions, those mutually perpendicular to *v* and the tangent vector to *A*, that contribute to *STw*. A key advantage of formulating *Tw* in terms of Φ and *STw* is that both Φ and *STw* give important information about DNA structure: Φ measures helical repeat, and *STw* describes the overall shape of supercoils.

Another measure of the shape of supercoils is of course *Wr*. Its sum with *STw* is another linking number with an important meaning. The linking number of *A* with A_ε is called the surface linking number, denoted *SLk*, because A_ε is obtained by moving along the normal to the surface. This quantity, like any linking number, is the sum of writhe and twist terms. Its writhe is just that of the axis *A*, and its twist is *STw*. Thus,

$$SLk = STw + Wr \qquad (3)$$

In many cases of interest, *SLk* is easily determined and can then be used to obtain *STw* or *Wr*. Four examples of values of *SLk* are illustrated in Figure 11.

SLk, like *Lk*, must have an integral value but is invariant under more limited conditions than *Lk*. It is a relative invariant like Φ because it is unchanged by what mathematicians call smooth deformations of its defining surface. For deformations to be smooth, i.e., to preserve the linking number of *A* and A_ε, two conditions must be met. First, the axis of the DNA must never leave the surface during the deformation. It may, however, slide along the surface. For example, if the capped cylinder shown in Figure 9B were allowed to expand, then the DNA wrapped around it could unwind while sliding along the surface and thereby reduce supercoiling. Both *SLk* and Φ would be unchanged by this operation. Second, the direction of the vector *v* along *A* must vary continuously; this will ensure that A_ε is a continuous curve and that *SLk* is always well defined. A smooth deformation does not introduce breaks or holes in the surface next to *A*. However, away from the curve *A*, the surface may be broken or discontinuously deformed without introducing any change in *SLk*. Smooth deformations are a subset of continuous deformations, which are unchanged as long as the axis is intact. *Lk* is invariant under more conditions than *SLk* because the only requirement for its constancy under deformation is that the deformation be continuous.

Figure 12 illustrates smooth deformations of spheroids (Fig. 12A),

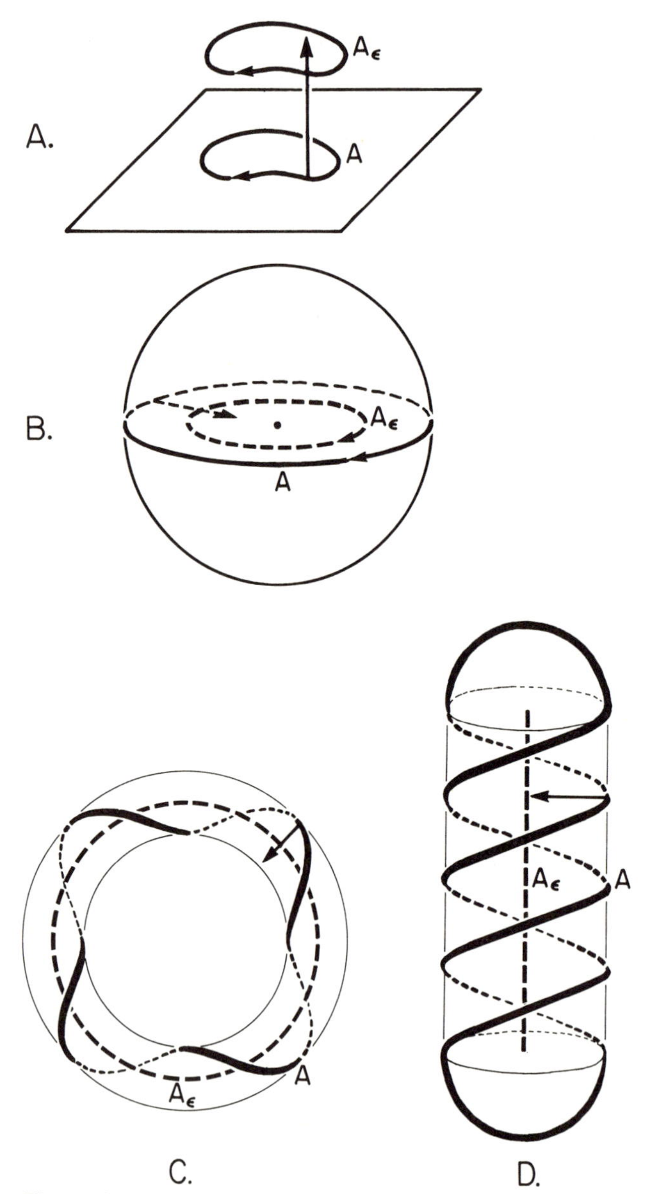

Figure 11 (See facing page for legend.)

sphere-like objects, and toroids (Fig. 12B), objects with a single hole such as a doughnut. Note that the shape of an object can be greatly changed under smooth deformation. An example of a nonsmooth deformation is the conversion of a DNA wrapped solenoidally on a torus to a DNA wrapped plectonemically on a capped cylinder, a transition that occurs when nucleosomal DNA sheds its bound histones. In this case, the DNA would have to be lifted off the surface of the torus and then wound about the surface of the capped cylinder. A saucer may be smoothly deformed into a cup without a handle but not into a cup with a handle or into a doughnut.

SLk has very different values for the two key types of supercoiled DNA. We have modeled DNA in solution as lying on a cylinder with hemispherical caps (Fig. 9B), a spheroid, and the DNA of a minichromosome as winding on a torus (Fig. 9A). $SLk = 0$ for a spheroid because the displacement curve A_ε can be entirely inside the spheroid and hence unlinked to A. This is clear from Figure 11D because A_ε can be pulled free of the superhelical DNA no matter how many times the DNA axis winds about the cylinder. Because $SLk = 0$, Equation 3 shows that $STw = -Wr$. In the minichromosome case, *SLk* is just the signed value of the number of toroidal turns (n) of the axis A (Fig. 11C). The sign of *SLk* is (+) if the winding is right-handed and (–) if it is left-handed.

The concept of *SLk* allows an important reinterpretation of the linking number of DNA. If we combine Equations 1, 2, and 3 successively, we obtain

$$\begin{aligned} Lk &= Tw + Wr \\ &= STw + \Phi + Wr \\ &= SLk + \Phi \end{aligned} \tag{4}$$

Figure 11 Examples of displacement curves and *SLk*. For any curve A lying on a surface, the displacement curve A_ε is formed by moving a small distance ε along the surface normal at each point on the curve. For planar curves as in part *A*, all the normal vectors can be chosen to point upward, and then A_ε is above A. Since the two curves are unlinked, *SLk* is zero. For curves on spherical surfaces, as in part *B*, the surface normals can be chosen to point inward, and A_ε is completely contained within the spherical surface. Thus, *SLk* is zero for this case as well. In parts *C* and *D*, the surface normals have also been chosen to point inward from the surface, and ε has been set equal to the radii of the surfaces on which the DNA is wound. A_ε is the torus axis in part *C* and the cylinder axis in *D*. In part *C*, the solenoidal curve is linked with its displacement curve by four right-handed turns, and *SLk* is +4. In part *D*, the displacement curve is not linked with the plectonemic curve, and *SLk* is zero.

Thus, *Lk* is the sum of two integers: the surface linking number and the winding number. Both are surface invariants whose key advantage over *Tw* and *Wr* is that they can be more readily measured in experimental situations. Φ can be determined from *h*, and *SLk* can be determined from the DNA geometry. For any closed DNA on a spheroid, $SLk = 0$ and therefore $Lk = \Phi$.

The division of *Lk* into *SLk* and Φ separates out the axis geometry component *SLk* from the winding of the DNA about the axis Φ. It is commonly thought that *Wr* and *Tw* accomplish the same subdivision, but as we have seen, *Tw* contains an axis geometry component *STw*. Equation 4 can be obtained from Equation 1 by subtracting *STw* from *Tw* to give Φ and then adding it to *Wr* to obtain *SLk*.

VIII. DESCRIPTORS OF SUPERCOILING

With the topological and geometric foundations laid, we now address the description of DNA supercoiling in both its plectonemic and solenoidal forms. The nonsupercoiled or relaxed state of DNA can be defined operationally as the equilibrium state after nicking and religation under a particular set of conditions. Prior to ligation, we presume that the nicked DNA is on the average planar. However, the total number of base pairs, *N*, will usually not be an integral multiple of the number of base pairs per turn *h*. As a result, the integral values of Φ required to close the DNA cannot be achieved without introducing some *Wr*. Alternatively, *h* could change a little to allow ligation. However, even with a small amount of *Wr*, the DNA will still be nearly planar, and *SLk* will be exactly zero. Because any potential *Wr* and the compensating *STw* will be small ($\leq 1/2$ in absolute value), they are conventionally ignored, and relaxed DNA is approximated as planar. Thus, the values of *SLk*, *Wr*, and *STw* for relaxed DNA, denoted SLk_0, Wr_0, and STw_0, are all taken to equal zero. Lk_0, Tw_0, Φ_0, and N/h_0 are therefore all equal. It is a common error to equate also the *Tw* of supercoiled DNA to N/h_0. This would be true only if *h* were unchanged by supercoiling and *STw* equaled zero for supercoiled DNA. Neither is correct.

Supercoiling can be introduced in two different but related ways. If a DNA free in solution is underwound, i.e., its *Lk* is less than Lk_0, the DNA generally assumes the plectonemic supercoiled form, and the number of supercoils is directly proportional to the linking difference, ΔLk, as defined by

$$\Delta Lk = Lk - Lk_0$$

It is often more convenient to describe the linking difference in terms of a length-independent descriptor σ, the specific linking difference, which is equal to $\Delta Lk/Lk_0$. Thus, the first way to introduce supercoiling is to reduce the linking number so that ΔLk and σ are negative numbers. DNA can also be supercoiled by a positive linking difference, and recent work implies a much more important role for positive supercoiling than previously thought (Lockshon and Morris 1983; Kikuchi and Asai 1984; Nadal et al. 1986; Wu et al. 1988). However, because its properties have not yet been well studied, we will generally restrict ourselves to negative supercoiling. The second common way in which supercoiling arises is by the wrapping of DNA around a protein complex, such as in a nucleosome (Richmond et al. 1984), the Tn*3* resolvase synaptic complex (Benjamin and Cozzarelli 1988), the intasome (Better et al. 1982), and the gyrasome (Liu and Wang 1978). Although wrapping per se does not change *Lk*, this is the expected result in vivo because topoisomerases can remove the compensating supercoils. Indeed, σ for eukaryotic DNA that is wrapped around nucleosomes is similar to that of bacterial DNA that is underwound by DNA gyrase (Bauer 1978).

Plectonemic and solenoidal supercoils can readily interconvert. This presumably occurs in vivo when the nucleosome structure is broken down and can be simulated in the laboratory when a length of rubber tubing solenoidally wound around a cylinder is freed of the constraint.

Changes in the amount and type of supercoiling can be accompanied by changes in many topological and geometric properties of DNA. Therefore, any of these properties can be used as a quantitative measure of supercoiling. The three most commonly used descriptors are ΔLk (or σ), *Wr*, and the number of superhelical turns about the superhelix axis, *n*. No one of these descriptors of supercoiling is universally useful; they are all valuable. In this section, we analyze these descriptors and indicate their advantages and limitations in describing the two types of supercoiling. In the next section, we compute the descriptors for two particular examples of supercoiled DNA.

A. ΔLk

The most widely used descriptor of supercoiling is ΔLk itself. Indeed, σ is frequently called superhelix density. ΔLk has the critical advantage that it can be measured experimentally with ease, accuracy, and sensitivity. Moreover, some of the major biological consequences of supercoiling, such as the opening of the double helix, are due to superhelical free energy that is a direct function of the linking deficit (Bauer and Vinograd 1970; Depew and Wang 1975; Pulleyblank et al. 1975). ΔLk, however, has limitations as a measure of supercoiling. It is often defined independently of DNA geometry, and yet supercoiling is a geometric

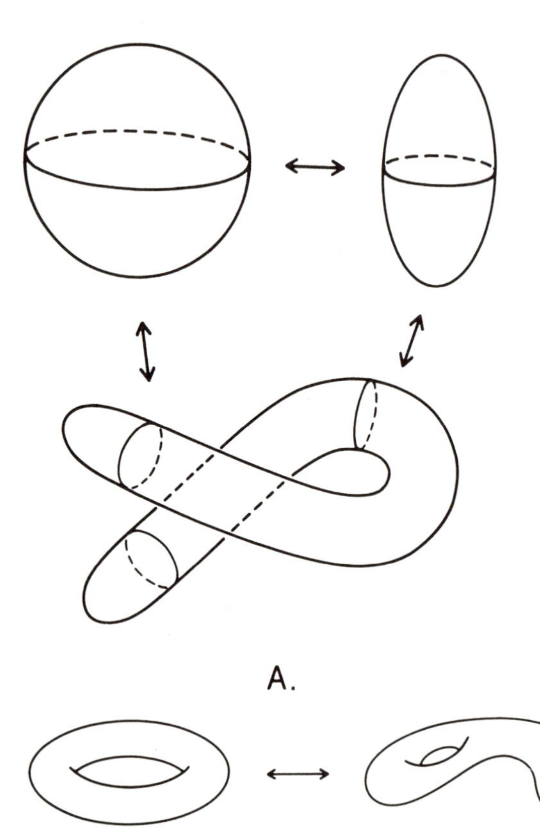

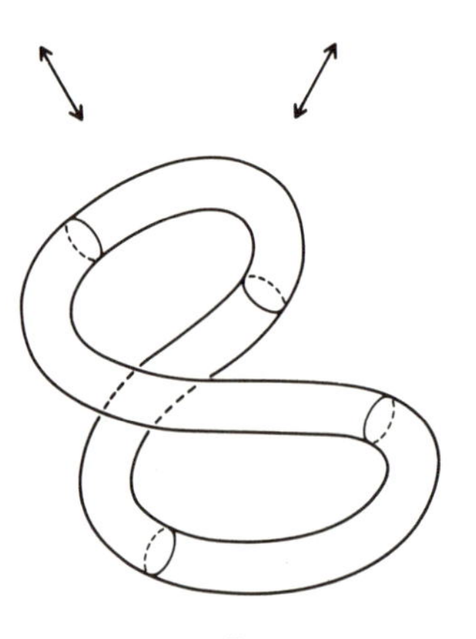

Figure 12 (See facing page for legend.)

concept. Thus, it is formally possible for a perfectly planar DNA molecule with Lk not equal to Lk_0 to be described nevertheless as "supercoiled" because it has a non-zero ΔLk. Moreover, unlike Wr and n, ΔLk cannot be computed from the axis shape.

In using ΔLk as a descriptor of supercoiling, one must be careful about the definition of Lk_0. Unfortunately, a spectrum of definitions is in use that differ in how much Lk_0 is allowed to vary as a function of experimental conditions. At one extreme, Lk_0 is viewed as a constant, usually the number of base pairs N divided by 10.5, the average value of the helical repeat under physiological ionic conditions. At the other extreme, Lk_0 is viewed as the equilibrium Lk after nicking and ligation in the particular set of experimental conditions in use. As experimental conditions such as temperature (Wang 1969; Depew and Wang 1975; Pulleyblank et al. 1975), ionic strength and type (Wang 1969; Anderson and Bauer 1978), and DNA ligand concentration (Keller 1975) change, Lk_0 changes. Thus, even in the absence of breakage and reunion, ΔLk will change, unlike Lk. Although a constant Lk_0 simplifies calculations and retains for ΔLk the topological invariance of Lk, the variable definition of Lk_0 makes ΔLk a more flexible and informative descriptor of supercoiling that can be measured under many experimental conditions.

We illustrate the effect of the different definitions of Lk_0 by considering a typical experiment involving supercoiling. A closed DNA is first bound by a ligand, such as ethidium bromide, that unwinds DNA. The resulting positive supercoils are relaxed by a topoisomerase, and the ligand is removed. This procedure is the most common laboratory method for generating negatively supercoiled DNA. By the constant definition of Lk_0, the DNA has a negative ΔLk once the topoisomerase acts; according to the variable definition of Lk_0, it has a negative ΔLk only after the ethidium bromide is removed. The latter interpretation is more in accord with the common-language definition of supercoiling because the ligand-bound DNA is a planar molecule, but the final ligand-free form is supercoiled. The situation is less clear cut with DNA that is supercoiled around proteins and then treated with a topoisomerase, such as in nucleosome formation. For experiments examining the binding of li-

Figure 12 Smooth deformations of spheroids and toroids. Spheroids (sphere-like objects) are shown in part *A* and toroids (torus-like objects) are in part *B*. A toroid can be generated by cutting a hole through a spheroid and sealing the cut edges. The changes in shape that are indicated by arrows are examples of smooth deformations. These deformations do not introduce or remove holes in the objects. Changes that convert spheroids to toroids or vice versa are non-smooth deformations.

gands to nucleosomal DNA, it is useful to consider the nucleosomal DNA as "relaxed" although it is supercoiled around the histones. In this case, the variable definition of ΔLk gives a good picture of superhelical free energy but a poor one of DNA geometry. Once again, we emphasize the need for a careful and consistent definition of terms.

B. *Wr*

Writhe has also been widely used to describe supercoiling. It is, like supercoiling, a geometric property of the DNA axis. *Wr* is directly proportional to *n* and equals zero for relaxed DNA. In the plectonemic model of negatively supercoiled DNA, for each side view perpendicular to the superhelix axis such as in Figure 13A, Wr_p is exactly the sum of the signed number of crossings. The average of Wr_p over all side views equals $-n$. However, *Wr* is the average of Wr_p over all views. As we illustrate below, the side views have only (–) crossings, whereas other views will show (+) crossings as well. *Wr* is therefore always less in absolute value than *n*. This is clear from a mathematical description of helical windings. If γ is the pitch angle of the plectonemic superhelix (Fig. 14)

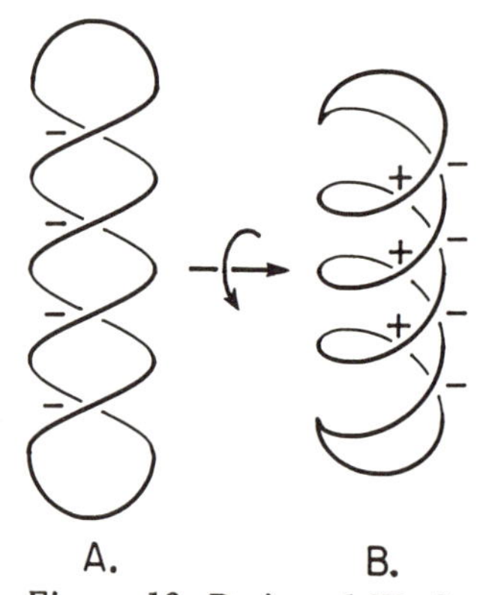

Figure 13 Projected *Wr* for plectonemically supercoiled DNA. Two different views of the same supercoiled molecule are shown. The curve represents the axis of the double helix. The view in part *B* is obtained by rotating the molecule about the horizontal axis indicated so that the top of the molecule is tipped toward the viewer. The view in part *A* has only the negative nodes ($Wr_p = -4$) which result from the crossing of distant DNA segments. Part *B* has these same negative nodes but in addition shows positive nodes contributed by crossings of nearby DNA segments. Thus, Wr_p for this view is -1. Because of this effect, the $|Wr|$ of plectonemic DNA is less than the number of supercoils.

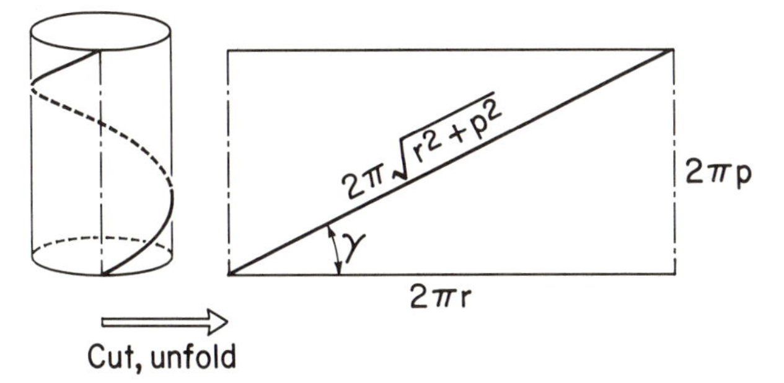

Figure 14 Mathematical description of helical winding. The drawing at the left represents one superhelical turn of DNA wrapped on a cylindrical surface with a radius r. If the surface is cut open and unfolded into a plane as shown on the right, the DNA forms a diagonal across the rectangle. The height of the rectangle is the superhelical pitch $2\pi p$, and the base of the rectangle is the circumference of the cylinder $2\pi r$. The length of the DNA is the diagonal. The superhelix winding angle γ is also illustrated.

$$Wr = -n \sin \gamma \tag{5}$$

Thus, writhe takes into account not only the number of superhelical turns, but also the pitch at which these turns are revolving about the plectonemic axis (Fuller 1971; White et al. 1988).

Wr can also be expressed very simply in terms of n and γ for a closed solenoidal superhelix. For example, if we consider a left-handed and therefore negative toroidal superhelix that is relatively long and slender so that we can approximate it as winding on n cylindrical segments, then

$$Wr \approx -n + n \sin \gamma \tag{6}$$

The consequences of this relationship are clear. As the pitch angle goes to 0°, so does $\sin \gamma$, and *Wr* is maximally negative. At the other extreme, *Wr* goes to zero as γ approaches 90° because sin 90° equals 1. Equation 6 is a quantitative expression of the relationship we described qualitatively above (Fig. 7); the absolute value of *Wr* for a solenoidal supercoil is maximal when it is nearly planar and diminishes as the winding angle increases.

Although the DNA can be viewed as winding helically for both plectonemic and solenoidal negative supercoiling, the handedness of the superhelices are opposite: right-handed for the plectonemic case and left-handed for the solenoidal. There is a simple explanation for this. The sign of a node depends on only two things: the relative orientation and the relative overlay of the crossing segments. Because the plectonemic supercoils are made by a doubling back of the helix axis and solenoidal coils are not, the relative orientation of the crossing segments is opposite. To achieve the same topological sign, the overlay must then be opposite, and in a superhelix, this can only be done by reversing handedness.

From Equations 5 and 6, it is clear that increasing γ has opposite effects on Wr in solenoidal and plectonemic supercoils. For (–) solenoidal supercoils, the magnitude of Wr decreases with increasing γ because of the decrease in the number of projections that contain the (–) crossings of the axis (Fig. 7). To understand the Wr of plectonemic supercoiling, we need to consider not only the usual side views (Fig. 13A), but also the nearly end-on views (Fig. 13B). In these projections, Wr_p is made up of two kinds of nodes: those from nearby DNA segments pointing in the same direction made by right-handed solenoidal coiling, and those made by the crossing of more distant segments that point in the opposite direction. The former are always (+), and the latter are always (–). Because the Wr of the positive solenoidal coils is small at high γ, the negative crossings dominate. In the limit, as γ goes to 90°, the solenoidal Wr goes to zero, and the total Wr approaches $-n$, the number of superhelical turns. Thus, for the plectonemic case, increasing γ increases the magnitude of Wr not by increasing the number of negative nodes but by decreasing the contribution of positive ones.

In conclusion, Wr is a smoothly varying geometric parameter, like supercoiling, that can often be computed readily if the superhelix is well defined. Its major limitation is that it is usually difficult to evaluate experimentally. We show below examples where the writhe of a regular superhelix was determined indirectly from other DNA parameters. It is also possible to use the Gauss integral to calculate Wr for a given structure directly, if the complete three-dimensional path of the axis is known.

C. *n*

The most direct, quantitative description of supercoiling comes from the general definition of supercoiling as the coiling of the DNA axis. This is simply n, or the number of times the DNA winds about the axis of supercoiling. n is therefore by itself unsigned. Uniform solenoidal supercoiling about an axis can be viewed for closed DNA as a winding about a torus whose axis is the supercoil axis. For solenoidal supercoiling, the absolute value of SLk, the surface linking number, is equal to n. SLk is equal to $+n$ for right-handed superhelices and $-n$ for left-handed superhelices. Because SLk is defined by a linking formula, small perturbations of the entire complex that do not remove the DNA from the torus surface or break the torus axis will leave this number unchanged; i.e., *SLk* is invariant under smooth deformation. Moreover, *n* here must be integral. We illustrate below the use of *SLk* as a solenoidal supercoiling descriptor by characterizing in detail the nucleosomal model where *SLk* has been accurately determined.

In the plectonemic supercoiling model (Fig. 9B), the number of superhelical turns also has a clear meaning. The DNA wraps solenoidally around the capped cylinder a total of *n* times; there *n*/2 coils up and *n*/2 coils down. In side views, as shown in Figure 9B, the number of crossings, $|Wr_p|$, is directly related to *n*. In the special case, where *n* is an integer, rotation of the superhelix about its axis leaves $|Wr_p|$ unchanged and equal to *n*. If *n* is not integral, then $|Wr_p|$ will differ from *n* by at most ±1 as the view changes with rotation. However, the average value of $|Wr_p|$ over all side views equals *n*. When DNA is viewed by electron microscopy, it lies with its long axis along the grid so that the number of nodes and therefore *n* can be experimentally measured as long as the structure is sufficiently regular that the axis is well-defined.

For the plectonemic supercoiling model, *SLk* is equal to zero and thus clearly does not equal *n*. Moreover, *n* is not a surface invariant in this instance. This can be illustrated using Figure 9B. By rotating the DNA lying on one cap about the superhelix axis, the number of supercoils *n* will change continuously. We can simultaneously adjust the dimensions of the capped cylinder so that the DNA never leaves the surface. Therefore, for plectonemic supercoiling, *n* can vary even under smooth deformation.

D. Summary

In summary, supercoiling can be quantified in three ways, ΔLk, *Wr*, and *n*. The more of these quantifiers that are known, the better specified is the structure. The choice of the supercoiling measure depends on what is known and what needs to be known. Clearly, *n* is the common-language meaning of supercoiling as the "coiling of a coil," and in the specific case of DNA represents the "coiling about the superhelical axis" (Bauer and Vinograd 1968). This choice is most useful mathematically in the solenoidal case, where $n = |SLk|$. *SLk* is well defined, quantized, a relative invariant directly related to surface topology, and the sum of two geometric quantities, *Wr* and *STw*, which change sensitively with supercoiling. For the plectonemic case, *n* is also an obvious measure of supercoiling, and it can be determined by electron microscopy. In some cases, however, to compute or measure *n*, considerable information about the structure must be known. *Wr* is usually even harder to evaluate because it depends on the superhelix pitch angle as well as n. Wr, however, is a pure expression of axis geometry that varies subtly and directly with changes in shape, unlike ΔLk and n. Wr is also an important measure of the overall hydrodynamic shape of supercoiled DNA. If the size and shape of the supercoils are unimportant or unknown, then ΔLk is clearly the measure of choice because it can be easily and precisely measured and many biological and energetic properties of supercoiling derive from ΔLk. n and Wr are much harder to measure but, unlike ΔLk, can usually be computed if the supercoil geometry is known.

IX. HISTORICAL PERSPECTIVE: THE MEANING OF $\alpha = \beta + \tau$

In the original formulation by Vinograd and colleagues of the DNA conservation equation (Vinograd and Lebowitz 1966; Bauer and Vinograd 1968, 1974), the result was given

$$\alpha = \beta + \tau \tag{7}$$

where α is the "topological winding number" or what we have called linking number; β is the "duplex winding number, . . . (or) the number of revolutions made by one strand about the duplex axis in the unconstrained (supercoiled) molecule," and τ is the "superhelix winding number, . . . (or) the number of revolutions made by the duplex about the superhelix axis." τ was also called the number of superhelical turns (Vinograd and Lebowitz 1966).

This equation was of fundamental importance in the early studies of supercoiling. It emphasized that the properties of supercoiling and the DNA double helix were interrelated and that α was a constant short of strand breakage. However, as first pointed out by Bauer (1978), the explication of this relationship was not mathematically rigorous and in some cases inconsistent. We will present two different reinterpretations of this equation in terms of precisely defined parameters.

First, it seems clear from the definition of Vinograd just cited, that τ should be interpreted as the signed number of superhelical turns or what we have termed $\pm n$. For solenoidally supercoiled DNA, this definition works very well. For such DNA, $\tau = SLk$. This equality provides justification for Vinograd's insight that τ is integral only for solenoidally supercoiled DNA (Bauer and Vinograd 1974). From Equation 4, β must then equal Φ. Thus, in Vinograd's original treatment, the number of times a backbone strand winds about the axis A must be taken with respect to a surface reference frame. Attempts to reconcile Equation 1 with Equation 7 by equating τ with Wr and β with Tw are incorrect although both pairs add up to *Lk*. For example, for left-handed solenoidal winding, Vinograd's definitions give $\tau = SLk = -n$, but from the discussion of *Wr* above (Eq. 6), $Wr = -n + n \sin \gamma$. Similarly, β cannot be equated with *Tw* because, as we have shown, $Tw = STw + \Phi = -n \sin \gamma + \Phi$ and Vinograd's definition requires that $\beta = \Phi$.

For plectonemically supercoiled DNA, setting τ equal to the signed value of *n* causes problems. β is then not a helical winding parameter, but the quantity

$$(Lk[1 + \sin \gamma] - Tw)/\sin \gamma$$

which is of unknown physical significance. Moreover, β cannot equal Φ for plectonemically supercoiled DNA because, instead, $Lk = \alpha = \Phi$. To equate τ with any precisely defined parameter of plectonemically supercoiled DNA, some part of the Vinograd treatment must be omitted.

Second, Wang has interpreted Vinograd's equation in terms of the way τ had been measured experimentally (Wang et al. 1983). Wang set $\beta = Lk_0$, and therefore $\tau = Lk - Lk_0 = \Delta Lk$. Thus, $\alpha = \beta + \tau$ becomes

$$Lk = Lk_0 + (Lk - Lk_0) \tag{8}$$

and Lk_0 is interpreted in the variable way as the equilibrium *Lk* after relaxation under the given experimental conditions. Because $\beta = Lk_0 = \Phi_0 = N/h_0$ and *N* is a constant, β is a direct measure of the winding of the strands of the double helix. Thus, the Vinograd equation then states that the linking number of DNA equals the sum of a superhelical parameter $\tau(\Delta Lk)$ and a double helical parameter $\beta(Lk_0)$.

This second interpretation also has limitations. For solenoidally supercoiled DNA, one must discard the identity of τ and *SLk*. Combining Wang's definition, $\tau = \Delta Lk$, and our formulation of the original Vinograd treatment, $\tau = SLk = Lk - \Phi$, we obtain $\Delta Lk = Lk - \Phi$ and $Lk_0 = \Phi_0 = \Phi$. However, we show in the next section that this is not true; Φ changes from Φ_0 as DNA is supercoiled. Previously, many thought that the B-type double helix had such great stability that *h* would be unchanged by supercoiling. If this had been correct, the two interpretations given of $\alpha = \beta + \tau$ would be consistent with each other for solenoidal supercoiling. The interpretation of ΔLk as τ also has the unfortunate consequence of reducing the classical Vinograd relationship to the somewhat trivial identity in Equation 8. The insight of two variable geometric entities adding up to a quantized topological parameter is lost. Given the confusion surrounding $\alpha = \beta + \tau$, we think it is best that this equation and the terms α, β, and τ not be used any longer.

X. GEOMETRY OF SUPERCOILED DNA

In this section, we calculate for plectonemically supercoiled DNA in solution and DNA solenoidally supercoiled in nucleosomes the values of the topological and geometric parameters that we have described. Some of the experimental data used in the calculations are recent (Boles et al. 1990) and await independent confirmation, and the precise values for some of the helical parameters of nucleosomal DNA are controversial (Morse and Simpson 1988; Klug and Travers 1989; White and Bauer 1989). We use the data nevertheless to develop a concrete picture of the structure of supercoiled DNA because we believe that the values are probably close to the true ones.

We first consider supercoiling that arises from a negative ΔLk. Figure 15A shows to scale an idealized average structure for a 4.6-kb supercoiled plasmid with $\sigma = -0.06$. A feature of this structure that we have not yet discussed is branching of the superhelix axis. Nearly all plectonemically supercoiled molecules display a branched structure in the electron microscope. Thus, in each molecule, there are one or more points where three plectonemic segments meet to form a Y junction. We show below that, given the assumption of uniform winding on a branched cylinder and the measurement of the length of the superhelix axis and the number of superhelical turns (Boles et al. 1990), it is simple to calculate all the other parameters we have discussed.

Over the entire range of σ studied (−0.02 to −0.12), the molecules observed had a branched plectonemic structure. The length of the superhelix axis, l, defined as the sum of the lengths of the axes of the segments of the branched superhelix, remained essentially constant at 41% of the length of the DNA. Because l can be no larger than 50% of total DNA length, plectonemically supercoiled molecules are very long and thin. This is evident from the drawing in Figure 15A. As supercoiling increases, the DNA winds more times around the virtual cylinders, and since the length of the DNA is constant, either l, r (radius of the cylindrical segments), or both must decrease. It turns out that l is independent of σ and that only r changes. An important consequence of the constancy of l is that, neglecting end effects, the superhelix winding angle γ (Fig. 9B) is constant. This is easy to prove. For our plectonemically supercoiled DNA model (Fig. 15A), the relationship between DNA length L, the superhelix axis length l, and γ, is, while neglecting the caps

$$\sin \gamma = 2l/L \tag{9}$$

Because the right-hand side is constant at 0.81, then $\gamma = 54°$. Perhaps the bending and twisting of the helix axis that accompanies supercoiling

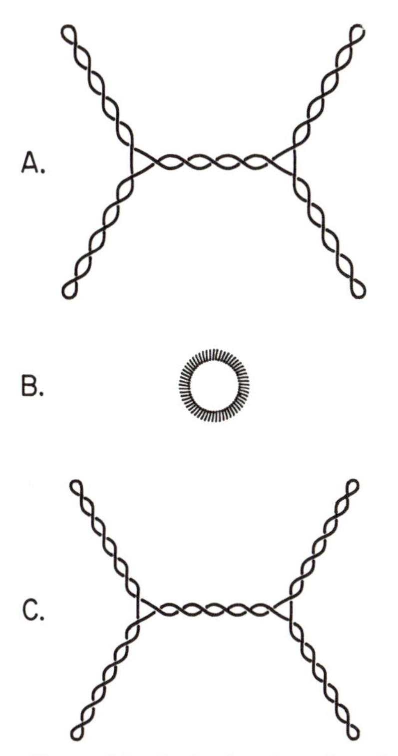

Figure 15 Scale drawings for plectonemic and solenoidal supercoiled DNA models. Parts *A* and *C* are scale drawings of plectonemic DNA in solution, and part *B* is a model for nucleosomal DNA. The value of σ is –0.06 in part *A* and –0.079 in parts *B* and *C*. In all cases, the length of the DNA is 15,500 Å (4.6 kb), and the diameter of the DNA is taken to be 22 Å. All dimensions are to scale except that the diameter of the DNA duplex in part *B* has been reduced by 50% for the sake of clarity. For the plectonemic DNA, the length of the plectonemic axis is 6300 Å, and the superhelix radius is 56 Å in part *A* and 43 Å in part *C*. For both, there are two branch points that result in a total of four ends. To model nucleosomal DNA, we imagine the DNA to be solenoidally supercoiled with nucleosomal geometry. This model is equivalent to a hypothetical case in which nucleosomes form a continuous end-to-end stack with no linker DNA. We have assumed 81-bp/superhelical turn, giving a total of 57 supercoils in 4.6 kb. The superhelix radius is 43 Å, the radius of the torus axis is 250 Å, and the length of the torus axis is 1600 Å. Notice the striking difference in compaction for the two forms of DNA supercoiling.

is energetically at a minimum when $\gamma = 54^{\circ}$. The geometric relationship between L, the length of the DNA, and the superhelix parameters for our model is given by the formula

$$L = 2\pi n([r^2 + p^2]^{1/2}) + E\pi r \qquad (10)$$

where E is the number of ends of the superhelical axis, $2\pi p$ is the superhelical pitch or the distance between adjacent superhelical turns, and r is the radius of the cylindrical segments. The two terms on the right of this expression correspond, respectively, to the winding around the cylinder and that around the hemispherical caps. This is easy to derive. Each wrap around an equator of a hemisphere is equal to πr, and thus the total length of the DNA on the caps is $E\pi r$. $2\pi([r^2 + p^2]^{1/2})$ is the formula for the length of one turn of a helix, and it is multiplied by n turns for winding up and down the cylinder. As shown in Figure 14, we derive this by slicing the cylinder longitudinally and flattening it out. Each turn then will be the diagonal of a rectangle of dimensions $2\pi r$ (circumference of the cylinder) and $2\pi p$ (pitch). From the Pythagorean theorem, the length of the diagonal is $([2\pi r]^2 + [2\pi p]^2)^{1/2}$, and for n helical turns, this becomes $n([\{2\pi\}^2]^{1/2})([r^2 + p^2]^{1/2}) = 2\pi n([r^2 + p^2]^{1/2})$.

p can be determined from l and n using the relationship

$$p = (l - Er)/\pi n \qquad (11)$$

which is also easily derived. The total length of the capped cylinder is l, so the length of the cylindrical part is just this length minus the total height of the all of the hemispherical caps or $l - Er$. There are $n/2$ windings up and $n/2$ down, and since the distance between successive windings is $2\pi p$, this same length can be expressed as $2\pi p(n/2)$. Equating the two expressions for the length of the uncapped cylinder gives us $\pi np = l - Er$. Equation 11 is then obtained by dividing each side by πn.

A simple linear relationship between n, the number of supercoils, and ΔLk was obtained (Boles et al. 1990)

$$n = -0.89\ \Delta Lk \qquad (12)$$

This result shows that for plectonemic DNA, $-\Delta Lk$ is a good approximation for the number of supercoils. Rewriting in terms of σ, the expression becomes

$$n = -0.89\ \sigma\ Lk_0 \qquad (13)$$

Substitutions of Equations 13 and 11 into Equation 10 enables the calculation of r as a function of σ. The resulting dependance of r on σ is shown in Figure 16. The value of r decreases approximately hyperbolically so that the radius drops from 150 Å to 55 Å between $\sigma = -0.02$ and

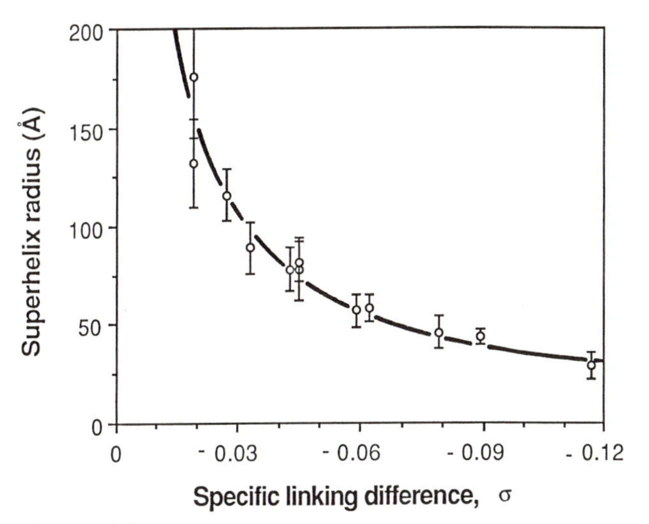

Figure 16 The relationship between plectonemic superhelix radius and σ. Experimentally determined values for the number of supercoils, the length of the superhelix axis, and the DNA length were used to calculate the superhelix radius as a function of σ. Note that at low values of σ the radius is much more sensitive to changes in supercoiling than at high values.

$\sigma = -0.06$ but decreases only to 50 Å at $\sigma = -0.08$. The shape of this curve offers a reason why the overall dimensions of supercoiled molecules, as reflected in properties such as sedimentation coefficient (Upholt et al. 1971) and electrophoretic mobility (Keller 1975; Shure and Vinograd 1976), change less and less as the DNA becomes more negatively supercoiled.

We can now determine the relationship of n and ΔLk to the third descriptor of supercoiling Wr. We stated above that, for plectonemic supercoiling

$$Wr = -n \sin \gamma \tag{5}$$

Since $\sin \gamma = 0.81$, we have

$$Wr = -0.81n \tag{14}$$

We can, alternatively, combine Equations 12 and 14 to get

$$Wr = 0.73\ \Delta Lk \tag{15}$$

From Equations 12, 14, and 15, we obtain the important result that all three descriptors of plectonemic supercoiling, ΔLk, Wr, and n, are directly proportional to each other. Moreover, once the value of one descriptor is known, the others are easily calculated using these Equations.

Because $\Delta Tw = \Delta Lk - Wr$, by substituting in Equation 15 we get

$$\Delta Tw = \Delta Lk - 0.72\ \Delta Lk = 0.28\ \Delta Lk$$

Thus, there is a constant proportion between Wr and ΔTw for plectonemic supercoiling. Each unit drop in linking number is accompanied by a 0.72 unit decrease in writhe and a 0.28 unit reduction in twist. These values are set by the value of γ in Equation 5. Since γ is constant with supercoiling, so is the ratio of Wr and Tw.

Because $SLk = 0$ for plectonemic supercoiling

$$STw = -Wr$$

and

$$STw = -0.72\ \Delta Lk$$

The values of five parameters for supercoiling (n, Wr, ΔTw, STw, and $\Delta\Phi$) as a function of ΔLk are plotted in Figure 17. In Figure 18 (upper panel) these values, ΔLk and Δh, are given for 1 kb of DNA, whereas in the lower panel they are shown for one superhelical turn.

We next consider the wrapping of DNA around core nucleosomes. This is the best-studied example of protein-stabilized supercoiling and also of the solenoidal form of supercoiling. Nuclease digestion experiments revealed that core particles contain 146 bp of DNA (Lutter 1979). Crystallographic data shows that this DNA wraps around an octamer of histones about 1.8 times in a left-handed helix with a pitch of 28 Å and a radius of 43 Å (Richmond et al. 1984). Analyses of circular viral minichromosomes give a ΔLk of about -1 per nucleosome (Shure and Vinograd 1976; Sogo et al. 1986). The interpretation of this value for ΔLk is uncertain because of the unknown disposition of linker DNA between nucleosomes, and therefore we do not use it in our calculations. Instead, we concern ourselves only with the DNA that is in contact with the histone octamer by using a DNA model that has a smooth, uninterrupted solenoidal winding with the same parameters as the nucleosome. Several lines of evidence suggest that the helical repeat of nucleosomal DNA, defined with respect to the surface of the histone octamer, is significantly below 10.5 bp/turn or the average value for free DNA (Rhodes and Klug

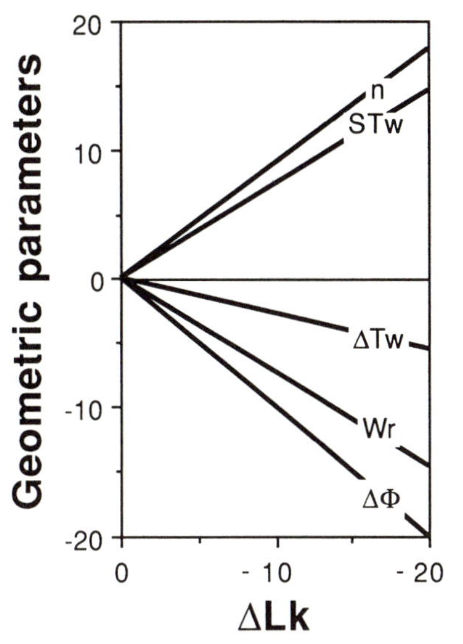

Figure 17 Geometric and topological parameters for plectonemic DNA in solution. The values were obtained as described in the text. The data shown are for the 4.6-kb plasmid illustrated in Fig. 15. The values shown for linking deficits below five ($|\Delta Lk| < 5$) are extrapolations from the experimental data.

1980; Peck and Wang 1981; Tullius and Dombroski 1985). The exact value is controversial, but we will assume a helical repeat of 10.0 bp/turn for our nucleosome model (Morse and Simpson 1988; Klug and Travers 1989; White and Bauer 1989). To facilitate comparisons with the plectonemic supercoiling case, we illustrate in Figure 15B the nucleosomal model for a circular DNA of the same length, 4.6 kb, and in Figure 15C we show plectonemically supercoiled DNA at the same σ value as the nucleosome model. The difference in the shape of the two forms of supercoiling is dramatic.

For this nucleosome model, we first calculate the ΔLk per supercoil. From Equation 4, for left-handed solenoidal DNA

$$\Delta Lk = SLk + \Delta\Phi$$

$$= -n + \Delta\Phi$$

At 81 bp/superhelical turn, our 4.6-kb model has a total of 57 left-handed supercoils. Assuming a helical repeat of nucleosomal DNA (h_{nuc}) of 10.0 bp/turn from the definition of $\Phi = N/h$, for one superhelical turn of 81 bp we have

$$\Delta\Phi = (N/h_{\mathrm{nuc}}) - (N/h_0)$$
$$= (81/10.0) - (81/10.5) = 0.39$$

Since our model has 57 supercoils, $\Delta\Phi$ for the whole plasmid is equal to 22. We can now calculate ΔLk. For the 4.6-kb plasmid

$$\Delta Lk = SLk + \Delta\Phi$$
$$= -57 + 22$$
$$= -35$$

or

$$\Delta Lk = -0.61n \qquad (16)$$

Because there are 1.8 turns/nucleosome, ΔLk equals −1.1 per nucleosome, or the number of supercoils is 1.6 times greater than the magnitude of ΔLk. The discrepancy between measured values of n and ΔLk has been widely discussed in the literature as the "linking number paradox" (Finch et al. 1977). In fact, it would be a remarkable coincidence, if ΔLk equaled n.

We calculate Wr using the relationship presented in Equation 6, $Wr = -n + n \sin\gamma$, which is valid when the size of the central hole of the torus is large compared with the radius of the superhelix, r. To evaluate $\sin\gamma$, we consider a single supercoil wrapped on a cylinder of nucleosomal dimensions. The superhelical pitch is $2\pi p$, and the circumference is $2\pi r$. We showed above (Fig. 14) that unfolding the cylinder into a rectangle reveals that the length of the DNA is the diagonal and equal to $2\pi([p^2 + r^2]^{1/2})$. Thus

$$\sin\gamma = p/([p^2 + r^2]^{1/2})$$

Substituting the nucleosomal values, $p = 28$ Å$/2\pi$ and $r = 43$ Å, we find that $\sin\gamma = 0.10$ ($\gamma = 6.3°$). Using these values in Equation 6, we obtain

$$Wr = -0.90n \qquad (17)$$

Substituting in Equation 16

$$Wr = 1.46\ \Delta Lk \qquad (18)$$

Once again, ΔLk, Wr, and n are directly proportional to each other.

To define the remaining parameters, we first rearrange Equation 3 to give

$$STw = SLk - Wr$$

Substituting the values of SLk and Wr in terms of ΔLk gives

$$STw = 1.63\ \Delta Lk - 1.46\ \Delta Lk = +0.17\ \Delta Lk$$

ΔTw is easily calculated from the values of Wr and ΔLk, using Equation 1. The result is

$$\Delta Tw = -0.46\ \Delta Lk$$

Notice here that ΔTw is positive and relatively large despite the large negative ΔLk. This explains why Wr is larger in magnitude than ΔLk (see Eq. 18).

σ for the nucleosome model equals –0.079 for one turn (σ = –0.61/(81/10.5). This value is different from that of protein-free DNA from minichromosomes of about –0.06 because of contributions of the linker DNA to ΔLk or perhaps because one of the assumptions about the nucleosome is a little off. The bar graphs on the right-hand side of Figure 18 summarize the structural parameters for our two model superhelices at this value of σ. In the upper panel, the values are given for a 1-kb length of DNA, whereas in the lower panel they are for one superhelical turn. At the left of each panel, we show the values for plectonemically supercoiled DNA at a σ of –0.06, a more commonly cited native level.

From an examination of the bar graphs in Figure 18 and the scale drawings (Fig. 15), one is immediately struck by the fact that the two types of supercoiling can cause the same change in a topological quantity, ΔLk, by entirely different means. In both cases, Wr is large, negative, and the major contributor to ΔLk. However, this is achieved for the left-handed solenoidal superhelices by having a small pitch ($2\pi p$) or equivalently a small winding angle (γ), but it is achieved for the right-handed plectonemic form by having high values for $2\pi p$ and γ. The values in our two cases are $2\pi p$ = 28 Å and $\gamma = 6^\circ$ for nucleosomal supercoiling and $2\pi p$ = 174 Å and $\gamma = 54^\circ$ for plectonemic supercoiling.

The difference in handedness of the two types of supercoiling also has an important effect on $\Delta\Phi$ and helical repeat h. Since B-type DNA is right-handed, the strand-axis vector v_{ac} (Fig. 10) always rotates in a

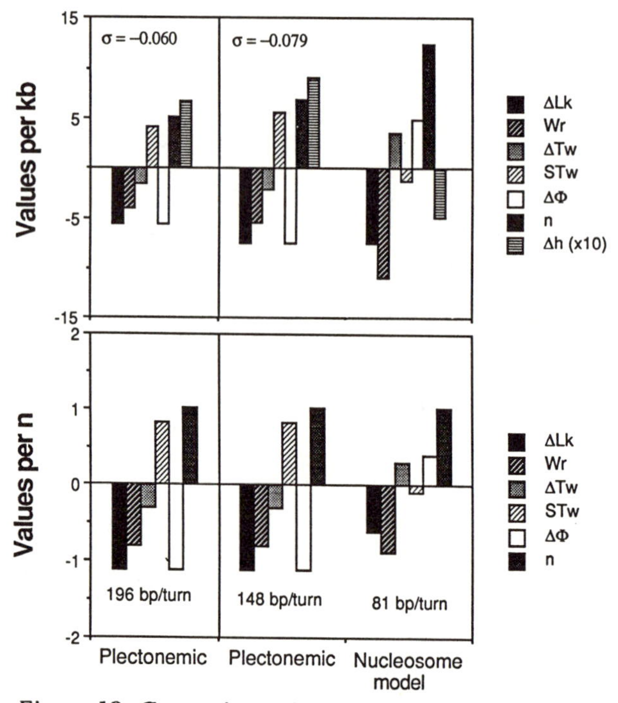

Figure 18 Comparison of parameters for plectonemic and solenoidally supercoiled DNA. Parameters are given for the three forms of the 4.6-kb plasmid shown in Fig. 15. In the upper panel, the parameters are reported for 1 kb of DNA length, whereas in the lower panel the values are given for one supercoil. In each panel, the data on the left are for plectonemically supercoiled DNA at a σ of –0.060 (Fig. 15A), the data in the center are for plectonemically supercoiled DNA at a σ of –0.079 (Fig. 15C), and at right the data are for the nucleosome model shown (Fig. 15B), which also has a σ of –0.079. The vertical separation between left and center portions of each panel emphasize the difference in σ. The amount of DNA per supercoil is given below the graphs in the lower panel. Notice that one supercoil of plectonemic DNA contains much more DNA than a nucleosomal supercoil.

right-handed fashion about the DNA axis A. However, the rotational sense of the surface normal ν about A is determined by the handedness of the supercoils. The rotation of ν about A is left-handed for left-handed supercoils and right-handed for right-handed coils. Therefore, if we start at a point on A where ν_{ac} and ν coincide and then move along A until they next meet again, ν_{ac} will have rotated through more base pairs than for relaxed DNA for a right-handed superhelix (i.e., $h > h_0$) and fewer base pairs for a left-handed superhelix (i.e., $h < h_0$). The actual values of h for our two models are 10.0 for the nucleosome and 11.4 for plectonemic DNA ($\sigma = -0.079$). Note that when DNA solenoidally supercoiled about histones is freed from protein, h changes from $<h_0$ to $>h_0$.

Our two models for supercoiling clearly illustrate the effect of superhelical pitch on the magnitude of *STw*. The nucleosome has very low values for both pitch and |*STw*| in contrast with the plectonemic case where both values are large. The reason for this correlation can be seen by considering the difference in the motion of the surface normal between the two models. As in the section defining *STw*, we consider the surface normal as an "observer" walking on the surface along the DNA axis. For the nucleosome case, the majority of the observer's movement is forward and backward as he traverses the DNA. He experiences very little lateral rotation about the DNA axis, and therefore |*STw*| is small. In contrast, for the plectonemic model where the pitch is large, the observer experiences large lateral rotations about the DNA axis as he walks, demonstrating that |*STw*| is large.

Surprisingly, *Tw* changes in opposite direction for the two models. The high positive $\Delta\Phi$ and low negative *STw* for the nucleosomal case lead to a positive ΔTw for the nucleosome. This contrasts with the plectonemic winding, where ΔTw is negative. Clearly, the often stated idea that supercoiling affects *Wr* but not *Tw* is wrong, and the magnitude and sign of the change in *Tw* depend on geometry.

XI. PERSPECTIVES

By way of summary, we contrast plectonemic and solenoidal negative supercoiling. (1) The plectonemic form is right-handed and the solenoidal form is left-handed. (2) This difference in handedness results in opposite changes of h, Φ, and *STw*; the superhelix pitch and the amount of DNA per supercoil determines the magnitude of these changes. (3) Plectonemic supercoils necessarily bring together sequences distant in the primary structure as well as neighboring sequences, whereas solenoidal supercoiling only affects the latter. Thus, the interaction of distant *cis*-acting sequences, as occurs in the regulation of transcription and site-specific recombination, should be promoted by plectonemic supercoiling (Wasserman and Cozzarelli 1986). (4) Branching is a natural property of plectonemic supercoiling but not of solenoidal supercoiling. Branching brings together and mutually intertwines three distant sequences, and this has been postulated to promote recombination reactions in which two crossover sites and an enhancer site are involved (Johnson et al. 1987; Kanaar et al. 1988). (5) Because of the effect of the crossing of distant segments on the writhe of plectonemically supercoiled

DNA, solenoidal and plectonemic forms that have identical superhelical pitch and radius will nonetheless differ in *Wr*, *Tw*, and σ. (6) The large pitch of negatively supercoiled DNA in solution results in very little compaction. The very low pitch of nucleosomal DNA results in effectively shortened DNA, which is important in chromosomal packing.

The differences between the two forms of supercoiling are all the more important given that these forms are probably in dynamic equilibrium in eukaryotic cells. DNA free in solution is plectonemic in form and is converted to a solenoidal form when folded into nucleosomes. Not all protein-bound supercoils are solenoidal; the structure will be dictated by the winding surface. The synaptic intermediate of the Tn*3* resolvase is a well-characterized plectonemic DNA/protein complex (Benjamin and Cozzarelli 1988, 1990).

We have stressed that a good index of DNA architecture encodes in a single number an important structural feature. We have described a number of good indices. The indices can be divided into those that describe the winding of the backbone about the helix axis (Φ and h), those that describe the winding of the helix axis in space (n, *Wr*, *STw*, *SLk*, γ, p, and r), and those that are a function of both (*Tw*, *Lk*, and ΔLk). One (*Lk*) is an integral invariant for closed DNA; some are invariant and quantized as long as their defining surface is smoothly deformed (Φ, h, and *SLk*), and others are continuously varying geometric properties (n, *Wr*, *Tw*, γ, p, and r). ΔLk can be viewed as topological, geometric, or some of both, depending on the definition of Lk_0.

Some of the indices are easily measured experimentally (*Lk*, Lk_0, and ΔLk), whereas others are readily calculated if the superhelix geometry is known (n, *Wr*, *SLk*, *STw*, p, and γ). We showed, however, that all the parameters could be determined by a combination of mathematics and experimentation. For example, once n, γ, and the number of branches were measured as a function of σ for plectonemically supercoiled DNA, the equations presented above allowed calculation of *Wr*, *Tw*, *STw*, p, and r. This wholesale return on such a modest experimental effort emphasizes the power of the quantitative approach to DNA supercoiling.

We pointed out that there are three main descriptors of DNA supercoiling: ΔLk, *Wr*, and n. At one time or another, each has been taken to be the "definition" of supercoiling, but we prefer instead to call them quantitative measures or descriptors. n, the number of turns around the superhelix axis, is the common-language meaning of the "coiling of a coil." *Wr* is a function of not only n, but also of the superhelix winding angle. ΔLk is a function not only of *Wr*, but also of *Tw*. Each descriptor has advantages and limitations. When all are known, supercoiling is best described.

Glossary of terms

Symbol	Meaning	Equivalence	Illustrative Figure
A	Curve representing DNA axis		8
a	Point on A		8
α	Topological winding number in Vinograd nomenclature	Lk	3, 5
A_ε	Displacement curve of A		11
β	Duplex winding number in Vinograd nomenclature	Φ (sol.)	
C	Curve representing one of the strands of duplex DNA		8
c	Point on C		8
E	Number of ends in branched plectonemic superhelix		15
Φ	Winding number	Lk - SLk, Tw - STw	10
γ	Superhelix pitch angle	$\frac{p}{\sqrt{r^2 + p^2}}$ $\sin^{-1}\left(2\frac{\ell}{L}\right)$ (plect.)	14
h	Helical repeat of duplex DNA (relative to surface normal)	$\frac{N}{\Phi}$	10

L	Length of DNA		
Lk	Linking number	$Wr + Tw$, $\Phi + SLk$	3, 5
Lk_0, h_0, Φ_0	Parameters for relaxed DNA		
N	Total number of base pairs		
n	Number of superhelical turns	$\left\lvert \frac{\sin \gamma}{Wr} \right\rvert$ (plect.), $\lvert SLk \rvert$ (sol.)	11C 11D
$2\pi p$	Pitch of a helix		14
r	Superhelix radius		14
σ	Specific linking difference	$\frac{\Delta Lk}{Lk_0}$	
SLk	Surface linking number	$Lk - \Phi$, Lk of A and A_ε, $\pm n$ (sol.)	11
STw	Surface twist	$Tw - \Phi$, $n \sin \gamma$ (plect.), $-n \sin \gamma$ (sol.)	10
τ	Number of superhelical turns in Vinograd nomenclature	SLk (sol.)	

Tw	Twist	$Lk - Wr$, $STw + \Phi$	8
$\mathbf{v}$	Surface normal unit vector		8
$\mathbf{v}_{ac}$	Strand-axis unit vector		
W	Curve representing one of the strands of duplex DNA		
Wr	Writhe	$Lk - Tw$, $SLk - STw$, $-n \sin \gamma$ (plect.), $-n + n \sin \gamma$ (sol.)	7
Wr_p	Projected writhe		6
ΔLk	$Lk - Lk_0$	$Wr + \Delta Tw$, $SLk + \Delta\Phi$	
ℓ	Length of superhelix axis		

The equivalences are for models for closed (-) supercoiled DNA given in the text. The solenoidal superhelix is assumed to be long and thin for the geometric terms. "Sol." and "plect." refer to solenoidal and plectonemic supercoils, respectively.

REFERENCES

Anderson, P. and Bauer, W.R. 1978. Supercoiling in closed circular DNA: Dependence upon ion type and concentration. *Biochemistry* **17:** 594–601.

Bauer, W. 1978. Structure and reactions of closed duplex DNA. *Annu. Rev. Biophys. Bioeng.* **7:** 287–313.

Bauer, W. and J. Vinograd. 1968. The interaction of closed circular DNA with intercalative dyes. I. The superhelix density of SV40 DNA in the presence and absence of dye. *J. Mol. Biol.* **33:** 141–171.

———. 1970. Interaction of closed circular DNA with intercalative dyes. II. The free energy of superhelix formation in SV40 DNA. *J. Mol. Biol.* **47:** 419–435.

———. 1974. Circular DNA. In *Basic principles in nucleic acid chemistry,* (ed. P.O.P. Ts'o), vol. 2, p. 265–303. Academic Press, New York.

Benjamin, H.W. and N.R. Cozzarelli. 1988. Isolation and characterization of the Tn3 resolvase synaptic intermediate. *EMBO J.* **7:** 1897–1905.

———. 1990. Geometric arrangements of the Tn3 resolvase sites. *J.Biol. Chem.* (in press).

Benyajati, C. and A. Worcel. 1976. Isolation, characterization, and structure of the folded interphase genome of *Drosophila melanogaster*. *Cell* **9:** 393–407.

Better, M., C. Lu, R.C. Williams, and H. Echols. 1982. Site-specific DNA condensation and pairing mediated by the int protein of bacteriophage lambda. *Proc. Natl. Acad. Sci.* **79:** 5837–5841.

Boles, T.C., J.W. White, and N.R. Cozzarelli. 1990. The structure of plectonemically supercoiled DNA. *J. Mol. Biol.* (in press).

Cozzarelli, N.R. 1980. DNA topoisomerases. *Cell* **22:** 327–328.

Crick, F.H.C. 1976. Linking numbers and nucleosomes. *Proc. Natl. Acad. Sci.* **73:** 2639–2643.

Depew, R.E. and J.C. Wang. 1975. Conformational fluctuations of DNA helix. *Proc. Natl. Acad. Sci.* **72:** 4275–4279.

Drlica, K. and A. Worcel. 1975. Conformational transitions in the *Escherichia coli* chromosome: Analysis by viscometry and sedimentation. *J. Mol. Biol.* **98:** 393–411.

Finch, J.T., L.C. Lutter, O. Rhodes, R.S. Brown, B. Rushton, M. Levitt, and A. Klug. 1977. Structure of nucleosome core particles of chromatin. *Nature* **269:** 29–36.

Fuller, F.B. 1971. The writhing number of a space curve. *Proc. Natl. Acad. Sci.* **68:** 815–819.

Gauss, K.F. 1833. Zur mathematischen Theorie der electrodynamischen Wirkungen. In *Werke* (ed. E.I. Schering), vol. 5, p. 605. Georg Olm Verlag, Hildesheim.

Glaubiger, D. and J.E. Hearst. 1967. Effect of superhelical structure on the secondary structure of DNA rings. *Biopolymers* **5:** 691–696.

Johnson, R.C., A.C. Glasgow, and M.I. Simon. 1987. Spatial relationship of the Fis binding sites for Hin recombinational enhancer activity. *Nature* **329:** 462–465.

Kanaar, R., P. van de Putte, and N.R. Cozzarelli. 1988. Gin-mediated DNA inversion: Product structure and the mechanism of strand exchange. *Proc. Natl. Acad. Sci.* **85:** 752–756.

Keller, W. 1975. Determination of the number of superhelical turns in simian virus 40 DNA by gel electrophoresis. *Proc. Natl. Acad. Sci.* **72:** 4876–4880.

Kikuchi, A. and K. Asai. 1984. Reverse gyrase: A topoisomerase which introduces positive superhelical turns into DNA. *Nature* **309:** 677–681.

Klug, A. and A.A. Travers. 1989. The helical repeat of nucleosome-wrapped DNA. *Cell* **56:** 10–11.

Liu, L.F. and J.C. Wang. 1978. *Micrococcus luteus* DNA gyrase: Active components and a model for its supercoiling of DNA. *Proc. Natl. Acad. Sci.* **75:** 2098–2102.

Lockshon, D. and D.R. Morris. 1983. Positively supercoiled plasmid DNA is produced by treatment of *Escherichia coli* with DNA gyrase inhibitors. *Nucleic Acids Res.* **11:** 2999–3017.

Lutter, L.C. 1978. Kinetic analysis of deoxyribonuclease I cleavages in the nucleosome core: Evidence for a DNA superhelix. *J. Mol. Biol.* **124:** 391–420.

———. 1979. Precise location of DNase I cutting sites in the nucleosome core determined by high resolution gel electrophoresis. *Nucleic Acids Res.* **6:** 41–56.

Maxwell, A. and M. Gellert. 1986. Mechanistic aspects of DNA topoisomerases. *Adv. Protein Chem.* **38:** 69–107.

Morse, R.H. and R.T. Simpson. 1988. DNA in the nucleosome. *Cell* **54:** 285–287.

Nadal, M., G. Mirambeau, P. Forterre, W.-D. Reiter, and M. Duguet. 1986. Positively supercoiled DNA in a virus-like particle of an archaebacterium. *Nature* **321:** 256–258.

Olins, A.L. and D.E. Olins. 1974. Spheroid chromatin units (ν bodies). *Science* **183:** 330–332.

Paulson, J.R. and U.K. Laemmli. 1977. The structure of histone-depleted metaphase chromosomes. *Cell* **12:** 817–828.

Peck, L.J. and J.C. Wang. 1981. Sequence dependence of the helical repeat of DNA in solution. *Nature* **292:** 375–378.

Pettijohn, D.E. and R. Hecht. 1974. RNA molecules bound to the folded bacterial genome stabilize DNA folds and segregate domains of supercoiling. *Cold Spring Harbor Symp. Quant. Biol.* **38:** 31–42.

Pulleyblank, D.E., M. Shure, D. Tang, J. Vinograd, and H.P. Vosberg. 1975. Action of nicking-closing enzyme on supercoiled and nonsupercoiled closed circular DNA: Formation of a Boltzmann distribution of topological isomers. *Proc. Natl. Acad. Sci.* **72:** 4280–4284.

Rhodes, D. and A. Klug. 1980. Helical periodicity of DNA determined by enzyme digestion. *Nature* **286:** 573–578.

Richmond, T.J., J.T. Finch, B. Rushton, D. Rhodes, and A. Klug. 1984. Structure of the nucleosome core particle at 7Å resolution. *Nature* **311:** 532–537.

Shure, M. and J.W. Vinograd. 1976. The number of superhelical turns in native virion SV40 DNA and minicol DNA determined by the band counting method. *Cell* **8:** 215–226.

Sogo, J.M., H. Stahl, T. Koller, and R. Knippers. 1986. Structure of replicating simian virus 40 minichromosomes: The replication fork, core histone segregation and terminal structures. *J. Mol. Biol.* **189:** 189–204.

Spengler, S.J., A. Stasiak, and N.R. Cozzarelli. 1985. The stereostructure of knots and catenanes produced by phage λ integrative recombination: Implications for mechanism and DNA structure. *Cell* **42:** 325–334.

Stonington, O.G. and D.E. Pettijohn. 1971. The folded genome of *Escherichia coli* isolated in a protein-DNA-RNA complex. *Proc. Natl. Acad. Sci.* **68:** 6–9.

Travers, A.A. and A. Klug. 1987. The bending of DNA in nucleosomes and its wider implications. *Philos. Trans. R. Soc. Lond. B* **317:** 537–561.

Tullius, T.D. and B.A. Dombroski. 1985. Iron(II) EDTA used to measure the helical twist along any DNA molecule. *Science* **230:** 679–681.

Upholt, W.B., H.B. Gray, and J. Vinograd. 1971. Sedimentation velocity behavior of closed circular SV40 DNA as a function of superhelix density, ionic strength, counterion and temperature. *J. Mol. Biol.* **62:** 21–38.

Vinograd, J. and J. Lebowitz. 1966. Physical and topological properties of circular DNA. *J. Gen. Physiol.* **49:** 103–125.

Vinograd, J., J. Lebowitz, and R. Watson. 1968. Early and late helix-coil transitions in closed circular DNA: The number of superhelical turns in polyoma DNA. *J. Mol. Biol.* **33:** 173–197.

Vinograd, J., J. Lebowitz, R. Radloff, R. Watson, and P. Laipis. 1965. The twisted circular form of polyoma viral DNA. *Proc. Natl. Acad. Sci.* **53:** 1104–1111.

Wang, J.C. 1969. Variation of the average rotational angle of the DNA helix and the superhelical turns of covalently closed cyclic lambda DNA. *J. Mol. Biol.* **43:** 25–39.

———. 1971. Interaction between DNA and an *Escherichia coli* protein ω. *J. Mol. Biol.* **55:** 523–533.

———. 1985. DNA topoisomerases. *Annu. Rev. Biochem.* **54:** 665–697.

———. 1986. Circular DNA. In *Cyclic polymers* (ed. J.A. Semlyen), p. 225–260. Elsevier Applied Science Publishers, London.

Wang, J.C., L.J. Peck, and K. Becherer. 1983. DNA supercoiling and its effects on DNA structure and function. *Cold Spring Harbor Symp. Quant. Biol.* **47:** 85–91.

Wasserman, S.A. and N.R. Cozzarelli. 1986. Biochemical topology: Applications to DNA recombination and replication. *Science* **232:** 951–960.

Watson, J.D. and F. Crick. 1953. A structure for deoxyribose nucleic acid. *Nature* **171:** 737–738.

White, J.H. 1969. Self-linking and the Gauss integral in higher dimensions. *Am. J. Math.* **91:** 693–728.

———. 1989. An introduction to the geometry and topology of DNA structure. In *Mathematical methods for DNA sequences* (ed. M.S. Waterman), p. 225–253. CRC Press, Boca Raton, Florida.

White, J.H. and W.R. Bauer. 1986. Calculation of the twist and the writhe for representative models of DNA. *J. Mol. Biol.* **189:** 329–341.

———. 1987. Superhelical DNA with local substructures: A generalization of the topological constraint in terms of the intersection number and the ladder-like correspondence surface. *J. Mol. Biol.* **195:** 205–213.

———. 1988. Applications of the twist difference to DNA structural analysis. *Proc. Natl. Acad. Sci.* **85:** 772–776.

———. 1989. The helical repeat of nucleosome-wrapped DNA. *Cell* **56:** 9–10.

White, J.H., N.R. Cozzarelli, and W.R. Bauer. 1988. Helical repeat and linking number of surface-wrapped DNA. *Science* **241:** 323–327.

Worcel, A. and E. Burgi. 1972. On the structure of the folded chromosome of *Escherichia coli*. *J. Mol. Biol.* **71:** 127–147.

Wu, H.Y., S.H. Shyy, J.C. Wang, and L.F. Liu. 1988. Transcription generates positively and negatively supercoiled domains in the template. *Cell* **53:** 433–440.

5

DNA Supercoiling and Unusual Structures

Maxim D. Frank-Kamenetskii
Institute of Molecular Genetics
Academy of Sciences of USSR
Moscow 123182, Union of Soviet Socialist Republics

I. INTRODUCTION

What are the structural consequences of DNA supercoiling? Until recently, the importance of the problem was generally underestimated because the available data gave relatively low absolute values for the degree of DNA supercoiling inside the cell.

Many people, I concluded, considered the dramatic changes in DNA structure at high levels of supercoiling as physicochemical tricks that were not directly relevant to biology. The only excuse for continuing the studies stemmed from the hope that the phenomena observed under those artificial conditions pointed to the ability of DNA to form unusual structures, other than the canonical Watson-Crick double helix, under the action of special proteins. The attitude has changed after the recent demonstration that a high local degree of supercoiling can occur as a consequence of the double-helical nature of DNA (Giaever and Wang 1988; Wu et al. 1988). We now understand that the overall degree of DNA supercoiling in the cell is much less relevant from a biological viewpoint and that the local transient degree of supercoiling may be much higher than the most ambitious physicochemist could ever dream of. Thus, we have to study the structural consequences of DNA supercoiling to un-

DNA Topology and Its Biological Effects

derstand the biological significance of DNA topology and to elucidate the structural potential of DNA.

Supercoiling may be introduced into linear DNA provided both ends are fixed so that the DNA strands cannot rotate with respect to one another. Throughout this chapter, however, we restrict ourselves to the case of closed circular DNA, which is commonly used in physicochemical studies.

Supercoiling is typical to closed circular DNA because of the topological linkage of the two complementary strands. The linking number, *Lk*, is the quantitative measure of this linkage: It is the algebraic number of times one strand crosses the surface stretched over the other strand. The *Lk* value is a topological invariant for closed circular DNA; there is no way it can be changed without introducing chain scissions. The number of supercoils can be defined as

$$\Delta Lk = Lk - N/\gamma_0 \tag{1}$$

where N is the number of base pairs in a DNA and γ_0 is the number of base pairs per turn of the double helix under given ambient conditions. If $\Delta Lk > 0$, DNA is called positively supercoiled; if $\Delta Lk < 0$, DNA is negatively supercoiled. Specific linking difference, or superhelical density, is defined as

$$\sigma = \gamma_0 \Delta Lk / N \tag{2}$$

where the value of γ_0 has been shown to be close to 10.5.

The first problem of the physical chemistry of closed circular DNA is to measure the ΔLk value. The second problem is to resolve the ΔLk value into its two geometric components

$$\Delta Lk = \Delta Tw + Wr \tag{3}$$

where ΔTw is the difference in the axial twist of either strand about the axis of the double helix, Wr is the writhe of the axis, which is determined by its spatial shape. The third problem is the energy of supercoiling and how it is distributed between twist and writhe.

We considered these three problems in the next section. Throughout it, we model the double helix as an elastic rod, i.e., as a homogeneous, isotropic flexible rod. This model includes one geometric parameter, the diameter of the cylinder, and two energetic parameters, the bending and torsional rigidities. Experimental data make it possible to estimate all of them.

The elastic rod model, however, is valid only at relatively low degrees of supercoiling. As negative superhelical density increases, the regular double-helix structure is interrupted by local distortions. The position and character of these distortions are strongly sequence-dependent. They usually encompass just a tiny portion of the molecule but nevertheless have a dramatic effect on the physicochemical properties of supercoiled DNA. Besides, these local distortions arouse a special interest because they take the form of unusual structures that could potentially play a significant role in DNA functions in the cell.

Special experimental techniques have been developed to study the unusual structures formed under superhelical stress. These techniques made it possible not only to demonstrate the formation of DNA structures known from traditional experiments, such as the Z form, but also to find completely nonorthodox structures with unusual base pairing, viz., the protonated H forms. We discuss these findings at length in the last section of this chapter.

II. GEOMETRY AND ENERGETICS OF SUPERCOILED DNA

In this section, we consider closed circular DNA at comparatively low superhelical densities, where it retains its normal double-helical structure and can be treated as an isotropic homogeneous flexible rod. Within the framework of this model, the double helix is characterized by two energy parameters: the bending and torsional rigidities. To what extent is this simple model applicable to real DNA?

Bending and torsional rigidities are sequence-dependent. Nevertheless, as long as these parameters are obtained from and used for sufficiently long DNA stretches composed of many helical turns, variations in these parameters should average out, and such helices may well be described in terms of the homogeneous model with averaged bending and torsional rigidities. One can expect the model to fail for some special DNA sequences or for extremely short helices. However, even small rings (Horowitz and Wang 1984; Frank-Kamenetskii et al. 1985; Shimada and Yamakawa 1988; Klenin et al. 1989) still comprise quite a few helix turns, let alone large ones (Depew and Wang 1975; Pulleyblank et al. 1975; Shure et al. 1977; Vologodskii et al. 1979a).

Our knowledge of the behavior of closed circular DNA at low superhelical densities is based on experiments on the equilibrium distribution of DNA rings over the linking number *Lk*. To obtain such distribution, one subjects a sample containing closed circular DNA molecules to relaxation by a DNA topoisomerase or, alternatively, to nicking and subsequent ligation. Then the sample is analyzed with the aid of agarose gel electrophoresis. Molecules with different *Lk* values, i.e., different topoisomers, move in the gel with different velocities. As a result, after staining the gel, one observes a series of bands corresponding to different topoisomers with the Gaussian distribution of intensities. The variance of the distribution, $<(\Delta Lk)^2>$, yields valuable information about closed circular DNA. However, to decipher the message one needs an adequate theoretical treatment of the flexible-rod model.

The theoretical treatment is based on the assumption that without topological constraints (i.e., in the presence of topoisomerases) fluctuations in twist and writhe (i.e., torsional and bending motions) occur independently from each other with the result that

$$<(\Delta Lk)^2> = <(\Delta Tw)^2> + <(Wr)^2> \quad (4)$$

The $<(\Delta Tw)^2>$ value is expressed in terms of the torsional rigidity parameter C

$$<(\Delta Tw)^2> = (hk_BT/4\pi^2C)N \quad (5)$$

where h is the distance between adjacent base pairs in the double helix (0.34 nm), N is the number of base pairs in the DNA ring under consideration, k_B is the Boltzmann constant, and T is the absolute temperature, which we assume to be 310°K, the typical temperature at which the experiments are done. Thus, the first term in the right-hand part of Equation 4 is completely determined by the torsional rigidity of the double helix. Since writhe depends only on the shape of the axis of the double helix, the second term in the right-hand part of Equation 4, $<(Wr)^2>$, depends on the bending rigidity of DNA or on the Kuhn statistical length b (see, e.g., Cantor and Schimmel 1980). There is no such simple universal relation between $<(Wr)^2>$ and the values of b and N as the relation between $<(\Delta Tw)^2>$ and the C and N values given by Equation 5. The value of $<(Wr)^2>$ is much more complicated. However, it is solvable because an equation exists for calculating Wr from the spatial shape of the axis of the DNA double helix. As a result, $<(Wr)^2>$ may be calculated by the methods of polymer statistics as a function of the number of Kuhn statistical segments $n = Nh/b$. Using Equation 4, one can then determine from the data on $<(\Delta Lk)^2>$ both rigidity parameters C and b. In turn, the knowledge of these parameters yields, within the framework of the flexible rod model, the complete description of DNA supercoiling.

The first attempt to fulfill this program was undertaken by Benham (1978). However, we showed that Benham made serious errors in calculating $<(Wr)^2>$ (Vologodskii et al. 1979a). In our calculations of the dependence of the variance of writhe, $<(Wr)^2>$, on the number of Kuhn statistical segments n, we modeled DNA as a free-joint polymer chain without excluded volume (Vologodskii et al. 1979a). The polymer chain was assumed to consist of infinitely thin straight segments of equal lengths connected by universal joints. This model seemed adequate if short chains were not considered and the effective diameter of DNA did not greatly exceed its geometrical value.

The free-joint model was clearly inappropriate for short molecules comprising a few Kuhn statistical lengths, i.e., below a few thousand base pairs. The detailed experimental studies of small DNA rings (Shore and Baldwin 1983; Horowitz and Wang 1984) stimulated analytical (Shimada and Yamakawa 1985) and Monte Carlo studies (Frank-Kamenetskii et al. 1985; Levene and Crothers 1986; Shimada and Yamakawa 1988; Klenin et al. 1989) of the $\langle (Wr)^2 \rangle$ value for the more adequate worm-like model.

These studies made clear that the free-joint model underestimates $\langle (Wr)^2 \rangle$ even for long chains. In this model, two adjacent segments make no contribution to the Wr because they always are planar. This is not the case for the more realistic worm-like chain model.

Shimada and Yamakawa (1988) and Klenin et al. (1989) calculated the $\langle (Wr)^2 \rangle$ value for the worm-like model. Figure 1 shows the results of these calculations of the variance of writhe without excluded volume $\langle (Wr)^2 \rangle_0$. The figure also shows the data on the $\langle (Wr)^2 \rangle_0$ value for the free-joint model. One can see that the free-joint model underestimates the $\langle (Wr)^2 \rangle_0$ value by 20%. The results obey the following equation (Klenin et al. 1989)

$$\langle (Wr)^2 \rangle_0 / n = (0.00385\, n + 0.113\, \mathrm{f}(n))/(1 + \mathrm{f}(n)),\ n \leq 30, \qquad (6)$$

where

$$\mathrm{f}(n) = 0.752n \exp(-6.242/n)$$

Experimental (Brian et al. 1981; Yarmola et al. 1985) and theoretical (Stigter 1977) data now show that the actual excluded volume effects in DNA at moderate ionic strengths are much more significant than expected on the basis of the geometrical diameter of DNA. The first attempt to allow for excluded volume effects in the calculation of DNA topological properties within the free-joint model was undertaken by Le Bret (1980). However, Le Bret significantly underestimated the excluded-volume effects (see Klenin et al. 1988). Because of electrostatic interactions, the effective diameter of the double helix, d, can exceed the geometrical diameter of DNA by several times (Stigter 1977; Brian et al. 1981; Yarmola et al. 1985; Klenin et al. 1988). Thus, one cannot neglect excluded volume effects while calculating the macromolecular properties of DNA. This was shown to be true for the topological characteristics of DNA even when the effective diameter coincided with the geometrical diameter of the double helix (Klenin et al. 1988; 1989). On the basis of these data, Klenin et al. (1989) concluded that accurate theoretical estimations of $<(Wr)^2>$ require use of the worm-like chain with excluded volume.

Klenin et al. (1989) calculated that the difference between $<(Wr)^2>$ and $<(Wr)^2>_0$ is chain-length-independent for $n \geq 5$ and obeys the equation

$$<(Wr)^2> = <(Wr)^2>_0/(1 + 5.5\, d/b) \qquad (7)$$

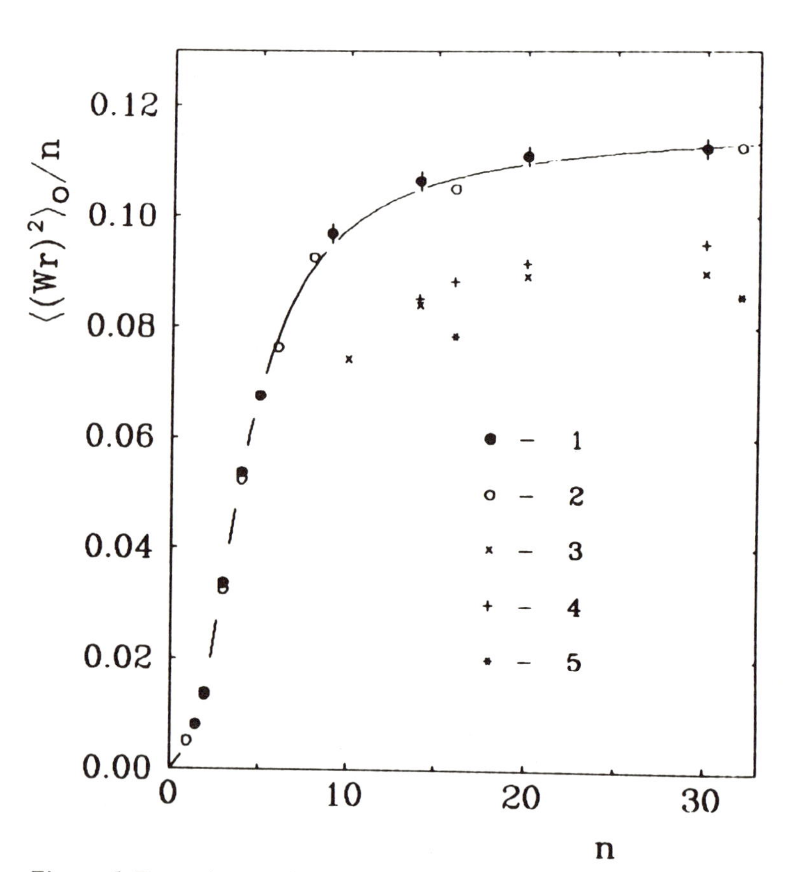

Figure 1 Dependence of the variance of writhe for chains without excluded volume, $<(Wr)^2>_0$, on the number of Kuhn statistical segments n. Data 1 and 2, obtained by Klenin et al. (1989) and Shimada and Yamakawa (1988) respectively correspond to the worm-like-chain model. Data 3–5 correspond to the free-joint model. The curve corresponds to Equation 6.

For $n < 5$, the excluded volume effects gradually decrease, becoming insignificant for $n \leq 1$ (Klenin et al. 1989). Equations 4–7 make it possible to compare theory and experiment (Depew and Wang 1975; Pulleyblank et al. 1975; Shure et al. 1977; Shore and Baldwin 1983; Horowitz and Wang 1984) using the most adequate worm-like model with excluded volume (Klenin et al. 1989).

There is a serious disagreement between the data of Shore and Baldwin (1983) and that of Horowitz and Wang (1984). Because the data of Shore and Baldwin are inconsistent with our theoretical results (Frank-Kamenetskii et al. 1985), we will ignore them in our comparison of theory and experiment.

According to Equations 3–7, the experimentally available quantity, $<(\Delta Lk)^2>$, depends on three parameters: the torsional rigidity C, the Kuhn statistical length b, and the effective diameter d. One can determine all three parameters by comparing theoretical and experimental data on the chain-length dependence of $<(\Delta Lk)^2>/n$ (Fig. 2) (see Klenin et al. 1989).

The C value is determined from the data on every small DNA rings. This yields $C = 3.0 \times 10^{-19}$ erg cm. The sharp increase of the $<(\Delta Lk)^2>/n$ value in the range of 2–5 segments practically does not depend on excluded-volume effects and makes it possible to determine the Kuhn statistical length b with high accuracy. This yields $b = 100$ nm, which is consistent with estimations by other methods (Hagerman 1981). Finally, the saturation level for $<(\Delta Lk)^2>/n$ at chain lengths yields an effective diameter d for DNA of 2 nm. It should be emphasized, however, that because the saturation level is a weak function of d (see Eq. 7), the d value is determined with relatively poor accuracy. We can only conclude that under the experimental conditions used by Depew and Wang (1975), Pulleyblank et al. (1975), Shure et al. (1977), and Horowitz and Wang (1984) (high sodium concentration and the presence of bivalent cations), the effective diameter of DNA is close to its geometrical value. The above data make it possible to answer many questions about closed circular DNA at a low degree of supercoiling. Specifically, they yield a reliable estimation of the superhelix energy, E, and its distribution between writhe and twist.

The expression for superhelix energy immediately follows from the fact that the equilibrium distribution of DNA topoisomers over the Lk value has a Gaussian shape with the variance $<(\Delta Lk)^2>$ shown in Figure 2

$$E = k_B T \, (\Delta Lk)^2 / 2 <(\Delta Lk)^2> \tag{8}$$

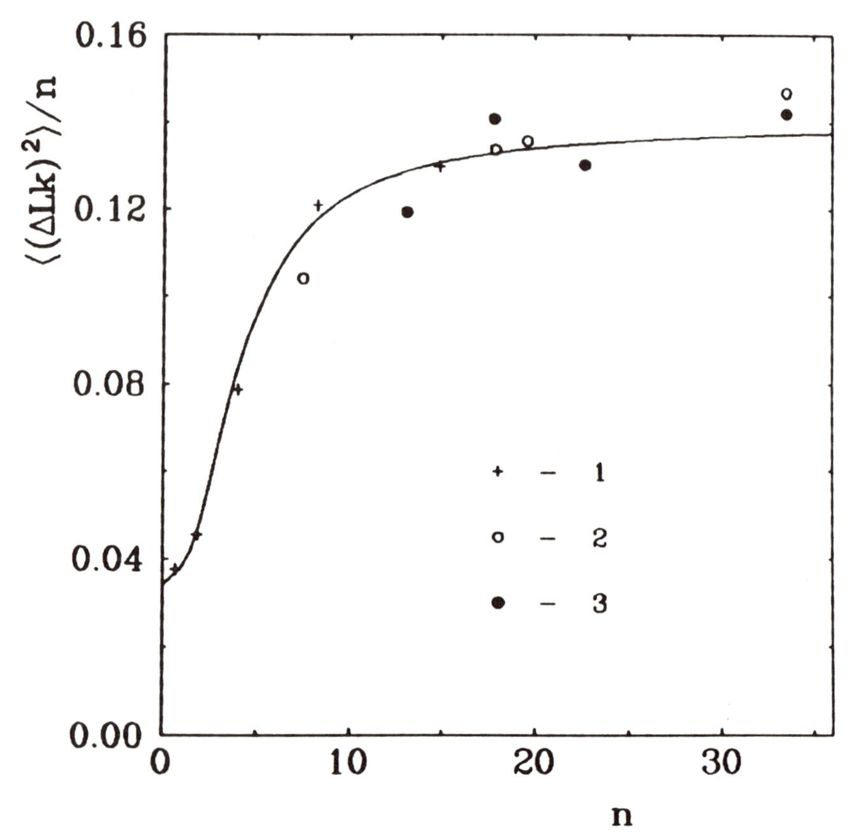

Figure 2 Dependence of the variance of the equilibrium distribution of closed DNA molecules over topoisomers, $<(\Delta Lk)^2>$, on the number of Kuhn statistical segments n. Data 1 are taken from Horowitz and Wang (1984), data 2 are from Depew and Wang (1975), and data 3 are from Pulleyblank et al. (1975) and Shure et al. (1977). The curve is calculated on the basis of Equations 3 to 7 for $C = 3 \times 10^{-19}$ erg cm, $b = 100$ nm, and $d = 2$ nm (Klenin et al. 1989).

From Equations 1 and 2, we obtain the more commonly used form of the equation (Bauer and Vinograd 1970; Hsieh and Wang 1975)

$$E = AN\sigma^2 \tag{9}$$

where $A = k_BTb(200h <(\Delta Lk)^2>/n)^{-1}$.

It follows from Figure 2 that the A value depends on N. For $N > 3000$, $A = 10RT$ (if E is measured in cal/mole). The A value increases with decreasing N and becomes three times as large for very small rings. In parallel with these changes, the distribution of superhelix energy between twist and writhe also changes. At large N, only one-third of the total superhelix energy is deposited in twist with the rest being in writhe, whereas at small N the entire energy is in twist.

It should be noted that the above estimations are based on the experimental data corresponding to neutralized DNA charge when the effective diameter of DNA is close to its geometrical value. At intermediate and low ionic strengths, the DNA effective diameter should be significantly larger (Stigter 1977; Brian et al. 1981; Yarmola et al. 1985). The above data indicate that this should lead to a drop of the $<(\Delta Lk)^2>$ value, which in turn should change superhelix energy (see Eqs. 8 and 9). Thus, the A factor in Equation 9 should increase with decreasing ionic strength (Klenin et al. 1989). An experimental verification of this prediction would be very important because in many cases the conformational transitions induced by superhelical stress are studied at low ionic strengths.

Not only the A value but the quadratic functional relation between G and σ may be challenged at high superhelix densities that typically occur in studies of conformational transitions. Unfortunately, this problem has not been studied theoretically because in the Monte Carlo procedure used by Klenin et al. (1989), chains with high Wr were extremely rare. The intriguing problems connected with highly supercoiled molecules require a special treatment and are still awaiting solution. (Note added in proof: We have recently developed a Monte Carlo procedure that makes it possible to simulate highly supercoiled chains and calculate the superhelical energy at high negative supercoiling [K.V. Klenin et al., in prep.].) Thus, two serious questions pertaining to superhelix energy remain unresolved. They are the following: Is the quadratic relation (Eqs. 8 and 9) valid at high superhelical densities, and to what extent is the superhelix energy ionic-strength-dependent? The answers to these questions are especially important for the quantitative analysis of DNA behavior under high superhelical stress, which will be considered in the next section.

III. EFFECTS OF SUPERCOILING ON DNA STRUCTURE

Despite its limitations, the standard expression for superhelix energy (Eq. 9) provides an important insight into the possible effects of DNA supercoiling. It suggests that, at a high superhelical density, considerable energy is accumulated in supercoiled DNA. As a result, one cannot expect the double helix to remain intact as superhelical density increases. Sooner or later, it should be cracked at the weakest site, and this local structural transition should be accompanied by a decrease of superhelical stress in the whole closed circular DNA molecule.

The questions that immediately arise are the following: What are the "weak" sites, and what are the structures they adopt under superhelical stress? I propose to answer these questions at some length in this section. However, before I do that, I have to discuss another problem: How can these local changes be studied?

These structural transitions embrace only a small portion of the DNA molecule. Moreover, the site adopts this structure only under superhelical stress, so one cannot cut it out of the molecule and study it separately. As a result, the whole arsenal of traditional structural methods, such as X-ray diffraction, nuclear magnetic resonance, and circular dichroism are less applicable. Thus, special approaches are required to study the unusual structures adopted by particular sequences under superhelical stress.

A. Methodology

1. Theoretical Studies

When direct physicochemical experiments fail, theory becomes especially important. The theoretical analysis of structural transitions under superhelical stress employs the following equation for the superhelix energy difference due to the transition (Vologodskii and Frank-Kamenetskii 1982, 1984; Peck and Wang 1983; Wang et al. 1983; Frank-Kamenetskii and Vologodskii 1984)

$$\Delta E = AN\,([\sigma + \kappa m/N]^2 - \sigma^2) \qquad (10)$$

Here m is the number of base pairs undergoing the transition into an alternative (non-B) conformation

$$\kappa = \gamma_0(\gamma_0^{-1} - \gamma_a^{-1}) \qquad (11)$$

where γ_0 and γ_a are the numbers of base pairs per turn for the B form and the alternative, respectively.

The most important feature of structural transitions in closed circular DNA, which immediately follows from Equation 10, is an unlimited correlation length encompassing the whole DNA molecule. In other words, structural changes in one site of the molecule affect structural changes in every other site, no matter how far removed along the chain. Thus, a consistent theoretical treatment requires a simultaneous allowance for all possible alternative structures throughout the closed circular DNA molecule. However, in a number of cases, people study specially prepared molecules that carry an insert undergoing a structural transition in the absence of any other transition in the rest of the molecule. In this case, a simple treatment is possible (Frank-Kamenetskii and Vologodskii 1984).

Consider an N-bp-long closed circular DNA carrying an M-bp-long insert that can adopt an alternative structure. The alternative structure is characterized by the parameter κ, the free energy ΔF of the transition of a base pair from B to the alternative form, which is assumed for simplicity to be the same for all M base pairs, and by the energy of nucleation F_n of the alternative structure inside B-DNA. The free energy change due to the transition of m bp (inside the M-bp-long insert) from B to the alternative form is

$$\Phi = \Delta E + m\Delta F + F_n \quad (12)$$

We are considering a transition under superhelical stress, i.e., with increasing negative superhelicity $-\sigma$. The transition point is determined from the condition $\Phi = 0$, which yields

$$-\sigma_{tr} = (m\kappa/2N) + (\Delta F/2A\kappa) + (F_n/2Am\kappa) \quad (13)$$

The dependence of σ_{tr} on m is shown in Figure 3. The m_0 and $\sigma^0{}_{tr}$ values are

$$m_0 = \kappa^{-1}(NF_n/A)^{1/2} \quad (14)$$

$$-\sigma^0{}_{tr} = (\Delta F/2A\kappa) + (F_n/AN)^{1/2} \quad (15)$$

It is clear from Figure 3 that for $M < m_0$ the whole insert of M bp flips into the alternative structure, and the superhelical density of transition is determined by equation 13 with M substituted for m. For $M > m_0$, only m_0 bp undergo the transition. The rest of the insert remains in the B form until the $-\sigma$ value is further increased. In the course of this elongation, the following relation holds true

$$\Delta F + 2A\kappa\,(\sigma + \kappa m/N) = 0 \quad (16)$$

This means that the value $(\sigma + \kappa m/N)$ is constant while the alternative structure propagates over the entire M-bp-long insert.

Although the above simple thermodynamic consideration is very useful in some cases, there is a more exact statistical/mechanical treatment (Anshelevich et al. 1979, 1988; Vologodskii et al. 1979b; Vologodskii and Frank-Kamenetskii 1982, 1984; Peck and Wang 1983).

This treatment has shown that the above equations need a significant correction when the position of the m-bp-long stretch that adopts the alternative structure is not fixed within the M-bp-long insert, as is generally the case (Frank-Kamenetskii and Vologodskii 1984; Vologodskii and Frank-Kamenetskii 1984). When the position is fixed the above equations are reliable.

2. *Two-dimensional Gel Electrophoresis*

Two-dimensional gel electrophoresis has proved to be an extremely powerful experimental technique for studying structural transitions under su-

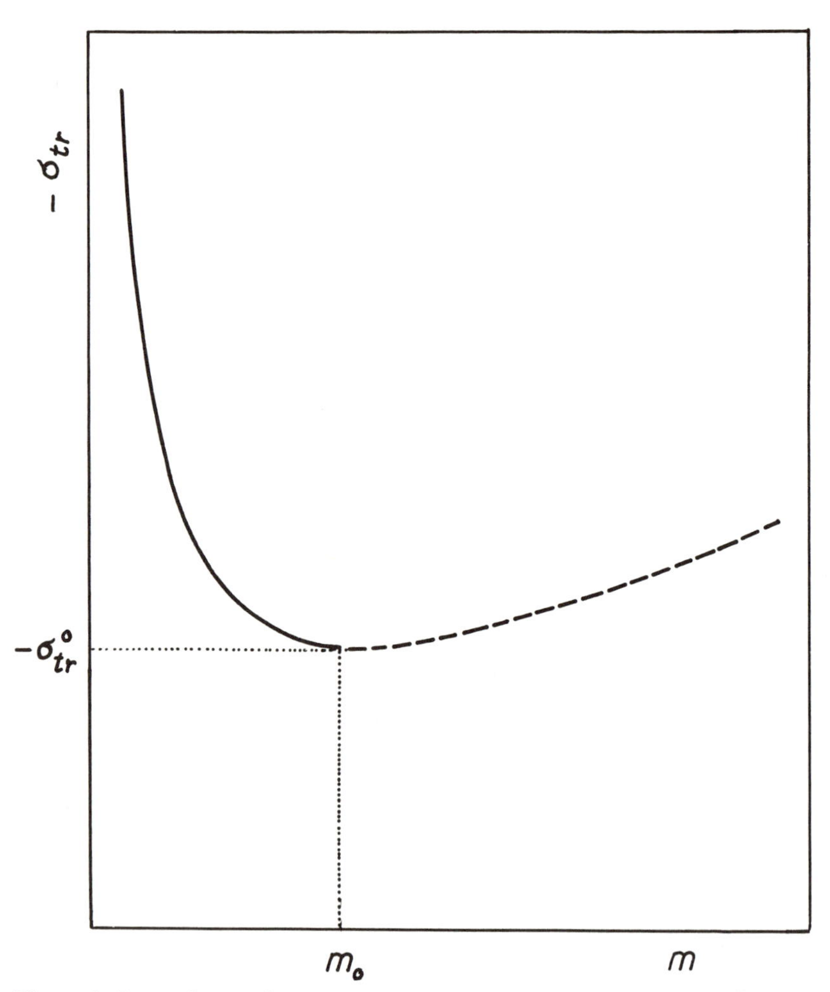

Figure 3 Dependence of σ_{tr} on m according to Equation 13. Values $\sigma^0{}_{tr}$ and m_0 correspond to the minimum.

perhelical stress (Haniford and Pulleyblank 1983; Lyamichev et al. 1983, 1985; Wang et al. 1983; Mirkin et al. 1987a,b). Up to a limit, the mobility in a gel of a closed circular DNA molecule depends on writhe because *Wr* is a measure of the shape of the molecule.

For closed circular DNA, the well-known relation is valid

$$Lk = Tw + Wr$$

If the double helix in the closed circular DNA experiences a structural transition accompanied by a decrease of the *Tw* value ($\delta Tw < 0$), the *Wr* value must be changed accordingly because $\delta Lk = 0$.

Any structural transition induced by negative superhelical stress should then lead to an increase of writhe. Since *Wr* is negative, its absolute value would decrease. The lower the absolute value of *Wr*, the slower the electrophoretic mobility. Thus, a structural transition should be accompanied by a drop in mobility.

To observe the mobility drop, one prepares, using topoisomerases, a sample consisting of DNA molecules with identical sequences but with a wide distribution over topoisomers. Then, the sample is subjected to gel electrophoresis in the first direction under chosen experimental conditions (Fig. 4, molecules run from top to bottom). After the completion of electrophoresis, the molecules are separated in the second direction (from left to right), after soaking the gel in a buffer containing an intercalating dye. The dye reduces negative supercoiling in all topoisomers. As a result, the alternative structures disappear, and the topoisomers move in the second direction according to their *Lk* values. Thus, a structural transition manifests itself in the two-dimensional patterns as the characteristic discontinuity clearly seen in Figure 4.

3. *Enzymatic and Chemical Probing*

Historically, alternative structures in supercoiled DNA were first detected with nucleases. The most commonly used enzyme is the S1 nuclease, which preferentially cleaves single-stranded regions. One should bear in mind that this enzyme is most effectively used at pH 4.5 and in the presence of Zn^{++}. There are single-strand-specific nucleases with neutral and even alkaline pH optima, but they are used only occasionally.

In the early 1980s, S1 nuclease probing at the sequence resolution level provided experimental indications of cruciforms in natural DNAs (Lilley 1980; Panayotatos and Wells 1981), the Z form in plasmids with

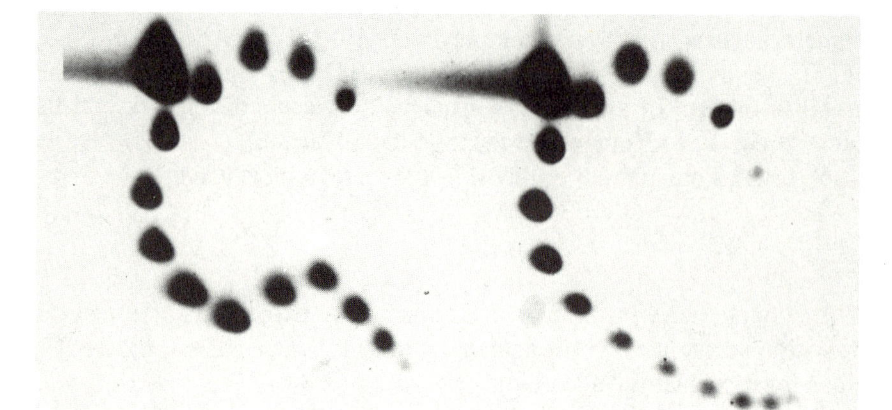

Figure 4 The two-dimensional gel electrophoresis pattern for a plasmid carrying the insert that undergoes a structural transition under superhelical stress (*left*). The right-hand pattern corresponds to the same molecule with a single-base-pair substitution inhibiting the transition over the range of superhelical densities under study. The large spot on both patterns corresponds to nicked molecules.

alternating purine/pyrimidine sequences (Singleton et al. 1982), and showed that something unusual happened in homopurine/homopyrimidine sequences (Hentschel 1982; Larsen and Weintraub 1982; Nickol and Felsenfeld 1983; Schon et al. 1983; Htun et al. 1984). The popularity of the method has declined in the mid-1980s, and it has been displaced by the much more informative techniques of two-dimensional gel electrophoresis and chemical probing.

Chemical probing is currently the most popular way of studying alternative structures formed under superhelical stress. This method came to the forefront in the mid-1980s when it was used successfully to detect Z form and cruciforms in supercoiled DNA (Lilley and Palecek 1984; Herr 1985; Johnston and Rich 1985; Furlong and Lilley 1986; Scholten and Nordheim 1986).

To apply this method, one subjects the plasmid carrying the insert under study to conditions favoring the formation of an alternative structure. Then a chemical agent is added that reacts with particular sites of the bases. A large arsenal of chemical probes with different specificities is available. The most popular ones are dimethylsulfate (DMS), which modifies guanines in both the double helix and in single-stranded regions, diethyl pyrocarbonate (DEP), which modifies primarily adenines in the single-stranded state and in the Z form but does not react with them in B-DNA, and osmium tetroxide (OT), which reacts with thymines in single strands. There are a number of less-specific probes, such as chloroacetaldehyde (Kohwi and Kohwi-Shigimatsu 1988) and potassium permanganate (Lyamichev et al. 1989).

To reveal the sites of modification at the sequence resolution, one uses one of two standard methods borrowed from the DNA-sequencing techniques. Most use the Maxam-Gilbert protocol, which consists of cutting out a convenient fragment with restriction enzymes, end-labeling it, treating the fragment with piperidine to cut it at modified sites, and separating the mixture thus obtained in a polyacrylamide gel saturated with urea (Herr 1985; Johnston and Rich 1985; Furlong and Lilley 1986; Scholten and Nordheim 1986; Hanvey et al. 1988a,b; Johnston 1988a,b; Kohwi and Kohwi-Shigimatsu 1988; Vojtiskova et al. 1988; Voloshin et al. 1988).

Very good results are also achieved by the Sanger-type approach when the insert under study is placed between two promoters with opposite orientation. RNA polymerase synthesizes labeled RNA chains from both strands. When it encounters a modified base, it stops. Polyacrylamide gel electrophoresis (PAGE) of the RNA molecules thus obtained localizes the sites of chemical modification of the alternative structure (Htun and Dahlberg 1988).

One should bear in mind the sensitivity of different methods to detect alternative structures. Two-dimensional gel electrophoresis is less sensitive than enzymatic and chemical probing because it requires that a considerable fraction of the DNA molecules in the sample carries the alternative structure. The same is true for DMS treatment. In contrast, probing by S1 nuclease, DEP, OT, and so forth, gives a positive result, even if only a tiny portion of DNA molecules in the sample has transiently adopted the alternative structure, because only cut molecules (or abortive transcripts) are detected when the radiolabel is used.

B. Unusual Structures in Supercoiled DNA

1. Cruciforms in Inverted Repeats

Sufficiently long inverted repeats adopt cruciform structures under superhelical stress. This was predicted theoretically (Hsieh and Wang 1975; Vologodskii et al. 1979b; Vologodskii and Frank-Kamenetskii 1982) and observed experimentally by enzymatic probing (Lilley 1980; Panayotatos and Wells 1981), two-dimensional gel electrophoresis (Courey and Wang 1983; Lyamichev et al. 1983), and chemical probing (Lilley and Palecek 1984; Furlong and Lilley 1986; Scholten and Nordheim 1986). Very large cruciforms were observed by electron microscopy (Gellert et al. 1979; Borst et al. 1984).

The thermodynamics of cruciform extrusion is determined almost exclusively by the nucleation energy, F_n, which is estimated to be 15–20 kcal/mole (Courey and Wang 1983; Panyutin et al. 1984; Naylor et al. 1986). For cruciforms $\kappa = 1$, it follows from Equation 14 that, for the commonly used plasmids of the pUC family with $N \approx 3000$, $m_0 = 100$.

Probably the most curious thing about cruciforms is their very slow kinetics. This was first noticed by Mizuuchi et al. (1982) for the case of their giant artificial inverted repeat. Then the very slow extrusion of cruciforms was demonstrated for shorter inverted repeats including natural ones (Courey and Wang 1983; Gellert et al. 1983; Lilley and Markham 1983; Lyamichev et al. 1983).

Vologodskii and Frank-Kamenetskii (1983) explained theoretically the slow relaxation of cruciforms. The overall process is described by first order kinetics

$$\text{B-DNA} \underset{k_2}{\overset{k_1}{\rightleftarrows}} \text{cruciform}$$

The most probable pathway for both the extrusion and the decay of a cruciform is via branch migration. By themselves, each step of a branch migration is energetically neutral since 2 bp merely change branches. As a result, branch migration is actually a random-walk process, as was first demonstrated by Thompson et al. (1976) for the free-ended cross. In negatively supercoiled DNA, the decay of a cruciform is a random walk against the gradient of superhelix energy. As a result, the decay rate constant k_2 increases exponentially with negative superhelical density. An accurate analysis shows (Vologodskii and Frank-Kamenetskii 1983) that the relaxation time

$$\tau = (k_1 + k_2)^{-1}$$

reaches its maximum at the superhelical density of the equilibrium transition point for the inverted repeat under study.

Our theoretical predictions have been completely confirmed in comprehensive experimental studies by Panyutin et al. (1984) and Courey and Wang (1988). In these studies, thermodynamically nonequilibrium DNA samples were prepared that did not carry any cruciform. These samples were incubated for different times at a given temperature, and their two-dimensional gel electrophoresis patterns were quantitatively studied at a low temperature.

Lilley and his colleagues have shown that inverted repeats highly enriched by A·T bp (such as $[AT]_n$) and normal inverted repeats embedded in regions highly enriched by A·T bp differ in relaxation behavior. They believe that in these cases cruciform extrusion does not occur via branch migration but rather through a large bubble as an intermediate state (for a concise review of these studies, see Lilley 1988).

2. *Z-form Extrusion under Superhelical Stress*

Negative supercoiling especially favors the formation of the left-handed Z form in DNA. Alternating purine/pyrimidine inserts ($[GC]_n$ and $[AC]_n$ but not $[AT]_n$, which form the cruciform rather than the Z form) undergo a structural transition into the Z form under superhelical stress. This has been shown by two-dimensional gel electrophoresis (Haniford and Pulleyblank 1983; Wang et al. 1983), anti-Z-antibody binding (Rich et al. 1984), and enzymatic (Singleton et al. 1982) and chemical (Herr 1985; Johnston and Rich 1985) probing. Since $\kappa = 1.8$ for the B–Z transition (Rich et al. 1984; see also Drew et al., this volume) and the nucleation energy F_n is in this case smaller than in the case of the cruciform, one could expect on the basis of Equation 14 comparatively low m_0 values. Indeed, Haniford and Pulleyblank (1983) observed the behavior that follows from Equations 14–16 for the $(AC)_{30}$ $(GT)_{30}$ stretch inserted into plasmid DNA. Theoretical treatment of these data (Frank-Kamenetskii and Vologodskii 1984; Vologodskii and Frank-Kamenetskii 1984) made it possible to estimate some of the energy parameters of the B–Z transition.

Two-dimensional gel electrophoresis supplemented by DNA tailoring made it possible to study the energetics of the B–Z transition in an arbitrary DNA sequence in quantitative terms (Ellison et al. 1985, 1986; Ho et al. 1986; Mirkin et al. 1987b). It turned out that the B–Z transition is described by the following energy parameters:

1. The energy difference between the B and Z forms for a base pair in the favorable conformation; purine in *syn* and pyrimidine (in the opposite strand) in *anti*. There are separate parameters for A·T and G·C base pairs, $\Delta F^{I}{}_{A}$ and $\Delta F^{I}{}_{G}$, respectively.
2. The energy difference between the B and Z forms for a base pair in the unfavorable conformation: purine in *anti* and pyrimidine (in the opposite strand) in *syn*. The parameters for A·T and G·C base pairs, $\Delta F^{II}{}_{A}$ and $\Delta F^{II}{}_{G}$, respectively.
3. The B–Z junction energy $F^{BZ}{}_{j} = F_n/2$.
4. The Z–Z junction energy $F^{ZZ}{}_{j}$, which appears every time when the chase of *syn–anti* alternation within the Z-form stretch changes.

The meaning of the energy parameters is explained in the following example:

nucleotide conformation in a selected strand	B	Z^s	Z^a	Z^s	Z^s	Z^a	Z^s	Z^a	B
sequence →	A	G	C	G	G	T	C	C	A
free energy of base pairs		ΔF^I_G	ΔF^I_G	ΔF^I_G	ΔF^I_G	ΔF^I_A	ΔF^{II}_G	ΔF^I_G	
free energy of junction	F^{BZ}_j				F^{ZZ}_j				F^{BZ}_j

Values for the various parameters are given in Table 1. Anshelevich et al. (1988) used the parameter values in Table 1 to calculate the probability of forming Z forms and cruciforms using a rigorous statistical/mechanical treatment. Some of the results are presented in Figure 5. Anshelevich et al. (1988) actually allowed for three DNA conformations: open (melted) base pairs, cruciform, and Z form. However, calculations showed that in ordinary DNA sequences the open state cannot compete

Table 1 Energy parameters of the B–Z transition

Parameter	Value in kcal/mole	Reference
ΔF^{I}_{G}	0.33	Peck and Wang (1983)
ΔF^{I}_{A}	1.15	Mirkin et al. (1987b)
ΔF^{II}_{G}	2.6	Ellison et al. (1985)
ΔF^{II}_{A}	3.6	Ellison et al. (1985)
F^{BZ}_{j}	5.2	Peck and Wang (1983)
F^{ZZ}_{j}	4.0	Mirkin et al. (1987b)

The parameter values are given for the tributyl phosphate buffer (Mirkin et al. 1987b).

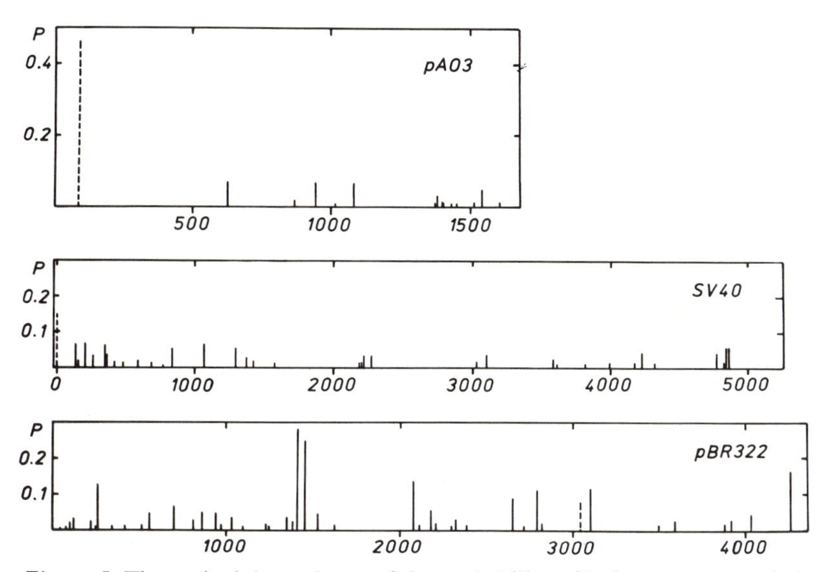

Figure 5 Theoretical dependence of the probability of Z-form segments (—) and cruciforms (- - -) on the nucleotide position for pAO3, SV40, and pBR322 DNA. The calculations corresponded to $\sigma = -0.06$ (Anshelevich et al. 1988).

with the other two conformations in natural DNA sequences under normal ambient conditions (room temperature and neutral pH).

For sequences such as $(GC)_n$ and $(AT)_n$, which can adopt both the Z conformation and the cruciform, theory predicts that their actual structure should depend on superhelical density and *n* value (see Fig. 6). These theoretical predictions await experimental verification.

3. *Protonated Structures in Supercoiled DNA*

The cruciform and the Z form were the structures that people had expected to observe in supercoiled DNA. Their actual observation has made it possible to study their energetics, kinetics, and other quantitative features, but it has not dramatically challenged our knowledge about the potential DNA structures.

A quite different situation arose with respect to a family of pH-dependent structures. They were discovered unexpectedly and then thoroughly studied using the whole arsenal of methods described above. The first sign that something unusual happened with homopurine/homopyrimidine sequences under superhelical stress was the increased sensitivity to S1 nuclease (Hentschel 1982; Larsen and Weintraub 1982; Nickol and Felsenfel 1983; Schon et al. 1983; Htun et al. 1984).

This stimulated a lot of speculation about the possible structural basis of the S1 hypersensitivity (Cantor and Efstratiadis 1984; Htun et al. 1984; Lee et al. 1984; Pulleyblank et al. 1985; for review, see Wells et al. 1988). At this stage, however, most people overlooked the fact that the S1 nuclease worked under acidic conditions and that the unusual structure it detected could be stabilized by protons.

Lyamichev et al. (1985) tackled the problem with the aid of two-dimensional gel electrophoresis. Our results carried three important messages. First, plasmid with homopurine/homopyrimidine insert $(AG)_n \cdot (CT)_n$ showed clear-cut discontinuities in two-dimensional gel electrophoresis patterns, and this indicated that the sequence actually adopted an alternative structure. Secondly, the superhelical density of the transition, σ_{tr}, proved to be strongly pH-dependent. Thirdly, the mobility drop was pH-independent and corresponded to that expected for non-interwound complementary strands throughout the $(AG)_n \cdot (CT)_n$ insert.

These findings led Lyamichev et al. (1985) to the conclusion that there was a structural transition into a DNA conformation unknown before. This proton-stabilized structure was termed the H form. We also

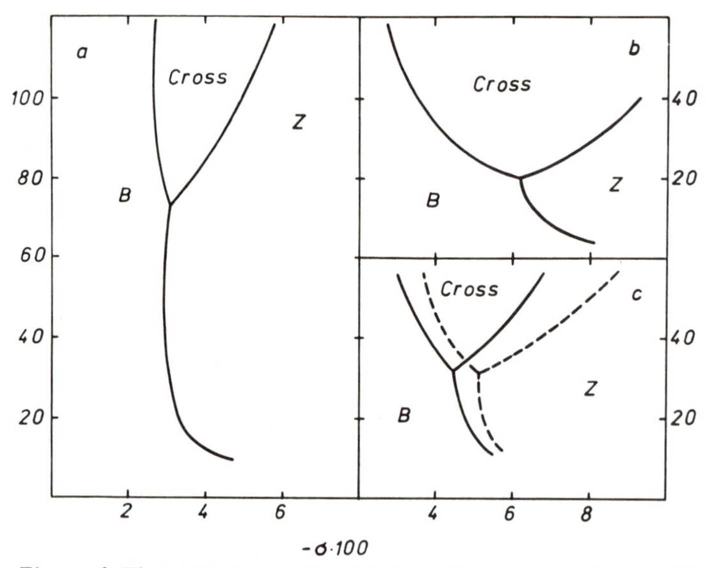

Figure 6 Theoretically predicted "phase diagrams" for inserts $(GC)_n$ (*a*), $(AT)_n$ (*b*), and $(CATG)_n$ (*c*) in 4000-bp-long supercoiled DNA (Anshelevich et al. 1988). The ordinate is the total insert length. The dashed line corresponds to 2100-bp-long DNA.

presented a theoretical treatment of the B–H transition in supercoiled DNA. This treatment corresponds to the above consideration (see Eq. 13) with

$$\Delta F = \Delta F_0 - RT \ln (1 + 10^{pK-pH})/r \tag{17}$$

Here r is the number of base pairs per protonation site in the insert, the pK value corresponds to the protonation site in the H form, ΔF_0 is the free energy per base pair of the insert for the formation of the unprotonated H form (the same structure with unoccupied proton-binding sites). Substituting Equation 17 for ΔF in Equation 13 and assuming that pH < pK, one obtains (Lyamichev et al. 1985, 1987)

$$\sigma_{tr} = (2.3RT/2A\kappa r)(pH_0 - pH) \tag{18}$$

The two-dimensional gel electrophoresis data indicated that $\kappa = 1$. As a result, the slope of the dependence of σ_{tr} on pH directly yielded the r value, which proved to be equal to 4, i.e., 1 protonation site/4 bp of the insert. This conclusion was astonishing since the sequence under study, $(GA)_n \cdot (TC)_n$, had a 2-bp repeat.

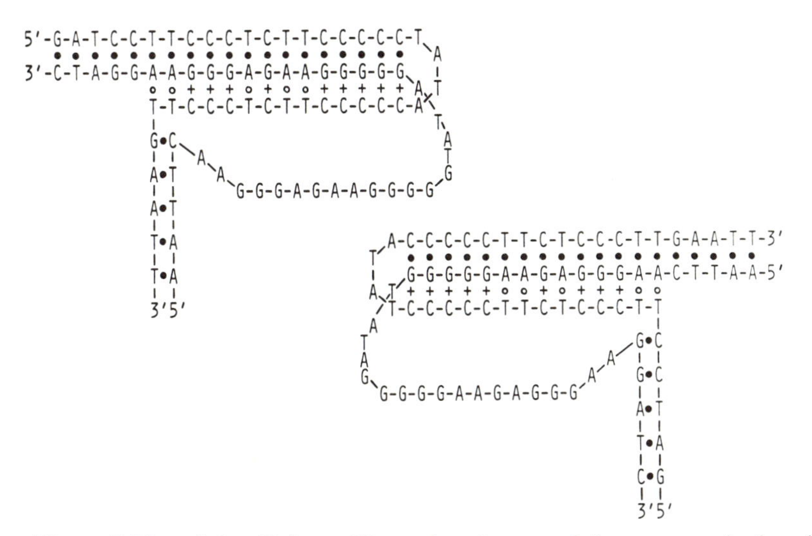

Figure 7 The triplex H form. The major element of the structure is the triplex that includes the Watson-Crick (•) duplex associated with a pyrimidine loop through Hoogsteen base pairing (o and +) (see Fig. 8). The two possible isomeric forms are shown. The data on chemical modification indicate that the upper conformation usually prevails.

In an attempt to solve this riddle, Lyamichev et al. (1986) arrived at the triplex H form (Fig. 7) as the most attractive hypothesis. This model easily explained the S1 hypersensitivity of homopurine/homopyrimidine inserts as a consequence of two single-stranded loops. The pH-dependence was explained in terms of the protonated CGC^+ base triads (Fig. 8). There are only two isomorphous base triads, CGC^+ and TAT (Fig. 8). This fact immediately led to the conclusion that, if the proposed model of the H form were correct, one would arrive at a very distinct sequence requirement for the H form. Namely, the H form required the homopurine/homopyrimidine mirror repeat, termed the H palindrome (Lyamichev et al. 1987). Since the central part of the homopurine/homopyrimidine tract is looped out in the H form (see Fig. 7), an H palindrome may carry a nonpalindromic (and even nonhomopurine/nonhomopyrimidine) insert in the middle.

To test this prediction, Mirkin et al. (1987a) designed and cloned in plasmids a series of sequences

5′-AAGGGAGAAXGGGGTATAGGGGYAAGAGGGAA-3′

where X and Y are either A or G. We expected supercoiling to readily induce a transition to the H form for X = Y, but the transition would be much more difficult or impossible for $X \neq Y$. This was exactly what

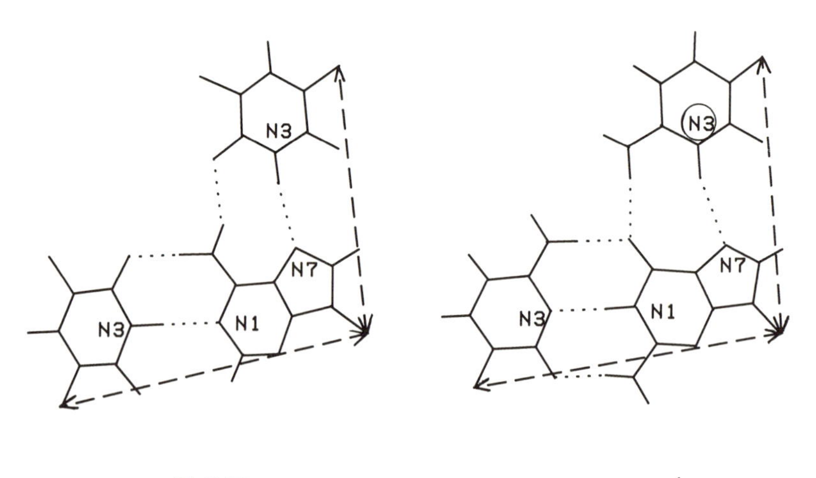

Figure 8 The TAT and CGC^+ base triads. They consist of the canonical Watson-Crick A·T and G·C base pairs with attached thymine and cytosine by the Hoogsteen pairing. The protonated site is encircled. The two base-triads are isomorphous; the respective distances between glycosyl bonds (→) are equal.

Mirkin et al. (1987a) observed with the help of two-dimensional gel electrophoresis. The inserts exhibited facile transitions into the H form for X = Y = G or X = Y = A, whereas the transition was much more difficult or impossible for the two nonpalindromes (X = A and Y = G or X = G and Y = A). The data of Mirkin et al. (1987a) therefore provided evidence that the H palindrome is the sequence requirement for the H conformation of DNA and left little if any doubt that the triplex model of the H form (Fig. 7) is correct.

The pK value of the CGC^+ base triad is between 7 and 8 (Howard et al. 1964; Lee et al. 1979). That is why Equation 18 fails in the alkaline pH region. Above pH 8, the σ_{tr} value becomes pH-independent (see Eq. 17). Under these conditions, one observes the triplex H form with unprotonated CGC base triad, which is stabilized exclusively by superhelical stress. This was experimentally observed for a very long homopurine/homopyrimidine insert (V.I. Lyamichev, unpubl.). This also explains the data of Evans and Efstratiadis (1986) who observed the hypersensitivity of a very long homopurine/homopyrimidine insert to the nuclease that cleaved single-stranded DNA at pH 9. Evans and Efstratiadis (1986) erroneously interpreted these data as a strong argument against an alternative structure stabilized by protons as a possible explanation of the single-stranded endonuclease hypersensitivity.

Chemical probing experiments with DEP, DMS, OT, and chloroacetaldehyde have confirmed completely the triplex H model (Hanvey et al. 1988a,b; Htun and Dahlberg 1988; Johnston 1988b; Kohwi and Kohwi-Shigimatsu 1988; Vojtiskova et al. 1988; Voloshin et al. 1988). They have shown that of the two possible isomeric structures in Figure 7, the one with the triplex at the 3′ end of the purine strand usually prevails (in this connection, see the recent paper by Htun and Dahlberg 1989).

Kohwi and Kohwi-Shigimatsu (1988) discovered that, in the presence of magnesium cations, the $(G)_n \cdot (C)_n$ insert adopts an alternative structure with the triplex formed by the unprotonated CGG base triads. Without magnesium, Kohwi and Kohwi-Shigimatsu (1988) observed a modification pattern completely consistent with the "canonical" H form in full agreement with the two-dimensional gel electrophoresis data of Lyamichev et al. (1987). The available data indicate that the "noncanonical" structure with the CGG base triads is observed only for homogeneous $(G)_n \cdot (C)_n$ tracts under special ambient conditions, whereas the canonical H form with the CGC^+ and TAT base triads is typical of an arbitrary H palindrome. Thus, the triplex H form is the first DNA structure discovered and studied exclusively within supercoiled DNA using the whole arsenal of special techniques described in the Methodology section of this chapter.

We have recently applied the same approach to the study of a structure that may be formed under superhelical stress by another type of sequence: a telomeric sequence (Lyamichev et al. 1989). We have chosen the typical motif $(G_4T_2)\cdot(A_2C_4)$ of *Tetrahymena* telomers (Weiner 1988).

As in the homopurine/homopyrimidine experiment, the first signal came from enzymatic probing (Budarf and Blackburn 1987). Two-dimensional gel electrophoresis revealed a strongly pH-dependent structural transition in the telomeric insert and indicated that in the structure under study the two complementary strands are not interwound (Lyamichev et al. 1989), just as in the case of the H form. However, it seemed very unlikely that this alternative structure was the triplex H form. Chemical modification experiments substantiated the doubts. In contrast with the case of homopurine/homopyrimidine inserts, the modification patterns for the telomeric insert proved to be symmetrical for both strands. We concluded that in the structure we observed, the C-rich strand forms a hairpin stabilized by non-Watson-Crick base pairs $C\cdot C^+$ and $A\cdot A^+$. We termed this novel DNA structure the (CA)-hairpin (see Fig. 9). The C-rich strand in the (CA)-hairpin is modified preferentially in the middle, indicating the the $C\cdot C^+$ and $A\cdot A^+$ pairs are stable at low pH. The protonated $C\cdot C^+$ and $A\cdot A^+$ pairs make the (CA)-hairpin stable at low pH. The G-rich strand is exclusively modified almost throughout the insert, indicating that it is virtually unstructured.

The (CA)-hairpin is stabilized primarily by $C\cdot C^+$ base pairs. Such pairs may be formed between both parallel and antiparallel strands (Fig. 10) under acidic conditions (Gray et al. 1987). Although several protonated pairing schemes of $A\cdot A^+$ base pairs have been proposed (Cantor and Schimmel 1980), we could not find in the literature an $A\cdot A^+$ pair isomorphous to the antiparallel $C\cdot C^+$ pair. Such an isomorphous $A\cdot A^+$ pair, as proposed by us, is shown in Figure 10. We assumed that protonation of the N1 site, which does not participate in the hydrogen bonding in one adenine, would stabilize the pair by favoring the formation of the hydrogen bond between C8 of the same adenine and N1 of the other one.

The asymmetry of two hairpins of the (CA)-hairpin was tested in a series of experiments on binding of the labeled oligonucleotides $C_3A_2C_4A_2C$ and $(G_4T_2)_2$ to a plasmid carrying the telomeric insert (Lyamichev et al. 1989). Neither oligonucleotide bound to nonsuperhelical DNA. However, $C_3A_2C_4A_2C$ but not $(G_4T_2)_2$ formed a strong complex with superhelical DNA at low pH. These data support our model of a stably base-paired, C-rich hairpin and a virtually unstructured G-rich strand.

The data supporting the (CA)-hairpin model are not absolutely unequivocal. Specifically, our data do not exclude the formation of an

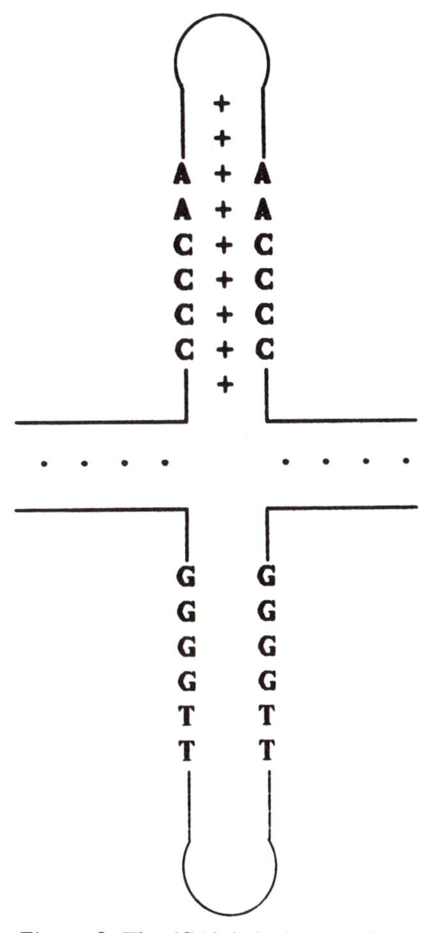

Figure 9 The (CA)-hairpin. The C-rich strand forms a hairpin stabilized by the noncanonical $C{\cdot}C^+$ and $A{\cdot}A^+$ base pairs (see Fig. 10). The available data indicate that the G-rich strand is unstructured.

altered triplex H form. If in such a noncanonical triplex, the two isomeric forms are present in comparable amounts, the symmetrical pattern of chemical modification would be explained. Thus, whereas for homopurine/homopyrimidine sequences the nature of the H form is well understood, establishment of the protonated structure adopted by telomeric sequences under superhelical stress requires further study.

IV. CONCLUDING REMARKS

Supercoiling is a powerful factor that forces the double helix to acquire structural forms it is reluctant to adopt. Do these structures have any

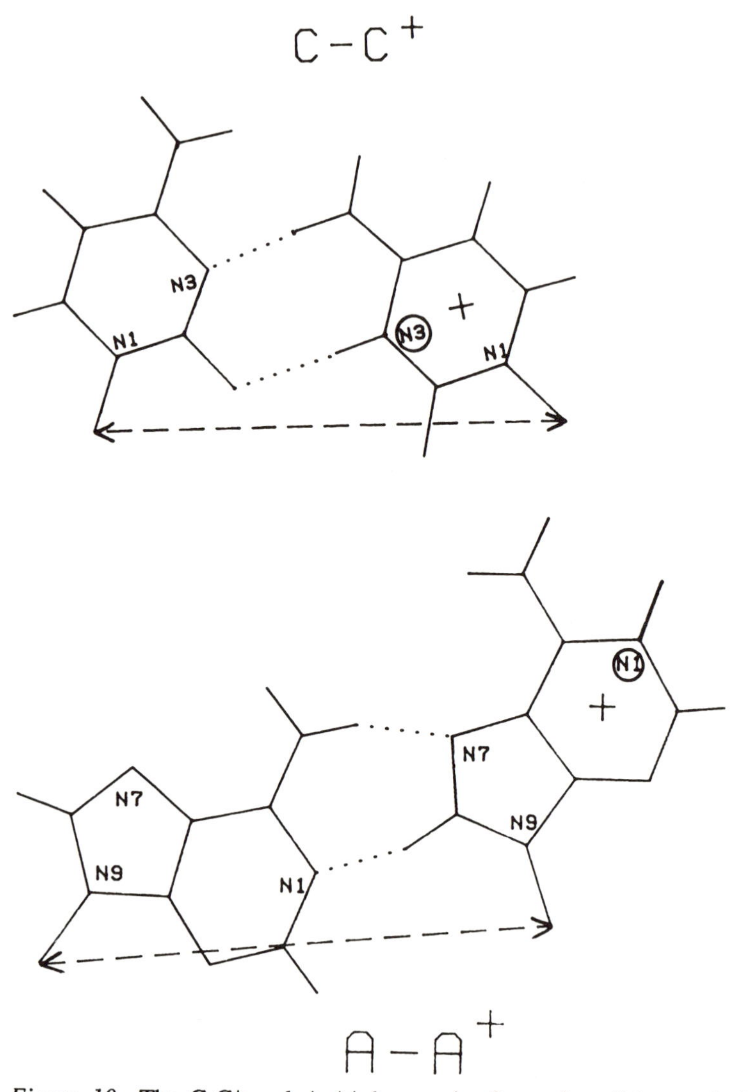

Figure 10 The C·C+ and A·A+ base pairs for antiparallel strands. The protonated sites are encircled. The base pairs are almost isomorphous (they have close distances between glycosyl bonds [→]).

physiological significance? This crucial question is deliberately left open. As is mentioned in the Introduction, the attitude toward this problem has changed dramatically after the discovery of transcriptional waves of supercoiling by J.C. Wang, L.F. Liu, and their co-workers. In my opinion, these findings made obsolete most of the approaches formerly used to uncover the biological role of unusual DNA structures.

After all, if such a fundamental phenomenon as transcriptional activation of supercoiling remained undisclosed until very recently, what do we know about the DNA structure inside the cell?

So far, we have had more than enough speculation, beliefs, and hints. I am sure that now many researchers are at the starting point of studies that will lead, within a couple of years, to dramatic progress in our knowledge of the biological implications of unusual DNA structures.

ACKNOWLEDGMENTS

I thank Yurii Lazurkin and Alex Vologodskii for a critical reading of the paper and Dmitrii Agrachev for some editorial suggestions.

REFERENCES

Anshelevich, V.V., A.V. Vologodskii, and M.D. Frank-Kamenetskii. 1988. A theoretical study of formation of DNA noncanonical structures under negative superhelical stress. *J. Biomol. Struct. Dyn.* **6:** 247–259.

Anshelevich, V.V., A.V. Vologodskii, A.V. Lukashin, and M.D. Frank-Kamenetskii. 1979. Statistical-mechanical treatment of violations of the double helix in supercoiled DNA. *Biopolymers* **18:** 2733–2744.

Bauer, W. and J. Vinograd. 1970. The interaction of closed circular DNA with intercalative dyes. II. The free energy of superhelix formation in SV40 DNA. *J. Mol. Biol.* **47:** 419–435.

Benham, C.A. 1978. The statistics of superhelicity. *J. Mol. Biol.* **123:** 361–370.

Borst, P., J.P. Overdulve, P.J. Weijers, F. Fase-Fowler, and M. van der Berg. 1984. DNA circles with cruciforms from *Isospora (Toxoplasma) gondii*. *Biochem. Biophys. Acta* **781:** 100–111.

Brian, A.A., H.L. Fritsch, and L.S. Lerman. 1981. Thermodynamics and equilibrium sedimentation analysis of the close approach of DNA molecules and a molecular ordering transition. *Biopolymers* **20:** 1305–1328.

Budarf, M. and E. Blackburn. 1987. S1-nuclease sensitivity of a double-stranded telomeric DNA sequence. *Nucleic Acids Res.* **15:** 6273–6292.

Cantor, C.R. and A. Efstratiadis. 1984. Possible structures of homopurine-homopyrimidine S1-hypersensitive sites. *Nucleic Acids Res.* **12:** 8059–8072.

Cantor, C.R. and P.R. Schimmel. 1980. *Biophysical chemistry*. W.H. Freeman, San Francisco.

Courey, A. and J.C. Wang. 1983. Cruciform formation is a negatively supercoiled DNA may be kinetically forbidden under physiological conditions. *Cell* **33:** 817–829.

———. 1988. Influence of DNA sequence and supercoiling on the process of cruciform formation. *J. Mol. Biol.* **202:** 35–43.

Depew, R.E. and J.C. Wang. 1975. Conformational fluctuations of DNA helix. *Proc. Natl. Acad. Sci.* **72:** 4275–4279.

Ellison, M.J., J. Feigon, R.J. Kelleher III, A.H.-J. Wang, J.F. Habener, and A. Rich. 1985. Sequence dependent energetics of the B-Z transition in supercoiled DNA containing non-alternating purine-pyrimidine sequences. *Biochemistry* **24:** 8320–8324.

———. 1986. An assessment of the Z-DNA forming potential of alternating dA-dT stretches in supercoiled plasmids. *Biochemistry* **25:** 3648–3655.

Evans, T. and A. Efstratiadis. 1986. Sequence-dependent S1 nuclease hypersensitivity of a heteronomous DNA duplex. *J. Biol. Chem.* **261:** 14771–14778.

Frank-Kamenetskii, M.D. and A.V. Vologodskii. 1984. Thermodynamics of the B-Z transition in superhelical DNA. *Nature* **307:** 481–482.

Frank-Kamenetskii, M.D., A.V. Lukashin, V.V. Anshelevich, and A.V. Vologodskii. 1985. Torsional and bending rigidity of the double helix from data on small DNA rings. *J. Biomol. Struct. Dyn.* **2:** 1005–1012.

Furlong, F.C. and D.M.J. Lilley. 1986. Highly selective chemical modification of cruciform loops by diethyl pyrocarbonate. *Nucleic Acids Res.* **14:** 3995–4007.

Gellert, M., M.H. O'Dea, and K. Mizuuchi. 1983. Slow cruciform transitions in palindromic DNA. *Proc. Natl. Acad. Sci.* **80:** 5545–5549.

Gellert, M., K. Mizuuchi, M.H. O'Dea, H. Ohmori, and J. Tomizawa. 1979. DNA gyrase and DNA supercoiling. *Cold Spring Harbor Symp. Quant. Biol.* **43:** 35–40.

Giaever, G.N. and J.C. Wang. 1988. Supercoiling of intracellular DNA can occur in eucaryotic cells. *Cell* **55:** 849–856.

Gray, D.M., R.L. Ratliff, V.P. Antao, and C.W. Gray. 1987. CD spectroscopy of acid-induced structures of polydeoxyribonucleotides: Importance of C C^+ base pairs. In *Structure and expression*, vol. 2: *DNA and its drug complexes* (ed. R.H. Sarma and M.H. Sarma), p. 147–166. Adenine Press, New York.

Hagerman, P.J. 1981. Investigation of the flexibility of DNA using transient electric birefringenece. *Biopolymers* **20:** 1503–1535.

Haniford, D.B. and D.E. Pulleyblank. 1983. Facile transition of poly[d(TG) d(CA)] into a left-handed helix in physiological conditions. *Nature* **302:** 632–634.

Hanvey, J.C., J. Klysik, and R.D. Wells. 1988a. Influence of DNA sequence on the formation of non-B right-handed helices in oligopurine-oligopyrimidine inserts in plasmids. *J. Biol. Chem.* **263:** 7386–7396.

Hanvey, J.C., M. Shimizu, and R.D. Wells. 1988b. Intramolecular DNA triplexes in supercoiled plasmids. *Proc. Natl. Acad. Sci.* **85:** 6292–6296.

Hentschel, C.C. 1982. Homocopolymer sequences in the spacer of sea urchin histone genes repeat are sensitive to S_1 nuclease. *Nature* **295:** 714–716.

Herr, W. 1985. Diethyl pyrocarbonate: A chemical probe for secondary structure in negatively supercoiled DNA. *Proc. Natl. Acad. Sci.* **82:** 8009–8013.

Ho, P.S., M.J. Ellison, G.J. Quigley, and A. Rich. 1986. A computer aid thermodynamic approach for predicting the formation of Z-DNA in naturally occurring sequences. *EMBO J.* **5:** 2737–2744.

Horowitz, D.S. and J.C. Wang. 1984. The torsional rigidity of DNA and the length dependence of the free energy of DNA supercoiling. *J. Mol. Biol.* **173:** 75–91.

Howard, F.B., F. Frazier, M.N. Lipsett, and H.T. Miles. 1964. Infrared demonstration of two- and three-strand helix formation between poly C and quanosine mononucleotides and oligonucleotides. *Biochem. Biophys. Res. Commun.* **17:** 93–102.

Hsieh, T.-S. and J.C. Wang. 1975. Thermodynamic properties of superhelical DNAs. *Biochemistry* **14:** 527–535.

Htun, H. and J.E. Dahlberg. 1988. Single strands, triple strands, and kinks in H-DNA. *Science* **241:** 1791–1796.

———. 1989. Topology and formation of triple-stranded H-DNA. *Science* **243:** 1571–1576.

Htun, H., E. Lund, and J.E. Dahlberg. 1984. Human U1 RNA genes contain an unusually sensitive nuclease S1 cleavage site within the conserved 3′ flanking region. *Proc. Natl. Acad. Sci.* **81:** 7288–7292.

Johnston, B.H. 1988a. Chemical probing of the B-Z transition in negatively supercoiled

DNA. *J. Biomol. Struct. Dyn.* **6:** 153–166.

———. 1988b. The S1-sensitive form of d(C-T)$_n$ d(A-G)$_n$: Chemical evidence for a three-stranded structure in plasmids. *Science* **241:** 1800–1804.

Johnston, B.H. and A. Rich. 1985. Chemical probes of DNA conformation: Detection of Z-DNA at nucleotide resolution. *Cell* **42:** 713–724.

Klenin, K.V., A.V. Vologodskii, V.V. Anshelevich, A.M. Dykhne, and M.D. Frank-Kamenetskii. 1988. Effect of excluded volume on topological properties of circular DNA. *J. Biomol. Struct. Dyn.* **5:** 1173–1185.

Klenin, K.V., A.V. Vologodskii, V.V. Anshelevich, V.Y. Klishko. A.M. Dykhne, and M.D. Frank-Kamenetskii. 1989. Variance of writhe for wormlike DNA rings with excluded volume. *J. Biomol. Struct. Dyn.* **6:** 707–714.

Kohwi, Y. and T. Kohwi-Shigimatsu. 1988. Magnesium ion-dependent triplex-helix structure formed by homopurine-homopyrimidine sequences in supercoiled plasmid. *Proc. Natl. Acad. Sci.* **85:** 3781–3785.

Larsen, A. and H. Weintraub. 1982. An altered DNA conformation detected by S1 nuclease occurs at specific regions in active chick globin chromatin. *Cell* **29:** 609–622.

Le Bret, M. 1980. Monte Carlo computation of supercoiling energy, the sedimentation constant, and the radius of gyration of unknotted and knotted circular DNA. *Biopolymers* **19:** 619–637.

Lee, J.S., D.A. Johnson, and A.R. Morgan. 1979. Complexes formed by (pyrimidine)$_n$ (purine)$_n$ DNAs in lowering the pH are three-stranded. *Nucleic Acids Res.* **6:** 3073–3091.

Lee, J.S., M.L. Woodsworth, L.J.P. Latimer, and A.R. Morgan. 1984. Poly(pyrimidine) poly(purine) synthetic DNAs containing S-methylcytosine form stable triplexes at neutral pH. *Nucleic Acids Res.* **12:** 6603–6614.

Levene, S.D. and D.M. Crothers. 1986. Topological distributions and the torsional rigidity of DNA. A Monte Carlo study of DNA circles. *J. Mol. Biol.* **189:** 73–83.

Lilley, D.M.J. 1980. The inverted repeat as a recognizable structural feature in supercoiled DNA molecules. *Proc. Natl. Acad. Sci.* **77:** 6468–6472.

———. 1988. DNA opens up—Supercoiling and heavy breathing. *Trends Genet.* **4:** 111–114.

Lilley, D.M.J. and A.F. Markham. 1983. Dynamics of cruciform extrusion in supercoiled DNA: Use of a synthetic inverted repeat to study conformational populations. *EMBO J.* **2:** 527–533.

Lilley, D.M.J. and E. Palecek. 1984. The supercoil-stabilized cruciform of ColE1 is hyper-reactive to osmium tetroxide. *EMBO J.* **3:** 1187–1192.

Lyamichev, V.I., S.M. Mirkin, and M.D. Frank-Kamenetskii. 1985. A pH-dependent structural transition in the homopurine-homopyrimidine tract in superhelical DNA. *J. Biomol. Struct. Dyn.* **3:** 327–338.

———. 1986. Structures of homopurine-homopyrimidine tracts in superhelical DNA. *J. Biomol. Struct. Dyn.* **3:** 667–669.

———. 1987. Structure of (dG)$_n$ (dC)$_n$ under superhelical stress and acid pH. *J. Biomol. Struct. Dyn.* **5:** 275–282.

Lyamichev, V.I., I.G. Panyutin, and M.D. Frank-Kamenetskii. 1983. Evidence of cruciform structures in superhelical DNA provided by two-dimensional gel electrophoresis. *FEBS Lett.* **153:** 298–302.

Lyamichev, V.I., S.M. Mirkin, O.N. Danilevkaya, O.N. Voloshin, S.V. Balatskaya, V.N. Dobrynin, S.A. Filippov, and M.D. Frank-Kamenetskii. 1989. Telomeric sequence under superhelical stress and low pH: A novel DNA structure with non-Watson-Crick pairing. *Nature* **339:** 634–637.

Mirkin, S.M., V.I. Laymichev, K.N. Drushlyak, V.N. Dobrynin, S.A. Filoppov, and M.D. Frank-Kamenetskii. 1987a. DNA H form requires a homopurine-homopyrimidine mirror repeat. *Nature* **330:** 495–497.

Mirkin, S.M., V.I. Lyamichev, V.P. Kumarev, V.F. Kobzev, V.V. Nosikov, and A.V. Vologodskii. 1987b. The energetics of the B-Z transition in DNA. *J. Biomol. Struct. Dyn.* **5:** 79–88.

Mizuuchi, K., M. Mizuuchi, and M. Gellert. 1982. Cruciform structures in palindromic DNA are favored by DNA supercoiling. *J. Mol. Biol.* **156:** 229–243.

Naylor, L.H., D.M.J. Lilley, and J.H. Van de Sande. 1986. Stress-induced cruciform foramtion in cloned d$(CATG)_{10}$ sequence. *EMBO J.* **5:** 2407–2413.

Nickol, J.M. and G. Felsenfeld. 1983. DNA conformation at the 5′ end of chicken adult β-globin gene. *Cell* **35:** 467–477.

Panayotatos, N. and R.D. Wells. 1981. Cruciform structures in supercoiled DNA. *Nature* **289:** 466–470.

Panyutin, I., V. Klishko, and V. Lyamichev. 1984. Kinetics of cruciform formation and stability of cruciform structure in superhelical DNA. *J. Biomol. Struct. Dyn.* **1:** 1311–1324.

Peck, L.J. and J.C. Wang. 1983. The energetics of B to Z transition in DNA. *Proc. Natl. Acad. Sci.* **80:** 6206–6210.

Pulleyblank, D.E., D.B. Haniford, and A.R. Morgan. 1985. A structural basis for S1 nuclease sensitivity of double-stranded DNA. *Cell* **42:** 271–280.

Pulleyblank, D.E., M. Shure, D. Tang, J. Vinograd, and H.-P. Vosberg. 1975. Action of nicking-closing enzyme on supercoiled and nonsupercoiled closed circular DNA: Formation of a Boltzmann distribution of topological isomers. *Proc. Natl. Acad. Sci.* **72:** 4280–4284.

Rich, A., A. Nordheim, and H.-J. Wang. 1984. The chemistry and biology of left-handed Z-DNA. *Annu. Rev. Biochem.* **53:** 791–846.

Scholten, P.M. and A. Norheim. 1986. Diethyl pyrocarbonate: A chemical probe for DNA cruciforms. *Nucleic Acids Res.* **14:** 3981–3993.

Schon, E., T. Evans, J. Welsh, and A. Efstratiadis. 1983. Conformation of promoter DNA: Fine mapping of S1-hypersensitive sites. *Cell* **35:** 837–848.

Shimada, J. and H. Yamakawa. 1985. Statistical mechanisms of DNA topoisomers: The helical worm-like chain. *J. Mol. Biol.* **184:** 319–329.

———. 1988. Moments for DNA topoisomers: The helical wormlike chain. *Biopolymers* **27:** 657–673.

Shore, D. and R.L. Baldwin. 1983. Energetics of DNA twisting. II. Topoisomer analysis. *J. Mol. Biol.* **170:** 983–1007.

Shure, M., D.E. Pulleyblank, and J. Vinograd. 1977. The problem of eukaryotic and prokaryotic DNA packaging and *in vivo* conformation posed by superhelix density heterogeneity. *Nucleic Acids Res.* **4:** 1183–1205.

Singleton, C.K., J. Klysik, S.M. Stirdivant, and R.D. Wells. 1982. Left-handed Z-DNA is induced by supercoiling in physiological ionic conditions. *Nature* **299:** 312–316.

Stigter, D. 1977. Interactions of highly charged colloidal cylinders with application to double-stranded DNA. *Biopolymers* **16:** 1435–1448.

Thompson, B.J., M.N. Camien, and R.C. Warner. 1976. Kinetics of branch migration in double-stranded DNA. *Proc. Natl. Acad. Sci.* **73:** 2299–2303.

Vojtiskova, M., S. Mirkin, V. Lyamichev, O. Voloshin, M. Frank-Kamenetskii, and E. Palecek. 1988. Chemical probing of the homopurine-homopyrimidine tract in supercoiled DNA at single-nucleotide resolution. *FEBS Lett.* **234:** 295–299.

Vologodskii, A.V. and M.D. Frank-Kamenetskii. 1982. Theoretical study of cruciform states in superhelical DNA. *FEBS Lett.* **143:** 257–260.

———. 1983. The relaxation time for a cruciform structure in superhelical DNA. *FEBS Lett.* **160:** 173–176.

———. 1984. Left-handed Z form in superhelical DNA: A theoretical study. *J. Biomol. Struct. Dyn.* **1:** 1325–1334.

Vologodskii, A.V., V.V. Anshelevich, A.V. Lukashin, and M.D. Frank-Kamenetskii. 1979a. Statistical mechanics of supercoils and the torsional stiffness of the DNA double helix. *Nature* **280:** 294–298.

Vologodskii, A.V., A.V. Lukashin, V.V. Anshelevich, and M.D. Frank-Kamenetskii. 1979b. Fluctuations in superhelical DNA. *Nucleic Acids Res.* **6:** 967–982.

Voloshin, O.N., S.M. Mirkin, V.I. Lyamichev, B.P. Belotserkovskii, and M.D. Frank-Kamenetskii. 1988. Chemical probing of homopurine-homopyrimidine mirror repeats in supercoiled DNA. *Nature* **333:** 475–476.

Wang, J.C., L.J. Peck, and K. Becherer. 1983. DNA supercoiling and its effects on DNA structure and function. *Cold Spring Harbor Symp. Quant. Biol.* **47:** 85–91.

Weiner, A.M. 1988. Eukaryotic nuclear telomers: Molecular fossils of the RNP world? *Cell* **52:** 155–158.

Wells, R.D., D.A. Collier, J.C. Hanvey, M. Shimizu, and F. Wohlrab. 1988. The chemistry and biology of unusual DNA structures adopted by oliopurine-oligopyrimidine sequences. *FASEB J.* **2:** 2939–2948.

Wu, H.-Y., S. Shyy, J.C. Wang, and L.F. Liu. 1988. Transcription generates positively and negatively supercoiled domains in the template. *Cell* **53:** 433–440.

Yarmola, E.G., M.I. Zarudnaya, and Y.S. Lazurkin. 1985. Osmotic pressure of DNA solutions and effective diameter of the double helix. *J. Biomol. Struct. Dyn.* **2:** 981–983.

6

Mechanistic Aspects of Type-I Topoisomerases

James J. Champoux
Department of Microbiology
School of Medicine
University of Washington
Seattle, Washington 98195

I. INTRODUCTION

Topoisomerases were first recognized by virtue of their ability to relax closed circular, negatively supercoiled DNA. The prokaryotic and eukaryotic enzymes were originally referred to as the ω protein (Wang 1971) and the DNA untwisting enzyme (Champoux and Dulbecco 1972), respectively. Because the relaxed structure of the closed DNA persists after removal of the protein, it is clear that the enzymes change the topological property called the linking number. In addition to altering the linking number, these enzymes were also shown to catalyze the interconversion of other kinds of topological isomers of DNA, and hence, they belong to the general class of enzymes now referred to as DNA topoisomerases. A change in the linking number of a closed circular

DNA Topology and Its Biological Effects
 0-87969-348-7/90 $1.00 + .00

DNA requires breakage and reclosure of one or both strands of the DNA. Topoisomerases are subdivided into two categories based on the number of strands broken during each cycle of breakage and reclosure. Type-I enzymes, the subject of this chapter (for recent reviews, see Maxwell and Gellert 1986; Sadowski 1986; Wang 1987), alter the linking number by introducing a transient single-strand break in duplex DNA, whereas type-II enzymes act by making transient double-strand breaks. Type-I topoisomerases have been found to change the linking number of the DNA by one (or possibly more), whereas type-II enzymes characteristically change the linking number in steps of two.

Since catalysis by the eubacterial and eukaryotic type-I enzymes proceeds without the need for an energy-donating cofactor, the temporary breaks introduced into the DNA must not be produced by simple hydrolysis of a phosphodiester bond. Thus, it was clear, even at the time of their discovery, that the overall enzymatic reaction involves a pathway other than simply the sum of the activities of a nuclease and a ligase.

The mechanism of the type-I enzymes is the subject of this chapter. Enzymes belonging to this class are ubiquitous, having been found in every prokaryotic or eukaryotic cell so far examined with the exception of sea urchin sperm (Poccia et al. 1978). In addition, they have been identified in both mitochondria (Fairfield et al. 1979; Brun et al. 1981) and chloroplasts (Siedlecki et al. 1983), and at least one virus (vaccinia) codes for its own type-I topoisomerase (Bauer et al. 1977; Shuman and Moss 1987). A related group of enzymes that share mechanistically some properties with the type-I enzymes includes site-specific recombinases, such as the bacteriophage λ integrase (Craig and Nash 1983), the Tn*3*-γδ resolvase (Krasnow and Cozzarelli 1983), the phage P1 Cre recombinase (Abremski et al. 1986), the yeast FLP recombinase (Andrews et al. 1985), and the bacteriophage Mu Gin invertase (Klippel et al. 1988), as well as the replication initiation-termination proteins of bacteriophage ϕX174 (*cis A* protein) (Ikeda et al. 1976; Eisenberg et al. 1977) and the filamentous phages, fd and M13 (gene-*II* protein) (Meyer and Geider 1979).

Because they have been so extensively studied, the *Escherichia coli* topoisomerase I and the eukaryotic topoisomerase I will be the main subjects of this chapter: the viral and organelle-associated enzymes will be mentioned only when they have been shown to exhibit some unusual property. Although the prokaryotic and eukaryotic enzymes share some properties, they are fundamentally different and are not evolutionarily related (Tse-Dinh and Wang 1986; D'Arpa et al. 1988). The enzymes from the two sources will be discussed separately, and their properties will be compared. Where information is available, the mechanisms of the site-specific recombinases and phage replication enzymes will be included for comparison with the topoisomerases.

II. REACTIONS CATALYZED BY TYPE-I TOPOISOMERASES

The type-I topoisomerases can promote the following four different kinds of topological interconversions of DNA in vitro: (1) Both the prokaryotic and eukaryotic type-I enzymes catalyze the relaxation of closed circular, negatively supercoiled DNA. The eukaryotic enzyme completely relaxes the DNA (Champoux and Dulbecco 1972; Pulleyblank et al. 1975), whereas the *E. coli* topoisomerase I can only partially remove the negative turns (Wang 1971). Although the eukaryotic enzyme can completely relax positive turns, the prokaryotic enzyme can only relax positive supercoils under special circumstances (see below). These limitations on the activity of the bacterial enzyme are consistent with its in vivo role of preventing excessive negative supercoiling by DNA gyrase (DiNardo et al. 1982; Pruss et al. 1982) or during transcription (Liu and Wang 1987; Wu et al. 1988; Brill and Sternglanz 1988; Giaever and Wang 1988). (2) The type-I enzymes can catalyze the intertwining and complete renaturation of two single-stranded, complementary DNA circles (Champoux 1977b; Kirkegaard and Wang 1978). (3) The *E. coli* topoisomerase I can catalyze the knotting of single-stranded circles (Liu et al. 1976). Despite numerous attempts under a variety of conditions, knotting of ssDNA by the eukaryotic enzyme has not been detected (M.D. Been and J.J. Champoux, unpubl.). (4) In the presence of a DNA aggregating agent such as spermidine, the type-I enzymes can catenate duplex circular DNA molecules, providing one of the molecules contains either a single-strand break or a gap (Tse and Wang 1980; Brown and Cozzarelli 1981; Badaracco et al. 1983; Tse et al. 1984). In a related reaction, the prokaryotic enzyme can knot a nicked or gapped duplex circle (Brown and Cozzarelli 1981; Dean et al. 1985). Knotting by the eukaryotic topoisomerase I has not been reported. As we will see below, the catenating and knotting reactions have some important implications for the mechanism of topoisomerization by the type-I enzymes.

III. INTERMEDIATES AND CHEMISTRY OF THE REACTION

A. Uncoupling of the Nicking and Closing Reactions

A variety of conditions have been discovered for both the prokaryotic and eukaryotic enzymes for producing breakage without reclosure. These aborted reactions have been useful in establishing the mechanism of the topoisomerase-catalyzed reactions.

1. Denaturants

E. coli topoisomerase I binds strongly to single-stranded DNA (ssDNA) and under the appropriate conditions can introduce knots into a circular ssDNA (Liu et al. 1976; Depew et al. 1978). With ssDNA as a substrate, addition of SDS, alkali, or in some cases proteinase-K to reactions lacking Mg^{++} results in breakage of the single strands with the concomitant covalent attachment of the enzyme to the 5′ end of the broken DNA (Depew et al. 1978; Kirkegaard et al. 1984). With negatively superhelical duplex DNA but not with relaxed closed circles, a covalent complex is similarly formed upon the addition of alkali or SDS (Liu and Wang 1979). Addition of proteinase-K to *Haemophilus gallinarum* topoisomerase I reactions lacking Mg^{++}, using a negatively supercoiled substrate, similarly leads to strand breakage (Shishido et al. 1983). Breakage of relaxed closed circles or purely linear duplexes in the presence of denaturants has not been detected, presumably because *E. coli* topoisomerase I requires at least a short single-stranded region for binding to the DNA (see below). Molecules that have a mixed structure where a region of single-stranded or partially single-stranded DNA is adjacent to a duplex region (gapped or nicked DNAs) have been used as substrates for the detergent-based breakage reaction (Kirkegaard et al. 1984; Dean and Cozzarelli 1985). Interestingly, the observed breaks are often found near the border between the single-stranded region and the duplex region of the molecule or in the intact strand opposite a nick. This observation suggests that a prerequisite for observing this kind of breakage is a region of DNA that exhibits both single- and double-stranded characters.

Using double-stranded DNA (dsDNA) as a substrate for a variety of different eukaryotic type-I topoisomerases, similar DNA/protein complexes have been identified after stopping the reactions with SDS, alkali, or low pH (Champoux 1976, 1977a; Trask and Muller 1983; Been et al. 1984b). The DNA need not be superhelical to observe breakage by eukaryotic topoisomerase I. Similar breakage is observed with single-stranded as a substrate (Been and Champoux 1980; Prell and Vosberg 1980; Edwards et al. 1982; Been et al. 1984b). In the case of the eukaryotic topoisomerase I, the enzyme is attached to the 3′ end rather than the 5′ end of the broken strand (Champoux 1977a, 1978).

There are two alternative hypotheses that may explain the breakage observed on the addition of a denaturant to a topoisomerase I reaction. First, the low level of nicked intermediates in the reaction could be trapped by the denaturation of the enzyme. If this is the correct explanation, the amount of observed breakage should reflect the proportion of enzyme molecules that are in the nicked state at any one instant. However, such an estimate is likely to be unreliable because one can never be confident that the trapping procedure is 100% efficient. The second hypothesis suggests that the denaturant itself triggers the enzyme to nick the DNA at the site where it is bound. In this case, the amount of breakage is presumably unrelated to the level of nicked intermediate in the reaction. It has not been possible to distinguish experimentally between these two hypotheses. In either case, it is generally believed that the breakage sites reflect the binding sites for the enzyme on the DNA.

It has not been possible in any system to demonstrate that the covalent complexes formed after addition of a detergent are true intermediates in the nicking-closing reaction. Apparently, the denaturation of the enzyme is, for all practical purposes, irreversible. Despite this problem, virtually all of the structural studies of the "nicked intermediate" (see below) have employed complexes formed after addition of detergents to the reactions. It is generally assumed that these complexes reflect the structure of the broken intermediate in the reaction.

Breakage of the target sequence on addition of detergents has also been observed for the λ integrase protein (Craig and Nash 1983), the yeast 2-μm circle FLP recombinase (Andrews et al. 1985), the phage P1 Cre recombinase (Abremski et al. 1986), and the phage Mu Gin invertase (Klippel et al. 1988). The λ integrase and the FLP recombinase are found attached to the 3′ end of the broken strands; the Cre and Gin enzymes are attached to 5′ ends. From these breakage experiments, it has proven possible to map to the nucleotide level the break sites that are generated during the site-specific recombination reactions catalyzed by this family of enzymes.

The formation of covalent DNA/protein complexes involving topoisomerases after treatment with detergents is reminiscent of the complexes formed with colE1 plasmid DNA (Blair and Helinski 1975; Guiney and Helinski 1975). Isolation of colE1 DNA by gentle procedures avoiding the use of ionic detergents yields a form of the plasmid DNA containing bound protein. Treatment of such structures with SDS causes nicking of the DNA and attachment of a protein to the 5′ end at the site of the nick. These DNA/protein complexes may reflect a process akin to the initiation of rolling circle replication by the *cis A* protein of ϕX174, and thus they may be important for the conjugal transfer of the DNA. In fact, current models for F-plasmid transfer during bacterial conjugation invoke the formation of such a complex as the initial step in the process (Willets and Wilkins 1984).

2. *Unusual Substrates or Reaction Conditions*

For most type-I topoisomerases, it has been possible to identify one or more unusual substrates or particular reaction conditions that cause the reaction to abort at the nicked stage. In some cases, it has been possible to adjust the conditions so that a subsequent closure reaction occurs, thus demonstrating a true intermediate status for the nicked complexes.

Abortive reactions have been demonstrated for the eukaryotic topoisomerase I using the following substrates and conditions: ssDNA in low salt (Been and Champoux 1980; Prell and Vosberg 1980; Halligan et al. 1982; Been et al. 1984b), short duplex oligonucleotides containing a recognition sequence for the enzyme (see below) (Champoux et al. 1984), and singly nicked, circular DNAs in the presence of polynucleotide kinase and ATP (McCoubrey and Champoux 1986). In the last case, breakage occurs in the intact strand opposite the nick, and the requirement for the kinase is to phosphorylate the free 5′-OH on the resulting linears and thereby prevent recircularization. With ssDNA as a template, it has been possible to show that most, if not all, of the cleavage occurs in regions of secondary structure (Been and Champoux 1984), indicating that at least a short region of duplex is probably required for the spontaneous breakage reaction. Consistent with this observation is the finding that single-stranded homopolymeric oligonucleotides up to a length of 20 are not cleaved by the eukaryotic enzyme (Tse-Dinh 1986).

What is the basis for the failure of the enzyme to complete the nicking-closing cycle with this set of substrates? With normal substrates, the duplex structure of the DNA holds the two ends to be rejoined in close proximity. However, with each of these abnormal substrates, juxtaposition of the two ends can easily be lost because of the unpairing of the DNA in the vicinity of the break. This view is supported by the observation that after breakage of circular ssDNA, the addition of salt or Mg^{++}, agents that stabilize base pairing, promotes recircularization of the DNA by the active complex (Been and Champoux 1981). With single-stranded substrates as well as with the duplex oligonucleotide substrate, it has been shown that the DNA/protein complexes formed in the breakage reaction can, under the appropriate conditions, catalyze the joining of the broken strand to the end of a different molecule containing a 5′-OH (Been and Champoux 1981; Halligan et al. 1982; Trask and Muller 1983; Champoux et al. 1984). Interestingly, the 5′ ends at internal nicks or on the ends of duplex or ssDNAs are equally efficient as acceptors in the reaction (Halligan et al. 1982). This intermolecular linking reaction has suggested a possible role for the type-I enzyme in illegitimate recombination in animal cells (Bullock et al. 1985).

E. coli topoisomerase I has been shown to cleave short single-stranded homopolymeric oligonucleotides (greater than 7 nucleotides in length) containing either dA or dT residues (Tse-Dinh et al. 1983). Cleavage of oligo(dC) and oligo(dG) is not detectable up to a chain length of 11 residues. Oligo$(dC)_{15}$ is cleaved but only very slowly. The cleaved oligonucleotides contain covalently bound topoisomerase that is active, since after addition of DNAs containing free 3′-OH, intermolecular joining can be demonstrated (Tse-Dinh 1986).

Interruption of the nicking-closing cycle has also been observed for two site-specific recombinases. Under most conditions, the Tn*3*-γδ resolvase requires Mg^{++} to carry out a complete cycle of site-specific recombination on a DNA molecule containing two appropriately oriented *res* sites. In the absence of Mg^{++}, breakage without reclosure occurs at *res* sites with simultaneous covalent attachment of the enzyme to the 5′ ends of the broken strands (Reed and Grindley 1981). This structure may be an intermediate in the reaction because addition of Mg^{++} allows recombination to go to completion. Using artificial substrates containing one or more mismatches in the crossover region, spontaneous breakage apparently also occurs for the bacteriophage λ integrase (Nash et al. 1987).

The A protein of ϕX174 initiates rolling circle replication by nicking the circular duplex DNA at the origin (Ikeda et al. 1976; Eisenberg et al. 1977; Eisenberg and Kornberg 1979). The protein becomes covalently attached to the 5′end at the site of the break and after one round of replication catalyzes the recircularization of the DNA. Although the enzyme possesses topoisomerase activity (Ikeda et al. 1976), normally the covalent DNA/enzyme complex is long-lived, and no special treatments or substrates are required to observe it. The analogous protein of bacteriophage fd, the gene-*II* protein, has also been shown to nick the DNA at the origin and to possess topoisomerase activity that is specific to fd DNA (Meyer and Geider 1979). However, in this case, it has not proven possible to identify a covalent complex between the protein and either end of the broken strand. Conceivably, the linkage in this particular case is easily hydrolyzed, precluding detection of the complex.

B. Structure of the Protein/DNA Intermediate and Energy Storage

As discussed above, breakage of a DNA strand by a type-I topoisomerase is always accompanied by the covalent attachment of the enzyme to a phosphate residue at one of the ends created at the break. The closure step of the reaction restores continuity to the DNA strand and releases functional enzyme. The structure of the covalent intermediate is in most cases consistent with the hypothesis that the energy required to resynthesize the DNA phosphodiester bond is stored in the bond formed between the enzyme and the broken strand.

Most topoisomerases, including all known type-II enzymes (see Hsieh, this volume), attach to the 5′ end of the broken strand. From the known enzymes that attach to the 3′ end of the broken strand—the eukaryotic topoisomerase I, the λ integrase, and the yeast 2-μm circle FLP recombinase—no obvious pattern emerges that suggests a possible functional significance to the different polarities of attachment. It seems likely therefore that the different polarities represent equivalent, independent solutions to the same problem.

What is the chemical nature of the linkage between the terminal phosphate residue on the DNA chain and the protein molecule? On the basis of the reactivity of the bond and direct chemical analysis, the DNA/protein linkages for both the prokaryotic and eukaryotic type-I enzymes and the FLP recombinase have been found to be phosphodiester bonds between the ends of the broken strands and tyrosine residues in the respective proteins (Tse et al. 1980; Champoux 1981; Gronostajski and Sadowski 1985). For *E. coli* topoisomerase I, Tyr-319 has been identified as the active site tyrosine (Lynn and Wang 1989), whereas the active site residue for the *Saccharomyces cerevisiae* topoisomerase I has been located to Tyr-727 (Eng et al. 1989; Lynn et al. 1989). In both cases, substitution of either phenylalanine or serine for the active site tyrosine resulted in an inactive enzyme, confirming that the tyrosine residues are essential for enzyme activity. On the basis of homology arguments, it is possible to identify Tyr-723 in the human topoisomerase I (D'Arpa et al. 1988) and Tyr-771 in the *Schizosaccharomyces pombe* enzyme as the active site tyrosines. In addition, it appears that Tyr-274 in the vaccinia virus topoisomerase I (Shuman and Moss 1987) is the active site residue (Lynn et al. 1989).

Chemically, breakage of the DNA by the enzyme is a transesterification reaction involving nucleophilic attack by the tyrosine hydroxyl in the active site on a phosphodiester bond in the DNA. Closure is similarly a transesterification reaction in which the attacking group is the hydroxyl group on the free end of the broken DNA strand. The reaction is readily reversible, providing that the free energy of hydrolysis of the protein/DNA phosphodiester bond is comparable to the free energy of hydrolysis of the DNA phosphodiester linkage. The free energy of hydrolysis of a phosphodiester bond in DNA has been estimated to be approximately –9 kcal/mole (Peller 1976). The tyrosine-adenylate linkage in AMP-glutamine synthetase, which is a structural analog of the DNA/tyrosine linkage in the topoisomerase intermediate, has a free energy of hydrolysis of approximately –10 kcal/mole (Holzer and Wohlhueter 1972). Thus, the change in free energy for the breakage reaction would be expected to be small, on the order of +1 kcal/mole, consistent with the reaction being freely reversible. This free energy change predicts that for a type-I topoisomerase reaction at equilibrium, less than 20% of the bound enzyme molecules will be in the nicked intermediate form. The observed breakage frequency for the eukaryotic topoisomerase I after stopping reactions with SDS or alkali (Champoux 1976) is consistent with this prediction, but for reasons mentioned above, this agreement should be viewed with caution.

Interestingly, the ϕX174 *cis A* protein contains two tyrosine residues, separated by three amino acids, each of which can form a covalent bond to the end of the broken DNA strand (van Mansfeld et al. 1984, 1986). It appears that successive cleavage events alternate between the tyrosine residues. Thus, after the first breakage event, each successive cleavage is chemically coupled to the recircularization of the previously broken strand (van Mansfeld et al. 1986).

The γδ resolvase and the phage Mu Gin invertase attach covalently to their respective DNA targets by way of a serine phosphodiester bond rather than through the usual tyrosine phosphodiester linkage found for all other topoisomerases (Reed and Moser 1984; Hatfull and Grindley 1986; Klippel et al. 1988). The free energy of hydrolysis of serine-phosphate is substantially lower than the value for tyrosine-phosphate with a value of approximately –3 kcal/mole (Guynn and Thames 1982). Therefore, it is difficult to explain the conservation of the DNA phosphodiester bond energy by the formation of the protein/DNA linkage. One possible resolution of this discrepancy is to imagine that the active site serine for these two enzymes is imbedded in an environment that destabilizes the phosphodiester bond between the serine and the end of the DNA chain. Structural studies on the enzyme may help solve this paradox.

IV. SPECIFICITY

A. DNA Supercoiling

The *E. coli* topoisomerase I exhibits a specificity for negatively supercoiled DNA and will not under normal circumstances relax positive supercoils (Wang 1971). Moreover, the enzyme is not capable of completely relaxing negative supercoils. These limitations on the activity of the enzyme are consistent with its role as an antagonist of the DNA gyrase (DiNardo et al. 1982; Pruss et al. 1982). Thus, the enzyme must act to prevent over-supercoiling of the DNA without completely relaxing the negative turns introduced by the gyrase.

The basis for this spectrum of specificities for the *E. coli* topoisomerase I appears to be a requirement for the formation of a region of DNA that is at least partially single-stranded. The enzyme has been shown to bind very tightly to ssDNA (Depew et al. 1978) and, as mentioned above, forms a cleavable complex with DNAs providing they contain at least a short single-stranded region (Depew et al. 1978; Dean et al. 1983; Kirkegaard et al. 1984; Dean and Cozzarelli 1985). Apparently, the ease of strand-separation in the vicinity of a single-strand break is sufficient to satisfy this requirement (Kirkegaard et al. 1984; Dean and Cozzarelli 1985). The free energy associated with high negative supercoiling destabilizes the helix sufficiently to facilitate the formation of a single-stranded region when the *E. coli* topoisomerase I binds to the DNA (Liu and Wang 1979). In fact, many of the nuclease S1 sensitive sites in a negatively supercoiled plasmid DNA are also found to be sites of detergent-based strand cleavage by the *H. gallinarum* topoisomerase I (Shishido et al. 1983). Perhaps the most dramatic demonstration of the requirement for single-stranded regions for catalysis by the bacterial topoisomerase I is the finding that the *E. coli* enzyme will relax positive supercoils providing the substrate DNA contains a single-stranded loop (Kirkegaard and Wang 1985).

The behavior of the eukaryotic topoisomerase I is quite different from its bacterial counterpart. The eukaryotic enzyme removes both negative and positive supercoils with equal efficiency (Champoux and Dulbecco 1972) and, moreover, completely relaxes the DNA to generate a topoisomer distribution equivalent to that obtained if the DNA is closed with ligase under the same reaction conditions (Pulleyblank et al. 1975). Eukaryotic topoisomerase I does not bind strongly to ssDNA and appears to require at least a short region of secondary structure within an otherwise single-stranded molecule to undergo the spontaneous breakage reaction described above (Been and Champoux 1984). Therefore, the evidence to date indicates that the eukaryotic enzyme interacts primarily with duplex DNA.

The question remains open whether the eukaryotic enzyme exhibits a preference for supercoiled over nonsupercoiled substrates. Evidence has been presented that negatively supercoiled DNA is a better substrate for topoisomerization (Camilloni et al. 1988) and for the SDS cleavage reaction than relaxed closed circular DNA (Muller 1985). Since the negatively supercoiled DNA was quickly relaxed under the conditions of these analyses, the effects must have been mediated at the level of the initial binding of the enzyme to the DNA. It would be interesting to know whether a similar binding preference also occurs with positively supercoiled DNA. If this kind of substrate specificity is confirmed, it could provide a basis for targeting the enzyme in vivo to supercoiled regions needing relaxation.

The site-specific recombinases vary with respect to their requirements for supercoiling. For example, both the λ integrase and the γδ resolvase require a negatively supercoiled substrate (Mizuuchi et al. 1978; Kikuchi and Nash 1979; Reed 1981), whereas the bacteriophage P1 Cre and yeast 2-μm FLP recombinases can act on a relaxed DNA as well as on a supercoiled substrate (Abremski et al. 1983; Vetter et al. 1983). For those recombinases that require a supercoiled substrate, the requirement may be related to the path of the DNA in the protein/DNA synaptic complex (Richet et al. 1986; Kanaar et al. 1988). Relaxation by Tn*3* resolvase has the same extremely high specificity requirements as recombination by this enzyme, namely, two directly repeated resolvase sites in a supercoiled molecule (Krasnow and Cozzarelli 1983). The ϕX174 gene-*A* protein is another example of a site-specific topoisomerase that requires a negatively supercoiled substrate DNA (Ikeda et al. 1979).

B. Nucleotide Sequence

The *E. coli* topoisomerase I appears to exhibit only limited specificity for nucleotide sequence in DNA. As mentioned above, single-stranded homopolymeric oligonucleotides of dA, dT, and to a lesser extent dC, are spontaneously broken by the enzyme. Only oligo(dG) appears to be refractory to breakage (Tse-Dinh et al. 1983). However, when reactions with ssDNA as a substrate are stopped with alkali or detergent, the enzyme is not found to break the DNA indiscriminately. Analysis of the nucleotides surrounding the break sites (enzyme attached at +1 position) reveals that the enzyme exhibits an obvious preference for a C residue 4 bases 5′ to the break site (–4 position) (Tse et al. 1980; Dean et al. 1983; Kirkegaard et al. 1984) and a bias against a G at the –1 position (Dean and Cozzarelli 1985). Most break sites can be accounted for by this set of rules, but other structural features may influence site selection by the enzyme. The same sequence preference has also been observed for SDS-induced breakage in the vicinity of nicks in duplex DNA (Dean and Cozzarelli 1985). The finding that the enzyme prefers to break DNAs at the border between single- and double-stranded regions (Kirkegaard et al. 1984; Dean and Cozzarelli 1985) suggests that secondary structure in ssDNA may be an additional determinant of the observed specificity.

Fine level mapping of sites in negatively supercoiled circular DNAs has not yet been achieved. Presumably, the sites of action would cluster within regions that are driven by negative supercoiling to have the most single-stranded character (Liu and Wang 1979; Shishido et al. 1983).

By comparison with the bacterial type-I enzyme, the eukaryotic type-I enzyme appears to exhibit a slightly higher degree of base sequence specificity. Using dsDNA as a substrate and the alkali or SDS stop methods, break sites have been mapped to the nucleotide level for the rat and wheat germ type-I enzymes (Been et al. 1984a). The break sites are found to map on the DNA in a nonrandom fashion with the strongest class of rat topoisomerase I sites conforming to the following consensus sequence:

–4 –3 –2 –1
5′-(A or T)-(G or C)-(A or T)-T-3′

The 3′T residue at the –1 position is the nucleotide to which the enzyme is attached. For the wheat germ enzyme, C residues are also found at the –4 position, otherwise the two enzymes have the same consensus sequence. There is no obvious preference for bases 3′ to the break site. Some versions of the consensus sequence (e.g., AGTT) appear to be invariably broken in vitro, whereas other versions (e.g., TCAT) are only occasionally broken. Consistent with the degeneracy of this consensus sequence, break sites for the enzyme are found every 7 to 10 bp on duplex DNA. A very similar consensus sequence has recently been found for the monkey cell enzyme acting on SV40 DNA in vivo (Porter and Champoux 1989a).

Some feature besides the sequence of 4 bases 5′ to the break site influences the frequency of breakage by the eukaryotic topoisomerase I when reactions are stopped with denaturants (Been et al. 1984a). Sites that conform to the consensus sequence are not all broken with the same efficiency, but there is no clear-cut correlation between breakage frequency and the various versions of the consensus sequence. The same sequence of 4 bases are often found to be broken with very different frequencies at different locations in a DNA molecule. Moreover, some break sites are found to deviate substantially from the consensus sequence. The ACTT site found within a repeated hexadecameric sequence near the *Tetrahymena* rDNA genes perhaps best illustrates the effects of context on the breakage frequency (Gocke et al. 1983; Andersen et al. 1985; Bonven et al. 1985). This particular site is not only a highly preferred site for breakage and relaxation by the *Tetrahymena* enzyme (Busk et al. 1987), but is also a preferred site for other eukaryotic type-I enzymes as well (Christiansen et al. 1987). The basis for these context effects is unclear, but it is worth noting that the observed frequency of breakage may not necessarily correlate with the strength of binding at a particular site. It is possible, for example, that the highest frequencies of breakage are found at sites where binding is weak but the closure reaction of the nicking-closing cycle is relatively slow. Likewise, good binding sites may exhibit relatively lower frequencies of breakage if, for example, the closure step is relatively fast. Break sites have also been mapped on single-stranded substrates (Edwards et al. 1982; Been and Champoux 1984), but the analysis of such sites is complicated by the finding that only those sites that are contained within regions of secondary structure are broken efficiently by the enzyme (Been and Champoux 1984).

Is the limited sequence specificity displayed by the prokaryotic and eukaryotic type-I enzymes of any biological significance? Given the low degree of specificity and the corresponding high frequency of breakage, it would appear that the sequence preference of the enzymes places very little limitation on the access of the enzymes to DNA. Indeed, it seems more likely that the enzymes have been designed by evolution to act on supercoiled DNA without regard to nucleotide sequence. If this speculation is true, then the observed site selection by the two enzymes is probably based on the best steric fit between the active site regions of the enzymes and the DNA strand to be broken.

V. INHIBITORS AND COVALENT MODIFICATIONS THAT AFFECT ACTIVITY

A. *E. coli* Topoisomerase I

As mentioned above, *E. coli* topoisomerase I forms a tight complex with single strands, and for this reason ssDNA is a potent competitive inhibitor of the relaxing activity of the enzyme (Wang 1971). It is likely, however, that the binding of the enzyme to single-stranded regions present in the cell during replication or repair is precluded by the single-strand binding protein (Srivenugopal and Morris 1986) or other factors. Thus, it is doubtful that this inhibition is a significant factor in vivo. No specific inhibitory compounds or drugs have been described for the bacterial topoisomerase I, and there is no information about possible covalent modifications that might modulate the activity in vivo.

B. Eukaryotic Topoisomerase I

1. ssDNA

Unlike the case with *E. coli* topoisomerase I, ssDNA does not appear to bind to the eukaryotic enzyme and is therefore not a competitive inhibitor. However, ssDNA is a breakage substrate for the enzyme, and the spontaneous formation of the protein/DNA complex results in inactivation of the enzyme (Been and Champoux 1980). Presumably, eukaryotic cells possess protective mechanisms to guard against breakage of single-stranded regions by topoisomerase I (see below).

2. *Small Molecules*

Ethidium and a series of other intercalators have been found to slow the rate of the relaxation of supercoiled DNA (Douc-Rasy et al. 1983, 1984; Pommier et al. 1987). Presumably, this effect is the result of an alteration in the structure of the substrate DNA, and it is noteworthy that these drugs do not enhance breakage of the DNA when reactions are stopped with denaturants (see below).

ATP inhibits the activity of the human topoisomerase I, provided that inorganic phosphate is present (Low and Holden 1985; Castora and Kelly 1986). Since 2′ and 3′ AMP are as effective inhibitors as ATP, hydrolysis of the nucleotide or phosphorylation of the protein are unlikely to be involved in the inhibition. The requirement for phosphate remains a puzzle. Whether this inhibition operates to regulate the activity of the enzyme in vivo remains to be determined.

The rat liver mitochondrial topoisomerase I has been found to be inhibited by relatively low concentrations of ethidium and also by the drug berenil (Fairfield et al. 1979). This pattern of inhibition apparently distinguishes the mitochondrial enzyme in this cell type from the nuclear topoisomerase I. In contrast with these findings in rat liver, the mitochondrial topoisomerase I in *Xenopus laevis* oocytes has been found to be indistinguishable from the nuclear enzyme with respect to sensitivity to these same two drugs (Brun et al. 1981).

Unlike ATP and intercalators, the cytotoxic alkaloid, camptothecin, appears to have a minimal effect on the relaxation activity of both the mammalian and wheat germ type-I enzymes at low concentrations, but its presence in a reaction in vitro greatly enhances the breakage observed when reactions are stopped with SDS (Hsiang et al. 1985; Champoux and Aronoff 1989). The formation of covalent complexes between topoisomerase I and DNA have also been observed when cells are treated with the drug and lysed with SDS (Gilmour and Elgin 1987; Champoux 1988). At higher drug concentrations in vitro, some inhibition of relaxation is observed with the mammalian enzymes (Hsiang et al. 1985), but little effect is observed with the wheat germ enzyme (Champoux and Aronoff 1989). Recently, actinomycin D has been reported to have the same effect as camptothecin (Trask and Muller 1988). At salt concentrations above 0.25 M, the eukaryotic topoisomerase I dissociates from DNA unless it is covalently attached (McConaughy et al. 1981). Addition of high salt to reactions containing camptothecin prior to terminating the reaction with SDS eliminates some but not all of the observed breakage (Champoux and Aronoff 1989). This residual salt-stable breakage suggests that the drug causes some long-lived or permanent nicking of the DNA even without exposure to a detergent. The topoisomerase I is likely to be the sole in vivo target of the drug camptothecin (Andoh et al. 1987; Eng et al. 1988; Nitiss and Wang 1988). Therefore, it is possible that the formation of permanent protein-associated breaks in vivo upon treatment of cells with camptothecin explains the rapid inhibition of DNA and RNA synthesis and the consequent cytotoxicity of the drug (Horowitz et al. 1971; Abelson and Penman 1973). If this is the basis for the cytotoxic effects of the drug, caution should be exercised in using camptothecin as an "inhibitor" of topoisomerase I in in vivo studies.

Since camptothecin has been used in several studies to map topoisomerase I to transcribing or replicating DNAs (Snapka 1986; Gilmour and Elgin 1987; Stewart and Schutz 1987; Champoux 1988; Zhang et al. 1988; Porter and Champoux 1989a), it is pertinent to inquire whether the drug affects the breakage specificity of the enzyme. Nearly all of the in vitro break sites observed in the absence of the drug are also observed in its presence (Kjeldsen et al. 1988; Champoux and Aronoff 1989). A few sites are only observed in the presence of the drug (Thomsen et al. 1987). Consistent with these observations is the finding that the consensus sequence for breakage by the wheat germ enzyme in the presence of the drug is the same as in its absence (Champoux and Aronoff 1989). Therefore, it appears that no gross changes in breakage specificity are caused by camptothecin. Interestingly, the drug enhances the breakage to different extents at different sites (Kjeldsen et al. 1988; Champoux and Aronoff 1989). At some sites, there is little or no enhancement. Typically, breakage is enhanced between 10- and 20-fold, but in a few cases enhancement is greater than 150-fold. The few sites that are broken only in the presence of the drug may not represent a change in specificity induced by camptothecin (Thomsen et al. 1987). Instead, they may simply be sites for which the enhancement factor is so great that the sites are not observable in the absence of the drug.

What is the basis for the effects of camptothecin on the eukaryotic topoisomerase I reaction? An increase in the binding affinity of the enzyme for DNA in the presence of the drug could explain the enhanced breakage when reactions are stopped with SDS. However, the drug does not appear to affect the processivity of the reaction (Champoux and

Aronoff 1989) or the equilibrium dissociation constant for the enzyme/DNA complex (Kjeldsen et al. 1988). It appears more likely that the drug interferes with the closure step in the nicking-closing cycle and thereby increases the amount of detergent-induced breakage. Different closure rates at different sites in the presence of the drug could account for the different enhancement factors at the different sites. Moreover, an exceptionally slow closure rate at some sites could explain the formation of the salt-stable breakage products (Kjeldsen et al. 1988; Champoux and Aronoff 1989). Consistent with these suggestions is the recent finding that the sites that are enhanced the greatest by the drug are the slowest to close in high salt in the presence of the drug (Porter and Champoux 1989b).

3. *Covalent Modifications*

Topoisomerase I isolated from Novikoff hepatoma cells is a phosphoprotein (Durban et al. 1983, 1985). Treatment of the enzyme with a phosphatase in vitro causes approximately a threefold loss of activity. Rephosphorylation in vitro using casein type-II protein kinase results in an approximately fivefold reactivation of the enzyme (Mills et al. 1982; Durban et al. 1983). At least one of the serine residues phosphorylated in vitro is also phosphorylated in vivo (Durban et al. 1985). It is not known whether these modest changes in enzyme activity by cycles of phosphorylation-dephosphorylation are physiologically significant.

The eukaryotic topoisomerase I has been found to be a good substrate for poly(ADP-ribose) synthetase (Ferro et al. 1983). Poly(ADP-ribosylation) of the protein inactivates the enzyme as measured by the relaxation assay (Jongstra-Bilen et al. 1983; Ferro and Olivera 1984). Activity is recovered upon removal of the poly(ADP-ribose) units by mild alkali treatment (Ferro et al. 1983). Since the poly(ADP-ribose) synthetase is activated by DNAs containing gaps and breaks, it has been speculated that under normal circumstances, poly(ADP-ribosylation) of the topoisomerase serves to protect single-stranded gaps generated in vivo by DNA damage or during replication and repair from spontaneous breakage by the enzyme (Ferro and Olivera 1984). Providing the modification is reversible, such a protective mechanism could be restricted to the immediate vicinity of the single-stranded region. The finding that treatment of cells with 3-aminobenzamide, an inhibitor of poly(ADP-ribose) synthetase, results in increased damage when cells are treated with camptothecin (Mattern et al. 1987) is consistent with a role for this modification in the regulation of topoisomerase I in vivo.

VI. MECHANISM OF RELAXATION: FREE ROTATION VERSUS STRAND PASSAGE

The two alternative mechanisms that have been proposed to explain the relaxation of DNA by type-I topoisomerases are shown schematically in Figure 1. In one mechanism (model I), the end of the broken strand that is not covalently attached to the enzyme is released from the active site region, permitting free and unlimited rotation about the phosphodiester bond in the unbroken strand. The free rotation hypothesis was considered the most likely explanation for the relaxation reaction until it was discovered that type-I enzymes are capable of catenating circular DNAs containing nicks or gaps (Tse and Wang 1980; Brown and Cozzarelli 1981). This observation led to the suggestion that in the normal reaction, the enzyme passes the unbroken strand through a gate provided by the bridge formed when the enzyme attaches to the two ends of the DNA at the transient break. This hypothesis will be referred to as the enzyme-bridging model (Fig. 1, model II) (Brown and Cozzarelli 1981; Wang 1982; Kirkegaard et al. 1984).

In the enzyme-bridging model, the attachment to one end would necessarily be covalent, whereas attachment to the other end would be noncovalent. This model nicely explains catenation. The enzyme-bridging model for topoisomerase I is also appealing because of the mechanistic parallels with the double-strand passage model proposed for type-II topoisomerases (see Hsieh, this volume).

Model II makes a very specific prediction that could, in principle, be used to distinguish between the two hypotheses. Since, according to model II, the enzyme holds on to both ends of the broken strand, there can be, at most, the relaxation of one turn for each cycle of nicking and closing. In model I, the enzyme is not so constrained, and many turns could be relaxed in each nicking-closing cycle. It has proven difficult to distinguish the two models based on this prediction since it has not been possible to show that over the range of conditions where the enzyme is active it invariably removes supercoils in steps of one.

Although it has not been possible to rigorously discriminate between these two alternative mechanisms, the weight of the evidence suggests that model II may hold for the *E. coli* enzyme, whereas model I could explain relaxation by the eukaryotic topoisomerase I (Brown and Cozzarelli 1981; Kirkegaard et al. 1984). Support for the enzyme-bridging model for the bacterial enzyme comes from the following observations. Using high salt and only moderately negatively supercoiled DNA to make the enzyme function distributively, Brown and Cozzarelli (1981) have shown that *E. coli* topoisomerase I removes supercoils in steps of one.(It is not known, however, whether under less restrictive conditions the en-

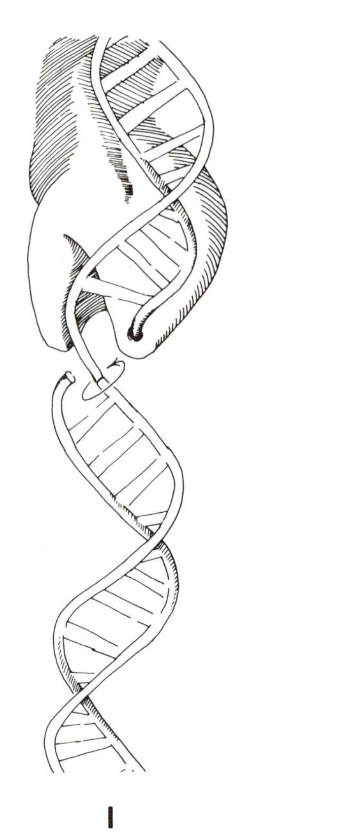

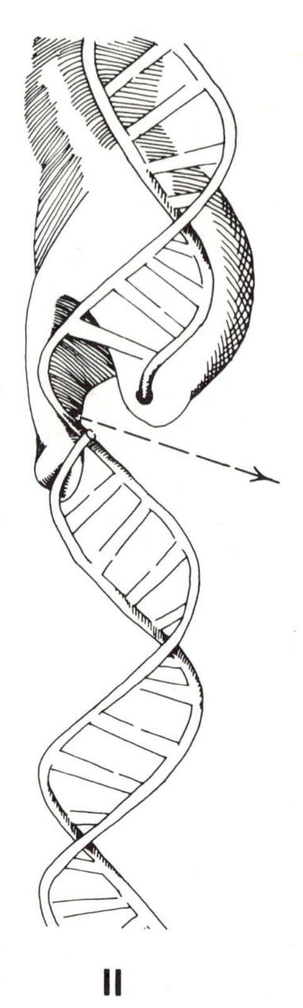

Figure 1 Drawings showing two alternative mechanisms for type-I topoisomerases. Model I depicts the free rotation model, and model II shows the enzyme-bridging model. See text for explanation.

zyme also works by a one-step mechanism.) Not only can the enzyme catenate gapped or nicked duplex circles, it can also tie single-stranded circles into knots (Liu et al. 1976). Moreover, gapped circles are much better substrates for catenation than nicked circles (Low et al. 1984; Dean and Cozzarelli 1985). In all of these reactions, it is likely that both ends of the broken strand are anchored to the enzyme, while the single strand (or double strand in the case of catenation) is passed through the break. Spontaneous breakage of single-stranded or nicked DNAs in the absence of denaturants has not been observed for the bacterial enzyme (except for short oligonucleotides), further suggesting that the enzyme holds onto both ends of the broken strand. That the enzyme is indeed interacting intimately with the DNA end not covalently attached to the protein follows from the observation that there is a strong bias for or against certain bases at positions –1 and –4 (enzyme attached at +1) (Kirkegaard et al. 1984; Dean and Cozzarelli 1985).

The following considerations favor the free rotation model for the eukaryotic enzyme. First, one does not necessarily need to postulate an enzyme bridge to explain the catenation reaction because an alternate pathway is possible. Since the eukaryotic enzyme can spontaneously break nicked circles in the strand opposite the preexisting nick to generate linears (McCoubrey and Champoux 1986), recircularization of the aggregated DNA could entrap other circles to generate catenanes. Second, there is no indication that the eukaryotic enzyme interacts with the end of the broken strand that is not covalently bound to the active site. The enzyme shows little or no preference for bases 3′ to the break site and spontaneously breaks a variety of substrates for which the recognition sequence is located so close to unpaired regions that juxtaposition of the two ends is lost following breakage (see above). Third, the failure to observe knotting of single-stranded circles and the fact that gapped circles are no better than nicked circles in the catenation reaction (Holden and Low 1985) are consistent with the free rotation model for the eukaryotic enzyme.

VII. SUMMARY

Elucidation of the mechanism for type-I topoisomerases has revealed a mechanism for temporarily breaking a DNA strand without sacrificing the energy of the phosphodiester bond. This basic mechanism not only applies to both prokaryotic and eukaryotic type-I topoisomerases, but also to type-II topoisomerases, the ssDNA phage replication proteins, and site-specific recombinases. The distinguishing feature of the recombinases is that during the lifetime of the break an exchange takes place so that, rather than restore continuity to the original strand, new DNA joints are formed. The prokaryotic and eukaryotic type-I enzymes may carry out topoisomerization of supercoiled DNA by fundamentally different pathways. Action of the prokaryotic enzyme occurs in a single-stranded region, the unpairing of which is facilitated by negative supercoiling, and therefore probably acts by an enzyme-bridging mechanism where the enzyme holds onto both ends of the broken strand. The eukaryotic enzyme, on the other hand, appears to act only on purely duplex DNA where the structure of the DNA ensures that rejoining can occur. In this case, if the enzyme does not firmly hold onto both ends of the broken strand, relaxation could occur by a free rotation mechanism.

ACKNOWLEDGMENTS

This work was supported by National Science Foundation Grant DMB-8603208. I thank Stephanie Porter and Andrew Ching for their helpful criticisms of the manuscript and Stephanie Porter for the drawings of the topoisomerization models.

REFERENCES

Abelson, H.T. and S. Penman. 1973. Induction of alkali labile links in cellular DNA by camptothecin. *Biochem. Biophys. Res. Commun.* **50:** 1048–1054.

Abremski, K., R. Hoess, and N. Sternberg. 1983. Studies on the properties of P1 site-specific recombination: Evidence for topologically unlinked products following recombination. *Cell* **32:** 1301–1311.

Abremski, K., A. Wierzbicki, B. Frommer, and R.H. Hoess. 1986. Bacteriophage P1 Cre-*loxP* site-specific recombination. Site-specific DNA topoisomerase activity of the Cre recombination protein. *J. Biol. Chem.* **261:** 391–396.

Andersen, A.H., E. Gocke, B.J. Bonven, O.F. Nielsen, and O. Westergaard. 1985. Topoisomerase I has a strong binding preference for a conserved hexadecameric sequence in the promoter region of the rRNA gene from *Tetrahymena pyriformis*. *Nucleic Acids Res.* **13:** 1543–1557.

Andoh, T., K. Ishii, Y. Suzuki, Y. Ikegami, Y. Kusunoki, Y. Takemoto, and K. Okada. 1987. Characterization of a mammalian mutant with a camptothecin-resistant DNA topoisomerase I. *Proc. Natl. Acad. Sci.* **84:** 5565–5569.

Andrews, B.J., G.A. Proteau, L.G. Beatty, and P.D. Sadowski. 1985. The FLP recombinase of the 2μ circle DNA of yeast: Interaction with its target sequences. *Cell* **40:** 795–803.

Badaracco, G., P. Plevani, W.T. Ruyechan, and L.M.S. Chang. 1983. Purification and characterization of yeast topoisomerase I. *J. Biol. Chem.* **258:** 2022–2026.

Bauer, W.R., E.C. Ressner, J. Kates, and J.V. Patzke. 1977. A DNA nicking-closing enzyme encapsidated in vaccinia virus: Partial purification and properties. *Proc. Natl. Acad. Sci.* **74:** 1841–1845.

Been, M.D. and J.J. Champoux. 1980. Breakage of single-stranded DNA by rat liver nicking-closing enzyme with the formation of a DNA-enzyme complex. *Nucleic Acids Res.* **8:** 6129–6142.

———. 1981. DNA breakage and closure by rat liver type I topoisomerase: Separation of the half-reactions by using a single-stranded DNA substrate. *Proc. Natl. Acad. Sci.* **78:** 2883–2887.

———. 1984. Breakage of single-stranded DNA by eukaryotic type 1 topoisomerase occurs only at regions with the potential for base-pairing. *J. Mol. Biol.* **180:** 515–531.

Been, M.D., R.R. Burgess, and J.J. Champoux. 1984a. Nucleotide sequence preference at rat liver and wheat germ type 1 DNA topoisomerase breakage sites in duplex SV40 DNA. *Nucleic Acids Res.* **12:** 3097–3114.

———. 1984b. DNA strand breakage by wheat germ type 1 topoisomerase. *Biochim. Biophys. Acta* **782:** 304–312.

Blair, D.G. and D.R. Helinski. 1975. Relaxation complexes of plasmid DNA and protein. I. Strand-specific association of protein and DNA in the relaxed complexes of plasmids colE1 and colE2. *J. Biol. Chem.* **250:** 8785–8789.

Bonven, B.J., E. Gocke, and O. Westergaard. 1985. A high affinity topoisomerase I binding sequence is clustered at DNAase I hypersensitive sites in *Tetrahymena* R-chromatin. *Cell* **41:** 541–551.

Brill, S.J. and R. Sternglanz. 1988. Transcription-dependent DNA supercoiling in yeast DNA topoisomerase mutants. *Cell* **54:** 403–411.

Brown, P.O. and N.R. Cozzarelli. 1981. Catenation and knotting of duplex DNA by type 1 topoisomeraes: A mechanistic parallel with type 2 topoisomerases. *Proc. Natl. Acad. Sci.* **78:** 843–847.

Brun, G., P. Vannier, I. Scovassi, and J.-C. Callen. 1981. DNA topoisomerase I from mitochondria of *Xenopus laevis* oocytes. *Eur. J. Biochem.* **118:** 407–415.

Bullock, P., J.J. Champoux, and M. Botchan. 1985. Association of crossover points with topoisomerase I cleavage sites: A model for nonhomologous recombination. *Science* **230:** 954–958.

Busk, H., B. Thomsen, B.J. Bonven, E. Kjeldsen, O.F. Nielsen, and O. Westergaard. 1987. Preferential relaxation of supercoiled DNA containing a hexadecameric recognition sequence for topoisomerase I. *Nature* **327:** 638–640.

Camilloni, G., E. Di Martino, M. Caserta, and E. Di Mauro. 1988. Eukaryotic DNA topoisomerase I reaction is topology dependent. *Nucleic Acids Res.* **16:** 7071–7085.

Castora, F.J. and W.G. Kelly. 1986. ATP inhibits nuclear and mitochondrial type I topoisomerases from human leukemia cells. *Proc. Natl. Acad. Sci.* **83:** 1680–1684.

Champoux, J.J. 1976. Evidence for an intermediate with a single-strand break in the reaction catalyzed by the DNA untwisting enzyme. *Proc. Natl. Acad. Sci.* **73:** 3488–3491.

———. 1977a. Strand breakage by the DNA untwisting enzyme results in covalent attachment of the enzyme to DNA. *Proc. Natl. Acad. Sci.* **74:** 3800–3804.

———. 1977b. Renaturation of complementary single-stranded DNA circles: Complete rewinding facilitated by the DNA untwisting enzyme. *Proc. Natl. Acad. Sci.* **74:** 5328–5332.

———. 1978. Mechanism of the reaction catalyzed by the DNA untwisting enzyme: Attachment of the enzyme to 3′-terminus of the nicked DNA. *J. Mol. Biol.* **118:** 441–446.

———. 1981. DNA is linked to the rat liver DNA nicking-closing enzyme by a phosphodiester bond to tyrosine. *J. Biol. Chem.* **256:** 4805–4809.

———. 1988. Topoisomerase I is preferentially associated with isolated replicating simian virus 40 molecules after treatment of infected cells with camptothecin. *J. Virol.* **62:** 3674–3683.

Champoux, J.J. and R. Aronoff. 1989. The effects of camptothecin on the reaction and the specificity of the wheat germ type I topoisomerase. *J. Biol. Chem.* **264:** 1010–1015.

Champoux, J.J. and R. Dulbecco. 1972. An activity from mammalian cells that untwists superhelical DNA-A possible swivel for DNA replication. *Proc. Natl. Acad. Sci.* **69:** 143–146.

Champoux, J.J., W.K. McCoubrey, Jr., and M.D. Been. 1984. DNA structural features that lead to strand breakage by eukaryotic type-I topoisomerase. *Cold Spring Harbor Symp. Quant. Biol.* **49:** 435–442.

Christiansen, K., B.J. Bonven, and O. Westergaard. 1987. Mapping of sequence-specific chromatin proteins by a novel method: Topoisomerase I on *Tetrahymena* ribosomal chromatin. *J. Mol. Biol.* **193:** 517–525.

Craig, N.L. and H.A. Nash. 1983. The mechanism of phage λ site-specific recombination: Site-specific breakage of DNA by Int topoisomerase. *Cell* **35:** 795–803.

D'Arpa, P., P.S. Machlin, H.R. Ratrie III, N.F. Rothfield, D.W. Cleveland, and W.C. Earnshaw. 1988. cDNA cloning of human DNA topoisomerase I: Catalytic activity of a 67.7-kDa carboxyl-terminal fragment. *Proc. Natl. Acad. Sci.* **85:** 2543–2547.

Dean, F.B. and N.R. Cozzarelli. 1985. Mechanism of strand passage by *Escherichia coli* topoisomerase I. The role of the required nick in catenation and knotting of duplex DNA. *J. Biol. Chem.* **260:** 4984–4994.

Dean, F.B., A. Stasiak, T. Koller, and N.R. Cozzarelli. 1985. Duplex DNA knots produced by *Escherichia coli* topoisomerase I. Structure and requirements for formation. *J. Biol. Chem.* **260:** 4975–4983.

Dean, F., M.A. Krasnow, R. Otter, M.M. Matzuk, S.J. Spengler, and N.R. Cozzarelli. 1983. *Escherichia coli* type-I topoisomerases: Identification, mechanism, and role in recombination. *Cold Spring Harbor Symp. Quant. Biol.* **47:** 769–784.

Depew, R.E., L.F. Liu, and J.C. Wang. 1978. Interaction between DNA and *Escherichia coli* protein ω. Formation of a complex between single-stranded DNA and ω protein. *J. Biol. Chem.* **253:** 511–518.

DiNardo, S., K.A. Voelkel, R. Sternglanz, A.E. Reynolds, and A. Wright. 1982. *Escherichia coli* DNA topoisomerase I mutants have compensatory mutations in DNA gyrase genes. *Cell* **31:** 43–51.

Douc-Rasy, S., A. Kayser, and G. Riou. 1983. A specific inhibitor of type I DNA-topoisomerase of *Trypanosoma cruzi:* Dimethyl-hydroxy-ellipticinium. *Biochem. Biophys. Res. Commun.* **117:** 1–5.

———. 1984. Inhibition of the reactions catalysed by a type I topoisomerase and a catenating enzyme of *Trypanosoma cruzi* by DNA-intercalating drugs. Preferential inhibition of the catenating reaction. *EMBO J.* **3:** 11–16.

Durban, E., M. Goodenough, J. Mills, and H. Busch. 1985. Topoisomerase I phosphorylation *in vitro* and in rapidly growing Novikoff hepatoma cells. *EMBO J.* **4:** 2921–2926.

Durban, E., J.S. Mills, D. Roll, and H. Busch. 1983. Phosphorylation of purified Novikoff hepatoma topoisomerase I. *Biochem. Biophys. Res. Commun.* **111:** 897–905.

Edwards, K.A., B.D. Halligan, J.L. Davis, N.L. Nivera, and L.F. Liu. 1982. Recognition sites of eukaryotic DNA topoisomerase I: DNA nucleotide sequencing analysis of topo I cleavage sites on SV40 DNA. *Nucleic Acids Res.* **10:** 2565–2576.

Eisenberg, S. and A. Kornberg. 1979. Purification and characterization of ϕX174 gene *A* protein. A multifunctional enzyme of duplex DNA replication. *J. Biol. Chem.* **254:** 5328–5332.

Eisenberg, S., J. Griffith, and A. Kornberg. 1977. ϕX174 *cistron A* protein is a multifunctional enzyme in DNA replication. *Proc. Natl. Acad. Sci.* **74:** 3198–3202.

Eng, W.-K., S.D. Pandit, and R. Sternglanz. 1989. Mapping of the active site tyrosine of eukaryote DNA topoisomerase I. *J. Biol. Chem.* **264:** 13373–13376.

Eng, W.-K., L. Faucette, R.K. Johnson, and R. Sternglanz. 1988. Evidence that DNA topoisomerase I is necessary for the cytotoxic effects of camptothecin. *Mol. Pharmacol.* **34:** 755–760.

Fairfield, F.R., W.R. Bauer, and M.Y. Simpson. 1979. Mitochondria contain a distinct DNA topoisomerase. *J. Biol. Chem.* **254:** 9352–9354.

Ferro, A.M. and B. Olivera. 1984. Poly(ADP-ribosylation) of DNA topoisomerase I from calf thymus. *J. Biol. Chem.* **259:** 547–554.

Ferro, A.M., N.P. Higgins, and B.M. Olivera. 1983. Poly(ADP-ribosylation) of a DNA topoisomerase. 1983. *J. Biol. Chem.* **258:** 6000–6003.

Giaever and Wang. 1988. Supercoiling of intracellular DNA can occur in eukaryotic cells. *Cell* **55:** 849–856.

Gilmour, D.S. and S.C.R. Elgin. 1987. Localization of specific topoisomerase I interactions within the transcribed region of active heat shock genes by using the inhibitor camptothecin. *Mol. Cell. Biol.* **7:** 141–148.

Gocke, E., B.J. Bonven, and O. Westergaard. 1983. A site and strand specific nuclease activity with analogies to topoisomerase I frames the rRNA gene of *Tetrahymena*. *Nucleic Acids Res.* **11:** 7661–7678.

Gronostajski, R.M. and P.D. Sadowski. 1985. The FLP recombinase of the *Saccharomyces cerevisiae* 2 μm plasmid attaches covalently to DNA via a phosphotyrosyl linkage. *Mol. Cell. Biol.* **5:** 3274–3279.

Guiney, D.C. and D.R. Helinski. 1975. Relaxation complexes of plasmid DNA and protein. III. Assocation of protein with the 5′ terminus of the broken DNA strand in the relaxed complex of plasmid colE1. *J. Biol. Chem.* **250:** 8796–8803.

Guynn, R.W. and H. Thames. 1982. Equilibrium constants under physiological conditions for the reactions of L-phosphoserine phosphatase and pyrophosphate: L-serine phosphotransferase. *Arch. Biochem. Biophys.* **215:** 514–523.

Halligan, B.D., J.L. Davis, K.A. Edwards, and L.F. Liu. 1982. Intra- and intermolecular strand transfer by HeLa DNA topoisomerase I. *J. Biol. Chem.* **257:** 3995–4000.

Hatfull, C.F. and N.D.F. Grindley. 1986. Analysis of γδ resolvase mutants *in vitro:* Evidence for an interaction between serine-10 of resolvase and site I of *res*. *Proc. Natl. Acad. Sci.* **83:** 5429–5433.

Holden, J.A. and R.L. Low. 1985. Characterization of a potent catenation activity of HeLa cell nuclei. *J. Biol. Chem.* **260:** 14491–14497.

Holzer, H. and R. Wohlhueter. 1972. (Glutamine synthetase) tyrosyl-O-adenylate: A new energy-rich phosphate bond. *Adv. Enzyme Regul.* **10:** 121–132.

Horowitz, S.B., C. Chang, and A.P. Grollman. 1971. Studies on camptothecin. I. Effects on nucleic acid and protein synthesis. *Mol. Pharmacol.* **7:** 632–644.

Hsiang, Y.-H., R. Hertzberg, S. Hecht, and L.F. Liu. 1985. Camptothecin induces protein-linked DNA breaks via mammalian DNA topoisomerase I. *J. Biol. Chem.* **260:** 14873–14878.

Ikeda, J.-E., A. Yudelevich, and J. Hurwitz. 1976. Isolation and characterization of the protein coded by gene *A* of bacteriophage ϕX174 DNA. *Proc. Natl. Acad. Sci.* **73:** 2669–2673.

Ikeda, J.-E., A. Yudelevich, N. Shimamoto, and J. Hurwitz. 1979. Role of polymeric forms of bacteriophage ϕX174 coded gene *A* protein in ϕXRFI DNA cleavage. *J. Biol. Chem.* **254:** 9416–9428.

Jongstra-Bilen, J., M.-E. Ittel, C. Niedergang, H.-P. Vosberg, and P. Handel. 1983. DNA topoisomerase I from calf thymus is inhibited *in vitro* by poly(ADP-ribosylation). *Eur. J. Biochem.* **136:** 391–396.

Kanaar, R., P. van de Putte, and N.R. Cozzarelli. 1988. Gin-mediated DNA inversion: Product structure and the mechanism of strand exchange. *Proc. Natl. Acad. Sci.* **85:** 752–756.

Kikuchi, Y. and H. Nash. 1979. Nicking-closing activity associated with bacteriophage λ *int* gene product. *Proc. Natl. Acad. Sci.* **76:** 3760–3764.

Kirkegaard, K. and J.C. Wang. 1978. *Escherichia coli* DNA topoisomerase I catalyzed linking of single-stranded rings of complementary base sequences. *Nucleic Acids Res.* **5:** 3811–3820.

———. 1985. Bacterial DNA topoisomerase I can relax positively supercoiled DNA containing a single-stranded loop. *J. Mol. Biol.* **185:** 625–637.

Kirkegaard, K., G. Plugfelder, and J.C. Wang. 1984. The cleavage of DNA by type-I DNA topoisomerases. *Cold Spring Harbor Symp. Quant. Biol.* **49:** 411–419.

Kjeldsen, E., S. Mollerup, B. Thomsen, B.J. Bonven, L. Bolund, and O. Westergaard. 1988. Sequence-dependent effect of camptothecin on human topoisomerase I DNA cleavage. *J. Mol. Biol.* **202:** 333–342.

Klippel, A., G. Mertens, T. Patschinsky, and R. Kahmann. 1988. The DNA invertase Gin of phage Mu: Formation of a covalent complex with DNA via a phosphoserine at amino acid position 9. *EMBO J.* **7:** 1229–1237.

Krasnow, M.A. and N.R. Cozzarelli. 1983. Site-specific relaxation and recombination by the Tn3 resolvase: Recognition of the DNA path between oriented *res* sites. *Cell* **32:** 1313–1324.

Liu, L.F. and J.C. Wang. 1979. Interaction between DNA and *Escherichia coli* DNA topoisomerase I. Formation of complexes between the protein and superhelical and nonsuperhelical duplex DNAs. *J. Biol. Chem.* **254:** 11082–11088.

———. 1987. Supercoiling of the DNA template during RNA transcription. *Proc. Natl. Acad. Sci.* **84:** 7024–7027.

Liu, L.F., R.E. Depew, and J.C. Wang. 1976. Knotted single-stranded DNA rings: A novel topological isomer of circular single-stranded DNA formed by treatment with *Escherichia coli* ω protein. *J. Mol. Biol.* **106:** 439–452.

Low, R.L. and J.A. Holden. 1985. Inhibition of HeLa cell DNA topoisomerase I by ATP and phosphate. *Nucleic Acids Res.* **13:** 6999–7014.

Low, R.L., J.M. Kaguni, and A. Kornberg. 1984. Potent catenation of supercoiled and gapped DNA circles by topoisomerase I in the presence of a hydrophilic polymer. *J. Biol. Chem.* **259:** 4576–4581.

Lynn, R.M. and J.C. Wang. 1989. Peptide sequencing and site-directed mutagenesis identify tyrosine 319 as the active site tyrosine of *Escherichia coli* DNA topoisomerase I. *Proteins* **6:** 231–239.

Lynn, R.M., M.A. Bjornsti, P.R. Caroh, and J.C. Wang. 1989. Peptide sequencing and site-directed mutagenesis identify tyrosine 727 as the active site tyrosine of *Saccharomyces cerevisiae* DNA topoisomerase I. *Proc. Natl. Acad. Sci.* **86:** 3559–3563.

Mattern, M.R., S.-M. Hong, H.F. Bartus, C.K. Mirabelli, S.T. Crooke, and R.D. Johnson. 1987. Relationship between the intracellular effects of camptothecin and the inhibition of DNA topoiomserase I in cultured L1210 cells. *Cancer Res.* **47:** 1793–1798.

Maxwell, A. and M. Gellert. 1986. Mechanistic aspects of DNA topoisomerases. *Adv. Protein Chem.* **38:** 69–107.

McConaughy, B.L., L.S. Young, and J.J. Champoux. 1981. The effect of salt on the binding of the eucaryotic DNA nicking-closing enzyme to DNA and chromatin. *Biochim. Biophys. Acta* **655:** 1–8.

McCoubrey, W.K., Jr. and J.J. Champoux. 1986. The role of single-strand breaks in the catenation reaction catalyzed by the rat type I topoisomerase. *J. Biol. Chem.* **261:** 5130–5137.

Meyer, T.F. and K. Geider. 1979. Bacteriophage fd gene II-protein. II. Specific cleavage and relaxation of supercoiled RF from filamentous phages. *J. Biol. Chem.* **254:** 12642–12646.

Mills, J.S., H. Busch, and E. Durban. 1982. Purification of a protein kinase from human Namalwa cells that phosphorylates topoisomerase I. *Biochem. Biophys. Res. Commun.* **109:** 1222–1227.

Mizuuchi, K., M. Gellert, and H. Nash. 1978. Involvement of super-twisted DNA in integrative recombination of bacteriophage lambda. *J. Mol. Biol.* **121:** 375–392.

Muller, M.T. 1985. Quantitation of eukaryotic topoisomerase I reactivity with DNA. Preferential cleavage of supercoiled DNA. *Biochim. Biophys. Acta* **824:** 263–267.

Nash, H.A., C.E. Bauer, and J.F. Gardner. 1987. Role of homology in site-specific recombination of bacteriophage lambda: Evidence against joining of cohesive ends. *Proc. Natl. Acad. Sci.* **84:** 4049–4053.

Nitiss, J. and J.C. Wang. 1988. DNA topoisomerase-targeting antitumor drugs can be studied in yeast. *Proc. Natl. Acad. Sci.* **85:** 7501–7505.

Peller, L. 1976. On the free-energy changes in the synthesis and degradation of nucleic acids. *Biochemistry* **15:** 141–146.

Poccia, D.L., D. LeVine, and J.C. Wang. 1978. Activity of a DNA topoisomerase (nicking-closing enzyme) during sea urchin development and the cell cycle. *Develop. Biology* **64:** 273–283.

Pommier, Y., J.M. Covey, D. Kerrigan, J. Markovits, and R. Pham. 1987. DNA unwinding and inhibition of mouse leukemia L1210 DNA topoisomerase I by intercalators. *Nucleic Acids Res.* **15:** 6713–6731.

Porter, S.E. and J.J. Champoux. 1989a. Mapping *in vivo* topoisomerase I sites on simian virus 40 DNA: Asymmetric distribution of sites on replicating molecules. *Mol. Cell. Biol.* **9:** 541–550.

———. 1989b. The basis for camptothecin enhancement of DNA breakage by eukaryotic topoisomerase I. *Nucleic Acids Res.* **17:** 8521–8532.

Prell, B. and H.-P. Vosberg. 1980. Analysis of covalent complexes formed between calf thymus DNA topoisomerase and single-stranded DNA. *Eur. J. Biochem.* **108:** 389–398.

Pruss, G.J., S.H. Manes, and K. Drlica. 1982. *Escherichia coli* DNA topoisomerse I mutants: Increased supercoiling is corrected by mutations near gyrase genes. *Cell* **31:** 35–42.

Pulleyblank, D.E., M. Shure, D. Tang, J. Vinograd, and H.-P. Vosberg. 1975. Action of nicking-closing enzyme on supercoiled and nonsupercoiled closed circular DNA: Formation of a Boltzmann distribution of topological isomers. *Proc. Natl. Acad. Sci.* **72:** 4280–4284.

Reed, R.R. 1981. Transposon-mediated site-specific recombination: A defined *in vitro* system. *Cell* **25:** 713–719.

Reed, R.R. and N.D.F. Grindley. 1981. Transposon-mediated site-specific recombination *in vitro:* DNA cleavage and protein-DNA linkage at the recombination site. *Cell* **25:** 721–728.

Reed, R.R. and C.D. Moser. 1984. Resolvase-mediated recombination intermediates contain a serine residue covalently linked to DNA. *Cold Spring Harbor Symp. Quant. Biol.* **49:** 245–249.

Richet, E., P. Abcarian, and H.A. Nash. 1986. The interaction of recombination proteins with supercoiled DNA: Defining the role of supercoiling in lambda integrative recombination. *Cell* **46:** 1011–1021.

Sadowski, P. 1986. Site-specific recombinases: Changing partners and doing the twist. *J. Bacteriol.* **165:** 341–347.

Shishido, K., N. Noguchi, and T. Ando. 1983. Correlation of enzyme-induced cleavage sites on negatively superhelical DNA between prokaryotic topoisomerase I and S_1 nuclease. 1983. *Biochim. Biophys. Acta* **740:** 108–117.

Shuman, S. and B. Moss. 1987. Identification of a vaccinia virus gene encoding a type I DNA topoisomerase. *Proc. Natl. Acad. Sci.* **84:** 7478–7482.

Siedlecki, J., W. Zimmermann, and A. Weissbach. 1983. Characterization of a prokaryotic topoisomerase I activity in chloroplast extracts from spinach. *Nucleic Acids Res.* **11:** 1523–1536.

Snapka, R.M. 1986. Topoisomerase inhibitors can selectively interfere with different stages of simian virus 40 DNA replication. *Mol. Cell. Biol.* **6:** 4221–4227.

Srivenugopal, K.S. and D.R. Morris. 1986. Modulation of the relaxing activity of *Escherichia coli* topoisomerase I by single-stranded DNA binding proteins. *Biochem. Biophys. Res. Commun.* **137:** 795–800.

Stewart, A.F. and G. Schutz. 1987. Camptothecin-induced *in vivo* topoisomerase I cleavages in the transcriptionally active typrosine aminotransferase gene. *Cell* **50:**

1109–1117.
Thomsen, B., S. Mollerup, B.J. Bonven, R. Frank, H. Blocker, O.F. Nielsen, and O. Westergaard. 1987. Sequence specificity of DNA topoisomerase I in the presence and absence of camptothecin. *EMBO J.* **6:** 1817–1823.
Trask, D.K. and M.T. Muller. 1983. Biochemical characterization of topoisomerase I purified from avian erythrocytes. *Nucleic Acids Res.* **11:** 2779–2800.
———. 1988. Stabilization of type I topoisomerase-DNA covalent complexes by actinomycin D. *Proc. Natl. Acad. Sci.* **85:** 1417–1421.
Tse, Y.-C. and J.C. Wang. 1980. *E. coli* and *M. luteus* DNA topoisomerase I can catalyze catenation or decatenation of double-stranded DNA rings. *Cell* **22:** 269–276.
Tse, Y.-C., K. Javaherian, and J.C. Wang. 1984. HMG17 protein facilitates the DNA catenation reaction catalyzed by DNA topoisomerases. *Arch. Biochem. Biophys.* **231:** 169–174.
Tse, Y.-C., K. Kirkegaard, and J.C. Wang. 1980. Covalent bonds between protein and DNA. Formation of phosphotyrosine linkage between certain DNA topoisomerases and DNA. *J. Biol. Chem.* **255:** 5560–5565.
Tse-Dinh, Y.-C. 1986. Uncoupling of the DNA breaking and rejoining steps of *Escherichia coli* type I DNA topoisomerase. *J. Biol. Chem.* **261:** 10931–10935.
Tse-Dinh, Y.-C. and J.C. Wang. 1986. Complete nucleotide sequence of the *topA* gene encoding *Escherichia coli* DNA topoisomerase I. *J. Mol. Biol.* **191:** 321–331.
Tse-Dinh, Y.-C., B.G.H. McCarron, R. Arentzen, and V. Chowdhry. 1983. Mechanistic study of *E. coli* DNA topoisomerase I: Cleavage of oligonucleotides. *Nucleic Acids Res.* **11:** 8691–8701.
van Mansfeld, A.D.M., P.D. Baas, and H.S. Jansz. 1984. Gene *A* protein of bacteriophage ϕX174 is a highly specific single-strand nuclease and binds via a tyrosyl residue to DNA after cleavage. *Adv. Exp. Med. Biol.* **179:** 221–230.
van Mansfeld, A.D.M., H.A.A.M. van Teeffelen, P.D. Bass, and H.S. Jansz. 1986. Two juxtaposed tyrosyl-OH groups participate in ϕX174 gene *A* protein catalysed cleavage and ligation of DNA. *Nucleic Acids Res.* **14:** 4229–4238.
Vetter, D., B.J. Andrews, L. Roberts-Batty, and P.D. Sadowski. 1983. Site-specific recombination of yeast 2-μm DNA *Proc. Natl. Acad. Sci.* **80:** 7284–7288.
Wang, J.C. 1971. Interaction between DNA and an *Escherichia coli* protein ω. *J. Mol. Biol.* **55:** 523–533.
———. 1982. DNA topoisomerases. In *Nucleases* (ed. S.M. Linn and R.J. Roberts), p. 41–57. Cold Spring Harbor Laboratory, Cold Spring Harbor, New York.
———. 1987. Recent studies of DNA topoisomerases. *Biochim. Biophys. Acta* **909:** 1–9.
Willets, N. and B. Wilkins. 1984. Processing of plasmid DNA during bacterial conjugation. *Microbiol. Rev.* **48:** 24–41.
Wu, H.-Y., S. Shyy, J.C. Wang, and L.F. Liu. 1988. Transcription generates positively and negatively supercoiled domains in the template. *Cell* **53:** 433–440.
Zhang, H., J.C. Wang, and L.F. Liu. 1988. Involvement of DNA topoisomerase I in transcription of human ribosomal RNA genes. *Proc. Natl. Acad. Sci.* **85:** 1060–1064.

7

Mechanistic Aspects of Type-II DNA Topoisomerases

Tao-shih Hsieh
Department of Biochemistry
Duke University Medical Center
Durham, North Carolina 27705

Escherichia coli DNA topoisomerase I and II (Wang and Liu 1979) were names given to the ω protein (Wang 1971) and DNA gyrase (Gellert et al. 1976b), respectively, to distinguish these two enzymes from the same organism. Later developments showed that these two enzymes differ in their mechanisms of action: The former breaks and rejoins one DNA strand at a time, and the latter breaks and rejoins a pair of strands in a duplex DNA in a somewhat concerted manner. This mechanistic difference provides the basis for classifying *E. coli* DNA topoisomerase I and other activities that break and rejoin one DNA strand at a time as type-I DNA topoisomerases and *E. coli* gyrase and other activites that transiently break duplex DNAs as type-II DNA topoisomerases (Brown and Cozzarelli 1979; Liu et al. 1980). Several reviews have appeared and covered the various stages of development in the field (Wang and Liu 1979; Cozzarelli 1980; Gellert 1981; Liu 1983; Wang 1985; Maxwell

 0-87969-348-7/90 $1.00 + .00

and Gellert 1986). This chapter summarizes the mechanistic aspects of type-II DNA topoisomerases and some of their biochemical properties that bear directly on their mechanism of action.

I. INTERCONVERSION OF DNA TOPOISOMERS

As a consequence of the transient breakage of DNA strands and the passage of duplex DNA segments through these breaks, a type-II DNA topoisomerase can alter the topology of DNA rings (Fig. 1). It is well established that covalently closed circular duplex DNA can serve as a substrate in these reactions, which reflects the double-stranded cleavage and rejoining mechanism of type-II topoisomerases.

A. Relaxation and Supercoiling

Among all known type-II topoisomerases, only bacterial DNA gyrase can introduce negative superhelical turns into the DNA substrate by using the chemical energy of ATP hydrolysis (Gellert et al. 1976a; Liu and Wang 1978a; Sugino and Bott 1980). The other type-II enzymes, including eukaryotic DNA topoisomerase II and bacteriophage T4 DNA topoisomerase, can relax both positively and negatively supercoiled DNA molecules but cannot drive a relaxed DNA into the supercoiled form. There has been no detailed kinetic analysis of the effect of the sign of DNA supercoiling on the relaxation rate by these enzymes. Bacterial DNA gyrase can readily remove positive supercoils in the presence of

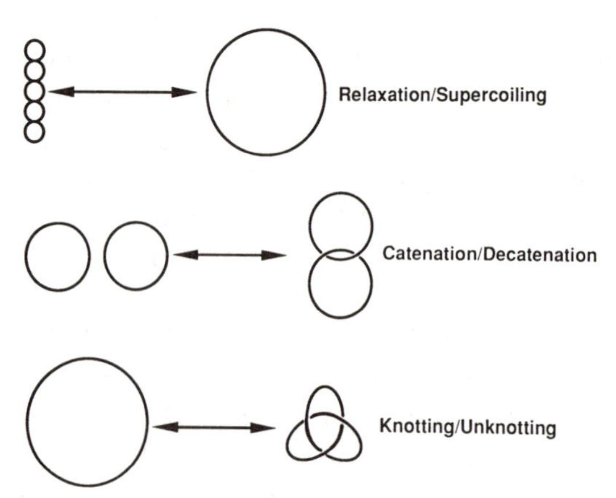

Figure 1 Schematic diagrams of DNA topoisomerization reactions mediated by type-II DNA topoisomerase.

ATP. It can also relax negatively supercoiled but not positively supercoiled DNA in the absence of ATP, but the relaxation reaction is much slower compared with the DNA supercoiling reaction (Higgins et al. 1978; Gellert et al. 1979). One intriguing observation about bacterial DNA gyrase is that the enzyme is able to catalytically relax positive supercoils in the presence of the nonhydrolyzable βγ-imido analog of ATP (Gellert et al. 1981). The mechanistic basis for this observation is as yet unknown.

A hallmark in the type-II DNA topoisomerase catalyzed relaxation or supercoiling reaction is that the linking number of the DNA ring changes in steps of two (Brown and Cozzarelli 1979; Hsieh and Brutlag 1980; Liu et al. 1980). This result substantiates the mechanism of action of topoisomerase II, the passage of a duplex DNA segment through a reversible double-stranded break (Crick 1976; Fuller 1978).

B. Catenation and Decatenation

DNA topoisomerase II mediates both the catenation and decatenation of double-stranded DNA (dsDNA) rings; the equilibrium position of which de-pends on the state of DNA condensation. If the DNA molecules are in an aggregated state, catenated networks of circles are formed in the presence of the enzyme; in the absence of reagents that affect DNA condensation, the catenanes can be resolved by the enzyme into separate circles. Effective condensation reagents in this reaction include DNA-binding proteins (Hsieh and Brutlag 1980; Goto and Wang 1982), histone H1 (Liu et al. 1980; Hsieh and Brutlag 1980), polycations such as spermidine and $Co^{3+}(NH_3)_6$ (Baldi et al. 1980; Kreuzer and Cozzarelli 1980; Krasnow and Cozzarelli 1982), and DNA topoisomerase II itself at a high concentration (Hsieh 1983).

Both the catenation and decatenation reactions have been used to assay topoisomerase II activity. The decatenation reaction is particularly interesting because one of the important biological functions of topoisomerase II is the resolution of newly replicated duplex daughter chromosomes that are multiply intertwined (Uemura and Yanagida 1984, 1986; DiNardo et al. 1984; Steck and Drlica 1984; Holm et al. 1985; Uemura et al. 1987; Yang et al. 1987). It has been suggested that the pulling force generated by the spindles during the anaphase provides the necessary directionality for the topoisomerase-II-mediated decatenation reaction in yeast cells (Holm et al. 1989). Using multiply intertwined catenated DNA dimers as the substrate, the biochemical properties of the in vitro decatenation reaction catalyzed by *E. coli* DNA gyrase have also been analyzed (Marians 1987).

C. Knotting and Unknotting

Introducing topological knots into a circular DNA can be viewed as an intramolecular equivalent of the catenation reaction. At a high enzyme concentration, both T4 DNA topoisomerase and *Drosophila* topoisomerase II can tie topological knots into circular DNA molecules, suggesting that these enzymes might play a role in bringing DNA segments together in addition to providing the requisite strand passage activity (Liu et al. 1980; Hsieh 1983). Some aspects in the DNA knotting reaction mediated by these two enzymes are different. The knotted DNA products are transient in the reaction by the T4 enzyme, and the reaction appears to require supercoiled DNA substrate. In the knotting reaction by the *Drosophila* enzyme, the DNA products are stable, and either supercoiled or nicked DNA can serve as substrate. These differences probably reflect a difference in the kinetic stability of the DNA/protein complexes. Similar to the decatenation reaction, type-II DNA topoisomerases can also untie topological knots. The knotted DNA molecules isolated from bacteriophage P4 capsids have been used as a convenient source of DNA substrate for assays of type-II DNA topoisomerases (Liu et al. 1981).

II. QUATERNARY STRUCTURE OF TYPE-II DNA TOPOISOMERASES

The apparently concerted double-stranded cleavage-rejoining mediated by these enzymes suggests that they are dimeric and that interactions between the identical halves of each enzyme are responsible for coordinating the cleavage-rejoining of the two DNA strands.

A. Bacterial DNA Gyrases

These enzymes are made of heterotypic tetramers, A_2B_2, with subunits A (~100 kD) and B (~90 kD) encoded by the gyrase *gyr*A and *gyr*B genes, respectively (Higgins et al. 1978; Liu and Wang 1978a; Mizuuchi et al. 1978; Klevan and Wang 1980; Orr and Staudenbauer 1982; Moriya et al. 1985). Both biochemical and genetic data demonstrate that the gyrase B subunit is involved in the ATP hydrolysis and energy transduction, and the gyrase A subunit is involved in the DNA breakage and reunion. An enzyme related to DNA gyrase, topoisomerase II′, has been isolated from *E. coli*, and it has a structure of $A_2B'_2$ (Brown et al. 1979; Gellert et al. 1979). The A subunit is identical to the gyrase A subunit, whereas the B′ subunit is a proteolytic fragment with 393 amino acid residues removed from the amino terminal of the gyrase B subunit (Adachi et al. 1987). DNA topoisomerase II′, unlike DNA gyrase, cannot supercoil

DNA. It can relax both positively and negatively supercoiled DNA, however, in an ATP-independent manner.

B. Bacteriophage T4 DNA Topoisomerase

Three genes encode bacteriophage T4 DNA topoisomerase: genes *39*, *52* and *60* (Liu et al. 1979; Stetler et al. 1979). The sizes of the genes *39*, *52* and *60* peptides based on the nucleotide sequences of the cloned genes are, respectively, 58 kD, 50 kD, and 18 kD (Huang 1986a,b; Huang et al. 1988). Gel filtration data suggest that T4 topoisomerase II is heterogeneous in size and that each holoenzyme contains at least two of each of the three subunits (Kreuzer and Huang 1983). The sequence homology between T4 gene *39* and bacterial *gyrB* suggests that gene *39* is involved in ATP hydrolysis (Huang 1986b). Similarly, sequence homology between T4 gene *52* and bacterial *gyrA* gene implicates that the gene *52* protein is the one involved in DNA breakage and reunion (Huang 1986a). The gene *52* subunit has indeed been found to form a covalent linkage with DNA in the DNA cleavage reaction mediated by T4 topoisomerase (Rowe et al. 1984).

The biochemical function of the gene *60* subunit in the T4 topoisomerase II is as yet unknown. However, the sequence of gene *60* protein contains stretches of homology with the carboxy-terminal regions of bacterial *gyrB*. The structure of T4 gene *60* also contains an intriguing, untranslated interruption; the interruption sequence appears to persist in the mature mRNA, suggesting that the translation machinery can skip it (Huang et al. 1988).

C. Eukaryotic Type-II Topoisomerases That Are Located in the Nucleus

All these enzymes are homotypic dimers with subunit molecular masses around 170 kD. These enzymes are likely to be closely related to each other at both the functional and structural level. Nucleotide sequencing of genes encoding DNA topoisomerase II from various organisms shows that there is interesting homology among these enzymes. The sequence information further indicates that the polypeptide of eukaryotic topoisomerase II forms at least three domains: an amino-terminal region with homology with the B subunit of bacterial gyrase, a central region with homology with the A subunit of gyrase, and a carboxy-terminal region characterized by clusters of charged amino acids (Lynn et al. 1986; Uemura et al. 1986; Wyckoff et al. 1989; see Appendix, this volume).

The functional relatedness of eukaryotic topoisomerase II from various organisms is supported by the observations that the topoisomerase II structural genes from *Saccharomyces cerevisiae* and *Schizosaccharomyces pombe* can genetically complement each other (Uemura et al. 1986) and that the *Drosophila* and human enzyme can genetically complement the deficiency of yeast *top2* (Wyckoff and Hsieh 1988; M. Tsai-Pflugfelder and J.C. Wang, unpubl.).

It is also interesting to note that a type-II DNA topoisomerase purified from *Trypanosoma cruzi* can carry out the topoisomerization reactions in the absence of ATP (Douc-Rasy et al. 1986). The cytological location of this enzyme and its structural relationship to other eukaryotic enzymes are as yet unknown. Recently, a DNA sequence from *Trypanosoma brucei* has been cloned based on its sequence homology with other topoisomerase II (Strauss and Wang 1990). Nucleotide sequence analysis of the cloned gene from *T. brucei* suggests it encodes a protein very similar to the nuclear DNA topoisomerases of other eukaryotes.

D. Kinetoplast Topoisomerase II

An ATP-dependent type-II DNA topoisomerase has been purified to homogeneity from the trypanosomatid *Crithidia fasciculata* (Melendy and Ray 1989). The isolated enzyme is a homodimer of 132-kD polypeptides, and immunocytochemical experiments have demonstrated the preferential localization of this enzyme at two distinct sites at the periphery of the kinetoplast (Melendy et al. 1988). Its enzymatic properties are very similar to the nuclear eukaryotic enzymes, however, and it should be interesting to establish whether the enzyme is distinct from the nuclear enzyme.

III. WRAPPING OF DNA AROUND TOPOISOMERASE II

Several lines of evidence suggest that a segment of DNA wraps around bacterial DNA gyrase molecules. Nuclease digestion experiments show that bacterial DNA gyrase can protect a DNA segment of about 110–160 bp from various nucleases (Liu and Wang 1978b; Klevan and Wang 1980; Fisher et al. 1981; Kirkegaard and Wang 1981; Morrison and Cozzarelli 1981). The DNase I footprint experiment revealed that the flanking regions of the gyrase-protected fragment, each about 50 bp in length, contain DNase-I-sensitive sites spaced roughly at 10 nucleotides intervals. The DNase I cleavage sites in the complementary DNA strands are staggered by about 2–4 nucleotides (Fisher et al. 1981; Kirkegaard and Wang 1981). These results are similar to the nuclease protection experiments with nucleosomes (see A. Travers and A. Klug, this volume), strongly suggesting the wrapping of DNA on the surface of gyrase molecules. In marked contrast with this protection against DNase I digestion, the binding of DNA gyrase does not prevent the alkylation of DNA by DMS, consistent with the notion that DNA is wrapped around the enzyme and is accessible to the solvent (Kirkegaard and Wang 1981).

In the absence of ATP, binding of bacterial gyrase molecules can increase the linking number of a DNA when a nicked DNA is sealed by DNA ligase (Liu and Wang 1978a). This observation is consistent with a change in the DNA writhe as a result of right-handed superhelical wrapping around a protein. DNA wrapping as assayed by its stoichiometric induction of positive supercoiling is sensitive to temperature; reducing the temperature from 25°C to 0°C results in a gradual loss of the right-handed superhelical wrapping (Gellert et al. 1981).

Electron microscopy also indicates that there is a reduction of DNA contour length by about 110–160 bp when the gyrase/DNA complex is formed (Kirchhausen et al. 1985). These results, when taken together, suggest that the formation of a gyrase/DNA complex involves a stretch of about 150-bp DNA wrapped around the enzyme by roughly one turn. As described earlier, the handedness of wrapping is right-handed, and this defined handedness may impose a directionality in the change of the linking number catalyzed by DNA gyrase. For example, the outside to inside passage of a DNA segment through a gyrase/DNA complex in which the DNA is wrapped around the enzyme in the right-handed manner will always result in the *reduction* of linking number by two (Fig. 2). The notion that DNA wrapping is important mechanistically because the supercoiling action of DNA gyrase is also supported by the observation that DNA gyrase cannot introduce negative supercoils into a DNA circle with a size smaller than 152 bp (Bates and Maxwell 1989). These small DNA circles, nevertheless, can serve as substrates for relaxation by gyrase. The size of these DNA circles may limit the extent of DNA wrapping by gyrase and thereby may interfere with its DNA supercoiling action; however, alternative interpretations are also plausible.

In contrast with the DNA/bacterial gyrase complex, there is no evidence for DNA wrapping around eukaryotic DNA topoisomerases II. The DNase I footprint experiment for the *Drosophila* topoisomerase II/DNA complex revealed a protected region of about 20–30 bp in length and no indication of DNase-I-sensitive sites spaced at an interval of 10 nucleotides (Lee et al. 1989b). The passage of DNA segments through a DNA gate for which there is no defined topological structure will only result in changing DNA topology toward the most thermodynamically stable state of DNA. This indeed characterizes all the reactions catalyzed by eukaryotic DNA topoisomerase II.

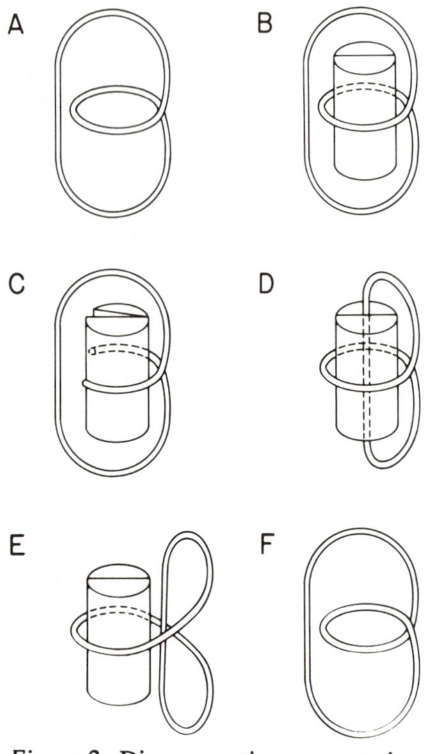

Figure 2 Diagrammatic representation of the DNA topological change during the gyrase supercoiling action (*A–F*). dsDNA and gyrase molecules are represented by the ribbon and the cylinder, respectively. The starting substrate for a cycle of gyrase supercoiling action is shown in *A* and *B*. The product of this reaction is given in *E* and *F*. The putative intermediates of the reaction are diagrammatically shown in *C* and *D*. Note the linking number change between *A*/*B* and that of *E*/*F* is –2.

Whereas DNA wrapped around the gyrase molecule contains the breakage-reunion site (DNA gate), the spatial relationship between the DNA segment to be translocated through this DNA gate and the wrapped DNA has not been determined. Since bacterial DNA gyrase can catalyze both the catenation and decatenation reactions, it is apparent that the translocated DNA segment does not have to be linked tandemly to the wrapped segment. It is also interesting to note that the results from electric dichroism experiments suggest a compaction in the gyrase/DNA complex upon addition of ATP or its analogs (Rau et al. 1987). It is possible that this structural transition might be a prerequisite for bringing together the DNA segment to be translocated through the DNA gate.

IV. ATP HYDROLYSIS

Except for the proteolytic product topoisomerase II′ of *E. coli* (Brown et al. 1979; Gellert et al. 1979) and partially purified activity from trypanosome (Douc-Rasy et al. 1986), all known type-II topoisomerases require ATP as well as its hydrolysis to ADP and orthophosphate for their catalytic activities in the topoisomerization reactions. The requirement for ATP hydrolysis is not restricted to type-II topoisomerases; "reverse gyrase," a type-I topoisomerase from archaebacteria that can introduce positive superhelical turns into DNA, also requires ATP (Kikuchi and Asai 1984; Forterre et al. 1985; see also, Kikuchi, this volume). ATP hydrolysis is obviously needed to provide the energy for bacterial gyrase and archaebacterial reverse gyrase to catalyze the thermodynamically uphill reactions. It is less apparent as to why ATP hydrolysis is required for other type-II topoisomerases since the reactions they catalyze follow a descending path into a thermodynamic valley.

Using the ATP analog in which the covalent bond between the β and γ phosphate groups cannot be hydrolyzed, there is only a limited negative supercoiling by bacterial gyrase (Sugino et al. 1978) and relaxation of supercoils by T4 topoisomerase II (Liu et al. 1979) and *Drosophila* topoisomerase II (Osheroff et al. 1983). These observations have led to the suggestion that the hydrolysis of ATP is necessary for reinitiating the catalytic cycle after the strand passage step (Peebles et al. 1979; Wang et al. 1981; Osheroff 1986).

The ATPase activity associated with type-II DNA topoisomerases is greatly stimulated by the presence of DNA. A preference for double-stranded over the single-stranded DNA (ssDNA) as an effector for ATPase has been demonstrated for DNA gyrase (Mizuuchi et al. 1978; Sugino and Cozzarelli 1980), T4 topoisomerase (Liu et al. 1979), and *Drosophila* topoisomerase II (Osheroff et al. 1983). Although the requirement in terms of the structure and sequence of the effector DNA is not stringent, there is a clear correlation between enzyme binding and the ability to stimulate ATPase. For DNA gyrase, relaxed DNA, the substrate for its supercoiling, forms a tighter complex with the enzyme than does negatively supercoiled DNA (Higgins and Cozzarelli 1982). Negatively supercoiled DNA is also less efficient than linear, nicked circular, or relaxed DNA in stimulating ATPase (Sugino and Cozzarelli 1980). Positive supercoiling of DNA around gyrase disfavors the binding of negatively supercoiled DNA. In an interesting contrast, the *Drosophila* topoisomerase II binds negatively supercoiled DNA with a higher affinity, and supercoiled DNA is a more efficient effector than linear or relaxed DNA (Osheroff et al. 1983). The length of the DNA fragment is also a factor in determining gyrase-binding affinity and efficiency in stimulating ATPase activity. For *Micrococcus luteus* gyrase, a 240-bp DNA fragment is more potent as an effector DNA than a 100-bp fragment (Klevan and Tse 1983). *E. coli* DNA gyrase appears to require a DNA fragment of at least 70 bp to serve as ATPase effector DNA and to form a tight enzyme/DNA complex (Maxwell and Gellert 1984). At much higher concentrations, DNA molecules shorter than 70 bp can stimulate ATPase activity and form a complex with DNA gyrase, thereby suggesting that a gyrase/DNA complex can form with more than one molecule of the shorter DNA fragment per enzyme (Maxwell and Gellert 1984).

The ATPase site is also the target for various inhibitors of topoisomerase II. The ATPase activity and the DNA supercoiling action of gyrase can be efficiently inhibited by the coumarin drugs coumermycin A1 and novobiocin (Mizuuchi et al. 1978; Sugino et al. 1978). At high concentrations, these antibiotics can also inhibit eukaryotic topoisomerase II (Hsieh and Brutlag 1980; Miller et al. 1981; Goto and Wang 1982).

V. DOUBLE-STRANDED DNA CLEAVAGE

E. coli DNA gyrase induces the double-stranded breakage of DNA if the reaction is carried out in the presence of an antibiotic like oxolinate or nalidixate and if it is terminated by adding SDS (Gellert et al. 1977; Sugino et al. 1977). This observation provided the earliest hint that DNA topoisomerization by this class of enzymes involves double-stranded breakage and rejoining.

The generation of topoisomerase-II-cleaved products usually requires the addition of a potent protein denaturant like an ionic detergent or strong alkali. Two additional features distinguish this reaction from those mediated by nucleases: the reversibility of the reaction and the covalent linkage of topoisomerase to the cleaved DNA. Prior to the addition of a protein denaturant, the dsDNA break by topoisomerase II can usually be reversed by one of the following treatments: proteinase, depletion of divalent cations, increasing the ionic strength in the reaction mixture, or shifting the incubation temperature (Gellert et al. 1977; Sugino et al. 1977; Liu et al. 1983; Sander and Hsieh 1983).

The structure of DNA at the cleavage site is identical for all type-II topoisomerases: a staggered double-stranded break with 4-nucleotide-long 5′ protruding ends and 3′ recessed ends with free hydroxyl groups (Morrison and Cozzarelli 1979; Sander and Hsieh 1983; Liu et al. 1983). The 5′ phosphoryl ends at the cleavage site are blocked because of their covalent attachment to the topoisomerase polypeptides. The linked protein moiety has been identified as the *gyrA* subunit in the case of bacterial gyrase (Sugino et al. 1980; Tse et al. 1980), gene *52* protein in T4 DNA topoisomerase (Rowe et al. 1984), and the 170-kD subunit in *Drosophila* topoisomerase II (Sander and Hsieh 1983). The covalent linkage is between a DNA phosphoryl group and a protein tyrosyl group (Tse et al. 1980; Rowe et al. 1984). Amino acid sequence around this DNA-linked tyrosyl residue has been determined for *E. coli* DNA gyrase (Horowitz and Wang 1987) and yeast topoisomerase II (Worland and Wang 1989). Sequence comparison of the type-II topoisomerases in this region clearly demonstrates significant homologies around the tyrosyl residue (Lynn et al. 1986; Tsai-Pflugfelder et al. 1988; Wyckoff et al. 1989; Strauss and Wang 1990).

In addition to incubation temperature, ionic strength, and divalent cation, the extent of DNA cleavage by topoisomerase II is also modulated by the size of DNA substrate, DNA sequence around the cleavage site, and the presence of ATP. The DNA sequence preference in the cleavage reaction will be addressed in the next section. The size of a DNA fragment is an important determinant in the cleavage reaction for bacterial gyrase: It requires a minimal length of DNA, which is correlated with the length of DNA wrapped around gyrase. For a DNA fragment of 77 bp (Morrison et al. 1980) or 34 bp (Fisher et al. 1986), there is no cleavage by bacterial DNA gyrase. In marked contrast, eukaryotic DNA topoisomerase II appears to be active on short DNA fragments for its cleavage reaction; calf thymus topoisomerase II can cleave a 74-bp DNA fragment at numerous sites (Liu et al. 1983). This is again correlated with the apparent lack of extensive DNA wrapping around the eukaryotic enzyme. ATP appears to stimulate the DNA gyrase cleavage reaction at some sites but not at others (Morrison et al. 1980; Kirkegaard and Wang 1981). The DNA cleavage by *Drosophila* topoisomerase II is stimulated by ATP (Sander and Hsieh 1983), and to a larger extent, by the ATP analog with $\beta\gamma$-nonhydrolyzable bond (Osheroff 1986). The presence of ATP also stimulates the cleavage of glucosylated/hydroxymethylcytosine-containing DNA by the bacteriophage T4 topoisomerase II, whereas its cleavage of unmodified DNA is only slightly affected by ATP (Kreuzer and Alberts 1984).

For bacterial DNA gyrase, the DNA cleavage reaction is greatly enhanced by the quinolone antibiotics (Gellert et al. 1977; Sugino et al. 1977). For the eukaryotic enzymes, enhancement of the cleavage reaction has been observed for a number of antitumor agents (Chen and Liu 1986; Potmesil and Ross 1987; Drlica and Franco 1988). The interactions between DNA topoisomerases and the drugs that target these enzymes are reviewed in Liu (this volume).

VI. SEQUENCE PREFERENCE IN INTERACTIONS BETWEEN DNA AND TYPE-II DNA TOPOISOMERASES

dsDNA cleavage by topoisomerase II provides a convenient means to monitor the sequence preference in DNA/enzyme interactions. Because the topoisomerase II cleavage sites in DNA constitute at least a subset of the enzyme-binding sites, they furnish information regarding the sequence determinants in the topoisomerase/DNA interactions. The sequence specificity in topoisomerase cleavage reactions is only moderate; depending on the experimental conditions, there can be numerous cleavage sites distributed along a DNA. For this reason, most of the attention has been focused on the strong cleavage sites. Both in vitro and in vivo mapping experiments have demonstrated that some of the strong eukaryotic topoisomerase II cleavage sites are located at the ends of the genes, particularly at the 5′ends (Udvardy et al. 1985, 1986; Riou et al. 1986; Rowe et al. 1986; Yang et al. 1987). The significance of the distribution of topoisomerase II cleavage sites with respect to the gene organization is not known.

It has been suggested that DNA attachment to the nuclear matrix/scaffold is mediated through topoisomerase II; the locations of the sites of DNA attachment to these structures appear to correlate with the topoisomerase II cleavage sites (Cockerill and Garrard 1986; Gasser and Laemmli 1986). In *E. coli,* DNA gyrase can be induced to generate approximately 45 double-stranded breaks in the chromosome, which is about the same as the number of putative topological domains in a bacterial nucleoid (Snyder and Drlica 1979).

Nucleotide sequence analysis of DNA topoisomerase II cleavage sites has provided the cleavage "consensus sequences" of bacterial DNA gyrase (Morrison and Cozzarelli 1979; Lockshon and Morris 1985), *Drosophila* topoisomerase II (Sander and Hsieh 1985), and chicken topoisomerase II (Spitzner and Muller 1988). These cleavage sequences are obtained from statistical analysis of the cleavage site sequences, and they are characterized by a high degree of degeneracies. However, the following two lines of evidence suggest that they can provide some information about determinants in the nucleotide sequence that are important in DNA/enzyme interactions. A synthetic DNA 21-mer containing the cleavage consensus sequence of *Drosophila* topoisomerase II was cloned into a plasmid vector. The *Drosophila* enzyme cleaves this sequence at precisely the position predicted by the consensus sequence. Furthermore, the DNase I footprint shows that the enzyme can protect a region of approximately 20–30 nucleotides surrounding the cleavage site (Lee et al. 1989b). Site-specific mutagenesis of a strong DNA gyrase cleavage site in pBR322 DNA showed that the cleavage efficiency is reduced by alterations of the consensus sequence (Fisher et al. 1986).

An interesting question about the DNA cleavage site is its relation to the essential topoisomerase II activities, such as DNA binding, strand passage, and DNA-dependent ATPase activity. The ATPase associated with *E. coli* gyrase probably does not require a specific sequence to serve as DNA cofactor since there is good ATPase activity in the presence of several simple DNA sequences like homopolymers and alternating copolymers (Sugino and Cozzarelli 1980). Furthermore, the kinetic specificity factor k_{cat}/K_m is about the same for ATPase activity monitored in the presence of several different DNA fragments (Maxwell and Gellert 1984). On the basis of the nitrocellulose filter binding assays and nuclease protection experiments, the gyrase strong cleavage sites correlate with their binding to the enzyme (Morrison et al. 1980; Fisher et al. 1981; Kirkegaard and Wang 1981; Morrison and Cozzarelli 1981). However, an interesting exception has been noted in which a strong DNA gyrase binding site is not cleaved (Kirkegaard and Wang 1981). The intergenic region of the divergent heat-shock protein (hsp) 70 genes in *Drosophila* contains clustered, strong cleavage sites for *Drosophila* topoisomerase II (Udvardy et al. 1985), and both nitrocellulose filter binding assay and DNase I footprint analysis have demonstrated that these sites are strong topoisomerase-II-binding sites as well (Sander et al. 1987; Lee et al. 1989b). The plasmid DNA containing the hsp70 intergenic sequence is relaxed by the *Drosophila* topoisomerase II faster than the DNA containing the hsp70-coding sequence by the *Drosophila* topoisomerase II, suggesting that the intergenic fragment is enriched in sites for strand passage (Sander et al. 1987).

VII. SINGLE-STRANDED DNA CLEAVAGE REACTION

The dsDNA cleavage reaction has played an essential role in the elucidation of the mechanism of type-II topoisomerases. However, cleavage in only one strand of duplex DNA has been observed for DNA gyrase in the presence of oxolinate (Gellert et al. 1977) and for eukaryotic topoisomerase II in the presence of antitumor agents (Chen et al. 1984; Yang et al. 1985; Muller et al. 1988) or with Ca^{++} substituting Mg^{++} in the reaction mixture (Osheroff and Zechiedrich 1987). Furthermore, single-stranded cleavage is greatly enhanced and double-stranded cleavage reaction is suppressed if the *Drosophila* topoisomerase II is terminated first by the addition of EDTA or if a divalent cation like Mn^{++}, Co^{++}, or Ca^{++} is present instead of Mg^{++} (Lee et al. 1989a).

Single-stranded cleavage in the duplex DNA is clearly mediated by topoisomerase II since the sequence specificity of this reaction closely parallels that of the double-stranded cleavage, and a topoisomerase II protomer is linked covalently to the 5′ phosphoryl end at the nick (Lee et al. 1989a). The ssDNA cleavage is expected to occur when the coordination between the two subunits in a topoisomerase II molecule is disrupted and the breakage of two DNA strands is no longer a concerted event. This coordination could be lost when the enzyme is poisoned by a topoisomerase-targeting drug. It is also possible that a divalent cation plays an important role in mediating the coupled action between the two halves of the enzyme. This intimate intersubunit coordination could be interrupted when Mg^{++} is chelated by a molar excess of EDTA or when it is substituted by Mn^{++}, Co^{++}, or Ca^{++}. Since most of the topoisomerase II cleavage sites lack dyadic symmetry in the DNA sequence, this asymmetry in the protein/DNA interaction could lead to different rates in the breakage and/or to the rejoining of two strands. Therefore, the single-stranded cleavage reaction suggests that, under the optimal conditions for topoisomerase II reactions, two DNA strands are cleaved and rejoined in a more or less concerted manner but the breaking-rejoining of each strand could be sequential. Although the biological significance of the single-stranded cleavage reaction is not yet clear, there is genetic evidence implicating the existence of the nicked intermediate generated by topoisomerase II in the process of acridine-induced mutagenesis in bacteriophage T4 (Ripley et al. 1988).

In a different reaction, topoisomerase II can also cleave ssDNA substrates (Liu et al. 1983; Kreuzer 1984). The cleavage sites in ϕX174 ssDNA by T4 DNA topoisomerase II are very close to the hairpin region, and this cleavage reaction is distinct from the cleavage reaction of duplex DNA substrates in that it is inhibited rather than stimulated by oxolinate and that its generation does not depend on the addition of a strong denaturant like SDS (Kreuzer 1984).

VIII. DNA STRAND TRANSFER REACTION

A combination between the two half-reactions in the breakage-rejoining cycle mediated by a DNA topoisomerase can in principle result in a strand transfer reaction between two DNA molecules. Type-I eukaryotic topoisomerase, being a single subunit enzyme, can indeed carry out an inter- and intramolecular strand transfer reaction when ssDNA is used as the substrate to initiate the formation of covalent complex between DNA and enzyme (Been and Champoux 1981; Halligan et al. 1982). Similar intermolecular strand transfer mediated by *E. coli* topoisomerase I was also observed between short oligodeoxynucleotide and DNA (Tse-Dinh 1984). For type-II topoisomerases, protein-protein interactions hold together the transient double-stranded break, and the strand transfer reaction via subunit exchange is not expected to be efficient. Ikeda et al. (1981) used a sensitive genetic assay to monitor illegitimate recombination between plasmid and bacteriophage λ DNA or between two genetically marked phage DNAs. They detected the rare recombination products generated from the in vitro strand transfer reaction mediated by bacterial DNA gyrase (Ikeda and Shiozaki 1984), phage T4 topoisomerase (Ikeda 1986), and calf thymus topoisomerase II (Bae et al. 1988). In all these recombination reactions, sequence homology is not required at the crossover junction. The crossover sites in the gyrase-mediated recombination events overlap with cleavage sites by gyrase (Ikeda et al. 1984). The illegitimate recombination generated through the strand transfer reaction by DNA gyrase is stimulated by oxolinate, and this stimulation is inhibited by coumermycin (Ikeda et al. 1981). Similarly, nonhomologous recombination mediated by phage T4 topoisomerase is enhanced by oxolinate (Ikeda 1986). It therefore appears that stimulation in the production of topoisomerase-mediated cleavage complex can increase the nonhomologous recombination events, consistent with the notion that they are generated by the strand transfer reaction of topoisomerase II.

IX. CONCLUDING REMARKS

The overall mechanism of DNA type-II topoisomerases, especially from the point of view of DNA topology, has been well established. A number of important issues remain, however, and will be the focus of future studies. Examples are the relation between the strong topoisomerase-mediated DNA cleavage sites and the efficiency of strand passage through them, the mechanistic role of ATP and its hydrolysis, the interaction between DNA topoisomerase II and the DNA segment to be translocated, the spatial relation between the translocated segment and the DNA segment containing the breakage/reunion gate, the biological significance of the sequence preferences, the mechanistic basis for the coupling of the action of two subunits, and the establishment of the mechanism and the biological significance of the strand transfer reaction. These issues also clearly illustrate the multifunctional nature of these enzymes. Despite the complexity of the system, tools are now available to carry out both genetic and biochemical analyses at the molecular level, and they should facilitate progress in answering these important questions.

ACKNOWLEDGMENTS

I thank many colleagues for helpful and stimulating discussions, especially Jim Wang for his assistance in revising the manuscript. The work from my laboratory is supported by National Institutes of Health grant GM-29006.

REFERENCES

Adachi, T., M. Mizuuchi, E.A. Robinson, E. Appella, M.H. O'Dea, M. Gellert, and K. Mizuuchi. 1987. DNA sequence of the *E. coli gyrB* gene: Application of a new sequencing strategy. *Nucleic Acids Res.* **15:** 771–783.

Bae, Y.S., I. Kawasaki, H. Ikeda, and L.F. Liu. 1988. Illegitimate recombination mediated by calf thymus DNA topoisomerase II in vitro. *Proc. Natl. Acad. Sci.* **85:** 2076–2080.

Baldi, M.I., P. Benedetti, E. Mattoccia, and G.P. Toccini-Valentini. 1980. In vitro catenation and decatenation of DNA and a novel eukaryotic ATP-dependent topoisomerase. *Cell* **20:** 461–467.

Bates, A.D. and A. Maxwell. 1989. DNA gyrase can supercoil DNA circles as small as 174 basepairs. *EMBO J.* **8:** 1861–1866.

Been, M.D. and J.J. Champoux. 1981. DNA breakage and closure by rat liver type I topoisomerase: Separation of the half-reactions by using a single-stranded DNA substrate. *Proc. Natl. Acad. Sci.* **78:** 2883–2887.

Brown, P.O. and N.R. Cozzarelli. 1979. A sign inversion mechanism for enzymatic supercoiling of DNA. *Science* **206:** 1081–1083.

Brown, P.O., C.L. Peebles, and N.R. Cozzarelli. 1979. A topoisomerase from *Escherichia coli* related to DNA gyrase. *Proc. Natl. Acad. Sci.* **76:** 6110–6114.

Chen, G.L. and L.F. Liu. 1986. DNA topoisomerases as therapeutic targets in cancer chemotherapy. *Annu. Rep. Med. Chem.* **21:** 257–262.

Chen, G.L., L. Yang, T.C. Rowe, B.D. Halligan, K.M. Tewey, and L.F. Liu. 1984. Nonintercalative antitumor drugs interfere with the breakage-reunion reaction of mammalian DNA topoisomerase II. *J. Biol. Chem.* **259:** 13560–13566.

Cockerill, P.N. and W.T. Garrard. 1986. Chromosomal loop anchorage of the kappa immunoglobulin gene occurs next to the enhancer in a region containing topoisomerase II sites. *Cell* **44:** 273–282.

Cozzarelli, N.R. 1980. DNA topoisomerases. *Science* **207:** 953–960.

Crick, F.H.C. 1976. Linking numbers and nucleosomes. *Proc. Natl. Acad. Sci.* **73:** 2639–2643.

DiNardo, S., K. Voelkel, and R. Sternglanz. 1984. DNA topoisomerase II mutant of *Saccharomyces cerevisiae:* Topoisomerase II is required for segregation of daughter molecules at the termination of DNA replication. *Proc. Natl. Acad. Sci.***81:** 2616–2620.

Douc-Rasy, S., A. Kayser, J.F. Riou, and G. Riou. 1986. ATP-independent type II topoisomerase from trypanosomes. *Proc. Natl. Acad. Sci.* **83:** 7152–7156.

Drlica, K. and R.J. Franco. 1988. Inhibitors of DNA topoisomerases. *Biochemistry* **27:** 2252–2259.

Fisher, L.M., H.A. Barot, and M.G. Cullen. 1986. DNA gyrase complex with DNA: Determinants for site-specific DNA breakage. *EMBO J.* **5:** 1411–1418.

Fisher, L.M., K. Mizuuchi, M.H. O'Dea, H. Ohmori, and M. Gellert. 1981. Site-specific interaction of DNA gyrase with DNA. *Proc. Natl. Acad. Sci.* **78:** 4165–4169.

Forterre, P., G. Mirambeau, C. Jaxel, M. Nadal, and M. Duguet. 1985. High positive supercoiling in vitro catalyzed by an ATP and polyethylene glycol-stimulated topoisomerase from *Sulfolobus acidocaldarius*. *EMBO J.* **4:** 2123–2128.

Fuller, F.B. 1978. Decomposition of the linking number of a closed ribbon: A problem from molecular biology. *Proc. Natl. Acad. Sci.* **68:** 3557–3561.

Gasser, S.M. and U.K. Laemmli. 1986. The organization of chromatin loops: Characterization of a scaffold attachment site. *EMBO J.* **5:** 511–518.

Gellert, M. 1981. DNA topoisomerases. *Annu. Rev. Biochem.* **50:** 879–910.

Gellert, M., L.M. Fisher, and M.H. O'Dea. 1979. DNA gyrase: Purification and catalytic properties of a fragment of gyrase B protein. *Proc. Natl. Acad. Sci.* **83:** 7152–7156.

Gellert, M., M.H. O'Dea, T. Itoh, and J.I. Tomizawa. 1976a. Novobiocin and coumermycin inhibit DNA supercoiling catalyzed by DNA gyrase. *Proc. Natl. Acad. Sci.* **73:** 4474–4478.

Gellert, M., K. Mizuuchi, M.H. O'Dea, and H.A. Nash. 1976b. DNA gyrase: An enzyme that introduces superhelical turns into DNA. *Proc. Natl. Acad. Sci.* **73:** 3872–3876.

Gellert, M., L.M. Fisher, H. Ohmori, M.H. O'Dea, and K. Mizuuchi. 1981. DNA gyrase: Site-specific interactions and transient double-strand breakage of DNA. *Cold Spring Harbor Symp. Quant. Biol.* **45:** 391–398.

Gellert, M., K. Mizuuchi, M.H. O'Dea, T. Itoh, and J.I. Tomizawa. 1977. Nalidixic acid resistance: A second genetic character involved in DNA gyrase activity. *Proc. Natl. Acad. Sci.* **74:** 4772–4776.

Goto, T. and J.C. Wang. 1982. Yeast DNA topoisomerase II. An ATP-dependent type II topoisomerase that catalyzes the catenation, decatenation, unknotting, and relaxation of double-stranded DNA rings. *J. Biol. Chem.* **257:** 5866–5872.

Halligan, B.D., J.L. Davis, K.A. Edwards, and L.F. Liu. 1982. Intra- and intermolecular strand transfer by HeLa DNA topoisomerase I. *J. Biol. Chem.* **257:** 3995–4000.

Higgins, N.P. and N.R. Cozzarelli. 1982. The binding of gyrase to DNA: Analysis by retention by nitrocellulose filters. *Nucleic Acids Res.* **10:** 6833–6847.

Higgins, N.P., C.I. Peebles, A. Sugino, and N.R. Cozzarelli. 1978. Purification of the subunits of *Escherichia coli* DNA gyrase and the reconstitution of its enzymatic activity. *Proc. Natl. Acad. Sci.* **75:** 1773–1777.

Holm, C., T. Stearns, and D. Botstein. 1989. DNA topoisomerase II must act at mitosis to prevent nondisjunction and chromosome breakage. *Mol. Cell. Biol.* **9:** 159–168.

Holm, C., T. Goto, J.C. Wang, and D. Botstein. 1985. DNA topoisomerase II is required at the time of mitosis in yeast. *Cell* **41:** 553–563.

Horowitz, D.S. and J.C. Wang. 1987. Mapping the active site tyrosine of *Escherichia coli* DNA gyrase. *J. Biol. Chem.* **262:** 5339–5344.

Hsieh, T. 1983. Knotting of circular duplex DNA by type II DNA topoisomerase from *Drosophila melanogaster*. *J. Biol. Chem.* **258:** 8413–8420.

Hsieh, T. and D. Brutlag. 1980. ATP-dependent DNA topoisomerase from *D. melanogaster* reversibly catenates duplex DNA rings. *Cell* **21:** 115–125.

Huang, W.M. 1986a. Nucleotide sequence of a type II DNA topoisomerase gene. Bacteriophage T5 gene 52. *Nucleic Acids Res.* **14:** 7379–7390.

———. 1986b. Nucleotide sequence of a type II DNA topoisomerase gene. Bacteriophage T4 gene 39. *Nucleic Acids Res.* **14:** 7751–7765.

Huang, W.M., S.-H. Ao, S. Casjens, R. Orlandi, R. Zeikus, R. Weiss, D. Winge, and M. Fang. 1988. A persistent untranslated sequence within bacteriophage T4 DNA topoisomerase gene 60. *Science* **239:** 1005–1012.

Ikeda, H. 1986. Bacteriophage T4 DNA topoisomerase mediates illegitimate recombination in vitro. Proc. Natl. Acad. Sci. **83:** 922–926.

Ikeda, H. and M. Shiozaki. 1984. Nonhomologous recombination mediated by *Escherichia coli* DNA gyrase: Possible involvment of DNA replication. *Cold Spring Harbor Symp. Quant. Biol.* **49:** 401–409.

Ikeda, H., I. Kawasaki, and M. Gellert. 1984. Mechanism of illegitimate recombination: Common sites for recombination and cleavage mediated by *E. coli* DNA gyrase. *Mol. Gen. Genet.* **196:** 546–549.

Ikeda, H., K. Moriya, and T. Matsumoto. 1981. In vitro study of illegitimate recombination: Involvment of DNA gyrase. *Cold Spring Harbor Symp. Quant. Biol.* **45:** 399–408.

Kikuchi, A. and K. Asai. 1984. Reverse gyrase—A topoisomerase which introduces positive superhelical turns into DNA. *Nature* **309:** 677–681.

Kirkegaard, K. and J.C. Wang. 1981. Mapping the topography of DNA wrapped around gyrase by nucleolytic and chemical probing of complexes of unique DNA sequences. Cell **23:** 721–729.

Kirchhausen, T., J.C. Wang, and S.C. Harrison. 1985. DNA gyrase and its complexes with DNA: Direct observation by electron microscopy. *Cell* **41:** 933–43.

Klevan, L. and Y.-C. Tse. 1983. Chemical modification of essential tyrosine residues in DNA topoisomerases. *Biochim. Biophys. Acta* **754:** 175–180.

Klevan, L. and J.C. Wang. 1980. A DNA-DNA gyrase complex containing 140 bp of deoxyribonucleic acid and an $\alpha_2\beta_2$ protein core. *Biochemistry* **19:** 5229–5234.

Krasnow, M.A. and N.R. Cozzarelli. 1982. Catenation of DNA rings by topoisomerases. Mechanism of control by spermidine. *J. Biol. Chem.* **257:** 2687–2693.

Kreuzer, K.N. 1984. Recognition of single-stranded DNA by the bacteriophage T4-induced type II topoisomerase. *J. Biol. Chem.* **259:** 5347–5354.

Kreuzer, K.N. and B.M. Alberts. 1984. Site-specific recognition of bacteriophage T4 DNA by T4 type II DNA topoisomerase and *Escherichia coli* DNA gyrase. *J. Biol. Chem.* **259:** 5339–5346.

Kreuzer, K.N. and N.R. Cozzarelli. 1980. Formation and resolution of DNA catenanes by DNA gyrase. *Cell* **20:** 245–254.

Kreuzer, K. and W.M. Huang. 1983. T4 DNA topoisomerase. In *Bacteriophage T4* (ed. C.K. Mathews et al.), pp. 90-96. American Society for Microbiology, Washington, D.C.

Lee, M.P., M. Sander, and T. Hsieh. 1989a. Single strand DNA cleavage reaction of duplex DNA by *Drosophila* topoisomerase II. *J. Biol. Chem.* **264:** 13510–13518.

———. 1989b. Nuclease protection by *Drosophila* DNA topoisomerase II: Enzyme/DNA contacts at the strong topoisomerase II cleavage sites. *J. Biol. Chem.* **264:** 21779–21787.

Liu, L.F. 1983. DNA topoisomerases-enzymes that catalyze the breaking and rejoining of DNA. *Crit. Rev. Biochem.* **15:** 1–24.

Liu, L.F. and J.C. Wang. 1978a. *Micrococcus luteus* DNA gyrase: Active components and a model for its supercoiling of DNA. *Proc. Natl. Acad. Sci.* **74:** 2098–2102.

———. 1978b. DNA-DNA gyrase complex: The wrapping of the DNA duplex outside the enzyme. *Cell* **15:** 979–984.

Liu, L.F., J.L. Davis, and R. Calendar. 1981. Novel topologically knotted DNA from bacteriophage P4 capsids: Studies with DNA topoisomerases. *Nucleic Acids Res.* **9:** 3979–3989.

Liu, L.F., C.C. Liu, and B.M. Alberts. 1979. T4 DNA topoisomerase: A new ATP-dependent enzyme essential for initiation of T4 bacteriophage DNA replication. *Nature* **281:** 456–461.

———. 1980. Type II DNA topoisomerases: Enzymes that can unknot a topologically knotted DNA molecule via a reversible double-strand break. *Cell* **19:** 697–707.

Liu, L.F., T.C. Rowe, L. Yang, K.M. Tewey, and G.L. Chen. 1983. Cleavage of DNA by mammalian DNA topoisomerase II. *J. Biol. Chem.* **258:** 15365–15370.

Lockshon, D. and D.R. Morris. 1985. Sites of reaction of *Escherichia coli* DNA gyrase on pBR322 in vivo as revealed by oxolinic acid-induced plasmid linearization. *J. Mol. Biol.* **181:** 63–74.

Lynn, R., G. Giaever, S.L. Swanberg, and J.C. Wang. 1986. Tandem regions of yeast DNA topoisomerase II share homology with different subunits of bacterial gyrase. *Science* **233:** 647–649.

Marians, K.J. 1987. DNA gyrase-catalyzed decatenation of multiple linked DNA dimers. *J. Biol. Chem.* **262:** 10362–10368.

Maxwell, A. and M. Gellert. 1984. The DNA dependence of the ATPase activity of DNA gyrase. *J. Biol. Chem.* **259:** 14472–14480.

———. 1986. Mechanistic aspects of DNA topoisomerases. *Adv. Protein Chem.* **38:** 69–107.

Melendy, T. and D.S. Ray. 1989. Novobiocin affinity purification of a mitochondrial type II topoisomerase from the trypanosomatid *Crithidia fasciculata. J. Biol. Chem.* **264:** 1870–1876.

Melendy, T., C. Scholine, and D.S. Ray. 1988. Localization of a type II DNA topoisomerase to two sites at the periphery of the kinetoplast DNA of *Crithidia fasciculata. Cell* **55:** 1083–1088.

Miller, K.G., L.F. Liu, and P.T. Englund. 1981. A homogeneous type II DNA topoisomerase from HeLa cell nuclei. *J. Biol. Chem.* **256:** 9334–9339.

Mizuuchi, K., M.H. O'Dea, and M. Gellert. 1978. DNA gyrase: Subunit structure and ATPase activity of the purified enzyme. *Proc. Natl. Acad. Sci.* **75:** 5960–5963.

Moriya, S., N. Ogasawara, and H. Yoshikawa. 1985. Structure and function of the region of the replication origin of the *Bacillus subtilis* chromosome. III. Nucleotide sequence of some 10,000 base pairs in the origin region. *Nucleic Acids Res.* **13:** 2251–2265.

Morrison, A. and N.R. Cozzarelli. 1981. Contacts between DNA gyrase and its binding site on DNA: Features of symmetry and asymmetry revealed by protection from nucleases. *Proc. Natl. Acad. Sci.* **78:** 1416–1420.

———. 1979. Site-specific cleavage of DNA by *E. coli* DNA gyrase. *Cell* **17:** 175–184.

Morrison, A., N.P. Higgins, and N.R. Cozzarelli. 1980. Interaction between DNA gyrase and its cleavage site on DNA. *J. Biol. Chem.* **255:** 2211–2219.

Muller, M.T., J.R. Spitzner, J.H. DiDonato, V.B. Mehta, K. Tsutsui, and K. Tsutsui. 1988. Single-strand DNA cleavages by eucaryotic topoisomerase II. *Biochemistry* **27:** 8369–8379.

Orr, E. and W.L. Staudenbauer. 1982. *Bacillus subtilis* DNA gyrase. Purification of subunits and reconstitution of supercoiling activity. *J. Bacteriol.* **151:** 524–527.

Osheroff, N. 1986. Eucaryotic topoisomerase II. Characterization of enzyme turnover. *J. Biol. Chem.* **261:** 9944–9950.

Osheroff, N. and E.L. Zechiedrich. 1987. Calcium promoted DNA cleavage by eucaryotic topoisomerase II: Trapping the covalent enzyme-DNA complex in an active form. *Biochemistry* **26:** 4303–4309.

Osheroff, N., E.R. Shelton, and D. Brutlag. 1983. DNA topoisomerase II from *Drosophila melanogaster:* Relaxation of supercoiled DNA. *J. Biol. Chem.* **258:** 9536–9543.

Peebles, C.L., N.P. Higgins, K.N. Kreuzer, A. Morrison, P.O. Brown, A. Sugino, and N.R. Cozzarelli. 1978. Structure and activities of *Escherichia coli* DNA gyrase. *Cold Spring Harbor Symp. Quant. Biol.* **43:** 41–52.

Potmesil, M. and W.E. Ross, eds. 1987. Proceedings of the First Conference on DNA Topoisomerases in Cancer Chemotherapy. *Natl. Cancer Inst. Monogr.* **4.**

Rau, D.C., M. Gellert, F. Thoma, and A. Maxwell. 1987. Structure of the DNA gyrase-DNA complex as revealed by transient electric dichroism. *J. Mol. Biol.* **193:** 555–569.

Riou, J.-F., E. Multon, M.-J. Vilarem, C.-J. Larsen, and G. Riou. 1986. In vivo stimulation by antitumor drugs of the topoisomerase II induced cleavage site in c-*myc* protooncogene. *Biochem. Biophys. Res. Commun.* **137:** 154–160.

Ripley, L.S., J.S. Dubins, J.G. deBoer, D.M. De Marini, A.M. Bogerd, and K.N. Kreuzer. 1988. Hotspot sites for acridine-induced frameshift mutations in bacteriophage T4 correspond to sites of action of the T4 type II topoisomerase. *J. Mol. Biol.* **200:** 665–680.

Rowe, T.C., K.M. Tewey, and L.F. Liu. 1984. Identification of the breakage-reunion subunit of T4 DNA topoisomerase. *J. Biol. Chem.* **259:** 9177–9181.

Rowe, T.C., J.C. Wang, and L.F. Liu. 1986. In vivo localization of DNA topoisomerase II cleavage sites on *Drosophila* heat shock chromatin. *Mol. Cell. Biol.* **6:** 985–992.

Sander, M. and T. Hsieh. 1983. Double strand DNA cleavage by type II DNA topoisomerase from *Drosophila melanogaster*. *J. Biol. Chem.* **258:** 8421–8428.

———. 1985. *Drosophila* topoisomerase II double-strand DNA cleavage: Analysis of DNA sequence homology at the cleavage site. *Nucleic Acids Res.* **13:** 1057–1072.

Sander, M., T. Hsieh, A. Udvardy, and P. Schedl. 1987. Sequence dependence of *Drosophila* topoisomerase II in plasmid relaxation and DNA binding. *J. Mol. Biol.* **194:** 219–229.

Snyder, M. and K. Drlica. 1979. DNA gyrase on the bacterial chromosome: DNA cleavage induced by oxolinic acid. *J. Mol. Biol.* **131:** 287–302.

Spitzner, J.R. and M.T. Muller. 1988. A consensus sequence for cleavage by vertebrate DNA topoisomerase II. *Nucleic Acids Res.* **16:** 5533–5556.

Steck, T.R. and K. Drlica. 1984. Bacterial chromosome segregation: Evidence for DNA gyrase involvement in decatenation. *Cell* **36:** 1081–1088.

Stetler, G.L., G.J. King, and W.M. Huang. 1979. T4 DNA-delay proteins required for specific DNA replication form a complex that has ATP-dependent DNA topoisomerase activity. *Proc. Natl. Acad. Sci.* **76:** 3737–3741.

Strauss, P.R. and J.C. Wang. 1990. The *Top2* gene of *Trypanosome brucei:* A single copy gene that shares extensive homology with other *Top2* genes encoding eucaryotic DNA topoisomerase II. *Mol. Biochem. Parasitol.* (in press).

Sugino, A. and K.F. Bott. 1980. *Bacillus subtilis* deoxyribonucleic acid gyrase. *J. Bacteriol.* **141:** 1331–1339.

Sugino, A. and N.R. Cozzarelli. 1980. The intrinsic ATPase of DNA gyrase. *J. Biol. Chem.* **255:** 6299–6306.

Sugino, A., N.P. Higgins, and N.R. Cozzarelli. 1980. DNA gyrase subunit stoichiometry and the covalent attachment of subunit A to DNA during DNA cleavage. *Nucleic Acids Res.* **8:** 3865–3874.

Sugino, A., C.L. Peebles, K.N. Kreuzer, and N.R. Cozzarelli. 1977. Mechanism of action of nalidixic acid: Purification of *E. coli* nalA gene product and its relationship to DNA gyrase and a novel nicking-closing enzyme. *Proc. Natl. Acad. Sci.* **74:** 4767–4771.

Sugino, A., N.P. Higgins, P.O. Brown, C.L. Peebles, and N.R. Cozzarelli. 1978. Energy coupling in DNA gyrase and the mechanism of action of novobiocin. *Proc. Natl. Acad. Sci.* **75:** 4838–4842.

Tsai-Pflugfelder, M., L.F. Liu, A.A. Liu, K.M. Tewey, J. Wang-Peng, T. Knutsen, K. Huebner, C.N. Croce, and J.C. Wang. 1988. Cloning and sequencing of cDNA encoding human DNA topoisomerase II and localization of the gene to chromosome region 17q21-22. *Proc. Natl. Acad. Sci.* **85:** 7177–7181.

Tse, Y.C., K. Kirkegaard, and J.C. Wang. 1980. Covalent bonds between protein and

DNA: Formation of phosphotyrosine linkage between certain DNA topoisomerases and DNA. *J. Biol. Chem.* **255:** 5560–5565.

Tse-Dinh, Y.C. 1984. Uncoupling of the DNA breaking and rejoining steps of *Escherichia coli* type I DNA topoisomerase. Demonstration of an active covalent protein-DNA complex. *J. Biol. Chem.* **259:** 10931–10935.

Udvardy, A., P. Schedl, M. Sander, and T. Hsieh. 1985. Novel partitioning of DNA cleavage sites for *Drosophila* topoisomerase II. *Cell* **40:** 933–941.

———. 1986. Topoisomerase II cleavage in chromatin. *J. Mol. Biol.* **191:** 231–246.

Uemura, T. and M. Yanagida. 1984. Isolation of type I and II DNA topoisomerase mutants from fission yeast: Single and double mutants show different phenotypes in cell growth and chromatin organization. *EMBO J.* **3:** 1737–1744.

———. 1986. Mitotic spindle pulls but fails to separate chromosomes in type II DNA topoisomerase mutants: Uncoordinated mitosis. *EMBO J.* **5:** 1003–1010.

Uemura, T., K. Morikawa, and M. Yanagida. 1986. The nucleotide sequence of the fission yeast DNA topoisomerase II gene: Structural and functional relationships to other DNA topoisomerases. *EMBO J.* **5:** 2355–2361.

Uemura, T., H. Ohkura, Y. Adachi, K. Morino, K. Shiozaki, and M. Yanagida. 1987. DNA topoisomerase II is required for condensation and separation of mitotic chromosomes in *S. pombe*. *Cell* **50:** 917–925.

Wang, J.C. 1971. Interaction between DNA and *Escherichia coli* protein ω. *J. Mol. Biol.* **55:** 523–533.

———. 1985. DNA topoisomerases. *Annu. Rev. Biochem.* **54:** 665–697.

Wang, J.C. and L.F. Liu. 1979. DNA topoisomerases: Enzymes that catalyze the concerted breaking and rejoining of DNA backbone bonds. In *Molecular genetics* (ed. J.H. Taylor), part 3, pp. 65–88. Academic Press, New York.

Wang, J.C., R.I. Gumport, K. Javaherian, K. Kirkegaard, L. Klevan, M.L. Kotewicz, and Y.-C. Tse. 1981. DNA topoisomerases. In *Mechanistic studies of DNA replication and genetic recombination* (ed. B.M. Alberts and C.F. Fox), pp. 769–784. Academic Press, New York.

Worland, S. and J.C. Wang. 1989. Inducible over-expression, purification and active site mapping of DNA topoisomerase II from yeast *Saccharomyces cerevisiae*. *J. Biol. Chem.* **264:** 4412–4416.

Wyckoff, E. and T. Hsieh. 1988. Functional expression of a *Drosophila* gene in yeast: Genetic complementation by DNA topoisomerase II. *Proc. Natl. Acad. Sci.* **85:** 6272–6276.

Wyckoff, E., D. Natalie, J.M. Nolan, M. Lee, and T. Hsieh. 1989. Structure of the *Drosophila* DNA topoisomerase II gene: Nucleotide sequence and homology among topoisomerase II. *J. Mol. Biol.* **205:** 1–14.

Yang, L., T.C. Rowe, E.M. Nelson, and L.F. Liu. 1985. In vivo mapping of DNA topoisomerase II-specific cleavage sites on SV40 chromatin. *Cell* **41:** 127–132.

Yang, L., M.S. Wold, J.J. Li, T.J. Kelly, and L.F. Liu. 1987. Roles of DNA topoisomerases in simian virus 40 DNA replication in vitro. *Proc. Natl. Acad. Sci.* **84:** 950–954.

8

Virus-encoded DNA Topoisomerases

Wai Mun Huang
Department of Cellular Viral and Molecular Biology
University of Utah Medical Center
Salt Lake City, Utah 84132

- **I. Introduction**
- **II. T4 DNA Topoisomerase**
  - A. General Description of T4 DNA Metabolism
  - B. Activities of Purified T4 DNA Topoisomerase
  - C. T4 Topoisomerase Is Sensitive to Antitumor Agents
  - D. DNA Sequence Arrangement of T4 Topoisomerase
  - E. Reconstitution of T4 Topoisomerase from Isolated Subunits
  - F. Regulation of T4 Topoisomerase Gene Expression
- **III. Vaccinia DNA Topoisomerase**
- **IV. Concluding Remarks**

I. INTRODUCTION

The ubiquity of DNA topoisomerases is now well established. These enzymes participate in many vital aspects of DNA metabolism including replication, transcription, recombination, and chromosomal segregation. In virus-infected prokaryotic and eukaryotic systems, the host DNA topoisomerases are usually recruited to perform the functions needed for virus growth and development. Two relatively large and complex double-stranded DNA (dsDNA) viruses, one in prokaryotes and one in eukaryotes, are known to encode their own topoisomerases that are required for viral growth and development. Bacteriophage T4 and its relatives encode a type-II ATP-dependent DNA topoisomerase, whereas vaccinia, a poxvirus, encodes a type-I enzyme. These two viral enzymes have properties distinct from the host enzymes. Their subunit structure and activity, the organization of their coding sequence, and their possible role in virus development are the subject of this chapter.

II. T4 DNA TOPOISOMERASE

A. General Description of T4 DNA Metabolism

Bacteriophage T4 is a large lytic virus of *Escherichia coli*. Its genome is 166-kb pairs in length. The viral DNA is terminally repetitious with 3%

DNA Topology and Its Biological Effects

redundancy. The genome is circularly permuted so that in different molecules any given gene lies at different distances from the end. Hence, the virus has a circular genetic map although the genome is a linear duplex DNA molecule. The viral DNA is modified with glucosylated hydroxymethylated cytosine in place of cytosine. These modifications protect the viral DNA from some host restriction endonucleases and may also provide an added avenue of control for the virus. T4 development follows a controlled temporal program, yet the process has built-in redundancy and alternate pathways. Within 3–5 minutes after infection, the synthesis of host DNA, mRNA, and protein is shut off. Therefore, it is not surprising that nearly half of the viral genome encodes functions in nucleotide metabolism and DNA replication that ensure the production of about 200 copies of the viral DNA within 30–35 minutes after infection. In fact, T4-infected cells have been one of the best sources of commonly used nucleic acid enzymes. The studies of T4 enzymes have provided many insights in establishing principles that govern DNA metabolism. Various aspects of T4 biology have been reviewed and compiled into one volume: *Bacteriophage T4* (Mathews et al. 1983). Developments since 1983 have been reviewed more recently (Mosig and Eiserling 1988). Because of the large demand for nucleotide precursors within a relatively short time, the supply of precursor is rate limiting for replication. T4 brings about the rapid rate of precursor biosynthesis by de novo synthesis, as well as by using virus-encoded and host enzymes to degrade the host DNA and recycle the nucleotides. These enzymes form a multienzyme complex that efficiently synthesizes and channels the precursor triphosphates into the T4 replication machinery (Mathews and Allen 1983). In support of the direct channeling and coupling of precursor biosynthesis and replication is the observation that a number of T4 proteins known to be involved in T4 replication are found associated with the large T4 precursor-synthesizing complex (Chiu et al. 1982).

T4 DNA replication can use exclusively virus-encoded proteins. The reconstituted replication system from purified T4 components has provided an elegant model for the formation of a specific multienzyme complex for processive DNA chain elongation and fork movement. The complex consists of T4 DNA polymerase, polymerase accessory proteins, helix-destabilizing protein, RNA priming proteins, and helicase. The enzyme complex forms an efficient "machine" that is capable of both leading and lagging strand synthesis at near in vivo rate (Nossal and Alberts 1983; Alberts 1984). On the other hand, T4 growth requires an efficient recombination process that is coupled to replication. The two processes are inseparable and equally essential for T4 growth (Mosig 1987). The two pathways share many T4 enzymes (Formosa and Alberts 1986). In addition to the large number of proteins needed for the virus-specified DNA metabolism, T4 also encodes a multisubunit DNA topoisomerase. Genetic evidence suggests that this topoisomerase has a profound effect on both DNA precursor biosynthesis (Wirak and Greenberg 1980) and the replication-recombination process (Mosig and Eiserling 1988).

The T4 DNA topoisomerase is specified by three genes, *39*, *52*, and *60*. Mutants in these genes do not synthesize active enzyme (Liu et al. 1979). These genes are located in the early region of the T4 genome, and their products are synthesized within 3–5 minutes after viral infection. In mutants defective in any one of these three genes, T4 DNA replication is initiated at a low level and rapid exponential replication does not occur until later times after infection. Because of the delay in the onset of exponential DNA replication in these mutants, the T4 DNA topoisomerase subunits were originally described as DNA delay proteins. The delay effect is accentuated by low temperature and varies according to the *E. coli* host. Hence, it is concluded that normal T4 development requires a functional viral topoisomerase (Yegian et al. 1971; Mufti and Bernstein 1974). Using permeable cellophane membrane disks to prepare highly concentrated cell extracts, it can be shown that T4-specific semiconservative DNA replication occurs and is dependent on the presence of the virus-encoded DNA topoisomerase (Huang 1983). T4 DNA replication in extracts prepared from T4-topoisomerase-deficient mutant-infected cultures is about 10–20% of wild-type virus-infected level. This level can be enhanced by exogenous purified T4 topoisomerase. Although a direct participation of T4 DNA topoisomerase in T4 DNA replication has been demonstrated, the precise role it plays in the replicative process remains to be established. By autoradiography, it was shown that in mutants defective in any one of the three genes, the rate of viral DNA chain elongation at each initiated fork is apparently comparable to that in wild-type T4 (McCarthy et al. 1976). This suggests that T4 topoisomerase is not directly involved in regulating the rate of DNA chain elongation, and a full complement of T4 replication proteins directly responsible for DNA chain elongation and fork movement is synthesized by topoisomerase defective mutants (Burke et al. 1983). However, the number of replication forks in the mutant infected cells is reduced (McCarthy et al. 1976). This is interpreted to mean that T4 topoisomerase is involved in the regulation of the number of replication forks that are initiated. Initiation in T4 topoisomerase mutant-infected cells is attributed to the host DNA gyrase because removal of gyrase activity either by mutation or by the addition of gyrase-specific antibiotics eliminates the residual activity, both in vivo and in vitro (McCarthy 1979; Huang 1983). The possible heterologous interactions between T4 topoisomerase subunits and host gyrase subunits, which are feasible in these single T4-topoisomerase-gene-defective mutant-infected cells, have not been investigated. However, in mutants where more than one T4 topoisomerase gene are defective, virtually no viral DNA replication is detected (Yegian et al. 1971; W.M. Huang, unpubl.). Hence, normal T4 DNA replication and development do require the presence of T4 DNA topoisomerase.

T4 initiates DNA replication bidirectionally from the internal portion of the viral genome. A number of strategies are apparently used to provide the 3′-OH end to prime DNA replication. The genetics of three suggested alternative modes of T4 DNA initiation have been reviewed recently (Mosig and Eiserling 1988). They included the use of (1) the host RNA polymerase to synthesize a specific short priming RNA, (2) recombination intermediates to provide DNA primers, and (3) a third mode that is independent of the host RNA polymerase and that uses a T4-specific promoter to synthesize an RNA primer. Under different experimental growth conditions, T4 replication can start at at least six regions of the T4 genome (Mosig and Eiserling 1988). It is believed that not all of them operate simultaneously, and the control of origin utilization is unknown. Because of the large size of the genome and the complexity of base modification, an in vitro replication system based on the use of any one of these identified origins for T4 replication has not been developed. This has hindered determination of how T4 topoisomerase fits into this replication-recombination process.

B. Activities of Purified T4 DNA Topoisomerase

T4 DNA topoisomerase is purified as a complex of three different polypeptides in equimolar ratio: a 60-kD peptide encoded by gene *39*, a 50-kD peptide encoded by gene *52*, and an 18-kD peptide encoded by gene *60*. The native enzyme is heterogeneous, sediments with a sedimentation coefficient of 6.5S, and has an average Stokes radius of 6.1 nm, as determined by gel filtration, corresponding to a globular protein of approximately 300,000 kD in molecular mass with a subunit structure of at least a dimer of the three polypeptides (Kreuzer and Huang 1983). The purified enzyme relaxes superhelical DNA by changing its linking number in steps of two (Liu et al. 1980); hence, it is a type-II DNA topoisomerase. The enzyme requires ATP or dATP for catalytic conversion of supertwisted DNA into partially relaxed topoisomers. Both negative and positive supercoils are effective substrates. This characteristic is different from bacterial DNA gyrase, a type-II enzyme that negatively supercoils DNA in an ATP-dependent reaction but that relaxes DNA in an ATP-independent reaction. The purified T4 enzyme is unable to introduce negative supertwists into closed circular DNA molecules. Thus, the catalytic activity of the T4 enzyme is very similar to the eukaryotic type-II topoisomerases (see Hsieh, this volume). In fact, it was the first known type-II DNA topoisomerase, other than DNA gyrase (Liu et al. 1979; Stetler et al. 1979). Despite these differences, type-II DNA topoisomerases share significant amino acid sequence homology, which will be discussed in detail in a later section.

Type-II DNA topoisomerases catalyze DNA strand passage reactions via a transient double-stranded break through which another segment of DNA is passed. In addition to the standard superhelical DNA relaxation reaction, T4 DNA topoisomerase, in the presence of ATP, catalyzes the decatenation of the interlocked closed circular network of kinetoplast DNA, as well as the unknotting of knotted phage P4 DNA (Liu et al. 1981). In the presence of DNA condensing agents such as polycationic polymin P, spermidine, or histone H1, DNA rings can be converted to knotted or interlocked circular species (Liu et al. 1980). Knotting and interlinking of closed circular DNA molecules are also observed at high concentrations of the T4 enzyme alone, suggesting that the T4 enzyme itself may be capable of condensing DNA (Liu et al. 1980). The enzyme has been useful for generating defined knots belonging exclusively to the twist family (Wasserman and Cozzarelli 1986).

The DNA strand passage reaction requires Mg^{++} ion. This divalent cation requirement cannot be replaced by Mn^{++}, Ca^{++}, or Co^{++}. In this reaction, the required ATP is hydrolyzed to ADP and P_i. Nonhydrolyzable ATP analogs such as ATPγS, adenylyl imidodiphosphate (p[NH]ppA), and adenylyl β,γ-methylene diphosphate (p[CH_2]ppA) are ineffective (Liu et al. 1979; Stetler et al. 1979). T4 DNA topoisomerase is also a DNA-dependent ATPase with an approximate K_m of 0.35 mM for ATP (G. Nicholson and W.M. Huang, unpubl.). Consistent with the use of dATP in DNA relaxation reaction, the T4 enzyme can also hydrolyze dATP. Single-stranded DNA (ssDNA) or dsDNA can be used as a cofactor in this ATPase reaction (G. Nicholson and W.M. Huang, unpubl.). As will be discussed below, the requirement for ATP hydrolysis in the catalytic reaction of DNA relaxation may be to change the conformation of the enzyme at the end of each cycle of concerted DNA breakage and rejoining. This notion is supported by the observation that a limited amount of DNA relaxation is observed in the absence of ATP if sufficient enzyme is used to provide one cycle of DNA relaxation (W.M. Huang, unpubl.).

Type-II DNA topoisomerases can cleave the substrate DNA to give a linearized duplex DNA with a free 3′-OH group and a blocked 5′-

phosphoryl group at each end. T4 topoisomerase performs this "induced cleavage" reaction in the absence of ATP and requires the addition of a protein denaturant such as SDS. Cleavable complex formation may be a reflection of a partial activity of strand passage in which both strands of the DNA duplex are transiently broken, and the protein denaturant traps this intermediate irreversibly by terminating the reaction without allowing the rejoining of the broken ends. In the cleaved product, the 52 protein subunit of the enzyme is covalently joined to the 5′- phosphoryl end via a tyrosyl phosphate linkage (Rowe et al. 1984). The enzyme cleaves pBR322 DNA or the replicative form of ϕX174 DNA at a number of specific locations dependent on the conditions of the reaction (Stetler 1980; Kreuzer and Alberts 1984). Although some sequence specificity is implied by the nonrandomness of the cleavage sites, the analysis of the locations did not provide strong rules for the recognition. Induced cleavage is stimulated by high concentrations (200–500 μg/ml) of oxolinic acid, a potent inhibitor of *E. coli* DNA gyrase. T4 topoisomerase is only marginally inhibited by quinolone derivatives such as oxolinic acid, and inhibition requires at least 50-fold more drug than needed for the inactivation of DNA gyrase (Stetler et al. 1979). Nevertheless, oxolinic acid provides a means to enhance the cleavage reaction for examination of the cleavage pattern of T4 DNA, which is clearly the true substrate for the enzyme. The analysis showed that glucosylation is partly responsible for increased specificity of the T4 topoisomerase in its reaction with T4 DNA. So far, a number of induced cleavage sites on the viral DNA have been recognized. They include sites of T4 recombination and mutational hot spots, as well as some presumed origins of DNA replication that were identified genetically (Kreuzer and Alberts 1984). In support of the enzyme's possible involvement in recombination, it is found that purified T4 DNA topoisomerase promotes illegitimate or nonhomologous recombination between two bacteriophage λ DNA molecules in vitro. The recombination event is ATP-independent and is stimulated by oxolinic acid, prompting the suggestion that although occurring at low frequency, this DNA strand exchange reaction is mediated by a mechanism that may involve the exchange of DNA linked subunits of T4 topoisomerase (Ikeda 1986, and this volume).

C. T4 Topoisomerase Is Sensitive to Antitumor Agents

In light of the similarity between the eukaryotic type-II DNA topoisomerase and the T4 enzyme, the susceptibilities of the T4 enzyme to antitumor drugs that inhibit eukaryotic type-II topoisomerase (Tewey et al. 1984; Drlica and Franco 1988) have been investigated. Two clinically important drugs, *m*-AMSA, (4′-9-acridinylamino)methanesulfon-*m*-anisidide) and mitoxanthrone, a synthetic anthracenedione, greatly stimulate the induced DNA cleavage by the T4 enzyme (Rowe et al. 1984; Huff et al. 1989; G. Harker and W.M. Huang, unpubl.). Furthermore, it was shown that under some conditions these drugs block viral growth, whereas the growth of the host *E. coli* is not effected (Huff et al. 1989). Thus, the simple prokaryotic T4 system is an easily manipulatable model system for the investigation of these antitumor drugs both as antiviral agents and as drugs targeted to type-II DNA topoisomerases. Since the basic enzymology of the T4 DNA topoisomerase is very similar to the eukaryotic type-II topoisomerases, the isolation and analyses of drug-resistant topoisomerase mutants will be much more easily done in the T4 system. An *m*-AMSA-resistant T4 topoisomerase has been isolated from *m*-AMSA-resistant virus-infected cells, unequivocally demonstrating that the in vivo target of this drug is the T4 topoisomerase (Huff et al. 1989). The resistant enzyme is unexpectedly different from the wild-type enzyme in a number of ways apparently mimicking an effect that *m*-AMSA has on the wild-type enzyme. The mutant enzyme has enhanced DNA cleavage activity, even in the absence of the drug, and the cleavage sites are different from that of the wild-type enzyme. Different antitumor drugs are likely to interact with DNA differently. The isolation of different drug-resistant topoisomerase mutants and their encoded enzymes is needed to distinguish between inhibitory effects caused by the direct interference of the enzyme and the distortion caused by drug/DNA interactions. In addition, antitumor drug-resistant mutants will provide a set of useful markers in the structure/function analysis of these proteins.

D. DNA Sequence Arrangement of T4 Topoisomerase

T4 DNA topoisomerase genes *39*, *60*, and *52* have been cloned into multicopy plasmids by virtue of their ability to rescue the corresponding T4 mutants. The nucleotide sequence of these genes has been determined, and the reading frames for the encoded proteins have been established by aligning the amino-terminal amino acids of the purified proteins with the DNA sequences. T4 gene *39* encodes a protein of 519 amino acids, gene *60* encodes a protein of 160 amino acids, and gene *52* encodes a protein of 441 amino acids (Huang 1986a,b; Huang et al. 1988). The DNA sequences and the derived amino acid sequences of the encoded proteins are given in the appendix to this volume. The most striking feature that emerges from the sequence analyses is that the T4 genes share significant amino acid sequence homology with both the eukaryotic type-II topoisomerases and the bacterial gyrases; they are distinct from the type-I topoisomerases of either group. The homology suggests that type-II enzymes are evolutionarily and structurally conserved despite their differences in subunit structure.

In eubacteria, type-II topoisomerases are represented by DNA gyrases, which are complexes of the products of *gyrA* and *gyrB* genes; they have a tetrameric subunit structure of A_2B_2. The sequence of these genes from two bacteria (*E. coli* and *Bacillus subtilis*) have been determined, and they are highly similar (Moriya et al. 1985; Adachi et al. 1987; Swanberg and Wang 1987; Yamagishi et al. 1987). The T4 *39* protein aligns with the amino-terminal portion of these *gyrB* proteins (Huang 1986b). The T4 *60* protein can be aligned with the carboxy-terminal end of the *gyrB* proteins. T4 *52* protein can be aligned with the amino-terminal half of the *gyrA* proteins with the highest similarity near the amino terminus (Huang 1986a). In eukaryotes, the type-II topoisomerases are homodimers of a single, large polypeptide. To date, the DNA sequences of four eukaryotic type-II topoisomerases have been published (yeast *Saccharomyces cerevisiae*, Giaever et al. 1986; yeast *Schizosaccharomyces pombe*, Uemura et al. 1986; human, Tsai-Pflugfelder et al. 1988; *Drosophila*, Wyckoff et al. 1989). They are homologous to sizes ranging from 1429 to 1530 amino acids. The subunit organization based on the amino acid similarities of all the sequenced type-II topoisomerases is summarized in Figure 1. The coding information of all type-II enzymes can be accounted for by positioning the prokaryotic enzyme subunits in tandem arrays along the eukaryotic counterpart. The *gyrB* subunit can be positioned at the amino terminus, whereas the *gyrA* subunit can be positioned at the carboxyl terminus (Lynn et al. 1986). Similarly, the three subunits of the T4 enzyme can be aligned with the eukaryotic enzyme according to the subunit order of the *39*, *60*, and *52* proteins. In *B. subtilis*, the two gyrase subunit genes are adjacent, but they are transcribed from separate promoters (Ogasawara et al. 1985). In *E. coli*, the two gyrase genes are far apart, separated by approximately 35 minutes of chromosomal distance. The *gyrB* gene of *E. coli* (encoding 804 amino acids) is larger than that of *B. subtilis* (638 amino acids) by 166 amino acids. By including the T4 gene *39* subunit of 519 amino acids in the analysis, it is clear that the first 500 amino acids of the three peptides are analogous. On the other hand, T4 gene *60* (encoding 160 amino acids) aligns with the carboxyl terminus of the *E. coli* and *B. subtilis gyrB* subunit. The fact that the carboxy-terminal portion of the *gyrB* gene corresponds to yet another gene in the T4 system implies that the *gyrB* gene product has two domains. The 166 amino acids unique to *E. coli* is located at the junction of the two *gyrB* domains. The amino acid

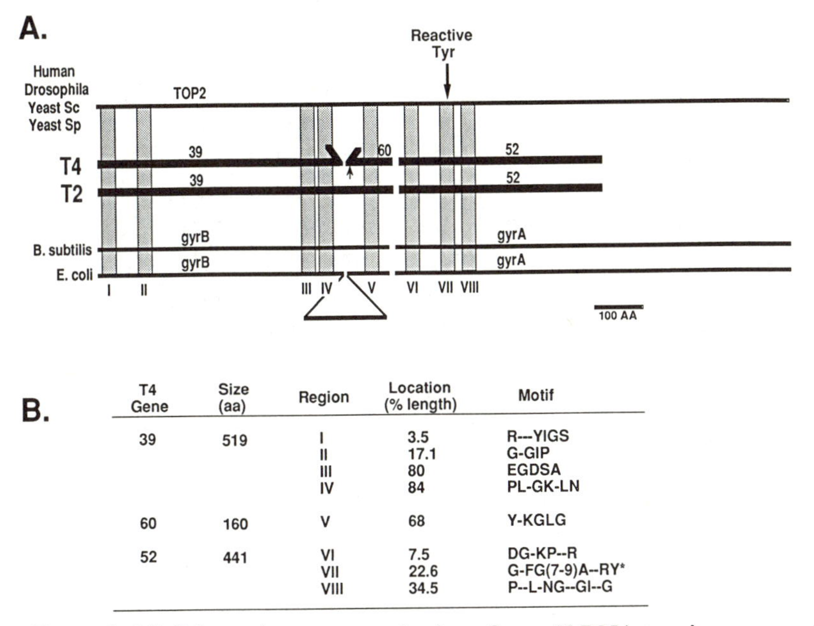

T4 Gene	Size (aa)	Region	Location (% length)	Motif
39	519	I	3.5	R---YIGS
		II	17.1	G-GIP
		III	80	EGDSA
		IV	84	PL-GK-LN
60	160	V	68	Y-KGLG
52	441	VI	7.5	DG-KP--R
		VII	22.6	G-FG(7-9)A--RY*
		VIII	34.5	P--L-NG--GI--G

Figure 1 (*A*) Schematic gene organization of type-II DNA topoisomerases. The primary sequence of the encoded proteins is depicted as parallel lines that are broken into subunits when appropriate. The orientation of amino-to-carboxyl termini is from left to right. Projections from the T4 proteins represent amino acids not found in T2. The small arrow under T4 gene *60* indicates the position of the 50-nucleotide untranslated interruption. See text for detailed comparison of the T2 and T4 proteins. The shaded regions indicate locations where amino acid motifs common to all eight proteins are found. (*B*) Description of common motifs found in type-II topoisomerases. The locations of the regions are described as the percentage of subunit length measured in amino acids starting from the initiation codon of each of the T4 proteins. Each dash in the motifs, presented as one letter amino acid codes, denotes a nonconserved amino acid. In region VII, the length of the nonconserved amino acid spacer is given in parenthesis, and the Y* represents the reactive tyrosine residue involved in the formation of covalent DNA/protein intermediates.

sequence alignment also suggests that the T4 gene *52* protein is analogous to the *gyrA* subunit of gyrase, but the gene *52* protein is only half the size of *gyrA* proteins. These regions can likewise be aligned with the carboxy-terminal half of the eukaryotic enzymes. The tyrosine residue that forms the DNA/protein intermediate has been located in two systems, *E. coli* and in *S. cerevisiae* (Horowitz and Wang 1987; Worland and Wang 1989). They are within the conserved amino acid motif VII (Fig. 1B) shared by all sequenced enzymes. This alignment provides a straightforward assignment of Tyr-116 of the T4 *52* subunit as the reactive tyrosine.

The partitioning of the coding information of functionally related type-II DNA topoisomerases into separate genes in prokaryotes may result from gene separation/fusion events during evolution. The boundaries that divide the eukaryotic topoisomerase into three domains analogous to the three genes of T4 topoisomerase bear no relationship to the locations of the three introns that interrupt the codons of the *Drosophila TOP2* gene (Wyckoff et al. 1989). The homology among the different species of type-II enzymes described above represents 32–55% identity in amino acid sequence when different pairs of topoisomerase genes are compared. It is not possible to define the precise boundaries of the different domains in the larger "fused" polypeptide or to identify the interdomain regions.

The most striking example of such a gene fusion/separation event is found in the topoisomerases of the closely related T even-phage family. The T2 and T6 topoisomerases are enzymatically indistinguishable from the T4 enzyme but have only two subunits (Huang et al. 1985). The gene *52* protein subunits are essentially identical genetically and biochemically in the three T phages. The larger T2 (and T6) gene *39* protein subunit is a fused protein equivalent to the combined products of T4 genes *39* and *60*. Recently, the complete DNA sequence for T2 gene *39* has been determined (A. Gibson et al., unpubl.). It encodes a protein of 605 amino acids. Its amino acid alignment is also included in Figure 1. The amino-terminal 473 amino acids of T2 and T4 gene *39* proteins are nearly identical with only six differences. On the other hand, the carboxy-terminal 129 amino acids of T2 *39* protein are the same as the carboxyl terminus of T4 *60* protein except for two differences. The alignment clearly identifies three T2-specific amino acids at the interdomain region. This short linker provides sufficient conformational flexibility to allow the proper folding and spatial juxtaposition of the two connected domains of the T2 *39* protein. The alignment also shows that the carboxy-terminal 46 amino acids of T4 gene *39*, as well as the amino-terminal 31 amino acids of T4 gene *60*, are unique to T4, and they may not be required for DNA topoisomerization reaction since they are absent in T2. The T2 topoisomerase, with the two subunits of 605 and 441 amino acids, is the smallest known topoisomerase that is capable of the dsDNA passage reaction.

Sequencing data from eight different sources have identified eight distinct blocks of conserved amino acids. They are scattered throughout the three genes of the T4 enzyme (Fig. 1A). These conserved protein sequences may be specifically related to the double-stranded breakage-rejoining activity characteristic of the type-II enzyme. These conserved amino acid motifs, together with their relative locations in the enzyme provide a useful diagnostic feature for future identifications of type-II topoisomerases from other organisms. The conserved sequences may also be used as guides for designing consensus protein or nucleic acid probes for studies of type-II DNA topoisomerases.

E. Reconstitution of T4 Topoisomerase from Isolated Subunits

The three subunits of T4 topoisomerase have been expressed independently and overproduced in *E. coli.* The purified subunit proteins possess partial activities relating to the type-II topoisomerization reaction (G. Nicholson and W. M. Huang, unpubl.). T4 *39* protein is a monomer under native conditions as determined by gel filtration analysis. It is an ATPase that hydrolyzes ATP to ADP and P_i. However, unlike the ATPase of the complete enzyme, it is only marginally stimulated by either dsDNA or ssDNA. It has no DNA relaxation activity. T4 *52* protein is a dimer. The purified *52* protein alone has a weak DNA relaxation activity. The reaction requires Mg^{++}, but it is independent of ATP. Furthermore, the *52* protein does not act catalytically. At the end of one round of cutting-rejoining reaction, the *52* protein is inactive. Purified *60* protein tends to form large aggregates under native conditions. It has neither ATPase activity nor DNA relaxation activity. The *60* protein does not stimulate the activities of *39* protein (ATPase) or *52* protein (DNA relaxation). Similarly, a mixture of T4 *39* protein and *52* protein has no enhanced activities relative to that of the single proteins. However, when the *39* and *60* proteins are added after *52* protein is rendered inactive by reaction with DNA, catalytic DNA relaxation activity characteristic of the complete enzyme can be demonstrated on an added DNA substrate. The reconstituted reaction also required ATP. With the 3-protein mixture, the ATPase activity is greatly stimulated by DNA. These experiments demonstrate that the complex ATP-dependent DNA strand passage reaction can be divided into a series of partial reactions carried out principally by the different units. A model can be formulated to account for the subunit reconstitution data: *52* protein is the DNA breaking-rejoining unit. At the end of one round of activity, it is left in an inactive state. If it is complexed with the *39* protein, through ATP hydrolysis, *52* protein may be converted back to the original state and is thus able to carry out a subsequent round of catalysis. *60* protein is apparently needed to hold the *39* and *52* subunits together via protein-protein interactions.

The partial activities of the T4 topoisomerase subunits have provided an explanation for the requirement of ATP hydrolysis in the energetically favored DNA relaxation reaction. It is probably used for protein conformational changes. In DNA gyrase-catalyzed reactions, ATP is not required for DNA relaxation reaction. ATP binding alone promotes a single round of supercoiling, introducing two negative supercoils into the DNA; ATP hydrolysis is needed for turnover (Sugino et al. 1978). Although both T4 *39* protein and the *gyrB* subunit of gyrase are ATPases, their DNA dependency is quite different. The *gyrA* subunit, which is twice the size of the T4 *52* protein, apparently does not have DNA breakage and rejoining activity in the absence of *gyrB* (Mizuuchi et al. 1984) although this functional domain is separately encoded as in T4. It would be of interest to determine whether separated peptide subdomains of the eukaryotic topoisomerases retain partial activities similar to the T4 system since their topoisomerase activities are more related.

F. Regulation of T4 Topoisomerase Gene Expression

The three gene products of T4 topoisomerase are coordinately synthesized in an approximately equimolar ratio shortly after infection. The adjacent genes *39* and *60*, separated by 782 nucleotides, are translated from one message (Brody et al. 1983), whereas *52* protein located 7-kb away is translated from a different message (Christensen and Young 1983). This is the only system where the coding information of a DNA topoisomerase is separately translated as three independent peptides. The translation of T4 gene *60* is most unusual in that it is a split gene whose transcript contains a 50 nucleotide interruption that is not removed before translation (Huang et al. 1988). This is an unprecedented form of translational regulation in which the translation machinery is thought to jump through an mRNA structure in order to decode the noncontiguous codons efficiently. It is of historical interest to note that the colinearity of the gene (and the implied functional message) with its corresponding peptide was first established in T4 25 years ago (Sarabhai et al. 1964). The exception to the colinearity rule was also discovered in T4. To date, it is the only known example of this kind. In the related T2 virus, the equivalent gene *60* information is covalently attached to the carboxy-terminal end of the preceding gene *39*, and the entire T2 gene *39* coding information is not interrupted. The T4 interruption occurs 15 amino acids downstream from the linker region that marks the separation of the T2 peptide into two domains (A. Gibson and W.M. Huang, unpubl.). The rationale for maintaining this mode of translational regulation in T4 is unknown. Whether there is any relationship between the unusual mode of translation used by T4 gene *60* and keeping gene *60* as a separate gene is also unknown. An obvious advantage of putting two genes together as one larger polypeptide is to ensure a coordinated transcription, translation, and immediate assembly into properly folded protein because the two domains are already together. On the other hand, the division of two domains into separately encoded genes may increase the diversity of domain utilization since one of the domains may be used in a separate pathway. Support for this notion is found in recent reports relating the effect of gene *39* to T4 nucleotide biosynthesis. Extragenic suppressors of a mutant in T4 ribonucleotide reductase (part of another multienzyme complex for DNA precursor biosynthesis) were found to be in gene *39*. This function of the T4 *39* protein subunit is apparently unrelated to the action of T4 topoisomerase since mutations in the other two subunits have no effect (Cook et al. 1988; Wirak et al. 1988). Because the independent translation of T4 gene *60* without the presence of T4 gene *39* protein is lethal to *E. coli* (W.M. Huang, unpubl.), one may speculate that the interruption may be used to coordinate protein synthesis by slowing down the progression of translation of this small protein sufficiently to ensure that the other subunits will be available for complex formation. Experiments are in progress to test this and other possibilities. Regardless of the outcome, the presence of this specialized mode of translation further illustrates the importance of the cellular requirement to regulate the level of topoisomerase and therefore the supercoiling of DNA.

Through site-directed mutagenesis and deletion analyses, the essential features of the exceptional mode of translation of T4 gene *60* have been deciphered. The 50-nucleotide interruption alone is insufficient. Approximately 90 nucleotides encoding 30 amino acids 5′ to the interruption plus one amino acid at the 3′ side are necessary for efficient ribosome-jumping (R. Weiss and W. M. Huang, unpubl.). Thus, approximately 140 nucleotides in the T4 gene *60* sequence is the minimal required length. The 50-nucleotide interruption together with its immediately surrounding sequence has the potential to assume a special secondary structure that brings together the separated codons (Huang et al. 1988). A model is described in Figure 2 and is based on the proposed secondary structure of the mRNA in this region. The transcribed mRNA itself may be folded into a stable stem-loop structure (Fig 2A). The amino-terminal half of gene *60* is marked by a stop codon following Gly-46. Translation to Gly-46 partially opens the stem and exposes the sequence complementary to the UGGA, a repeated sequence also found immediately 5′ to the codon coding for the next amino acid, Leu-47, located 50 nucleotides downstream. The required stop codon next to the Gly-46 may act as a pause point to slow down the progression of translation. The length of the remaining sequence in the side arm region (Fig. 2B) is less stringently regulated. So far, it appears that the direct repeat is one of the most

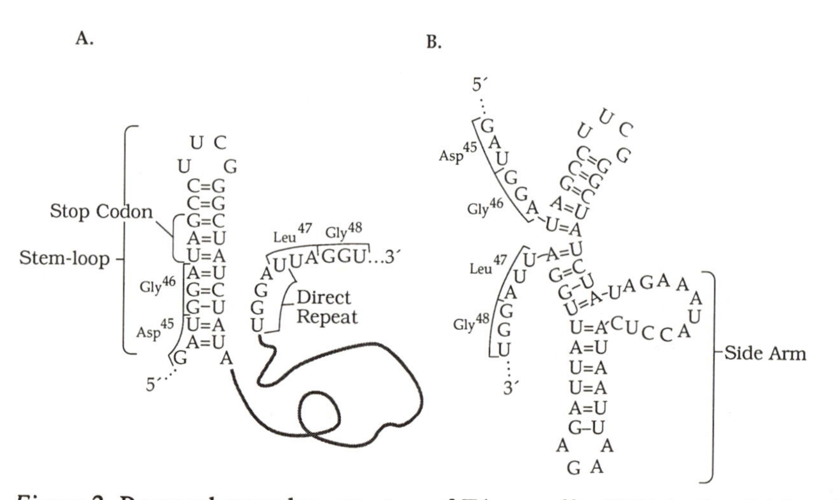

Figure 2 Proposed secondary structure of T4 gene *60* mRNA in the vicinity of the interrupted region. The pairing described in the stem of the stem-loop structure shown in *A* can be alternately folded as shown in *B* as a result of pairing with the two different halves of the direct repeat sequence. The adjacent codons Gly-46 and Leu-47 are brought to juxtaposition so that the ribosome may skip over the sequences in between. See text for a more detailed explanation of the other features. (Adapted from Huang et al. 1988.)

important features of this particular mode of translation. It provides the same codon, GGA, for the elongating peptidyl-tRNA to "land," so it can bypass the interrupted sequence and continue translation 50 nucleotides downstream. The role of the second required component involving sequences upstream of the interruption is not understood. It may be needed to provide added stability to the basic alternate pairing scheme, or it may interact with the translation machinery itself.

III. VACCINIA DNA TOPOISOMERASE

Vaccinia virus, a member of the poxvirus family, is one of the largest and most complex DNA viruses with a genome size of 185 kb. It has two distinguishing features: (1) At the termini of the linear genome, the two strands are covalently joined to form hairpin loops (Geshelin and Burns 1974; Baroudy and Moss 1982); and (2) it replicates its DNA exclusively in the cytoplasm of eukaryotic cells (for a recent review, see Moss 1985). During maturation of the virus, a number of virus- and host-encoded enzymes are encapsidated in the virion. They include a full complement of enzymes needed for the synthesis and processing of functional viral mRNAs that are normally polyadenylated, capped, and methylated. In addition, a DNA topoisomerase is also encapsidated. Because many essential replication enzymes, such as the viral DNA polymerase or DNA ligase, are apparently not in the virion although all enzymes needed for viral transcription are in the virion, and in light of the unusual organization of the DNA structure, it was postulated that the vaccinia topoisomerase may be involved in viral transcription (Bauer et al. 1977). Direct evidence for this is lacking. Other experiments suggest that it may be involved in the replicative process (Poddar and Bauer 1986). Insertion mutagenesis has shown that the vaccinia DNA topoisomerase gene is required for viral growth in cell culture (Shuman et al. 1989) although its exact role remains to be defined. The enzyme has been purified from viral core particles, as well as from *E. coli* cells containing a recombinant plasmid (Foglesong and Bauer 1984; Shaffer and Traktman 1987; Shuman et al. 1988). It is a type-I enzyme with a single polypeptide of about 32 kD that is only one-third the size of a typical eukaryotic type-I enzyme. It relaxes both positively and negatively supercoiled DNA in an ATP-independent reaction. Characteristic of eukaryotic type-I activity, the vaccinia enzyme can also be induced to form a cleavable protein/DNA complex via the 3′-phosphoryl end of one strand of the DNA. It is interesting to note that novobiocin, coumermycin A1, and to a lesser extent nalidixic acid, all reagents that inhibit type-II DNA topoisomerases, inhibit the DNA relaxation activity and cleavable complex formation of the vaccinia enzyme (Shaffer and Traktman 1987). Similarly, berenil, an inhibitor of mitochondrial type-II trypanosome topoisomerase (Melendy and Ray 1989), is also effective as an inhibitor. The vaccinia topoisomerase synthesized in *E. coli* is resistant to camptothecin, a specific inhibitor of cellular eukaryotic type-I topoisomerases. Various assays suggest that the enzyme does not bind camptothecin (Shuman et al. 1988). This observation, however, leaves open the question of how vaccinia infection is inhibited by camptothecin (Horwitz et al. 1972).

The recent determination of the DNA sequence of the vaccinia topoisomerase gene confirmed that it is part of the viral genome, and it encodes a protein of 314 amino acids (Shuman and Moss 1987). It shares a region of amino acid sequence similarity with three type-I enzymes, namely those from *S. cerevisiae*, *S. pombe*, and humans, but not with *E. coli* topoisomerase I (Lynn et al. 1989). Reasonable homology (26% identity) occurs near amino acids 116 through 228. Downstream from this homologous region, Tyr-274 has been implicated to be the active tyrosine residue that forms the phosphotyrosine in the covalent intermediate. This assignment is based on a weaker similarity of sequences surrounding the vaccinia Tyr-274 and those of yeast Tyr-727, the active tyrosine of yeast topoisomerase I. Furthermore, by mutating the *S. cerevisiae* enzyme at locations near its active Tyr-727 to match the environment of the vaccinia enzyme, full yeast enzymatic activity is retained (Lynn et al. 1989). Vaccinia topoisomerase represents a simple model system of a eukaryotic type-I enzyme with characteristics unique to the viral system.

IV. CONCLUDING REMARKS

Viral DNA topoisomerases encoded by T2, T4, and vaccinia are useful model systems to study the structure, function, and mechanism of type-II and type-I enzymes primarily because they are smaller and more amenable as genetic systems than their host counterparts. Because of the amino acid sequence homology within each class of topoisomerases, the viral enzymes have provided valuable demarcation lines in assigning functional domains in the larger cellular enzymes. They exhibit properties common to the typical DNA topoisomerases of each class, yet they possess unique properties especially in their interactions with topoisomerase-specific drugs. Why these viruses should spend their precious coding information on an entirely new topoisomerase instead of just modifying the host enzyme for their special needs is a mystery. The biochemistry and the enzymology of these enzymes are well in hand, but a missing link is the lack of information concerning the structures of their in vivo substrates. Much remains to be learned concerning their physiological functions. Since these genes are now cloned and sequenced and the encoded proteins can be purified in large quantities, detailed information concerning their three-dimensional structure, interaction with DNA, and consequences of their interactions with pharmacologically important drugs that use them as targets should be forthcoming.

ACKNOWLEDGMENTS

I thank Sherwood Casjens for patiently reading this manuscript and members of my laboratory for stimulating discussions. Support is provided by a grant from National Institutes of Health.

REFERENCES

Adachi, T., M. Mizuuchi, E. Robinson, E. Appella, M. O'dea, M. Gellert, and K. Mizuuchi. 1987. DNA sequence of the *E. coli* gyrB gene: Application of a new sequencing strategy. *Nucleic Acids Res.* **15:** 771–783.

Alberts, B.M. 1984. The DNA enzymology of protein machines. *Cold Spring Harbor Symp. Quant. Biol.* **49:** 1–12.

Baroudy, B.M. and B. Moss. 1982. Incompletely base-paired flip-flop terminal loops link the two DNA strands of the vaccinia virus genome into one uninterrupted polynucleotide chain. *Cell* **28:** 315–324.

Bauer, W.R., E.C. Ressner, J. Kates, and J.V. Patzke. 1977. A DNA nicking-closing enzyme encapsidated in vaccinia virus: Partial purification and properties. *Proc. Natl. Acad. Sci.* **74:** 1841–1845.

Brody, E., D. Rabussay, and D. Hall. 1983. Regulation of transcription of prereplicative genes. In *Bacteriophage T4* (ed. C.K. Mathews et al.), p. 174–183. American Society for Microbiology, Washington, D.C.

Burke, R.L., T. Formosa, K.S. Cook, A.F. Seasholtz, J. Hosoda, and H. Moise. 1983. Use of two-dimension polyacrylamide gels to identify T4 prereplicative proteins. In *Bacteriophage T4* (ed. C.K. Mathews et al.), p. 321–326. American Society for Microbiology, Washington, D.C.

Chiu, C.S., K.S. Cook, and G.R. Greenberg. 1982. Characterization of bacteriophage T4-induced complex synthesizing deoxyribonucleotides. *J. Biol. Chem.* **257:** 15087–15097.

Christensen, A.C. and E.T. Young. 1983. Characterization of T4 transcripts. In *Bacteriophage T4* (ed C.K. Mathews et al.), p. 184-188. American Society for Microbiology, Washington, D.C.

Cook, K.S., D.O. Wirak, A.F. Seasholtz, and G.R. Greenberg. 1988. Effect of bacteriophage T4 DNA topoisomerase gene 39 on level of β chain of ribonucleoside diphosphate reductase in T4 *nrdB* mutant. *J. Biol. Chem.* **263:** 6202–6208.

Drlica, K. and R. Franco. 1988. Inhibitors of DNA topoisomerases. *Biochemistry* **27:** 2254–2258.

Foglesong, P.W. and W.R. Bauer. 1984. Effects of ATP and inhibitory factors on the activity of vaccinia virus type I topoisomerase. *J. Virol.* **49:** 1–8.

Formosa, T. and B.M. Alberts. 1986. DNA synthesis dependent on genetic recombination: Characterization of a reaction catalyzed by purified bacteriophage T4 proteins. *Cell* **47:** 793–806.

Geshelin, P. and K. Burns. 1974. Characterization and location of the naturally occurring cross-links in vaccinia virus DNA. *J. Mol. Biol.* **88:** 785–796.

Giaever, G., R. Lynn, T. Goto, and J. Wang. 1986. The complete nucleotide sequence of the structural gene *Top2* of yeast DNA topoisomerase II. *J. Biol. Chem.* **261:** 12448–12454.

Horowitz, D.S. and J.C. Wang. 1987. Mapping the active site tyrosine of *Escherichia coli* DNA gyrase. *J. Biol. Chem.* **262:** 5339–5344.

Horwitz, S.B., C.K. Chang, and A.P. Grollman. 1972. Antiviral action of camptothecin. *Antimicrob. Agents Chemother.* **2:** 395–401.

Huang, W.M. 1983. T4 DNA replication on cellophane disks. In *Bacteriophage T4* (ed. C.K. Mathews et al.), p. 97–102. American Society for Microbiology, Washington, D.C..

———. 1986a. The 52-protein subunit of T4 DNA topoisomerase is homologous to the gyrA-protein of gyrase. *Nucleic Acids Res.* **14:** 7379–7389.

———. 1986b. Nucleotide sequence of a type II DNA topoisomerase gene: Bacteriophage T4 gene 39. *Nucleic Acids Res.* **14:** 7751–7765.

Huang, W.M., L. Wei, and S. Casjens. 1985. Relationship between bacteriophage T4 and T6 DNA topoisomerases. *J. Biol. Chem.* **260:** 8973–8977.

Huang, W.M., S. Ao, S. Casjens, R. Orlandi, R. Zeikus, R. Weiss, D. Winge, and M. Fang. 1988. A persistent untranslated sequence within bacteriophage T4 DNA topoisomerase gene 60. *Science* **239:** 1005–1012.

Huff, A., J. Leatherwood, and K. Kreuzer. 1989. Bacteriophage T4 DNA topoisomerase is the target of antitumor agent 4′-(9-acridinylamino)methanesulfon-m-anisidide (m-AMSA) in T4-infected *E. coli*. *Proc. Natl. Acad. Sci.* **86:** 1307–1311.

Ikeda, H. 1986. Bacteriophage T4 DNA topoisomerase mediates illegitimate recombination in vitro. *Proc. Natl. Acad. Sci.* **83:** 922–926.

Kreuzer, K.N. and B.M. Alberts. 1984. Site-specific recognition of bacteriophage T4 DNA by T4 type II DNA topoisomerase and *Escherichia coli* DNA gyrase *J. Biol. Chem.* **259:** 5339–5346.

Kreuzer, K.N. and W.M. Huang. 1983. T4 DNA topoisomerase. In *Bacteriophage T4* (ed. C.K. Mathews et al.), p. 90–96. American Society for Microbiology, Washington, D.C.

Liu, L.F., J.L. Davis, and R. Calendar. 1981. Novel topologically knotted DNA from bacteriophage P4 capsids: Studies with DNA topoisomerases. *Nucleic Acids Res.* **9:** 3979–3989.

Liu, L.F., C.C. Liu, and B.M. Alberts 1979. T4 DNA topoisomerase: A new ATP-dependent enzyme essential for initiation of T4 bacteriophage DNA replication. *Nature* **281:** 456–461.

———. 1980. Type II DNA topoisomerases: Enzymes that can unknot a topologically knotted DNA molecule via a reversible double-strand break. *Cell* **19:** 697–707.

Lynn, R., M. Bjornsti, P. Caron, and J.C. Wang. 1989. Peptide sequencing and site-directed mutagenesis identify tyrosine-727 as active site tyrosine of *Saccharomyces cerevisiae* DNA topoisomerase I. *Proc. Natl. Acad. Sci.* **86:** 3559–3565.

Lynn, R., G. Giaever, S. Swanberg, and J.C. Wang. 1986. Tandem regions of yeast DNA topoisomerase II share homology with different subunit of bacterial gyrase. *Science* **233:** 647–649.

Mathews, C.K. and J.A. Allen. 1983. DNA precursor biosynthesis. In *Bacteriophage T4* (ed. C.K. Mathews et al.), p. 59–70. American Society for Microbiology, Washington, D.C.

Mathews, C.K., E.M. Kutter, G. Mosig, and P.B. Berget, eds. 1983. *Bacteriophage T4*. American Society for Microbiology, Washington, D.C.

McCarthy, D. 1979. Gyrase-dependent initiation of bacteriophage T4 DNA replication: Interaction of *Escherichia coli* gyrase with novobiocin, courmermycin and phage DNA-delay gene products. *J. Mol. Biol.* **127:** 265–283.

McCarthy, D., C. Minner, H. Bernstein, and C. Bernstein. 1976. DNA elongation rates and growing point distributions of wild type T4 and DNA-delay amber mutant. *J. Mol. Biol.* **106:** 963–981.

Melendy, T. and D.S. Ray. 1989. Novobiocin affinity purification of a mitochondrial type II topoisomerase from the trypanosomatid *Crithidia fasciculata*. *J. Biol. Chem.* **264:** 1870–1876.

Mizuuchi, K., M. Mizuuchi, M. O'dea, and M. Gellert. 1984. Cloning and simplified purification of *Escherichia coli* gyrase A and B proteins. *J. Biol. Chem.* **259:** 9199–9201.

Moriya, S., N. Ogasawara, and H. Yoshikawa. 1985. Structure and function of the region of the replication origin of the *Bacillus subtilis* chromosome. III. Nucleotide sequence of some 10,000 base pairs in the origin region. *Nucleic Acids Res.* **13:** 2251–2265.

Mosig, G. 1987. The essential role of recombination in phage T4 growth. *Annu. Rev. Genet.* **21:** 347–371.

Mosig, G. and F. Eiserling. 1988. Phage T4 structure and metabolism. In *The bacteriophages* (ed. R. Calendar), p. 521–606. Plenum Press, New York.

Moss, B. 1985. Replication of poxviruses. In *Virology* (ed. B. Fields et al.), p. 685–703. Raven Press, New York.

Mufti, S. and H. Bernstein. 1974. The DNA-delay mutants of bacteriophage T4. *J. Virol.* **14:** 860–871.

Nossal, N.G. and B.M. Alberts. 1983. Mechanism of DNA replication catalyzed by purified T4 replication proteins. In *Bacteriophage T4* (ed. C.K. Mathews et al.), p. 71–78. American Society for Microbiology, Washington, D.C.

Ogasawara, N., S. Moriya, and H. Yoshikawa. 1985. Structure and function of the region of the replication origin of the *Bacillus subtilis* chromosome. IV. Transcription of the *oriC* region and repression of DNA gyrase genes and other open reading frames. *Nucleic Acids Res.* **13:** 2267–2279.

Poddar, S.K. and W.R. Bauer. 1986. Type I topoisomerase activity after infection of enucleated, synchronized mouse L cells by vaccinia virus. *J. Virol.* **57:** 433–437.

Rowe, T., K. Tewey, and L. Liu. 1984. Identification of the breakage-reunion subunit of T4 DNA topoisomerase. *J. Biol. Chem.* **259:** 9177–9181.

Sarabhai, A.S., A.O.W. Stretton, and S. Brenner. 1964. Colinearity of the gene with the polypeptide chain. *Nature* **201:** 13–17.

Shaffer, R. and P. Traktman. 1987. Vaccinia virus encapsidates a novel topoisomerase with the properties of a eukaryotic type I enzyme. *J. Biol. Chem.* **262:** 9309–9315.

Shuman, S. and B. Moss. 1987. Identification of a vaccinia virus gene encoding a type I DNA topoisomerase *Proc. Natl. Acad. Sci.* **84:** 7478–7482.

Shuman, S., M. Golder, and B. Moss. 1988. Characterization of vaccinia virus DNA topoisomerase I expressed in *Escherichia coli*. *J. Biol. Chem.* **263:** 16401–16407.

———. 1989. Insertional mutagenesis of the vaccinia virus gene encoding a type I DNA topoisomerase: Evidence that the gene is essential for virus growth. *Virology* **170:** 302–306.

Stetler, G. 1980. "The DNA-delay proteins of bacteriophage T4." Ph.D. thesis, University of Utah, Salt Lake City.

Stetler, G., G. King, and W.M. Huang. 1979. T4 DNA-delay proteins, required for specific DNA replication, form a complex that has ATP-dependent topoisomerase activity. *Proc. Natl. Acad. Sci.* **76:** 3737–3742.

Sugino, A., N.P. Higgins, P.O. Brown, C.L. Peebles, and N.R. Cozzarelli. 1978. Energy coupling in DNA gyrase and the mechanism of action of novobiocin. *Proc. Natl. Acad. Sci.* **75:** 4838–4842.

Swanberg, S. and J. Wang. 1987. Cloning and sequencing of the *E. coli* gyrA gene encoding for the A subunit of DNA gyrases. *J. Mol. Biol.* **197:** 729–736.

Tewey, K., T. Rowe, L. Yang, B. Halligan, and L. Liu. 1984. Adriamycin-induced DNA damage mediated by mammalian DNA topoisomerase II. *Science* **226:** 466–468.

Tsai-Pflugfelder, M., L. Liu, A. Liu, K. Tewey, J. Whang-Peng, T. Knutsen, K. Huebner, C. Croce, and J. Wang. 1988. Cloning and sequencing of cDNA encoding human DNA topoisomerase II and localization of the gene to chromosome region 17q21-22. *Proc. Natl. Acad. Sci.* **85:** 7177–7181.

Uemura, T., K. Morikawa, and M. Yanagida. 1986. The nucleotide sequence of the fission yeast DNA topoisomerase II gene: Structural and functional relationships to other DNA topoisomerases. *EMBO J.* **5:** 2355–2361.

Wasserman, S.A. and N.R. Cozzarelli. 1986. Biochemical topology: Applications to DNA recombination and replication. *Science* **232:** 951–960.

Wirak, D.O. and G.R. Greenberg. 1980. Role of bacteriophage T4 DNA-delay gene products in deoxyribonucleotide synthesis. *J. Biol. Chem.* **255:** 1896–1904.

Wirak, D.O., K.S. Cook, and G.R. Greenberg. 1988. Defect in synthesis of deoxyribonucleotides by a bacteriophage T4 *nrdB* mutant is suppressed on mutation of T4 DNA topoisomerase gene. *J. Biol. Chem.* **263:** 6193–6201.

Worland, S.T. and J.C. Wang. 1989. Inducible overexpression, purification and active site mapping of DNA topoisomerase II from the yeast *Saccharomyces cerevisiae*. *J. Biol. Chem.* **264:** 4412–4416.

Wyckoff, E., D. Natalie, J. Nolan, M. Lee, and T. Hsieh. 1989. Structure of the *Drosophila* DNA topoisomerase II gene: Nucleotide sequence and homology among topoisomerases II. *J. Mol. Biol.* **205:** 1–13.

Yamagishi, J., H. Yoshida, M. Yamayoshi, and S. Nakamura. 1987. Nalidixic acid-resistant mutations of the *gyrB* gene of *Escherichia coli*. *Mol. Gen. Genet.* **204:** 367–373.

Yegian, C.D., M. Mueller, G. Selzer, V. Russo, and F.W. Stahl. 1971. Properties of DNA-delay mutants of bacteriophage T4. *Virology* **46:** 900–919.

9

Reverse Gyrase and Other Archaebacterial Topoisomerases

Akihiko Kikuchi
Mitsuibishi Kasei Institute of Life Sciences
Machida-shi, Tokyo 194, Japan

There is controversy over whether or not prokaryotes should be divided into two separate kingdoms: eubacteria and archaebacteria. Archaebacteria have several distinct features, but most of these might be a consequence of their growth in extreme habitats, such as at high temperature for acidothermophiles, in high salt for extreme halophiles, and in an obligate anaerobic environment for methanogens. It is a difficult task to put these three archaebacteria into a single group, but they share common properties, such as sequence similarity in rRNA, high complexity of RNA polymerase subunits, and cell-wall structures devoid of peptideglycans. Figure 1 shows the proposed phylogenetic relationship, based on rRNA sequences, of archaebacteria to eubacteria or eukaryotes (Woese 1987).

It has often been suggested that in some aspects archaebacteria are more closely related to eukaryotes than prokaryotes (e.g., Zillig et al. 1985), although nuclear structures have never been shown in archaebacterial cells. An important issue in the study of archaebacteria is how chromosome structure is maintained under such extreme conditions, such as 75°C for acidothermophiles or 3–4 M KCl for halophiles. Under these conditions many proteins from other kingdoms dissociate from DNA, and the DNA itself would be expected not to have the canonical B-form configuration. To keep chromosomal structure intact, histone-like pro-

 0-87969-348-7/90 $1.00 + 00

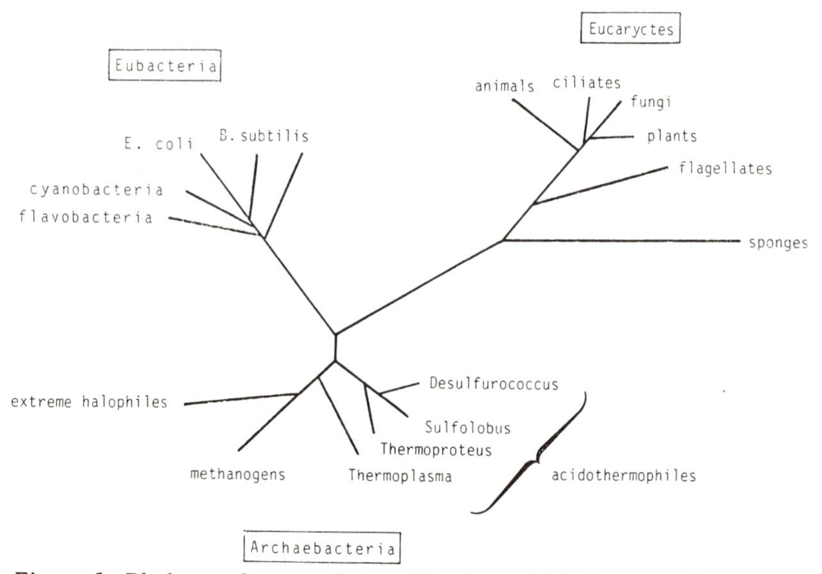

Figure 1 Phylogenetic tree of three kingdoms. This is a modification of the original figure based on rRNA sequence comparisons (Woese 1987).

teins or other DNA-binding proteins might play important roles (Notbohn 1982; Kimura et al. 1984; Kikuchi et al. 1986). Accordingly, specific topoisomerases might be needed to change local constraints on DNA and to construct tight DNA/protein complexes in obnoxious conditions. One of the so-called histone-like proteins identified in the acidothermophilic archaebacterium *Thermoplasma* has an amino acid sequence analogous to Hu protein of bacteria, but it is not directly related to eukaryotic histones (DeLange et al. 1981). It is worthwhile nevertheless to determine what kinds of topoisomerases are present in archaebacteria that function well at high temperature or at high salt concentrations.

I. TOPOISOMERASE OF *SULFOLOBUS*

Sulfolobus acidocaldarius is a relatively well-characterized acidothermophilic archaebacterium (Stetter and Zillig 1985). It is one of the smallest free-living organisms; its genome size is at most one-third of ordinary bacterial genomes (Doolittle 1985). Kikuchi and Asai (1984) and Mirambeau et al. (1984) were the first to identify topoisomerases in *Sulfolobus* extracts. Both groups concluded that little topoisomerase activity was detected in the absence of Mg^{++} and ATP, conditions under which abundant type-I topoisomerase activity is found in eukaryotes. In the presence of ATP and Mg^{++}, negatively supercoiled DNA was relaxed at 75°C but not below 55°C. Good activity was detected up to 95°C. The temperature range of the enzymatic activity is just what is expected from the growth temperature of *Sulfolobus*, from 70°C to 90°C.

The novel aspect of this topoisomerase activity is that not only was negatively supercoiled DNA relaxed, but also positive supercoils were introduced (Kikuchi and Asai 1984). On careful scrutiny of the migration of topoisomers produced by this enzyme, we initially observed a slight stagger of the ladders when compared with negatively supercoiled ones; this equivocally suggested the formation of positive supercoils. Since the catalytic introduction of positive supercoils was unprecedented, we verified that the product was really positively supercoiled by two methods. First, the reaction product had a higher buoyant density than relaxed DNA in CsCl ethidium bromide density gradient centrifugation. Second, positively supercoiled topoisomers were separated from negatively supercoiled ones by two-dimensional agarose gel electrophoresis (Fig. 2). Ethidium bromide or chloroquine in the second dimension separated negatively and positively superhelical topoisomers (Wu et al. 1988).

The introduction of positive supercoils was shown to be catalytic, not the stoichiometric reaction of a conventional topoisomerase and a DNA-binding protein as suggested in the mechanism of positive supercoiling by an altered form of DNA gyrase (Brown et al. 1979). We also showed that the product of the enzymatic reaction free of bound proteins was by itself positively supercoiled, demonstrating that it could be relaxed by yeast topoisomerase I but not by *Escherichia coli* topoisomerase I. Because the direction of the enzymatic reaction was just opposite to DNA gyrase, we proposed the name of reverse gyrase for this new enzyme (Kikuchi and Asai 1984).

II. PROPERTIES OF REVERSE GYRASE

Nakasu and Kikuchi (1985) purified reverse gyrase to homogeneity. The enzyme was composed of a single polypeptide with a molecular weight of 120,000. Ten consecutive amino-terminal amino acids were determined. The existence of other accessory proteins for supercoiling in our purified preparation was completely ruled out. Forterre et al. (1985) independently purified the enzyme, and Nadal et al. (1988) found a single 128,000 subunit. These investigators also showed that the addition of 10% polyethylene glycol 6000 significantly enhanced reverse gyrase activity. Contrary to our observation that nucleoside triphosphates are essential to enzymatic activity (Kikuchi and Asai 1984; Nakasu and Kiku-

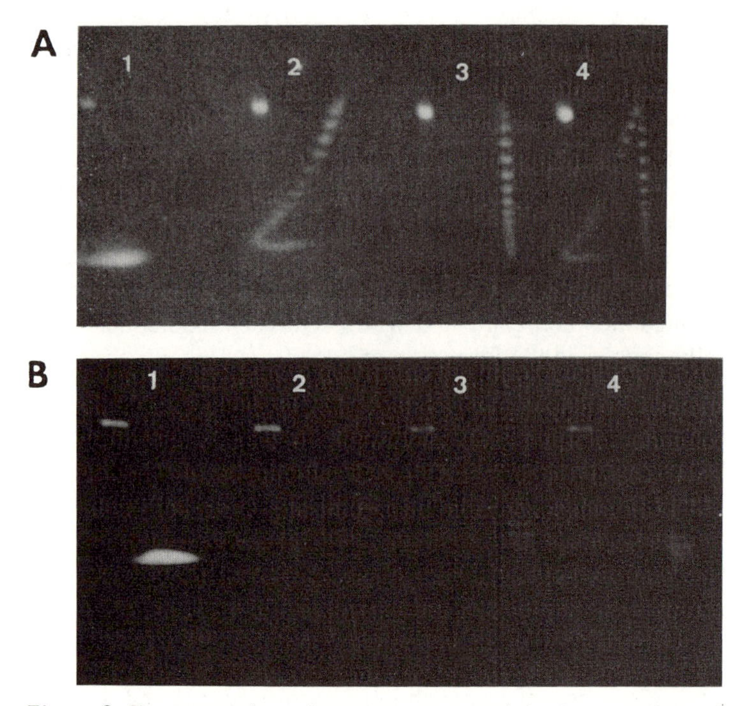

Figure 2 Demonstration of reverse gyrase activity by two-dimensional agarose gel electrophoresis. (*A*) Separation of positively and negatively supercoiled topoisomers. Gel electrophoresis in the second dimension (left to right) was performed perpendicular to the first dimension (top to bottom) after soaking the gel in the buffer (89 mM Tris base, 89 mM boric acid, 2.5 mM EDTA·Na_2) containing 3 μg/ml chloroquine. The samples were (*1*) negatively supercoiled pBR322 (control without enzymes), (*2*) negatively supercoiled topoisomers prepared using the type-II topoisomerase of *Sulfolobus*, (*3*) positively supercoiled topoisomers prepared by reverse gyrase, and (*4*) a mixture of samples 2 and 3. (*B*) Demonstration of reverse gyrase activity by two-dimensional gel electrophoresis using 0.02 μg/ml ethidium bromide in the second dimension. The molar ratio of enzyme to substrate (1.8 nM pBR322) was 0.0 (sample *1*), 2.3 (sample *2*), 9.0 (sample *3*), and 23.0 (sample *4*).

chi 1985), Forterre et al. (1985) claimed that their most purified fraction alone relaxed negatively supercoiled DNA in the absence of any nucleoside triphosphates. The discrepancy of the ATP requirement for relaxation of negative supercoils between two *Sulfolobus* strains is yet to be solved.

Slesarev (1988) purified reverse gyrase from another extremely thermophilic anaerobic archaebacterium, *Desulfurococcus amylolyticus*. This enzyme is a single polypeptide with a molecular weight of 135,000 and is active from 65°C to 100°C. ATP is required for both relaxation of negative supercoils and the introduction of positive supercoils.

Reverse gyrase changes the linking number of a single topoisomer in steps of one (Fig. 3). This was shown by Nakasu and Kikuchi (1985) and by Zivanovic et al. (1986). Our previous claim of a change in linking number of 2 (Kikuchi and Asai 1984) was an error caused by using a partially purified enzyme that was contaminated with a type-II topoisomerase. The unique topoisomers extracted from an agarose gel might have contained an inhibitor of reverse gyrase but not of the type-II enzyme.

Nakasu and Kikuchi (1985) observed that closed circular DNA was nicked by stoichiometric amounts of reverse gyrase in the absence of either ATP or Mg^{++} when the reaction was terminated by the addition of SDS. This suggests that reverse gyrase caused transient nicks during topoisomerization as expected for a type-I DNA topoisomerase.

A. Role of ATP

Nakasu and Kikuchi (1985), using the most purified reverse gyrase, and Forterre et al. (1985) showed that only 10 μM ATP was required for

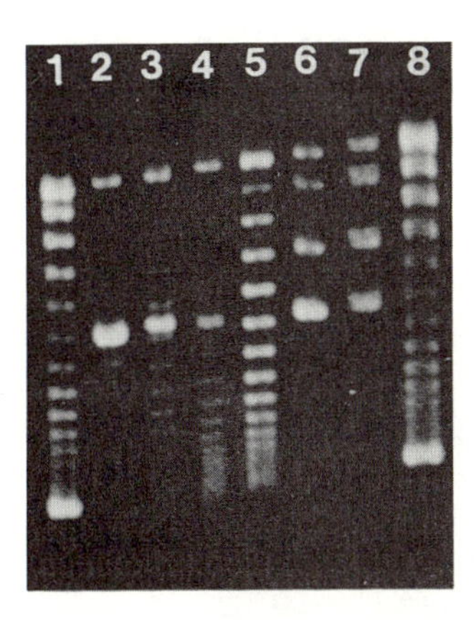

Figure 3 Linking number change by reverse gyrase and the type-II topoisomerase from *Sulfolobus* (Nakasu and Kikuchi 1985). The lanes contained the following: (*1* and *8*) negatively supercoiled topoisomers of pBR322 as markers, (*2*) a single topoisomer substrate (negatively supercoiled), (*3*) the substrate incubated with reverse gyrase for 1 min (most of the reaction products were positively supercoiled.), (*4*) the substrate incubated with reverse gyrase for 3 min (all the reaction products were positively supercoiled), (*5*) positively supercoiled topoisomers of pBR322 as markers, (*6*) the substrate incubated with type-II topoisomerase from *Sulfolobus* for 1 min, and (*7*) the substrate incubated with type-II topoisomerase from *Sulfolobus* for 3 min.

reverse gyrase. Other ATP-requiring topoisomerases, including the *Sulfolobus* type-II enzyme, required much more ATP (on the order of 1 mM).

Shibata et al. (1987) showed that the ATPase activity was strictly DNA dependent and intrinsic to reverse gyrase. ATP and dATP were hydrolyzed in the presence of DNA, but GTP, UTP, and CTP were not. These nonhydrolyzed nucleoside triphosphates, however, were good cofactors for the relaxation of negative supercoils but not for introduction of positive supercoils (Fig. 4). Therefore, NTP hydrolysis is required for the topoisomerase reaction only when the process is energetically unfavorable, that is, when there is the introduction of positive superhelical turns into DNA. AMP·PCP and AMP·PNP strongly interfered with the topoisomerase reaction, whereas ATPγS was significantly hydrolyzed by reverse gyrase and promoted positive supercoiling. The ATPase reaction was more active in the presence of single-stranded DNA although it strongly inhibited topoisomerization (Shibata et al. 1987).

Hydrolysis of ATP by reverse gyrase was quite efficient, and almost all the ATP was converted to ADP after 60 minutes at 75°C. The initial rate of ATP hydrolysis, monitored by high-performance liquid chromatography (HPLC) (Fig. 5), was much slower than that of topoisomerase reaction; that is, no more than 10% of ATP was hydrolyzed before the positive supercoiling reaction was completed. It is not known whether ATP hydrolysis facilitated the relaxation of negative supercoils or how

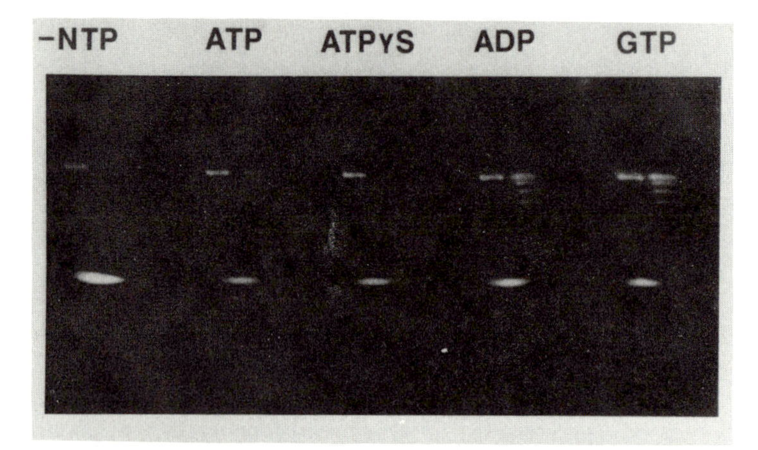

Figure 4 Effect of nucleoside triphosphates on reverse gyrase. Of the nucleotides indicated at the top of each lane, 10 μM was added in each reaction, containing 1.25 nM reverse gyrase and 1.8 nM negatively supercoiled pBR322. Topoisomerase products were analyzed by two-dimensional gel electrophoresis using ethidium bromide in the second dimension, as in Fig. 1B.

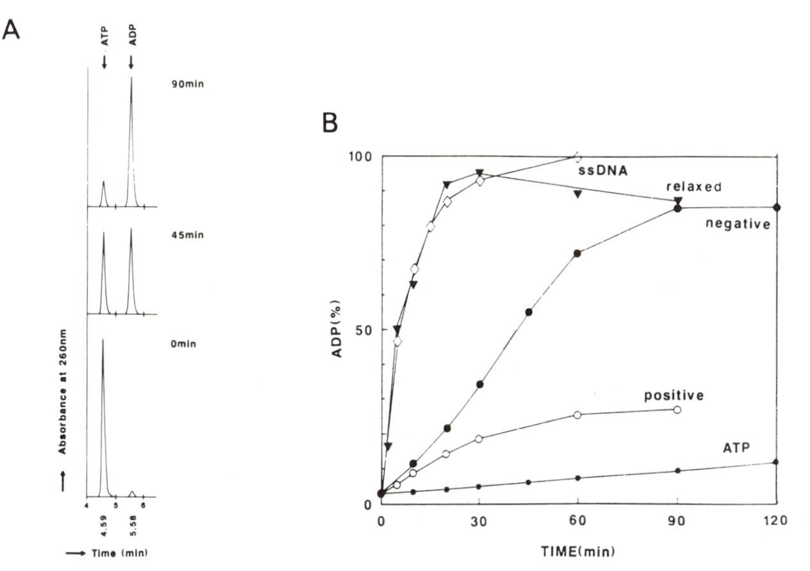

Figure 5 Quantitative analysis of ATPase activity of reverse gyrase using HPLC equipped with a C18 column. (*A*) HPLC elution profile. Reverse gyrase reaction was arrested at the time indicated, and the reaction mixture was directly applied to HPLC. (*B*) Kinetics of DNA-dependent ATP hydrolysis of reverse gyrase. The kinetics of ATP hydrolysis was strictly dependent on the presence of DNA and on its topological form.

many molecules of ATP are required to make one positive superhelical turn.

B. Negative Supercoiling Is Induced by Binding of Reverse Gyrase

Nicked circular DNA was incubated with reverse gyrase at 75°C in the absence of ATP, and the nicks were sealed by DNA ligase at 16°C where no reverse gyrase reaction takes place. The ligated DNA acquired negative superhelical turns upon removal of bound enzyme. About three negative superhelical turns were introduced per seven molecules of reverse gyrase. The reaction depended on the incubation temperature in the same fashion as supercoiling (Fig. 6). Similar results were obtained independently by Jaxel et al. (1989).

Addition of ATP abolished the acquisition of negative superhelicity, whereas nonhydrolyzed nucleoside triphosphates such as AMP·PCP, AMP·PNP, and GTP did not have an effect. ADP and ATPγS also strongly inhibited this binding reaction. This suggests that ATP hy-

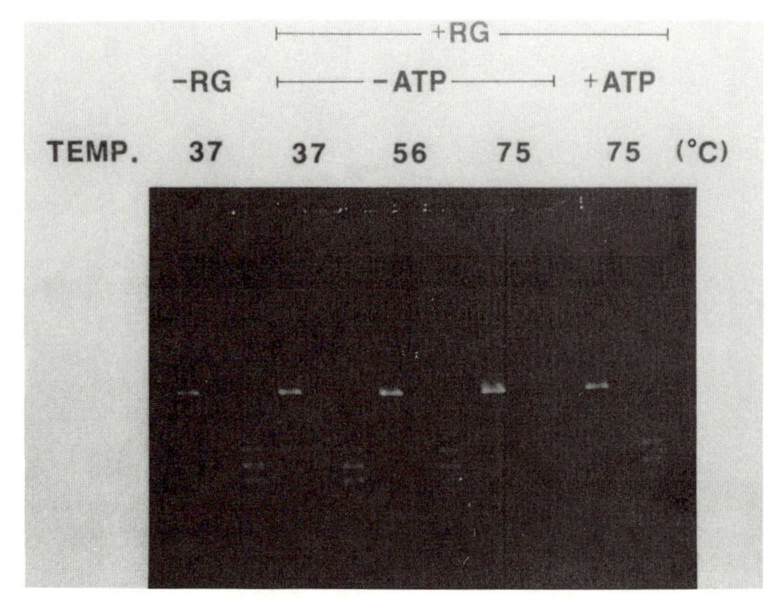

Figure 6 Acquisition of negative superhelical turns by binding of reverse gyrase. Nicked pBR322 was incubated with reverse gyrase at the temperature indicated on the top of each column. The nicks were sealed by DNA ligase at 16°C, and the change of superhelicity was analyzed by two-dimensional gel electrophoresis. Because of the difference in temperature and salt concentration in the reaction mixture and the gel, pBR322, which was nicked and resealed in the absence of reverse gyrase, gained a few positive superhelical turns (first column). When reverse gyrase was added at 75°C, the sealed topoisomers were less positively supercoiled (fourth column). Addition of ATP (last column) abolished the reduction of positive superhelicity.

drolysis might induce the release of reverse gyrase from DNA for recycling and that an ADP form of enzyme might effect the release. This mechanism is reminiscent of the GTPase cycle for many G proteins, including elongation factor Tu, and it was suggested as the mechanism for DNA gyrase by Sugino et al. (1978).

C. Positively Supercoiled Plasmid DNA in *Sulfolobus*

Since the positive supercoiling reaction is dominant in *Sulfolobus* extracts, it was important to determine the topological forms of DNA inside the cell. Yeats et al. (1982) found a lemon-shaped bacteriophage-like particle in one species of *Sulfolobus*. This particle contained positively supercoiled, closed circular DNA (Nadal et al. 1986). This DNA was also identified in a plasmid state after exposure to UV irradiation, presumably by excision from the chromosome. The plasmid was also positively supercoiled although less so than the DNA in the phage-like particles. These data imply that reverse gyrase is functional inside *Sulfolobus* to introduce positive superhelicity.

Although this is the only report of the natural occurrence of positively supercoiled DNA in vivo, positive supercoil stress must result transiently during processes such as DNA replication and transcription. Now, it is well-documented that the arrest of DNA gyrase activity by antibiotics causes the accumulation of positive supercoils in vivo (Lockshon and Morris 1983; Wu et al. 1988). Care should be taken in interpreting the appearance of positive supercoils in vivo. Positively supercoiled plasmid in a *Sulfolobus* cell might only result from the failure to remove positive superhelical constraints by perturbing other topoisomerases during the extraction of plasmids (e.g., prolonged UV exposure, cooling of culture, and change of pH).

III. PHYLOGENETIC STUDIES OF REVERSE GYRASE

We prepared antiserum against reverse gyrase and searched for cross-reacting material in extracts of *E. coli, Thermus thermophilus*, yeast, fruit fly, and *Halobacterium halobium*. None was found, and only *Sulfolobus* extracts gave positive results in a Western blot; it was a single band with a molecular weight of 120,000 (Kikuchi et al. 1986).

Collin et al. (1988) surveyed for reverse gyrase activity among several eubacterial species and many species of archaebacteria. Reverse gyrase was found only among sulfur-dependent acidothermophilic archaebacteria, including *Desulfurococcus* and *Thermoproteus*. In their preliminary report, it is noted that *Thermoplasma*, another acidothermophilic archaebacteria, did not have reverse gyrase activity and nor did halobacteria, methanogens, and any other thermophilic eubacteria.

In *Thermoplasma acidophilum*, Forterre et al. (1989) observed another efficient type-I topoisomerase. This enzyme had similar chromatographic properties as *Sulfolobus* reverse gyrase, but it relaxed negatively but not positively supercoiled DNA in an ATP-independent manner, as found for ordinary bacterial type-I topoisomerases.

IV. WHY REVERSE GYRASE AT ALL?

Liu and Wang (1987) suggested that important superhelical constraints in bacterial cells were caused by the movement of the transcription (and translation) machinery. To remove these constraints, two topoisomerases were needed: one, like topoisomerase I, to relax negative superhelical tension; and the other, like DNA gyrase, to remove positive superhelical stress. In this context, the reverse gyrase might correspond essentially to the standard topoisomerase I of eubacteria, because it is a type-I topoisomerase and it has a similar molecular size. We have no idea then why an ATP hydrolyzing system was added to reverse gyrase. The introduction of positive supercoils might be because the enzyme does not recognize the usual nearly relaxed end point of reaction but goes past it toward positive supercoiling. Reverse gyrase is useful in studying the physicochemical properties of positively supercoiled DNA (Zivanovic et al. 1986; Goulet et al. 1987; Brahms et al. 1989).

V. TYPE-II TOPOISOMERASE FROM *SULFOLOBUS*

In crude extracts of *Sulfolobus,* we detect a topoisomerase activity in addition to reverse gyrase. Reverse gyrase introduces positive supercoils and relaxes negative but not positive supercoils. We sought other enzymes that relaxed positive supercoils or introduced negative supercoils, which could function to balance the superhelical state of DNA in the cell (Liu and Wang 1987; Wu et al. 1988). Also, since reverse gyrase is a type-I topoisomerase and therefore incapable of decatenating or unknotting intact DNA, a type-II enzyme must be present for this purpose.

In our preliminary investigations (Kikuchi and Asai 1984), we identified a type-II topoisomerase activity and purified it using the unknotting of P4 knotted DNA as an assay (Fig. 7). The most purified fraction contained proteins with molecular weights of 40,000 and 60,000. These cosedimented during glycerol gradient centrifugation and comigrated through gel filtration in the presence of 1 M NaCl or 6 M urea with the unknotting and negative supercoil relaxation activities (Kikuchi et al. 1986). As shown in Figure 3, this topoisomerase changed the linking number in steps of two (Nakasu and Kikuchi 1985). This type-II topoisomerase required Mg^{++} and 1 mM ATP (or dATP) at 75°C; no reaction took place below 55°C (Kikuchi et al. 1986). During purification, a gyrase-like activity of negative supercoiling found in crude enzyme sources was completely lost, and it might have been a concerted reaction of this type-II topoisomerase and DNA-binding proteins. The purified enzyme relaxed both negative and positive supercoils, unknotted P4 knotted DNA, and catenated covalently closed circle DNA molecules. The unknotting reaction was ten times more efficient than the relaxation of negatively supercoiled pBR322.

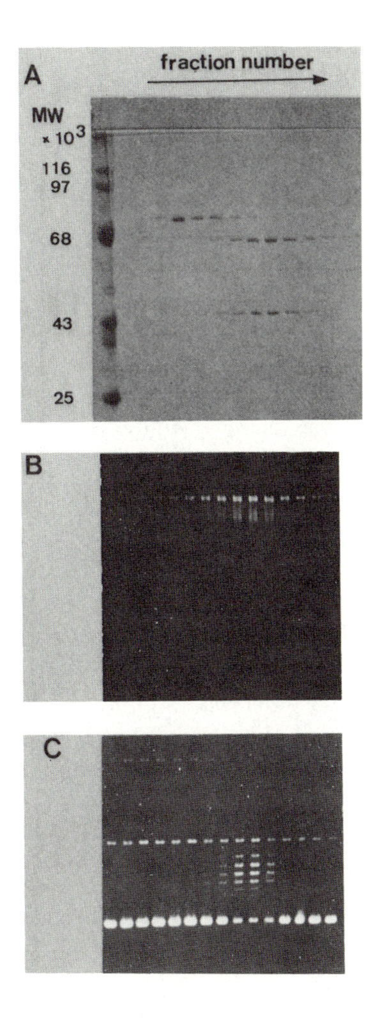

Figure 7 Activity of the type-II topoisomerase of *Sulfolobus* purified through DNA cellulose column chromatography. The type-II topoisomerase activity was purified successively with hydroxylapatite, heparine sepharose, and DNA cellulose column chromatography (Kikuchi et al. 1986). (*A*) Protein profile of a part of the DNA cellulose column fractions (left to right) as assayed by SDS-PAGE. In the active fraction, only two major bands were detected by Coomassie brilliant blue staining. The first lane contains molecular weight markers. (*B*) Unknotting activity of P4 knots by each fraction. (*C*) Relaxation of pBR322 by each fraction.

Reverse gyrase and this type-II topoisomerase are probably the two major topoisomerases in *Sulfolobus*, and there is roughly ten times more reverse gyrase than the type-II topoisomerase (Kikuchi et al. 1986). As a consequence, DNA in *Sulfolobus* cells might be positively supercoiled in vivo. However, there are many factors that influence the topological state of DNA in cells, and global positive supercoiling in vivo must be demonstrated directly.

VI. NEGATIVELY SUPERCOILED PLASMID DNA IN *HALOBACTERIUM*

Thus far, no topoisomerase has been detected from extreme halophiles. Plasmids extracted from an extreme halophile, *H. halobium,* were negatively supercoiled (Weidinger et al. 1979). In vivo investigations used specific antibiotics against known topoisomerases in eubacterial or eukaryotic cells. The topology of plasmids was changed after extensive treatment with the drugs, which suggests the presence of a topoisomerase in *Halobacterium.*

Forterre et al. (1986) and Sioud et al. (1987b) found that certain halophiles were as sensitive to etoposide and teniposide as eukaryotic cells and that a fraction of plasmid DNA extracted from drug-treated cells suffered DNA strand breaks (Sioud et al. 1987a; Sioud and Forterre 1989). Sioud et al. (1988c) also found that *Halobacterium* and *Methanosarcina* were sensitive to the coumarin and quinolone antibiotics effective against DNA gyrase in eubacteria. Sioud et al. (1988b) observed that on addition of novobiocin to *Halobacterium,* small multicopy plasmids in the cell acquired highly positive superhelical turns. They suggested that a topoisomerase (possibly DNA gyrase) might function to keep the plasmid DNA negatively supercoiled and that inhibition by novobiocin unmasked mechanisms to accumulate positive superhelical turns including perhaps reverse gyrase. Recently the same investigators (Sioud et al. 1988a) found that, after prolonged treatment of *Halobacterium* with novobiocin, the same plasmid turned to be a single-stranded circle, and they suggested that novobiocin affected DNA replication. The target and the mode of action of each drug are not necessarily the same among eukaryotes, prokaryotes, and archaebacteria, particularly in the high salt concentration or high growth temperature needed for survival of the archaebacteria. The effect of drugs must be carefully interpreted.

It is premature to conclude what kind of topoisomerases exist in halophiles. The biochemical studies of halobacterial topoisomerases might be strongly hindered by the 3–4 M of monovalent cation in cellular extracts.

ACKNOWLEDGMENTS

I thank K. Asai and K. Yasui for expert technical assistance, Drs. S. Nakasu and T. Shibata for collaborations and helpful discussions, and

Dr. Yoshiko Kikuchi for her critical reading of this manuscript. I gratefully acknowledge Drs. P. Forterre and D. Musgrave for providing comments and preprints about their recent progress. This work was supported in part by grant-in-aid from the Ministry of Education, Science, and Culture of Japan. I thank Etsuko Ueno for her invaluable help in manuscript preparation.

REFERENCES

Brahms, S., S. Nakasu, A. Kikuchi, and J.G. Brahms. 1989. Structural changes in positively and negatively supercoiled DNA. *Eur. J. Biochem.* **184:** 297–303.

Brown, P.O., C.L. Peebles, and N.R. Cozzarelli. 1979. A topoisomerase from *Escherichia coli* related to DNA gyrase. *Proc. Natl. Acad. Sci.* **76:** 6110–6114.

Collin, R.G., H.W. Morgan, D.R. Musgrave, and R.M. Daniel. 1988. Distribution of reverse gyrase in representative species of eubacteria and archaebacteria. *FEMS Microbiol. Lett.* **55:** 235–240.

DeLange, R.J., L.C. Williams, and D.C. Searcy. 1981. Histone-like protein (HTa) from *Thermoplasma acidophilum*. II. Complete amino acid sequence. *J. Biol. Chem.* **256:** 905–911.

Doolittle, W.F. 1985. Genome structure in archaebacteria. In *The bacteria* (ed. C.R. Woese and R.S. Wolfe), vol. VIII, pp. 545–560 Academic Press, New York.

Forterre, P., C. Elie, M. Sioud, and A. Hamal. 1989. Studies on DNA polymerase and topoisomerases in archaebacteria. *Can. J. Microbiol.* **35:** 228–233.

Forterre, P., G. Mirambeau, C. Jaxel, M. Nadal, and M. Duguet. 1985. High positive supercoiling *in vitro* catalyzed by an ATP and polyethylene glycol-stimulated topoisomerase from *Sulfolobus acidocaldarius*. *EMBO J.* **4:** 2123–2128.

Forterre, P., M. Nadal, C. Elie, G. Mirambeau, C. Jaxel, and M. Duguet. 1986. Mechanisms of DNA synthesis and topoisomerization in archaebacteria—Reverse gyration *in vitro* and *in vivo*. *Syst. Appl. Microbiol.* **7:** 67–71.

Goulet, I., Y. Zivanovic, and A. Prunell. 1987. Helical repeat of DNA in solution. The V curve method. *Nucleic Acids Res.* **15:** 2803–2821.

Jaxel, C., M. Nadal, G. Mirambeau, P. Forterre, M. Takahashi, and M. Duguet. 1989. Reverse gyrase binding to DNA alters the double helix structure and produces single-stranded cleavage in the absence of ATP. *EMBO J.* **8:** 3135–3139.

Kikuchi, A. and K. Asai. 1984. Reverse gyrase—A topoisomerase which introduces positive superhelical turns into DNA. *Nature* **309:** 677–681.

Kikuchi, A., T. Shibata, and S. Nakasu. 1986. Reverse gyrase and DNA supercoiling in *Sulfolobus*. *Syst. Appl. Microbiol.* **7:** 72–78.

Kimura, M., J. Kimura, P. Davie, R. Reinhardt, and J. Dijk. 1984. The amino acid sequence of a small DNA binding protein from the archaebacterium *Sulfolobus solfataricus*. *FEBS Lett.* **176:** 176–178.

Liu, L.F. and J.C. Wang. 1987. Supercoiling of the DNA template during transcription. *Proc. Natl. Acad. Sci.* **84:** 7024–7027.

Lockshon, D. and D.R. Morris. 1983. Positively supercoiled plasmid DNA is produced by treatment of *Escherichia coli* with DNA gyrase inhibitors. *Nucleic Acids Res.* **11:** 2999–3017.

Mirambeau, G., M. Duguet, and P. Forterre. 1984. ATP-dependent DNA topoisomerase from the archaebacterium *Sulfolobus acidocaldarius*. Relaxation of supercoiled DNA at high temperature. *J. Mol. Biol.* **179:** 559–563.

Nadal, M., G. Mirambeau, P. Forterre, W.-D. Reiter, and M. Duguet. 1986. Positively supercoiled DNA in a virus-like particle of an archaebacterium. *Nature* **321:** 256–258.

Nadal, M., C. Jaxel, C. Portemer, P. Forterre, G. Mirambeau, and M. Duguet. 1988. Reverse gyrase of *Sulfolobus:* Purification to homogeneity and characterization. *Biochemistry* **27:** 9102–9108.

Nakasu, S. and A. Kikuchi. 1985. Reverse gyrase: ATP-dependent type I topoisomerase from *Sulfolobus*. *EMBO J.* **4:** 2705–2710.

Notbohn, H. 1982. Low angle X-ray scattering analysis of the *Thermoplasma acidophilum* nucleoprotein subunit. *Biochim. Biophys. Acta* **696:** 223–225.

Shibata, T., S. Nakasu, K. Yasui, and A. Kikuchi. 1987. Intrinsic DNA-dependent ATPase activity of reverse gyrase. *J. Biol. Chem.* **262:** 10419–10421.

Sioud, M. and P. Forterre. 1989. Ciprofloxacin and etoposide (VP16) produce a similar pattern of DNA cleavage in a plasmid of an archaebacterium. *Biochemisty* **28:** 3638–3641.

Sioud, M., P. Forterre, and A.-M. de Recond. 1987a. Effects of the antitumor drug VP16 (etoposide) on the archaebacterial *Halobacterium* GRB 1.7 kb plasmid *in vivo*. *Nucleic Acids Res.* **15:** 8217–7234.

Sioud, M., G. Baldacci, P. Forterre, and A.-M. de Recondo. 1987b. Antitumor drugs inhibit the growth of halophilic archaebacteria. *Eur. J. Biochem.* **169:** 231–236.

———. 1988a. Novobiocin induces accumulation of a single strand of plasmid pGRB-1 in the archaebacterium *Halobacterium* GRB. *Nucleic Acids Res.* **16:** 7833–7842.

Sioud, M., G. Baldacci, A.-M. de Recondo, and P. Forterre. 1988b. Novobiocin induces positive supercoiling of small plasmids from halophilic archaebacteria *in vivo*. *Nucleic Acids Res.* **16:** 1379–1391.

Sioud, M., O. Possot, C. Elie, L. Sibold, and P. Forterre. 1988c. Coumarin and quinolone action in archaebacteria: Evidence for the presence of DNA gyrase-like enzyme. *J. Bacteriol.* **170:** 946–953.

Slesarev, A.I. 1988. Positive supercoiling catalysed *in vitro* by ATP-dependent topoisomerase from *Desulfurococcus amylolyticus*. *Eur. J. Biochem.* **173:** 395–399.

Stetter, H.O. and W. Zillig. 1985. Thermoplasma and the thermophilic sulfur dependent archaebacteria. In *The bacteria* (ed. C.R. Woese and R.S. Wolfe), vol. VIII, pp. 85–170. Academic Press, New York.

Sugino, A., N.P. Higgins, P.O. Brown, C.L. Peebles, and N.R. Cozzarelli. 1978. Energy coupling in DNA gyrase and the mechanism of action of novobiocin. *Proc. Natl. Acad. Sci.* **75:** 4838–4842.

Weidinger, G., G. Klotz, and W. Goebel. 1979. A large plasmid from *Halobacterium halobiun* carrying genetic information to gas vacuole formation. *Plasmid* **2:** 377–386.

Woese C.R. 1987. Bacterial evolution. *Microbiol Rev.* **51:** 221–271.

Wu, H.-Y., S. Shyy, J.C. Wang, and L.F. Liu. 1988. Transcription generates positively and negatively supercoiled domains in the template. *Cell* **53:** 433–440.

Yeats, S., P. McWilliam, and W. Zillig. 1982. A plasmid in the archaebacterium *Sulfolobus acidocaldarius*. *EMBO J.* **1:** 1035–1038.

Zillig, W., R. Schnabel, and H.O. Stetter. 1985. Archaebacteria and the origin of the eucaryotic cytoplasm. *Curr. Top. Microbiol. Immunol.* **114:** 1–18.

Zivanovic, Y., I. Goulet, and A. Prunell. 1986. Properties of supercoiled DNA in gel electrophoresis. The V-like dependence of mobility on topological constraint. DNA-matrix interactions. *J. Mol. Biol.* **192:** 645–660.

10

Genetics of DNA Topoisomerases

Mitsuhiro Yanagida
Department of Biophysics
Faculty of Science
Kyoto University
Kyoto 606, Japan

Rolf Sternglanz
Department of Biochemistry
State University of New York
Stony Brook, New York 11794

I. INTRODUCTION

This chapter concentrates on DNA topoisomerase genes and mutants from *Escherichia coli* and from the two yeasts, *Saccharomyces cerevisiae* and *Schizosaccharomyces pombe*. It is in these species that the greatest amount of work has been done.

II. *E. COLI* DNA TOPOISOMERASE MUTANTS

E. coli has three topoisomerases, DNA topoisomerase I, originally called ω protein (Wang 1971), DNA gyrase (Gellert et al. 1976), and DNA topoisomerase III (Dean et al. 1983; Srivenugopal et al. 1984; DiGate and Marians 1988). Topoisomerase I is a monomer of about 97 kD. DNA gyrase (topoisomerase II) is a tetramer of the A_2B_2 type; the molecular masses of the A and B subunits are about 97 kD and 90 kD, respectively.

DNA Topology and Its Biological Effects

Topoisomerase III is a monomer of 74 kD. Topoisomerase I and gyrase from other eubacteria are similar in size and enzymatic properties to their *E. coli* counterparts. Topoisomerase III has not yet been identified in other bacterial species.

The *E. coli* gyrase genes were known even before gyrase was discovered; they were defined by mutations causing resistance to two antibacterial drugs, nalidixic acid and coumermycin. The genes, initially called *nalA* and *cou* and now called *gyrA* and *gyrB*, code for the A and B subunits of gyrase, respectively. In *E. coli*, these two genes are far apart, at 48 and 83 minutes, respectively, on the genetic map. In *Bacillus subtilis*, the genes are adjacent and within a few kilobases of the origin of DNA replication (Moriya et al. 1985). The gyrase genes from *B. subtilis* and *E. coli* have been cloned and sequenced (Moriya et al. 1985; Yamagishi et al. 1986; Adachi et al. 1987; Swanberg and Wang 1987).

It is now known that the GyrA subunit contains the site for breakage and rejoining of the DNA; it forms a transient covalent intermediate with DNA. The quinolone family of antibacterial drugs, exemplified by nalidixic acid, oxolinic acid, and norfloxacin, prolongs the lifetime of this intermediate, effectively trapping the GyrA subunit on DNA. On the other hand, drugs such as coumermycin and novobiocin bind to the GyrB subunit and compete with ATP for binding to this subunit, thereby inhibiting the enzyme. Occasionally, nalidixic-acid-resistant mutations are found in the *gyrB* gene. Two of these mutations have been sequenced and found to cause single amino acid changes in the central region of the GyrB protein (Yamagishi et al. 1986).

One *gyrA* temperature-sensitive (*ts*) conditional-lethal mutant has been isolated (Kreuzer and Cozzarelli 1979) and several *ts gyrB* mutants are known (Gellert et al. 1979; Orr et al. 1979; Filutowicz and Jonczyk 1983). Inactivation of gyrase in these mutants or treatment of cells with gyrase inhibitors leads to a large decrease in negative supercoiling or an increase in DNA linking number. This holds true for plasmid and viral DNAs as well as for the bacterial chromosome itself. Thus, gyrase is involved in the negative supercoiling of prokaryotic DNA and is important for DNA transactions that depend on supercoiling. These include initiation of DNA replication, replication fork movement, initiation of transcription from some promoters, transcriptional elongation (see below), and a number of site-specific recombination processes.

The first *E. coli* DNA topoisomerase I mutants were identified by screening a collection of heavily mutagenized temperature-sensitive mutants with an enzymatic assay for DNA topoisomerase I. Two mutants, one with 1% and the other with 10% of the normal level of activity, were detected (Sternglanz et al. 1981). Genetic crosses showed that the temperature sensitivity of the original strains did not cosegregate with the mutation leading to the enzymatic defect and that strains harboring only the topoisomerase I mutation were not temperature sensitive for growth. The mutations were mapped to 28 minutes on the *E. coli* genetic map, near *trp* and *cysB*. Other experiments strongly suggested that the locus defined by these mutations is the structural gene for DNA topoisomerase I, now called *topA*. Strains with deletions of the *cysB* region had been isolated previously, and five of these strains had no detectable topoisomerase I activity (Sternglanz et al. 1981). These results demonstrated that this enzyme is not essential for viability. Subsequent work showed that *E. coli topA* mutants are viable only because of compensatory mutations in other genes. This point is discussed below.

The proximity of the *topA* gene to *trp* and *cysB* led to an investigation of whether the *Salmonella typhimurium supX* gene, known to map in this region, could be *topA*. *supX* mutants had been isolated many years earlier as extragenic suppressors of a promoter mutation called *leu500* (Mukai and Margolin 1963; Dubnau and Margolin 1972). Whereas *leu500* mutants are phenotypically Leu$^-$, *leu500 supX* mutants are Leu$^+$ and thus can be easily selected. In fact, the *cysB topA* deletions described above were originally isolated as *supX* mutants in *Salmonella* and subsequently transferred to *E. coli*. *Salmonella* and *E. coli supX* mutants were found to lack DNA topoisomerase I, and thus, it was concluded that either *supX* is the structural gene for the enzyme or a nearby regulatory gene (Sternglanz et al. 1981; Trucksis and Depew 1981). The subsequent evidence that *supX* amber mutants produce truncated topoisomerase I proteins proved that the *supX* locus is *topA*, the structural gene for the enzyme (Margolin et al. 1985).

As mentioned above, early studies showed that *E. coli topA* mutants were viable and grew only slightly more slowly than wild-type strains. The mutants have few noticeable phenotypes. Some of them have a more rapid rate of induction and a higher level of catabolite-sensitive enzymes such as β-galactosidase and tryptophanase. This phenotype is attributed to an increase in negative superhelicity of the DNA. Other mutants exhibit a large decrease in the frequency of Tn*5* transposition. Further analysis revealed a paradox. Despite the absence of significant growth defects for *topA* deletion mutants, it is very difficult to move such deletions to other strains, for example, by P1 transduction. In such experiments, transductants carrying *topA* deletions are found with a very low frequency and grow extremely slowly. They readily acquire secondary mutations, however, that allow them to grow normally (DiNardo et al. 1982; Pruss et al. 1982; Raji et al. 1985). These compensatory mutations occur in several genes. The best understood ones are those in *gyrA* or *gyrB* that lead to a lower level of gyrase activity in the cell. These results have led to the concept that the relative levels of topoisomerase I and gyrase determine the level of DNA superhelicity in the cell. Mutations in *topA* lead to an increase in negative superhelicity that is deleterious to the cell. The compensatory mutations reduce the level of gyrase and hence bring the superhelicity back to a normal or lower than normal level. Measurements of plasmid supercoiling in various *topA* mutant strains support these ideas (Pruss et al. 1982). Some of the phenotypes initially observed for *topA* deletions, such as the lower frequency of Tn*5* transposition and the inability to plate phage Mu, are actually due to the gyrase compensatory mutations rather than the *topA* mutation itself.

Not all compensatory mutations are in gyrase genes. For example, Raji et al. (1985) mapped several such mutations to a locus called *toc* (*t*opoisomerase *o*ne *c*ompensatory) near *tolC* at 66 minutes on the genetic map. These mutations do not change the level of gyrase activity. They appear to be gene duplications that in an unknown way compensate for the absence of topoisomerase I (Dorman et al. 1989). Other compensatory genes not at *gyrA*, *gyrB*, or *toc* have been identified but not mapped or otherwise characterized. Apparently, *E. coli* can compensate for the loss of topoisomerase I in several different ways, many of which are not understood.

In contrast with *E. coli, Salmonella topA* (*supX*) deletion mutants appear to survive and grow well without compensatory mutations. However, these strains can acquire modifier mutations that improve growth (Overbye and Margolin 1981; Richardson et al. 1984). Plasmids from *Salmonella topA* mutants are more negatively supercoiled than normal, whereas plasmids from strains with the additional modifier mutations have a normal degree of superhelicity (Richardson et al. 1984). Thus, the situation is reminiscent of the one described for *E. coli,* but excess negative superhelicity appears to be less deleterious for *Salmonella* than it is for *E. coli*. The suppression of the *leu500* promoter mutation, as well as other effects on gene expression caused by *Salmonella topA* mutations (Overbye et al. 1983), can be rationalized by the increased negative superhelicity. The *leu500* mutation is known to be an A-T to G-C transition in the −10 region of the promoter for this gene. Presumably, increased negative supercoiling allows this mutant promoter to function.

As described above, the relative levels of topoisomerase I and gyrase determine the degree of negative supercoiling in the cell. Mutants or drugs that inhibit one or the other enzyme significantly affect plasmid supercoiling (Pruss 1985; Pruss and Drlica 1986; Lockshon and Morris 1983). It is now apparent that these effects are coupled to transcription (Wu et al. 1988). During transcriptional elongation, a domain of positive supercoiling arises in the DNA ahead of the moving RNA polymerase, and a domain of negative supercoiling arises behind it (Liu and Wang 1987). Gyrase relaxes the positive supercoils, and topoisomerase I partially relaxes the negative ones. Thus, the combination of transcription and topoisomerase activity leads to the observed level of supercoiling. It is possible that some of the poorly understood compensatory mutations (that compensate for *topA* mutations; see above) are in genes coding for proteins that are part of the transcriptional apparatus.

E. coli topoisomerase III has been purified recently to homogeneity (DiGate and Marians 1988). Its enzymatic properties are fairly similar to those of topoisomerase I in that it relaxes negative but not positive supercoils, binds to single-stranded DNA (ssDNA), requires Mg^{++}, is active at high temperatures (52°C), and is very effective at decatenating catenated double-stranded DNA (dsDNA) rings as long as they have single-stranded regions. However, its cleavage pattern on a defined ssDNA differs from that of topoisomerase I (Dean et al. 1983). The gene for topoisomerase III (*topB*) has been cloned and sequenced (DiGate and Marians 1989). The sequence indicates that topoisomerase III, although smaller than topoisomerase I, shows considerable homology over a stretch of 308 amino acids in the middle of both proteins (24% identity and 46% similarity over this region). This region includes the active site tyrosine of topoisomerase I (Lynn and Wang 1989). The proteins are dissimilar at their amino- and carboxy-terminal ends. On the basis of the known restriction map of the *E. coli* genome, the *topB* gene was localized to 38.7 minutes on the genetic map. Interestingly, a strain exists that has a deletion from 38.4 minutes to 39 minutes and thus presumably has a deletion of the *topB* gene. It is not yet known if this strain actually lacks topoisomerase III activity or if it has compensatory mutations. It is intriguing that the *topA* and *topB* genes flank and are just outside the *E. coli* DNA replication fork terminators, T1 and T2. More work with better characterized *topB* mutants is needed to elucidate the biological roles of topoisomerase III and its relationship with the other topoisomerases of *E. coli*.

III. YEAST DNA TOPOISOMERASE MUTANTS

All eukaryotes have two major topoisomerases: a type-I enzyme called DNA topoisomerase I, first isolated from mouse nuclei (Champoux and Dulbecco 1972), and a type-II enzyme, DNA topoisomerase II (Baldi et al. 1980; Hsieh and Brutlag 1980; Liu et al. 1980). Since their initial identification, these two enzymes have been isolated from many eukaryotic species. It is clear from these studies that DNA topoisomerase I from all species from yeast to mammals is quite similar in size, subunit structure, and enzymatic properties. The same holds true for DNA topoisomerase II. Most of the remaining sections of this chapter are devoted to topoisomerase I and II mutants and genes from the budding yeast *S. cerevisiae* and the fission yeast *S. pombe*, each of which has a well-defined genetic system. These two yeasts are very distant evolutionarily, and thus, conserved features in them often can be extrapolated to higher eukaryotes.

Very recently, the gene for a third topoisomerase has been identified in the yeast *S. cerevisiae* (Wallis et al. 1989); it will be described later in this chapter. Mitochondrial topoisomerases have been reported for several species, and it is plausible that they are distinct from the nuclear enzymes. Because no genes or mutants for these enzymes have been identified, they will not be discussed in this chapter.

Yeast DNA topoisomerase mutants were initially isolated by biochemical screening of mutagenized cells of *S. cerevisiae* (Thrash et al. 1984) and *S. pombe* (Uemura and Yanagida 1984). Extracts of cells grown from individual members of a collection of *S. cerevisiae* temperature-sensitive mutants were assayed for topoisomerase activity; type-I activity was assayed by the relaxation of a supercoiled plasmid in a buffer without Mg^{++} and type-II activity by decatenation. Because the temperature-sensitive mutants were heavily mutagenized, the deficiency of a particular enzyme was not necessarily related to the temperature-sensitive phenotype. A *top1* mutant with a heat-labile DNA topoisomerase I was obtained, and tetrad analysis showed that the temperature sensitivity of the enzyme was unrelated to the temperature-sensitive growth defect (Thrash et al. 1984). It is now known that DNA topoisomerase I is nonessential for growth (see below). A temperature-sensitive DNA topoisomerase II mutant, *top2*, was also found; in this case, a single recessive mutation was responsible for both the heat-labile topoisomerase II activity and the growth defect (DiNardo et al. 1984).

In the work with *S. pombe*, mass screening of 1600 individual extracts of mutagenized cells yielded two mutants with heat-labile topoisomerase I activity (Uemura and Yanagida 1984). The mutations were mapped to a single locus, *top1*. These mutants show no phenotype and grow and sporulate normally. A *top1* mutant was crossed with an *end1* mutant deficient in a Mg^{++}-activated endonuclease, which cleaves dsDNA. The resulting double mutant *top1-end1*, which grows normally at 22°C to 36°C, was mutagenized further, and more than 600 temperature-sensitive mutants that grow at 26°C but not at 36°C were examined. Extracts of these temperature-sensitive mutants were then assayed for DNA topoisomerase II at 39°C. The Mg^{++}- and ATP-dependent relaxation of supercoiled DNA could be used as the assay because of the genetic elimination of the interfering DNA topoisomerase I and endonuclease activities. Three of the temperature-sensitive mutants showed heat-labile ATP-dependent relaxing activity. An unknotting assay confirmed the thermal inactivation of the type-II DNA topoisomerase in these mutants. Two additional *ts top2* mutants were obtained by rescreening the same temperature-sensitive collection (Hirano et al. 1986). Tetrad analyses indicated that all the five mutations mapped at an identical locus, designated *top2*. The *ts top2* mutation in *S. cerevisiae* and *S. pombe* is linked to the temperature-sensitive growth phenotype for each of the mutants, showing that the inactivation of the DNA topoisomerase II is responsible for the temperature-sensitive block of cell growth.

A cold-sensitive (*cs*) *S. pombe top2-250* mutant was obtained by examining a large number of cold-sensitive mutant strains for the cytological "cut" phenotype exhibited by all the *ts top2* mutants (Uemura et al. 1987b). These cold-sensitive mutants grow at 36°C but not at 22°C. The *cs top2* mutation was mapped to the same locus identified by the *ts top2* mutants. Mg^{++}- and ATP-dependent DNA topoisomerase II activity is cold-sensitive in the cold-sensitive mutant extract. The *cs top2* mutant is hypersensitive to the topoisomerase II inhibitor, adriamycin (100 µg/ml), at permissive temperature, and this phenotype cosegregates with the cs^- phenotype (T. Uemura and M. Yanagida, unpubl.).

A number of ts^+ revertants were obtained from *S. pombe top2-191* and *top2-342* at a frequency of 1×10^{-8}; all were intragenic revertants (T. Uemura, unpubl.). Similarly, all the cs^+ revertants from *top2-250* (at a frequency of 1×10^{-6}) were also intragenic.

Construction and properties of other *top1* and *top2* mutants involved the use of the cloned genes. (The cloning of the topoisomerase I and II genes is described below.) Holm et al. (1985) isolated five *S. cerevisiae ts top2* mutants by in vitro mutagenesis of a plasmid carrying the *TOP2* locus. Insertion and deletion *top1* mutants of *S. cerevisiae* (Goto and Wang 1985; Thrash et al. 1985) were constructed by the one step gene replacement method of Rothstein (1983). The same method was used in the disruption of one copy of the *TOP2* gene in a diploid strain; tetrad analysis using this diploid mutant showed that the single-copy *TOP2* gene is essential (Goto and Wang 1984).

In the work with *S. pombe*, a deletion and insertion *top1* mutant was also constructed (Uemura et al. 1987a). The disrupted *top1* mutant grows and sporulates, just as is the case for *S. cerevisiae*. Also, the disruption of one copy of the *S. pombe* $top2^+$ gene in a diploid strain followed by tetrad analysis shows that the disrupted haploid is lethal (K. Shiozaki and

M. Yanagida, unpubl.). A series of plasmids, carrying the topoisomerase II gene missing different carboxy-terminal regions, was constructed, and a large carboxy-terminal domain nonessential for relaxing activity was defined. These carboxy-terminal deletions complement *top2* null mutants as well as the *ts* and *cs top2* mutants (K. Shiozaki and M. Yanagida, unpubl.).

IV. CLONING OF YEAST DNA TOPOISOMERASE GENES

For the cloning of the *S. cerevisiae TOP1* gene, Thrash et al. (1985) used three methods: mass screening of a *top1* mutant transformed with a plasmid-borne yeast gene library for the presence of the enzyme in cell extracts, the screening of the same collection of transformants for the Mak$^+$ phenotype (see below), and the screening of transformants of the *ts top-1 top2-1* double mutant for an increase in colony size at a permissive temperature. Independently, Goto and Wang (1985) isolated the *TOP1* gene by immunoscreening of *E. coli* infected with a phage λgt11 yeast gene library. The *S. pombe top1*$^+$ gene was cloned by hybridization of a cosmid library of *S. pombe* genomic DNA with the *S. cerevisiae TOP1* sequence (Uemura et al. 1987a).

The *S. cerevisiae TOP2* gene was isolated by immunochemical screening (Goto and Wang 1984). Independently, it was also cloned by complementing a *ts top2* mutant with a plasmid carrying the *TOP2* gene (Voelkel-Meiman et al. 1986). The *S. pombe top2*$^+$ gene was isolated by transformation of *ts top2* mutant and by plaque hybridization of a phage λ library of *S. pombe* genomic DNA, using the *S. cerevisiae TOP2* gene as a probe (Uemura and Yanagida 1986; Uemura et al. 1986).

V. CHROMOSOME MAPPING OF YEAST *TOP1* AND *TOP2* LOCI

In both *S. cerevisiae* and *S. pombe*, single genes exist for DNA topoisomerase I and II, respectively. The *S. cerevisiae top1* locus is linked to the centromere of chromosome XV, near a *mak1* gene previously known for its role in the maintenance of a double-stranded killer RNA (Thrash et al. 1984). Further biochemical and genetic analyses indicated that *top1* and *mak1* are allelic. The *S. cerevisiae top2* locus is on the left arm of chromosome XIV between *met4* and its centromere (DiNardo et al. 1984; Holm et al. 1985). In *S. pombe*, the *top2* locus is tightly linked to *leu1* on the long arm of chromosome II (Uemura and Yanagida 1984, 1986). The *top1* locus is between *nuc1* and *lys4* on the long arm of chromosome II (Uemura et al. 1987a).

VI. NUCLEOTIDE SEQUENCES OF YEAST DNA TOPOISOMERASE GENES

The *S. cerevisiae TOP1* gene nucleotide sequence contains an open reading frame that codes for a 769-amino-acid polypeptide (Thrash et al. 1985). The predicted molecular mass (90,020 kD) is in good agreement with the molecular mass of purified *S. cerevisiae* topoisomerase I (Goto et al. 1984). The *TOP1* gene can be expressed in *E. coli*; a relaxing activity characteristic of the eukaryotic type-I enzyme was detected in extracts of *E. coli* cells containing the yeast *TOP1* sequence on a multicopy plasmid. Deletion mutagenesis has also shown that the amino-terminal domain of *S. cerevisiae* topoisomerase I is nonessential for the catalytic activity of the enzyme (Bjornsti and Wang 1987).

The nucleotide sequence of the *S. pombe* $top1^+$ gene shows a hypothetical coding frame interrupted by two short introns (59 and 46 bp), encoding an 812-residue polypeptide (predicted molecular mass is 94,000 kD), 43 residues longer than and 47% homologous to *S. cerevisiae* DNA topoisomerase I (Uemura et al. 1987a). The *S. cerevisiae* and *S. pombe* sequences are least homologous in the two large, very hydrophilic domains near the amino terminus and the carboxyl terminus. The sequences show no significant homology with bacterial DNA topoisomerase I. An eightfold increase of type-I enzyme activity was obtained in *S. pombe* cells containing a multicopy plasmid carrying the $top1^+$ gene. An antiserum made against a $top1^+$ fusion protein detects a protein in extracts of *S. pombe* with a molecular mass identical with that expected from the sequence (K. Shiozaki and M. Yanagida, unpubl.).

Nucleotide sequences of *S. cerevisiae* and *S. pombe TOP2* genes have been determined (Giaever et al. 1986; Uemura et al. 1986). The *S. cerevisiae* gene has an open reading frame coding for 1429 amino acids (164 kD). The *S. pombe* gene encodes a hypothetical protein of 1431 residues (162 kD). The *S. pombe* gene might encode a 54-residue longer protein (168 kD), however, if a hypothetical intron is present at the amino-terminal end (T. Uemura and M. Yanagida, unpubl.). A recent determination of the amino-terminal residues of purified topoisomerase II shows that the latter is indeed the case (K. Shiozaki and M. Yanagida, unpubl.). An antiserum made against a $top2^+$-*lacZ* fusion protein detected a 165-kD protein in *S. pombe* extracts, and the amount increased in cells transformed with a multicopy plasmid carrying the *top2*$^+$ gene (T. Uemura and M. Yanagida, unpubl.). In transformants with plasmids carrying carboxy-terminal-deleted *top2*$^+$ genes, the antiserum also showed polypeptides with lower molecular masses (T. Uemura et al., unpubl.).

The predicted amino acid sequences of budding and fission yeast topoisomerase II show a 49% overall homology. This is identical with the value obtained from a comparison of the predicted DNA topoisomerase I amino acid sequences between the two yeasts but is low when compared with the values obtained for other known proteins from the two distantly related yeasts (e.g., 83% for histone H2A, 73% for β-tubulin, and 62% for the protein kinase encoded by *CDC28* and *cdc2*$^+$). The cloned *S. pombe top2*$^+$ gene can complement the *S. cerevisiae top2* mutation, and conversely, the *S. cerevisiae TOP2* gene can complement *S. pombe top2* mutations (Uemura et al. 1986). These results indicate that the DNA topoisomerase II genes of the two yeasts are functionally exchangeable despite considerable differences in the amino acid sequences of the enzymes. Interestingly, the cloned *Drosophila* or human *TOP2* gene can also complement *S. cerevisiae top2* mutants when expressed in yeast (Wyckoff and Hsieh 1988; M. Tsai-Pflugfelder and J.C. Wang, unpubl.).

A weak but significant homology with the sequences of bacterial DNA gyrase subunits (Moriya et al. 1985) was discovered in the yeast topoisomerase II sequences (Lynn et al. 1986; Uemura et al. 1986). The amino-terminal half domain of the yeast sequences is homologous to the ATP-binding bacterial *gyrB* subunit. The central to the latter part of the yeast sequences is homologous to the amino-terminal domain of the catalytic bacterial *gyrA* subunit (it is more homologous to T4 phage topoisomerase subunit gp52 than to the *gyrA* subunit). These homologies suggest a possible gene fusion of the bacterial DNA gyrase genes during evolution to form a eukaryotic DNA topoisomerase II gene. The yeast sequences have higher homology with the *gyrB* subunit than with the *gyrA*. This is consistent with the fact that yeast DNA topoisomerase II is inhibited by coumermycin but not by nalidixic acid or oxolonic acid (Goto et al. 1984). The yeast topoisomerase II sequences have no significant homology with the type-I sequences.

DNA topoisomerase II consists of three domains, namely the amino-terminal half domain (perhaps ATP binding), the central to the latter part (DNA-binding, active site), and the carboxy-terminal highly hydrophilic (nonessential) domains. DNA topoisomerase II of the *cs top2* mutant (Uemura et al. 1987b) contains an amino acid substitution (Gly to Asp) in the amino-terminal domain, whereas the site of the mutation for a *ts top2* mutant was located in the central DNA-binding domain (K. Morino et al., unpubl.).

Very recently, evidence for a third topoisomerase gene in *S. cerevisiae* has been presented. Wallis et al. (1989) found a mutation that causes a fivefold increase in recombination between directly repeated δ sequences (δ sequences flank the yeast transposon Ty*1*). They called the mutation *edr* for *e*nhanced *d*elta *r*ecombination. Strains with the *edr* mutation grow slowly, and this phenotype enabled them to clone the gene. Sequencing of the gene revealed an open reading frame coding for a protein of 656 amino acids that, surprisingly, showed good homology with *E. coli* DNA topoisomerase I. Specifically, 21.5% of the residues are identical, and 39.1% are identical or conserved, comparing the entire yeast protein with the first 596 amino acids of the *E. coli* enzyme. Hence, the gene was renamed *TOP3*. A more recent three-way comparison among the putative yeast *TOP3* gene product and *E. coli* DNA topoisomerases I and III shows that all three proteins are about equally related to each other with about 40% homology in each pair-wise comparison (A. Rothstein, unpubl.). Given the homology with the *E. coli* enzymes, it is not surprising that the yeast *TOP3* protein shows no homology with any eukaryotic topoisomerase.

Strains with a *top3* null mutation are viable but grow at 50% of the wild-type rate (Wallis et al. 1989). Furthermore, *top1-top3* and *top2-top3* mutants grow more slowly than any of the single mutants. Interestingly, expression of *E. coli* DNA topoisomerase I from a plasmid in yeast complements the growth defect of a *top3* mutant. It should be emphasized, however, that to date no enzymatic activity corresponding to this gene has been detected. Thus, the evidence that the *TOP3* gene indeed codes for a topoisomerase is the sequence homology, the synergistic effect on growth rate of *top1-top3* and *top2-top3* double mutants, and the complementation of the *top3* growth defect by *E. coli* DNA topoisomerase I. The in vivo roles of this new topoisomerase and why *top3* mutants show increased recombination remain to be determined.

VII. ANALYSIS OF YEAST DNA TOPOISOMERASE MUTANTS

A. Studies with *top1* Mutants

DNA topoisomerase I exists in the nuclei of all eukaryotes, and it is thought to be involved in a variety of nuclear functions including replication, transcription, and recombination. DNA topoisomerase I in both *S. cerevisiae* and *S. pombe*, however, proved to be nonessential for viability. *top1* mutants are viable, and disruption or deletion of the gene causes no lethal phenotype although the generation time of DNA topoisomerase-I-deficient mutants is 20% longer than that of wild-type (Thrash et al. 1984, 1985; Uemura and Yanagida 1984, 1986; Goto and Wang 1985; Uemura et al. 1987a). One hypothesis to explain why the topoisomerase I gene is dispensable is that DNA topoisomerase II can substitute for topoisomerase I (Uemura and Yanagida 1984; see Section VII.D.).

It is not known whether DNA topoisomerase I is also dispensable in higher eukaryotes. Although the topoisomerase I inhibitor, camptothecin, kills mammalian cells and mammalian camptothecin-resistant mutants contain drug-resistant DNA topoisomerase I (Andoh et al. 1987), this does not necessarily mean that DNA topoisomerase I is essential for the survival of mammalian cells. Recent experiments showed that both *S. cerevisiae* and *S. pombe* are also sensitive to camptothecin, but *top1* mutants (*ts* or deletion) are found to be resistant to the drug. This indicates that the lethality is caused by the drug-induced aberrant enzyme action and not by the absence of DNA topoisomerase I activity (Eng et al. 1988; Nitiss and Wang 1988). As will be described in a later section, DNA topoisomerase I is dispensable in yeast only if DNA topoisomerase II activity is abundant.

S. cerevisiae top1 mutants do have the so-called *mak* phenotype (Wickner 1986); i.e., they are unable to maintain M_1 killer RNA, the double-stranded linear RNA found in a virus-like particle in the cytoplasm of many yeast strains. The reason for the inability of *top1* mutants to maintain this RNA is unknown. It could be that DNA topoisomerase I is involved directly in the metabolism of this cytoplasmic RNA or more likely that the effect is a secondary one caused by the somewhat deleterious effect that the absence of DNA topoisomerase I has on the cell (as evidenced by the slower growth rate of *top1* mutants.)

An unexpected phenotype recently found for *S. cerevisiae top1* mutants is that they exhibit greatly enhanced mitotic recombination within the rDNA cluster. This locus consists of 200 tandemly repeated 9-kb units (Christman et al. 1988).

The increased recombination is specific for the rDNA cluster; the frequency of recombination at other duplicated loci is unchanged in *top1* strains. Interestingly, *ts top2* mutants grown at a semipermissive temperature also show increased recombination specific to the rDNA cluster. These results have led to the suggestion that both DNA topoisomerases I and II are required to suppress recombination in this region of the genome (Christman et al. 1988). This will be discussed in more detail in Section VII. D.

B. Studies with *top2* Mutants

As described in previous sections, DNA topoisomerase II is essential for viability. Conditional lethal *top2* mutants show characteristic defective phenotypes at the restrictive temperature. The *S. cerevisiae ts top2* mutants accumulate replicated circular DNA plasmids as multiply-intertwined catenated dimers, indicating that DNA topoisomerase II is required for the segregation of circular DNA molecules after replication (DiNardo et al. 1984). In the *S. pombe top2* mutants that contain heat-sensitive DNA topoisomerase II, a highly uniform cell-cycle-dependent phenotype is observed (Uemura and Yanagida 1984). Nuclear division is blocked, and chromosomes are neither condensed nor segregated. Consistent with the nuclear division block in the *S. pombe top2* mutants, the *S. cerevisiae* mutants die during mitosis at the restrictive temperature (Holm et al. 1985); the onset of inviability coincides with the time of mitosis, and inviability is prevented by nocodazole, a tubulin inhibitor that blocks mitosis by inhibiting spindle formation. The *S. pombe top2* mutants also die during mitosis (Uemura and Yanagida 1986). These studies indicate that DNA topoisomerase II is required at this stage of the cell cycle.

Further studies of the *S. pombe top2* mutants (Uemura and Yanagida 1986) showed that the *top2* mutant cells produce abnormal chromosomes at the time of mitosis at the restrictive condition. The chromosomes are transiently pulled into filamentous structures by the elongating mitotic spindle but are not separated. Therefore, a primary defect in *top2* mutants may be the formation of aberrant mitotic chromosomes inseparable by the force generated by the spindle apparatus. DNA and RNA synthesis continue if cytokinesis is blocked by a *cdc11* mutation; in the *top2-cdc11* double mutant under nonpermissive conditions, an enlarged nucleus containing several times the normal haploid DNA content is observed.

An attempt was made to determine at which mitotic stage DNA topoisomerase II becomes essential. For this purpose, *cs-cs* and *cs-ts* double mutants in the genes coding for β-tubulin (*nda3*$^{+}$) and DNA topoisomerase II were constructed, and temperature-shift experiments were performed (Uemura et al. 1987b). The ATP-dependent relaxing activity of the *cs top2* gene product is cold-sensitive in vitro, and this cold sensitivity is reversible. Single *top2* mutants are not appropriate for examining the blocked stages during mitosis because the mutant phenotypes are obscured by other mitotic events, including spindle dynamics, septum formation, and cytokinesis, that take place at the nonpermissive temperature (Uemura and Yanagida 1984, 1986). On the other hand, temperature-shift experiments using *nda3-top2* double mutants show that DNA topoisomerase II is required continuously for mitotic chromosome changes (Uemura et al. 1987b). For example, a *nda3*^{cs}-*top2*^{cs} mutant at the restrictive temperature (20°C) is arrested at a stage similar to mitotic prophase, but the long, entangled chromosomes can condense and separate on the shift to the permissive temperature (36°C). If spindle formation is prevented at the permissive temperature by the addition of nocodazole, the chromosomes condense but do not separate. Thus, topoisomerase II is required for the final condensation of chromosomes. Moreover, pulse-shift experiments with a *top2*^{ts}-*nda3*^{cs} strain show that topoisomerase I is also required for chromatid disjunction in anaphase. Inactivation of topoisomerase II and reactivation of β-tubulin after chromosome condensation allow normal spindle formation but result in "streaked" chromosomes.

In *S. cerevisiae*, Holm et al. (1989) looked for the consequences expected by the attempted segregation of still-tangled chromosomes. The frequency of chromosome loss is substantially elevated in *top2*/*top2* diploid mutant cells kept at the restrictive temperature for one generation. However, only a minor increase in the amount of chromosome breakage is observed. Chromosome loss rather than the breakage may be a major cause of the inviability observed when *top2* cells undergo mitosis at the restrictive temperature.

C. DNA Topoisomerase II and Genetic Control of Chromosome Separation

Uncoordinated mitosis, i.e., the occurrence of spindle dynamics and cytokinesis in the absence of chromosome separation, is revealed in *S. pombe top2* mutants (Uemura and Yanagida 1986). The steps controlled by DNA topoisomerase II are not coordinated with spindle formation. The *top2*$^{+}$ gene product appears to become essential after the cells have passed the stages that commit the cell to the spindle-formation pathway. In all the previously isolated *cdc* and *cs nda* (tubulin) mutants in *S. pombe*, the arrest of nuclear division is always accompanied by the blocking of cytokinesis (Nurse et al. 1976; Toda et al. 1983). The phenotype of *top2* mutants that undergo cytokinesis in the absence of nuclear division, however, is not exceptional among the mutants defective in nuclear division; ten complementation groups have been identified among the *ts cut* mutants that show a cytological phenotype similar to *top2* (Hirano et al. 1986). The other *cut* gene products might also be implicated in chromosome separation. Extracts of the *cut* mutant cells, however, contain a normal level of topoisomerase II activity. The uncoupling of the chromosome pathway from the spindle (and cytokinesis) pathway must be a general phenomenon since mutations in more than 14 genes, including *top2*, *cut1-cut10*, and *dis1-dis3* (Ohkura et al. 1988), cause such a phenotype.

Little is known at the molecular level about the role of DNA topoisomerase II in chromosome separation. Catenating-decatenating and/or knotting-unknotting activity has been postulated to be required for condensation and separation of chromosomal DNAs (Cozzarelli 1980; Hsieh and Brutlag 1980; Sundin and Varshavsky 1981; DiNardo et al. 1984; Uemura and Yanagida 1986). In fact, catenated circular DNAs accumulate in *top2* cells at the restrictive temperature (DiNardo et al. 1984; Uemura and Yanagida 1986). Thus, it is tempting to assume that sister chromatids are topologically interlocked after replication so that they can be held together until anaphase, at which time they are separated by the combined action of topoisomerase II and the mitotic spindle (e.g., Murray and Szostak 1985; Holm et al. 1989). However, studies on the structure of a small minichromosome in various *S. cerevisiae* cell-cycle mutants (Koshland and Hartwell 1987) show that sister minichromosome molecules are not topologicaly interlocked with each other in the majority of cells arrested after S phase but before anaphase. The sister molecules are properly segregated when the cell cycle block is removed. Therefore, they need not remain topologically intertwined until anaphase to be properly segregated. If these studies can be extrapolated to large linear chromosomes, then topological interlocking apparently is not the primary force holding sister chromatids together.

D. Studies with *top1-top2* Double Mutants

The *top1-top2* double mutants of *S. pombe* and *S. cerevisiae* show phenotypes strikingly different from those of single *top1* and *top2* mutants (Uemura and Yanagida 1984; Goto and Wang 1985; Brill et al. 1987). *S. pombe* double mutant cells incubated at the restrictive temperature are quickly arrested irrespective of the stage in the cell cycle, producing a dramatically altered nuclear chromatin region (designated ring phenotype). The nuclear chromosome domain becomes a hollow bowl-like structure, in contrast with the hemispherical structure seen in wild-type interphase cells. Further studies indicated that DNA, RNA, and protein synthesis are diminished at the restrictive condition (Uemura and Yanagida 1986).

S. cerevisiae top1-top2 mutants also stop growth rapidly at the nonpermissive temperature. Whereas mitotic blocks can prevent killing of *ts top2* cells at the restrictive temperature, the same treatments are ineffective in preventing cell death of a *top1-top2* double mutant (Goto and Wang 1985). DNA and rRNA synthesis are drastically inhibited in a *S.*

cerevisiae top1-top2 mutant at the restrictive temperature, but the rate of poly(A)-terminated mRNA synthesis is reduced only threefold, and tRNA synthesis remains relatively normal (Brill et al. 1987). Studies with synchronized cells showed that DNA synthesis stops abruptly at the restrictive temperature. These results suggest that DNA topoisomerase activity is required as a swivel for DNA unwinding during fork movement and for the transcription of rRNA. The synthesis of some specific mRNAs seemed to be normal in a *top1-top2* mutant. For example, the rate and extent of induction from the *GAL1* promoter is completely normal even when the strain is induced 30 minutes after a shift to the nonpermissive temperature (Brill and Sternglanz 1988).

A preferential reduction of rRNA synthesis is also found in the *S. pombe top1-top2* mutant (Yamagishi and Nomura 1988). rRNA synthesis in the *top1-top2* mutant is only about 4% of that of the wild-type, whereas RNA synthesis other than rRNA in the mutant is about 70% of that in the wild-type. Only a small decrease in the ribosomal protein synthesis results from the shift to a restrictive temperature. Thus, in both yeast species, topoisomerase activity is much more important for rRNA synthesis than for mRNA synthesis. rRNA synthesis may be uniquely sensitive to topoisomerase depletion because the rDNA locus consists of a tandem array of heavily transcribed genes (Brill et al. 1987; Hirano et al. 1989).

An interesting example of the effect of topoisomerase depletion on the rDNA locus in *S. cerevisiae* has been reported recently. Kim and Wang (1989) found that, in a particular *top1-top2* mutant grown at the permissive temperature, over half of the rDNA is present as extrachromosomal rings containing one 9-kb unit of the rDNA gene or tandem repeats of it. Expression of a plasmid-borne *TOP1* or *TOP2* gene leads to integration of the extrachromosomal rDNA rings back into the chromosomal rDNA cluster. If a topoisomerase gene is expressed from the regulated *GAL1* promoter, then repression of the gene by addition of glucose to the strain leads to reappearance of the extrachromosomal rDNA rings. Presumably, this excision-integration phenomenon is related to the hyper-recombination at the rDNA locus seen in *top1* or *top2* mutants (Christman et al. 1988; see section above), although it should be emphasized that these observations were with single mutants rather than a double mutant. An interpretation of all these results is that transcription of the rDNA locus is inhibited somewhat in *top1* or *top2* single mutants and more so in double mutants, and the superhelical stress that arises in the DNA template during rRNA transcription in these mutants leads to DNA structures that are recombinogenic.

Examination of the degree of supercoiling of plasmids isolated from

S. cerevisiae topoisomerase mutants has provided evidence recently that topoisomerases are involved in relieving stress that arises during mRNA and rRNA synthesis. Brill and Sternglanz (1988) found that plasmids isolated from *top1-top2* mutants grown at the restrictive temperature are hypernegatively supercoiled. This supercoiling is more pronounced with highly transcribed genes and longer trancripts and is seen even in *top1* single mutants if transcription starts from very strong promoters such as the *GAL1-10* promoter or the one for the 35S rRNA precursor. It appears that the supercoiling arises during transcriptional elongation in the absence of topoisomerase I relaxing activity. These results have been interpreted in terms of the twin domain model for transcriptional elongation (Liu and Wang 1987) in which a positively supercoiled domain is generated ahead of the moving RNA polymerase and a negatively supercoiled domain arises behind it. The results suggest that topoisomerase I and, to a lesser extent, topoisomerase II are responsible for relieving the resulting torsional and flexural stress. In *top1* mutants and even in *top1-top2* mutants, it appears that another topoisomerase can relax both supercoiled domains but preferentially the positive domain, leading to plasmids with the observed high net negative supercoiling. An alternate interpretation is that the hypernegative supercoiling is not due to a bias in the rates of removal of negatively and positively supercoiled domains but is due to a bias in the binding of certain proteins (e.g., nucleosomes) to the differentially supercoiled domains (Giaever and Wang 1988). The question of which topoisomerase is responsible for the observed change in linking number remains unanswered; it could be a form of topoisomerase II that remains active in the *ts top1-top2* mutant at the restrictive temperature, or conceivably, another topoisomerase.

Strong support for the twin domain model for transcriptional elongation in yeast comes from experiments in which *E. coli* DNA topoisomerase I is expressed in a yeast *top1-top2* mutant at the restrictive temperature. The yeast 2-μm plasmid becomes positively supercoiled under these conditions (Giaever and Wang 1988). The explanation for this result is that since the *E. coli* enzyme can relax only negative supercoils, and since the two major yeast topoisomerases are inactivated, transcription of genes on the plasmid leads to net positive supercoiling. At first sight, it seems difficult to reconcile this result with the hypernegative supercoiling described above (Brill and Sternglanz 1988). The accumulation of positive supercoils in the strain suggests that no enzyme other than topoisomerases I and II can effectively remove positive supercoils. Furthermore, because positive supercoils appear only when the *E. coli* enzyme is expressed, it seems that in yeast no enzyme other than topoisomerases I and II can effectively relax negative supercoils. Probably, the overexpression of the *E. coli* enzyme in yeast leads to far more relaxing activity than the low level of activity remaining in a *top1-top2* mutant at the restrictive temperature.

To explain the different phenotypes among *top1*, *top2*, and *top1-top2* mutants, it has been postulated that DNA topoisomerase II can substitute for DNA topoisomerase I (Uemura and Yanagida 1984; Goto and Wang 1985: Brill et al. 1987); DNA topoisomerase II may have several in vivo functions, some of which are redundant because DNA topoisomerase I can play a similar role. The one essential role of DNA topoisomerase II is in chromosome condensation and segregation in mitosis through unknotting and/or decatenating the chromosomal DNA. Therefore, *top1* mutants are viable due to the functional substitution by DNA topoisomerase II. The defect in DNA topoisomerase II blocks chromosome condensation or segregation specifically because events in other parts of the cell cycle can be advanced with the complementing activity of DNA topoisomerase I. In the double *top1-top2*, the relaxing activity shared by DNA topoisomerase I and II is absent so that DNA replication and rRNA transcription are inhibited. Different *top1-top2* mutants, namely *top1*^{ts}-*top2*^{ts}, *top1*^{null}-*top2*^{ts}, *top1*^{ts}-*top2*^{cs}, and *top1*^{null}-*top2*^{cs} (the last one was lethal at any temperature), were constructed (Uemura et al. 1987a). Analysis of the mutants indicated that the total level of the relaxing activity by DNA topoisomerase I and II determines the growth rate of cells. The cells appear to grow poorly or even die when their level of relaxing activity is below a certain critical level. Thus, the DNA topoisomerase I gene becomes essential when the activity of DNA topoisomerase II is not abundant.

ACKNOWLEDGMENTS

Work done in the laboratory of M.Y. was supported by grants from the Ministry of Education, Science and Culture, and the Mitsubishi Foundation. Research from the laboratory of R.S. was supported by the National Institutes of Health grant GM-2822O.

REFERENCES

Adachi, T., M. Mizuuchi, E.A. Robinson, E. Appella, M.H. O'Dea, M. Gellert, and K. Mizuuchi. 1987. DNA sequence of the *E. coli gyrB* gene: Application of a new sequencing strategy. *Nucleic Acids Res.* **15:** 771–783.

Andoh, T., K. Ishii, Y. Suzuki, Y. Ikegami, Y. Kusunoki, Y. Takemoto, and K. Okada. 1987. Characterization of a mammalian mutant with a camptothecin-resistant DNA topoisomerase. *Proc. Natl. Acad. Sci.* **84:** 5565–5569.

Baldi, M.I., P. Benedetti, E. Mattoccia, and G.P. Tocchini-Valentini. 1980. In vitro catenation and a novel eucaryotic ATP-dependent topoisomerase. *Cell* **20:** 461–467.

Bjornsti, M. and J.C. Wang. 1987. Expression of yeast DNA topoisomerase I can complement a conditional-lethal DNA topoisomerase I mutation in *Escherichia coli*. *Proc. Natl. Acad. Sci.* **84:** 8971–8975.

Brill, S.J. and R. Sternglanz. 1988. Transcription-dependent DNA supercoiling in yeast DNA topoisomerase mutants. *Cell* **54:** 403–411.

Brill, S., S. DiNardo, K. Voelkel-Meiman, and R. Sternglanz. 1987. Need for DNA topoisomerase activity as a swivel for DNA replication and for transcription of ribosomal RNA. *Nature* **326:** 414–416.

Champoux, J.J. and R. Dulbecco. 1972. An activity from mammalian cells that untwists superhelical DNA—A possible swivel for DNA replication. *Proc. Natl. Acad. Sci.* **69:** 143–146.

Christman, M.F., F.S. Dietrich, and C.R. Fink. 1988. Mitotic recombination in the rDNA of *S. cerevisiae* is suppressed by the combined action of DNA topoisomerase I and II. *Cell* **55:** 413–425.

Cozzarelli, N.R. 1980. DNA gyrase and the supercoiling of DNA. *Science* **207:** 953–960.

Dean, F., M.A. Krasnow, R. Otter, M.M. Matzuk, S.J. Spengler, and N.R. Cozzarelli. 1983. *Escherichia coli* type-I topoisomerases: Identification, mechanism, and role in recombination. *Cold Spring Harbor Symp. Quant. Biol.* **47:** 769–777.

DiGate, R.J. and K.J. Marians. 1988. Identification of a potent decatenating enzyme from *Escherichia coli*. *J. Biol. Chem.* **263:** 13366–13373.

———. 1989. Molecular cloning and DNA sequence analysis of *Escherichia coli topB*, the gene encoding topoisomerase III. *J. Biol. Chem.* **264:** 17924–17930.

DiNardo, S., K. Voelkel, and R. Sternglanz. 1984. DNA topoisomerase II mutant of *Saccharomyces cerevisiae:* Topoisomerase II is required for segregation of daughter molecules at the termination of DNA replication. *Proc. Natl. Acad. Sci.* **81:** 2616–2620.

DiNardo, S., K.A. Voelkel, R. Sternglanz, A.E. Reynolds, and A. Wright. 1982. *Escherichia coli* DNA topoisomerase I mutants have compensatory mutations in DNA gyrase genes. *Cell* **31:** 43–51.

Dorman, C.J., A.S. Lynch, N. Ni Bhriain, and C.F. Higgins. 1989. DNA supercoiling in *Escherichia coli: topA* mutations can be suppressed by DNA amplifications involving the *tolC* locus. *Mol. Microbiol.* **3:** 531–540.

Dubnau, E. and P. Margolin. 1972. Suppression of promoter mutations by the pleiotropic *supX* mutations. *Mol. Gen. Genet.* **117:** 91–112.

Eng, W., L. Faucette, R.K. Johnson, and R. Sternglanz. 1988. Evidence that DNA topoisomerase I is necessary for the cytotoxic effects of camptothecin. *Mol. Pharmacol.* **34:** 755–760.

Filutowicz, M. and P. Joncyzk. 1983. The *gyrB* gene product functions in both initiation and chain polymerization of *Escherichia coli* chromosome replication: Suppression of the initiation deficiency in *gyrB*-ts mutants by a class of the *rpoB* mutations. *Mol. Gen. Genet.* **191:** 282–287.

Gellert, M., L.M. Fisher, and M.H. O'Dea. 1979. DNA gyrase: Purification and catalytic properties of a fragment of gyrase B protein. *Proc. Natl. Acad. Sci.* **76:** 6289–6293.

Gellert, M., K. Mizuuchi, M. O'Dea, and H. Nash. 1976. DNA gyrase: An enzyme that introduces superhelical turns into DNA. *Proc. Natl. Acad. Sci.* **73:** 3872–3876.

Giaever, G.N. and J.C. Wang. 1988. Supercoiling of intracellular DNA can occur in eukaryotic cells. *Cell* **55:** 849–856.

Giaever, G., R. Lynn, T. Goto, and J.C. Wang. 1986. The complete nucleotide sequence of the structural gene TOP2 of yeast DNA topoisomerase II. *J. Biol. Chem.* **261:** 12448–12454.

Goto, T. and J.C. Wang. 1984. Yeast DNA topoisomerase II is encoded by a single-copy, essential gene. *Cell* **36:** 1073–1080.

———. 1985. Cloning of yeast *TOP1*, the gene encoding DNA topoisomerase I, and construction of mutants defective in both DNA topoisomerase I and DNA topoisomerase II. *Proc. Natl. Acad. Sci.* **82:** 7178–7182.

Goto, T., P. Laipis, and J.C. Wang. 1984. The purification and characterization of DNA topoisomerases I and II of the yeast *Saccharomyces cerevisiae*. *J. Biol. Chem.* **259:** 10422–10429.

Hirano, T., S. Funahashi, T. Uemura, and M. Yanagida. 1986. Isolation and characterization of *Schizosaccharomyces pombe cut* mutants that block nuclear division but not cytokinesis. *EMBO J.* **5:** 2973–2979.

Hirano, T., G. Konoha, T. Toda, and M. Yanagida. 1989. Essential roles of the RNA polymerase I largest subunit and DNA topoisomerases in the formation of fission yeast nucleolus. *J. Cell Biol.* **108:** 243–253.

Holm, C., T. Stearns, and D. Botstein. 1989. DNA topoisomerase II must act at mitosis to prevent nondisjunction and chromosome breakage. *Mol. Cell. Biol.* **9:** 159–168.

Holm, C., T. Goto, J.C. Wang, and D. Botstein. 1985. DNA topoisomerase II is required at the time of mitosis in yeast. *Cell* **41:** 553–563.

Hsieh, T. and D. Brutlag. 1980. ATP-dependent DNA topoisomerase from *D. melanogaster* reversibly catenates duplex DNA rings. *Cell* **21:** 115–125.

Kim, R.A. and J.C. Wang. 1989. A subthreshold level of DNA topoisomerases leads to the excision of yeast rDNA as extrachromosomal rings. *Cell* **57:** 975–985.

Koshland, D. and L.H. Hartwell. 1987. The structure of sister minichromosome DNA before anaphase in *Saccharomyces cerevisiae*. *Science* **238:** 1713–1716.

Kreuzer, K.N. and N.R. Cozzarelli. 1979. *Escherichia coli* mutants thermosensitive for deoxyribonucleic acid gyrase subunit A: Effects of deoxyribonucleic acid replication, transcription and bacteriophage growth. *J. Bacteriol.* **140:** 424–435.

Liu, L.F. and J.C. Wang. 1987. Supercoiling of the DNA template during transcription. *Proc. Natl. Acad. Sci.* **84:** 7024–7027.

Liu, L.F., C. Liu, and B.M. Alberts. 1980. Type II DNA topoisomerases: Enzymes that can unknot a topologically knotted DNA molecule via a reversible double-strand break. *Cell* **19:** 697–707.

Lockshon, D. and D.R. Morris. 1983. Positively supercoiled plasmid DNA is produced by treatment of *Escherichia coli* with DNA gyrase inhibitors. *Nucleic Acids Res.* **11:** 2999–3017.

Lynn, R.M. and J.C. Wang. 1989. Peptide sequencing and site-directed mutagenesis identify tyrosine-319 as the active site tyrosine of *Escherichia coli* DNA topoisomerase I. *Protein* **6:** 231–239.

Lynn, R., G. Giaever, S.L. Swanberg, and J.C. Wang. 1986. Tandem regions of yeast DNA topoisomerase II share homology with different subunits of bacterial gyrase. *Science* **233:** 647–649.

Margolin, P., L. Zumstein, R. Sternglanz, and J.C. Wang. 1985. The *Escherichia coli supX* locus is *topA*, the structural gene for DNA topoisomerase I. *Proc. Natl. Acad. Sci.* **82:** 5437–5441.

Moriya, S., N. Ogasawara, and H. Yoshikawa. 1985. Structure and function of the region of the replication origin of the *Bacillus subtilis* chromosome. II. Nucleotide sequence of some 10,000 base pairs in the origin region. *Nucleic Acids Res.* **13:** 2251–2265.

Mukai, F.H. and P. Margolin. 1963. Analysis of unlinked suppressors of an 0° mutation in *Salmonella. Proc. Natl. Acad. Sci.* **50:** 140–148.

Murray, A.W. and J. Szostak. 1985. Chromosome segregation in mitosis and meiosis. *Annu. Rev. Cell Biol.* **1:** 289–315.

Nitiss, J. and J.C. Wang. 1988. DNA topoisomerase-targeting antitumor drugs can be studied in yeast. *Proc. Natl. Acad. Sci.* **85:** 7501–7505.

Nurse, P., P. Thuriaux, and K. Nasmyth. 1976. Genetic control of the cell division cycle in the fission yeast *Schizosaccharomyces pombe. Mol. Gen. Genet.* **146:** 167–178.

Ohkura, H., Y. Adachi, N. Kinoshita, O. Niwa, T. Toda, and M. Yanagida. 1988. Cold-senstive and caffeine supersensitive mutants of the *Schizosaccharomyces pombe* genes implicated in sister chromatid separation during mitosis. *EMBO J.* **7:** 1465–1473.

Orr, E., N.F. Fairweather, I.B. Holland, and R.H. Pritchard. 1979. Isolation and characterization of a strain carrying a conditional lethal mutation in the *cou* gene of *Escherichia coli* K-12. *Mol. Gen. Genet.* **177:** 103–112.

Overbye, K.M. and P. Margolin. 1981. Role of the *supX* gene in ultraviolet light-induced mutagenesis in *Salmonella typhimurium. J. Bacteriol.* **146:** 170–178.

Overbye, K.M., S.K. Basu, and P. Margolin. 1983. Loss of DNA topoisomerase I activity alters many cellular functions in *Salmonella typhimurium. Cold Spring Harbor Symp. Quant. Biol.* **47:** 785–791.

Pruss, G.J. 1985. DNA topoisomerase I mutants. Increased heterogeneity in linking number and other replicon-dependent changes in DNA supercoiling. *J. Mol. Biol.* **185:** 51–63.

Pruss, G.J. and K. Drlica. 1986. Topoisomerase I mutants: The gene on pBR322 that encodes resistance to tetracycline affects plasmid DNA supercoiling. *Proc. Natl. Acad. Sci.* **83:** 8952–8956.

Pruss, G.J., S.H. Manes, and K. Drlica. 1982. *Escherichia coli* DNA topoisomerase mutants: Increased supercoiling is corrected by mutations near gyrase genes. *Cell* **31:** 35–42.

Raji, A., D.J. Zabel, C.S. Laufer, and R.E. Depew. 1985. Genetic analysis of mutations that compensate for loss of *Escherichia coli* DNA topoisomerase I. *J. Bactiol.* **162:** 1173–1179.

Richardson, S.M.H., C.F. Higgins, and D.M.J. Lilley. 1984. The genetic control of DNA supercoiling in *Salmonella typhimurium. EMBO J.* **3:** 1745–1752.

Rothstein, R.J. 1983. One-step gene disruption in yeast. *Methods Enzymol.* **101:** 202–211.

Srivenugopal, K.S., D. Lockshon, and D.R. Morris. 1984. *Escherichia coli* DNA topoisomerase III: Purification and characterization of a new type I enzyme. *Biochemistry* **23:** 1899–1906.

Sternglanz, R., S. DiNardo, K.A. Voelkel, Y. Nishimura, Y. Hirota, K. Becherer, L. Zumstein, and J.C. Wang. 1981. Mutations in the gene coding for *Escherichia coli* DNA topoisomerase I affect transcription and transposition. *Proc. Natl. Acad. Sci.* **78:** 2747–2751.

Sundin, O. and A. Varshavsky. 1981. Arrest of segregation leads to accumulation of highly intertwined caternated dimers: Dissection of the final stages of SV40 DNA replication. *Cell* **25:** 659–669.

Swanberg, S.L. and J.C. Wang. 1987. Cloning and sequencing of the *Escherichia coli gyrA* gene coding for the A subunit of DNA gyrase. *J. Mol. Biol.* **197:** 729–736.

Thrash, C., A.T. Bankier, B.G. Barrell, and R. Sternglanz. 1985. Cloning, characterization, and sequence of the yeast DNA topoisomerase I gene. *Proc. Natl. Acad. Sci.* **82:** 4374–4378.

Thrash, C., K. Voelkel, S. DiNardo, and R. Sternglanz. 1984. Identification of *Saccharomyces cerevisiae* mutants deficient in DNA topoisomerase I activity. *J. Biol. Chem.* **259:** 1375–1377.

Toda, T., K. Umesono, A. Hirata, and M. Yanagida. 1983. Cold-sensitive nuclear division arrest mutants of the fission yeast *Schizosaccharomyces pombe*. *J. Mol. Biol.* **168:** 251–270.

Trucksis, M. and R.E. Depew. 1981. Identification and localization of a gene that specifies production of *Escherichia coli* DNA topoisomerase I. *Proc. Natl. Acad. Sci.* **78:** 2164–2168.

Uemura, T. and M. Yanagida. 1984. Isolation of type I and II DNA topoisomerase mutants from fission yeast: Single and double mutants show different phenotypes in cell growth and chromatin organization. *EMBO J.* **3:** 1737–1744.

———. 1986. Mitotic spindle pulls but fails to separate chromosomes in type II DNA topoisomerase mutants: Uncoordinated mitosis. *EMBO J.* **5:** 1003–1010.

Uemura, T., K. Morikawa, and M. Yanagida. 1986. The nucleotide sequence of the fission yeast DNA topoisomerase II gene: Structural and functional relationship to other DNA topoisomerases. *EMBO J.* **5:** 2355–2361.

Uemura, T., K. Morino, S. Uzawa, K. Shiozaki, and M. Yanagida. 1987a. Cloning and sequencing of *Schizosaccharomyces pombe* DNA topoisomerase I gene, and effect of gene disruption. *Nucleic Acids Res.* **15:** 9727–9739.

Uemura, T., H. Ohkura, Y. Adachi, K. Morino, K. Shiozaki, and M. Yanagida. 1987b. DNA topoisomerase II is required for condensation and separation of mitotic chromosomes in *S. pombe*. Cell **50:** 917–925.

Voelkel-Meiman, K., S. DiNardo, and R. Sternglanz. 1986. Molecular cloning and genetic mapping of the DNA topoisomerase II gene of *Saccharomyces cerevisiae*. *Gene* **42:** 193–199.

Wallis, J.W., G. Chrebet, G. Brodsky, M. Rolfe, and R. Rothstein. 1989. A hyper-recombination mutation in *S. cerevisiae* identifies a novel eukaryotic topoisomerase. *Cell* **58:** 409–419.

Wang, J.C. 1971. Interaction between DNA and an *Escherichia coli* protein ω. *J. Mol. Biol.* **55:** 523–533.

Wickner, R.B. 1986. Double-stranded RNA replication in yeast: The killer system. *Annu. Rev. Biochem.* **55:** 373–395.

Wu, H.-Y., S. Shyy, J.C. Wang, and L.F. Liu. 1988. Transcription generates positively and negatively supercoiled domains in the template. *Cell* **53:** 433–440.

Wyckoff, E. and T. Hsieh. 1988. Functional expression of a *Drosophila* gene in yeast: Genetic complementation of DNA topoisomerase II. *Proc. Natl. Acad. Sci.* **85:** 6272–6276.

Yamagishi, J., H. Yoshida, M. Yamayoshi, and S. Nakamura. 1986. Nalidixic acid-resistant mutations of the *gyrB* gene of *Escherichia coli*. *Mol. Gen. Genet.* **204:** 367–373.

Yamagishi, M. and M. Nomura. 1988. Deficiency in both type I and type II DNA topoisomerase activities differentially affect rRNA and ribosomal protein synthesis in *Schizosaccharomyces pombe*. *Curr. Genet.* **13:** 305–314.

11

DNA Replication: Topological Aspects and the Roles of DNA Topoisomerases

James C. Wang
Department of Biochemistry and Molecular Biology
Harvard University
Cambridge, Massachusetts 02138

Leroy F. Liu
Department of Biological Chemistry
Johns Hopkins University School of Medicine
Baltimore, Maryland 21205

The advent of the double-helix structure of DNA raised immediately the question of how the intertwined strands could come apart during replication (Delbruck 1954; Delbruck and Stent 1957). The problem was initially viewed as a kinetic one; as the strands separate, the unreplicated portion ahead of the replication fork would have to rotate rapidly around its helical axis. A rate of about 10,000 revolutions per minute was estimated based on the rate of replication in bacteria.

The problem of unraveling the parental strands during semiconservative replication became a topological one when the entire genome of the bacterium *Escherichia coli* was found to be in the form of a double-stranded ring. A duplex DNA ring can be viewed as two multiply-linked

DNA Topology and Its Biological Effects

single-stranded rings of complementary nucleotide sequences, and the unlinking of such a pair is impossible without disrupting, at least transiently, the continuity of the strands (Cairns 1963, 1964; Cairns and Davern 1967). Therefore, an enzyme or a group of enzymes must serve the role of a "swivel" to allow the unraveling of the intertwined single-stranded DNA (ssDNA) rings. The situation is not very different for long linear chromosomes organized into multiple loops, each of which appears to define a topological domain (for review, see Gross and Garrard 1988).

The topological problem described above for strand separation is basically one for the elongation step of replication. Because the linking number between the parental strands of a DNA ring or loop decreases continuously during replication, a swivel is required at most if not all times during the elongation of the new DNA chains.

There are also unique topological problems during the initiation and termination steps of replication. In the initiation step, the topology of a DNA ring or loop containing *ori* has strong effects on the binding of proteins to *ori*. In the termination step, a pair of interlocked progeny molecules would form if the parental strands are unpaired and come apart before they are completely untwined. Decatenation of this newly replicated progeny pair would be necessary before their segregation into separate cells.

In the next three sections, we discuss some aspects of DNA topology during the various stages of replication and the roles of DNA topoisomerases. For discussions on the more general aspects of DNA replication and comprehensive coverage of the literature, several excellent monographs and reviews should be consulted (Kornberg 1980, 1982, 1988a,b; Nossal 1983; McMacken and Kelly 1986; Kelly 1988; Kornberg and Baker 1990).

I. INITIATION OF REPLICATION

In vitro studies of several well-characterized systems indicate that the priming of replicative DNA synthesis is often preceded by the assembly of sequence-specific DNA-binding proteins at the origin of replication, *ori*, and the unpairing of a stretch of duplex DNA in that region (Dodson et al. 1985, 1986, 1987; Baker et al. 1986, 1987; Dean et al. 1987a,b; Stahl and Knippers 1987; Wold et al. 1987). The topology of a DNA can have strong effects on the binding of proteins and other molecules to the DNA as well as the unpairing of complementary strands (Vinograd et al. 1968; Bauer and Vinograd 1970; Davidson 1972; Hsieh and Wang 1975; Wang et al. 1983; Gellert and Nash 1987; Wang and Giaever 1988).

A. DNA Topology and the Binding of Proteins Involved in the Initiation of Replication

In general, if the binding of a protein to a DNA segment untwists the DNA double helix or imposes a negative spatial writhe on the helical axis of the segment, the binding will be more favorable in a negatively supercoiled than a relaxed DNA. Similarly, if the interaction overwinds the DNA or imposes a positive writhe, the process will be more favorable in a positively supercoiled DNA. Quantitative estimates of the effects of supercoiling on the binding constants can be made based on the free energy of DNA supercoiling (Davidson 1972; Hsieh and Wang 1975; Wang et al. 1983). In the case of the binding of the *DnaA* protein to *oriC*, the origin of replication of *E. coli*, the complex is more stable if *oriC* is present in a negatively supercoiled DNA (Fuller and Kornberg 1983). The same is true for complex formation between the λ initiation protein O and *ori*λ, the origin of replication of phage λ (Dodson et al. 1985, 1986).

B. Topological Constraints Imposed by DNA Looping

When proteins bound to separate sites along a DNA are brought together, they close a DNA loop in between. Similar to the effects of supercoiling in a DNA ring, the looping of a short segment of DNA by proteins may also introduce torsional and flexural stresses in the DNA that may in turn affect structural transitions within the loop and/or interactions between the DNA loop and other proteins (for discussion, see Wang and Giaever 1988). It is noteworthy that, if a DNA loop is strained, the proteins holding the loop are also strained. In principle, the reactivity of a protein can be affected by the strain imposed by the DNA; however, no experimentally established paradigm is yet available.

C. DNA Topology and the Initiation of Strand Separation

An important step in the initiation of replication is the unpairing of DNA strands in the *ori* region. This process is strongly influenced by the topology of the DNA segment containing *ori*.

In several in vitro systems, probing of the DNA structure by the use of single-strand-specific nucleases indicates that the unpairing of a short DNA segment within *ori* is an early event in the initiation process. Binding of phage λ protein O to *ori*λ, for example, makes a 60-bp AT-rich region adjacent to the O-binding site sensitive to single-strand-specific nuclease S1 or P1 (Schnos et al. 1988). This O-protein-induced structural change does not require ATP or any other high-energy cofactor but is strictly dependent on the DNA being negatively supercoiled. Similarly, the binding of the *E. coli* replication initiator *DnaA* protein to its binding sites in *oriC* renders an adjacent AT-rich region sensitive to S1 or P1 (Bramhill and Kornberg 1988). The *DnaA*-protein-induced reaction requires ATP, dATP, or CTP; no other protein is needed to confer nuclease sensitivity to the adjacent AT-rich sequence.

Significantly, the *DnaA*-protein-induced strand-opening reaction is sensitive to temperature. Little strand separation is detectable below 20°C. Once the open complex is formed, however, it remains stable at 0°C (Bramhill and Kornberg 1988). This steep temperature dependence is probably related to the high negative enthalpy accompanying the disruption of base pairs and may be a common feature of protein/DNA interactions that unpair DNA. The unwinding of DNA by RNA polymerase, for example, is strongly dependent on temperature; once an open complex is formed, it reverts to the closed complex very slowly when incubated at a low temperature (Wang et al. 1977; Kirkegaard et al. 1983). It is also likely that the temperature threshold can be lowered by an increase in the degree of negative supercoiling of the DNA.

The importance of strand separation in the initiation of replication is further strengthened by the identification of several yeast autonomously replicating sequences (ARS) as sites that unwind readily and are therefore sensitive to single-strand-specific nucleases when they are present in negatively supercoiled DNAs (Umek and Kowalski 1987, 1988). The ease of unwinding at an ARS appears to correlate with its function as a replication origin in vivo; it was shown that an ARS element near the histone H4 gene can be substituted by a biologically unrelated sequence with DNA unwinding properties similar to functional ARS (Umek and Kowalski 1988).

The elegant studies on the initiation of replication of the plasmid colE1 provide an additional paradigm of the effects of DNA topology on the initiation of replication. Here, the initiation step involves, in succession, the formation of an RNA-DNA hybrid, cleavage of the RNA in the hybrid by RNase H, and extension of the 3′ end of the severed RNA by DNA polymerase I (Itoh and Tomizawa 1980). The probability of RNA-DNA hybrid formation at *ori* is dependent on the structure of the initiator RNA, the transcription of which started some 550 bp upstream of *ori*. At least in purified systems, negative supercoiling of the DNA is necessary for the initiation of replication. It is plausible that the negatively supercoiled state of DNA may facilitate the unpairing of the DNA strands and the formation of the RNA-DNA hybrid.

D. Mechanisms of DNA Supercoiling

In the examples described above, macromolecular interactions and DNA supercoiling act synergistically to effect the unpairing of a short stretch of duplex DNA. Is intracellular DNA supercoiled, and what cellular processes might be responsible for DNA supercoiling?

Until recently, the prevailing view had been that in bacteria intracellular DNA was negatively supercoiled due to the action of DNA gyrase. In eukaryotes, the lack of a gyrase-type DNA supercoiling activity had cast doubt on the involvement of DNA supercoiling in biological processes. These views have been modified significantly with the introduction of the twin-supercoiled-domain model of transcription (Liu and Wang 1987).

Figure 1 illustrates two types of situations in which transcription may cause the supercoiling of the template. In Figure 1a, the RNA polymerase R undergoing transcription is shown to interact with an upstream DNA-binding protein X. This interaction anchors the polymerase; as RNA synthesis proceeds, the DNA is being threaded through R to form a negatively supercoiled loop between X and R (Wang 1985, 1987). The extent of negative supercoiling of this loop can be very high: The geometry of the DNA double helix predicts that one negative supercoil is introduced every 10.5 bp transcribed.

Anchoring of the polymerase does not necessarily involve a direct contact between the polymerase and a DNA-bound factor. If R and a point S are both attached to a cellular structure, transcription introduces positive supercoils into the template between S and R if S is downstream from R as illustrated in Figure 1a; negative supercoils would be introduced between S and R if S is upstream of R.

It is also plausible that the cellular milieu imposes a sufficiently large viscous drag on the transcriptional ensemble, which includes the polymerase, the nascent RNA, and RNA-associated proteins, so that the rotation of the DNA around its axis becomes energetically more favorable than rotating the entire ensemble around DNA (Liu and Wang 1987). In that case, as illustrated in Figure 1b, positive supercoiling of the DNA template ahead of R and negative supercoiling of the DNA template behind R would again occur as transcription proceeds.

Experimental data of *E. coli* (Wu et al. 1988) and yeast (Giaever and Wang 1988) support the notion that transcription is a major factor and under some conditions probably the predominant factor in the supercoiling of intracellular DNA. In contrast to supercoiling by bacterial gyrase, which progressively lowers the linking number of an entire topological domain, supercoiling by transcription involves no change in linking number, and equal numbers of positive and negative supercoils are simultaneously generated. The DNA topoisomerases may act differentially in relaxing the oppositely supercoiled regions and cause a net increase or decrease in the linking number of a topological domain.

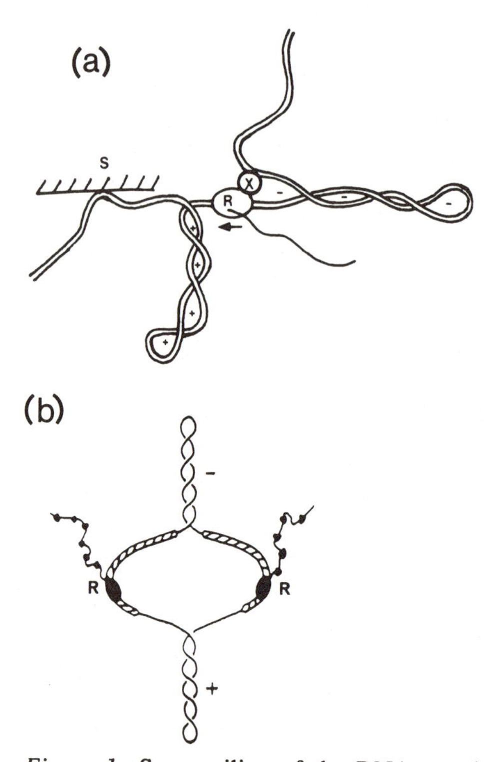

Figure 1 Supercoiling of the DNA template by transcription. Two plausible models are illustrated. (*a*) The transcription machinery R is assumed to be associated with an upstream DNA-bound factor X. As RNA synthesis commences, DNA is pulled through the polymerase, and it rotates as it translocates because of its helical geometry. A negatively supercoiled loop is thus formed in the region between R and the upstream factor X. Negative supercoiling of the template will also occur if R and an upstream point on the DNA are both in contact with a cellular structure (represented by S in the figure), and positive supercoiling will occur if R is in contact with a downstream point through a DNA-bound factor or if both R and the downstream point are attached to a cellular structure S. (*b*) It is assumed that the viscous drag against the rotation of the transcriptional ensemble around the DNA is greater than that against rotating the DNA around its helical axis. Two transcriptional ensembles oriented oppositely on a circular template are illustrated in the figure; positive supercoils are generated in the region ahead of the polymerases, and negative supercoils are generated in the region behind the polymerases. ([*a*] Reprinted, with permission, from Wang 1988; [*b*] reprinted, with permission, from Wu et al. 1988.)

Transcription is one of the most important processes but not the only one that affects DNA supercoiling. Translocations of other proteins along

DNA can similarly induce localized supercoiling (Liu and Wang 1978; Mizuuchi et al. 1978; Sugino et al. 1978), and the possibility was raised (Wang 1983) that DNA supercoiling might result from the movement of a multiprotein assembly, termed the primosome, that is involved in the initiation of DNA synthesis (Low et al. 1981). Electron microscopy also suggests that the ATP-dependent translocation of the type-I restriction enzyme *EcoK* may form a tightly supercoiled loop (Yuan et al. 1980).

Recently, it was shown that SV40 T antigen yielded a positively supercoiled DNA in the presence of *E. coli* DNA topoisomerase I. This indicates that the well-known helicase action of the protein may involve the formation of positively and negatively supercoiled domains as the protein tracks along DNA (Yang et al. 1989).

E. Roles of DNA Topoisomerases during Initiation

As discussed above, supercoiling of intracellular DNA may occur in both prokaryotes and eukaryotes. In eubacteria, such as *E. coli*, DNA gyrase appears to be required in the initiation of DNA replication. Presumably, the enzyme facilitates the initiation step by its negative supercoiling of DNA (Cozzarelli 1980; Gellert 1981; Vosberg 1985; Wang 1985). As described above, negative supercoiling of the DNA could facilitate the assembly of proteins that are involved in the initiation step, promote the unpairing of the strands either before the action of a helicase, such as *DnaB* protein, or jointly with such an activity; in the case of the plasmids colE1 and pBR322, negative supercoiling might also stabilize the initiator RNA-DNA hybrid.

In purified systems, the role of DNA gyrase during initiation appears to correlate with its DNA negative supercoiling activity. In the nine protein system for λ DNA replication (Mensa-Wilmot et al. 1989), for example, initiation and limited DNA synthesis was observed in the absence of DNA gyrase as long as a negative supercoiled DNA containing *oriλ* was used as the template. Curiously, the initiation of phage T4 DNA replication in vivo is delayed in T4 DNA topoisomerase mutants, although the phage enzyme, similar to eukaryotic DNA topoisomerase II, does not catalyze the negative supercoiling of DNA (see Huang, this volume). A plausible explanation of this observation is as follows. On the basis of the estimated rate of transcription and the turnover number of *E. coli* DNA gyrase, in regions undergoing rapid transcription, gyrase may be viewed more appropriately as an enzyme that removes positive supercoils generated by the translocating ensemble rather than as a negative supercoiling activity (Liu and Wang 1987). Such a role can be fulfilled by phage T4 DNA topoisomerase as well.

In purified systems, *E. coli* DNA topoisomerase I affects the specificity of initiation at *E. coli oriC*, and at plasmid pBR322 *ori*, probably because of its modulation of DNA supercoiling (Kaguni and Kornberg 1984; Minden and Marians 1985). In vivo, the deletion of the gene *topA* encoding DNA topoisomerase I suppresses a *dnaA* temperature-sensitive mutation *ts46*. Whether this suppression is due to the effect of Δ*topA* on the expression of *dnaA* or on the interaction between *oriC* and the mutant *DnaA* protein is not known.

In contrast with the dependence of the initiation of replication on DNA gyrase in bacteria, initiation of replication of most if not all yeast replicons does not seem to require the eukaryotic analog of gyrase; yeast DNA topoisomerase II is indispensable only during mitosis (for review, see Yanagida and Wang 1987). When both yeast DNA topoisomerase I and II are inactivated, nascent DNA chains in the size range of several kilobases are still being synthesized (Kim and Wang 1989).

F. Transcriptional Activation of DNA Replication

Transcriptional activation of *ori* is a phenomenon that was initially observed in the replication of phage λ (Dove et al. 1969). Transcription in the vicinity of *ori*λ is required for the initiation of replication. This requirement is responsible for the blockage of λ replication by λ repressor, which prevents transcription from promoters P_R and P_L located more than 1000 bp from *ori*λ. Whereas transcripts initiating at P_R pass through *ori*λ, mutations that create promoters near *ori*λ but direct transcription away from it, such as ri^c5b, are known to escape the block of replication by λ repressor (Furth et al. 1982). The coupling between transcription and initiation of replication is also well-established for *E. coli* (Lark 1972; Bagdasarian et al. 1977; Atlung 1984).

Several mechanisms are plausible for the activation of replication by transcription. A transcript passing through *ori* may form an RNA-DNA hybrid that, as exemplified by the replication of colE1, may serve as the primer after its cleavage by RNase H. Alternatively, as suggested by biochemical experiments with *oriC* plasmids, RNA-DNA hybrid formation may free a stretch of ssDNA and thus facilitate the assembly of an initiation complex (Baker and Kornberg 1988). The presence of a short single-stranded region near *oriC*, whether upstream of or downstream from the start of replicative DNA chain initiation, may, for example, permit the *DnaA*-mediated entry of the *DnaB* helicase. A third mechanism is that transcription near *ori* may affect the nucleoprotein structure in that region to facilitate initiation of replication. In vitro replication at *E. coli oriC* requires transcription when the histone-like protein HU is present at a relatively high level (Ogawa et al. 1985); the same is true for the replication of λ DNA in vitro (Mensa-Wilmot et al. 1989). How HU represses replicative initiation and how transcription counteracts the action of HU are unclear. For the purified *E. coli* replication system, however, it is known that, instead of adding HU, the lowering of the temperature or negative superhelicity also elicits the dependence of initiation on transcription. The common denominator of all these conditions appears to be the stabilization of the DNA double helix (Baker and Kornberg 1988).

Whether a common mechanism is responsible for transcriptional activation in various in vivo systems is unknown. Any process that affects the superhelicity in the *ori* region is expected to have a major effect on initiation, and it seems plausible that localized supercoiling by transcription may affect the initiation of replication in some cases. In this connection, it is noteworthy that genetic data point to the possibility that *E. coli* RNA polymerase might interact directly with proteins that are involved in the initiation of replication at *oriC* (Bagdasarian et al. 1977; Atlung 1984; Baker and Kornberg 1988) and *ori*λ (McKinney and Wechsler 1983). If this is true, RNA synthesis may strongly affect the topology of the DNA in between the binding sites of replication initiator proteins and promoters nearby (see Fig. 1a).

II. THE SWIVEL REQUIREMENT DURING REPLICATIVE DNA CHAIN ELONGATION

As a replication fork advances in a covalently closed DNA ring, the parental strands unwind, and the DNA becomes positively supercoiled. Eukaryotic DNA topoisomerase I and II and bacterial gyrase are known to remove positive supercoils readily. Bacterial DNA topoisomerase I, on the other hand, normally relaxes only negative supercoils (Wang 1971); positive supercoils are efficiently removed only if there is a single-stranded region in the DNA (Kirkegaard and Wang 1985). A second type-I DNA topoisomerase found in *E. coli,* DNA topoisomerase III, also relaxes negatively supercoiled DNA preferentially (Dean et al. 1983).

In a number of purified systems or cell-free extracts for the replication of DNA, DNA gyrase or one of the eukaryotic topoisomerases was shown to be sufficient for relieving the topological constraint during elongation. In *E. coli,* several temperature-sensitive mutants of DNA gyrase have been shown to block initiation and termination but not elongation (for review, see Gellert 1981; Wang 1985). However, rapid cessation of replication upon shifting the temperature to nonpermissive conditions was observed for one gyrase *ts* mutant (Filutowicz and Jonczk

1983); these investigators suggested that gyrase might be the sole swivelase during elongation in that particular strain. Whether *E. coli* DNA topoisomerase I or III generally participates in the elongation step is unknown. The existence of deletion mutants of DNA topoisomerase I shows that this enzyme is not necessary for replication. In vivo, DNA topoisomerase III appears to be inefficient in removing negative supercoils; intracellular λ DNA remains negatively supercoiled when DNA gyrase is inactivated in the absence of DNA topoisomerase I but in the presence of DNA topoisomerase III (Bliska and Cozzarelli 1987).

In two distantly related yeasts *S. cerevisiae* and *S. pombe*, DNA synthesis monitored by the incorporation of labeled uracil into DNA is much reduced by the inactivation of both topoisomerase I and II but not by the inactivation of either (Uemura and Yanagida 1986; Brill et al. 1987; Kim and Wang 1989). In synchronously or asynchronously grown *S. cerevisiae Δtop1 top2 ts* cells but not in *TOP1*$^{+}$ *top2 ts* or *Δtop1 TOP2*$^{+}$ cells, elongation of DNA strands larger than 5 kilonucleotides is blocked at the nonpermissive temperatures (Kim and Wang 1989). These results indicate that in eukaryotes either DNA topoisomerase I or II can fulfill the role of a swivel during the elongation step. Recently, a gene, *TOP3*, has been identified in *S. cerevisiae* that most likely encodes a second type-I DNA topoisomerase (Wallis et al. 1989). Mutations in this gene increase recombination between short repeated sequences, termed δ sequences, and show a reduction in growth rate. Interestingly, whereas *S. cerevisiae* and *E. coli* DNA topoisomerase I share no sequence homology, *S. cerevisiae TOP3* contains an open reading frame capable of encoding a protein homologous to *E. coli* DNA topoisomerase I. The poor growth phenotype of *top3* mutants but not the hyper-recombination phenotype is complemented by the expression of *E. coli topA* in yeast.

The role of the *TOP3* gene product in yeast DNA replication is unknown. Investigations of the linking numbers of plasmids in yeast topoisomerase mutants indicate that the major DNA relaxation activities in vivo are DNA topoisomerases I and II (Saavedra and Huberman 1986; Giaever and Wang 1988). Saavedra and Huberman (1986) showed that for yeast cells with active DNA topoisomerase I or II, the linking number of intracellular 2-μm DNA decreases upon increasing the temperature of the culture as one might expect from the temperature dependence of the helical twist of DNA. No linking number adjustment occurs upon a shift in temperature, however, if both enzymes are inactivated.

In an experimental test of the twin-supercoiled-domain model of transcription in yeast, Giaever and Wang (1988) have shown that positive supercoils accumulate in intracellular 2-μm DNA in the absence of yeast DNA topoisomerases I and II but in the presence of *E. coli* DNA topoisomerase I. This demonstrates that yeast DNA topoisomerase III can not be very efficient in removing the positive supercoils. Furthermore, because the accumulation of positive supercoils in this experiment is strictly dependent on the presence of the *E. coli* enzyme, one can deduce that no yeast DNA topoisomerase can specifically and efficiently remove negative supercoils in vivo; otherwise, the *E. coli* enzyme should be dispensable in this experiment.

In principle, a DNA topoisomerase can act either ahead of or behind the replication fork (Champoux and Been 1980). Avemann et al. (1988) suggested that DNA topoisomerase I might be preferentially present in the vicinity of the replication fork of SV40 DNA. When replicative intermediates of SV40 DNA in camptothecin-treated cells infected with the virus were examined by electron microscopy, it was found that the replication branches of a large fraction of DNA molecules were broken near the growth points, converting the θ-shaped molecules to σ-shaped ones.

The drug camptothecin acts on eukaryotic DNA topoisomerase I by converting the enzyme/DNA complex to an abortive one, which on exposure to a protein denaturant yields a covalent enzyme/DNA complex. A DNA strand is cleaved by the topoisomerase in the process, and the protein becomes covalently linked to the 3′-phosphoryl end of the broken DNA strand (see Champoux; Hsieh; Liu; all this volume). If camptothecin-induced broken ends of DNA are assumed to mark the positions of DNA topoisomerase I, then the high frequency of the broken replicative branches in the electron microscopy experiment just described would imply a preferential association of the topoisomerase with the replication fork. In the same series of experiments, nicks induced by DNA topoisomerase I were also detected in the unreplicated portions of the replicative forms of the viral DNA from camptothecin-treated cells.

It is plausible, however, that the high frequency of σ-shaped DNA molecules in the above example does not represent a preferential association of eukaryotic DNA topoisomerase I with the replication fork but is due instead to the formation of a broken branch when a replication fork moves toward a camptothecin-trapped topoisomerase molecule. In this view, the point where an enzyme molecule is entrapped resembles a single-stranded break on the replicating DNA.

Avemann et al. (1988) reported, however, that the addition of aphidicolin, which blocks replicative fork movement, has little effect on the formation of the σ-shaped DNA molecules. In mammalian cells (L.F. Liu, unpubl.) as well as in yeast (J. Nitiss and J.C. Wang, unpubl.), aphidicolin appears to prevent the killing of cells by camptothecin; one simple interpretation of these observations would be that aphidicolin prevents the camptothecin/DNA-topoisomerase-I-induced fork breakage. Further refinement of the experiments are needed to clarify whether the formation of the σ-shaped structures is related to fork movement and whether these structures are involved in the pathway of cell-killing by camptothecin.

III. THE TERMINATION STEP

A. Resolution of Interwrapped Duplex DNA Molecules following the Complete Unpairing of the Parental Strands

As mentioned above, when two converging replication forks merge, the complete unpairing of the stretch of unreplicated DNA segment in between may not coincide with the complete unraveling of the intertwined strands. A pair of newly replicated DNA rings may therefore end up in the form of a multiply-intertwined catenane, and their segregation would require decatenation by a topoisomerase. For linear chromosomes, the problem of unlinking the sibling molecules after a round of replication is basically the same as that of a ring because of the extreme length of a chromosome, the presence of multiple replicons, and the organization of a linear chromosome into loops.

Dimeric catenanes as replicative intermediates of DNA rings have been known for some time, and plausible enzymatic pathways for their resolution have been discussed previously (Gefter 1975; Sogo et al. 1976; for review, see Wasserman and Cozzarelli 1986). The findings that catenanes can be decatenated by a type-II (Kreuzer and Cozzarelli 1980; Liu et al. 1980) or type-I topoisomerase (Tse and Wang 1980; Brown and Cozzarelli 1981) points immediately to the plausible involvement of these enzymes in the segregation process.

Whereas both types of DNA topoisomerases can decatenate interlocked dsDNA rings, decatenation by a type-I enzyme requires the presence of a preexisting nick or gap in at least one of the pair of component rings in a dimeric catenane (Tse and Wang 1980; Brown and Cozzarelli 1981; Low et al. 1984). Therefore, once the component rings in a newly replicated catenane are converted to the covalently closed form, the type-II enzyme becomes the only activity that can decatenate them.

The mechanistic difference between the two types of DNA topoisomerases helps to distinguish two types of pathways for the segregation of the linked siblings. In the path shown in Figure 2, the complete unpairing of the parental strands is followed rapidly by the completion of the synthesis and covalent closure of the progeny strands, yielding a multiply-intertwined dimeric catenane with covalently closed component

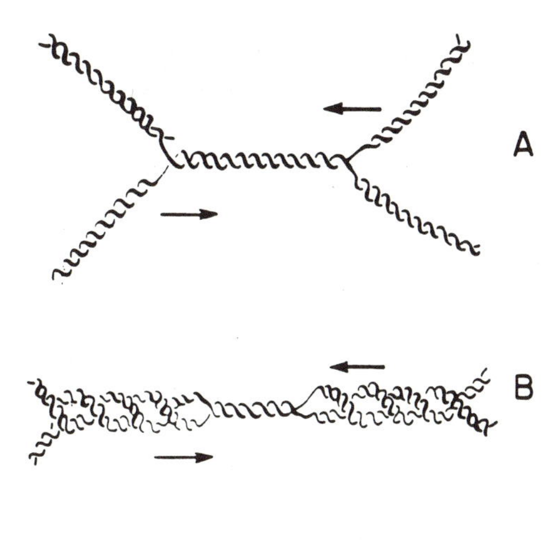

Figure 2 The formation of a pair of multiply-intertwined dsDNA molecules at the end of replication. As the two converging replication forks approach each other (*A*), the unraveling of the parental strands may be incomplete at the time when the parental strands have been completely unpaired and the synthesis of the progeny strands has been completed. In such a case, a pair of multiply-intertwined progeny molecules are formed (*C*). Because no single-stranded gap or nick is present in the progeny molecules, a type-II DNA is necessary for their separation. See the text for an alternative pathway. (Reprinted, with permission, from Varshavsky et al. 1983.)

rings. The final segregation process in this case requires the decatenation of the siblings by a type-II DNA topoisomerase. Alternatively, completion of the synthesis and covalent closure of the progeny strands may occur after the complete unpairing of the parental strands, yielding a pair of interlocked sibling molecules containing single-stranded gaps or nicks. In the latter case, decatenation of the gapped or nicked siblings can be accomplished by either a type-I or type-II topoisomerase.

The above discussion suggests that the relative rates of unpairing and unraveling of the parental strands prior to the merging of two converging forks and that of completion of progeny strand synthesis could determine the predominant pathway of final segregation. The pathway shown in Figure 2 was found to be the predominant one for SV40 DNA replication when cells infected with the virus were exposed to a hypertonic medium (Sundin and Varshavsky 1980, 1981). The covalently closed dimeric catenanes were shown to convert readily to separate monomeric closed circles, presumably through the action of DNA topoisomerase II, upon adjusting the medium to an isotonic one. Covalently closed dimeric catenanes were also found to accumulate in a cell-free system on the addition of VM26, an inhibitor of eukaryotic DNA topoisomerase II, or when the cell extract was immunologically depleted of DNA topoisomerase II (Yang et al. 1987).

Covalently closed yeast 2-μm plasmid catenanes were found to accumulate when cells of a temperature-sensitive DNA topoisomerase II mutant were heated to a nonpermissive temperature (DiNardo et al. 1984). Studies with both *S. cerevisiae* and *S. pombe* show that DNA topoisomerase II is essential during mitosis, which is consistent with the notion that the enzyme is required in the separation of intertwined chromosomal DNA siblings during mitosis (for review, see Yanagida and Wang 1987; see also Yanagida and Sternglanz, this volume). In *E. coli* it has also been shown that catenanes formed by λ integrase are readily decatenated by DNA gyrase in vivo (Bliska and Cozzarelli 1987).

The results cited above provide strong evidence that DNA topoisomerase II is involved in the final segregation of newly replicated siblings in a number of systems. It is plausible, however, that a type-I topoisomerase is also normally involved. In SV40-infected cells grown in isotonic media, for example, the predominant species of newly replicated DNA was found to be gapped nonomers, indicating that DNA topoisomerase II may not be obligatory in the decatenation reaction (Weaver et al. 1985). In a purified system for the replication of pBR322, it was shown that *E. coli* DNA topoisomerase I could segregate the replicating progeny molecules when the enzyme is present at a high concentration (Minden and Marians 1986). *E. coli* DNA topoisomerase III, a type-I topoisomerase, has been found to be efficient in the final segregation of newly replicated pBR322 in a purified system (DiGate and Marians 1988). Replication intermediates in the form of dimeric catenanes with at least one nicked or gapped component ring were also observed in a number of other prokaryotic and eukaryotic systems (for example, see Sogo et al. 1976 and references therein).

B. Unraveling of the Last Stretch of Unreplicated DNA Segment

In the above section, the discussion was focused on the events following the complete unpairing of the parental strands between the two converging forks. It appears that as the converging forks are near each other, the unraveling of the intertwined parental strands between them becomes more difficult. When SV40-infected cells are exposed to hypertonic conditions, late Cairns structures in which a short segment of the parental DNA remains intact accumulate in addition to the covalently closed catenated dimers (Weaver et al. 1985). Similarly, inhibition of SV40-infected cells by an inhibitor of eukaryotic DNA topoisomerase II (VM-26) also leads to the accumulation of the late Cairns structures (Richter et al. 1987). These results suggest a role for DNA topoisomerase II in the maturation of the late replicative intermediates. The kinetic barrier in the unraveling of the last stretch of unreplicated DNA segment might be due to the inaccessibility of the intact parental duplex to DNA topoisomerase I when the merging forks are very close to each other.

V. CONCLUDING REMARKS

For the faithful preservation of genetic information, nature apparently chose to minimize the number of breaks in DNA, and a vast amount of information is usually contained in a small number of chromosomes of extreme lengths. Yet, to separate the intertwined strands during replication, breaks must be frequently introduced into them. The DNA topoisomerases appear to be the best solution to this dilemma because the breaks they introduce are transient and are protected by the enzymes. Biochemical and genetic studies in the past two decades have demonstrated the importance of DNA topology in the various steps of the replication cycle and established the roles of the DNA topoisomerases in them.

Although much progress has been made in understanding the roles of the DNA topoisomerases in replication, many questions remain. For example, it is unclear whether an intrinsic safety device is built into the topoisomerases, so that a transient break in a DNA strand generated by an enzyme does not become a permanent one. How is the accidental formation of a break in one or both of the DNA strands prevented when a replication fork encounters a type-I or type-II topoisomerase ahead of it? As described earlier in this chapter and in Liu (this volume), there is evidence that, in the presence of drugs that interfere with the DNA-strand-rejoining activity of the topoisomerases, DNA replication may indeed cause DNA breakage. How do the enzymes normally avoid such disasters? As another example, we have little information regarding the coordination of DNA topoisomerases with other proteins involved in replication. Large assemblies containing DNA topoisomerase and other activities involved in replication have been reported previously (Noguchi et al. 1983), but the functional and mechanistic aspects of such associations remain unknown.

In a paper entitled, "Genetic Implications of the Structure of Deoxyribonucleic Acid," Watson and Crick (1953) stated, "Since the two chains in our model are intertwined, it is essential for them to untwist if they are to separate. . . . It is well known from microscopic observation that much coiling and uncoiling occurs during mitosis, and although it is on a much larger scale, it probably reflects similar processes on a molecular level. Although it is difficult at the moment to see how these processes occur without everything getting tangled, we do not feel that this objection will be insuperable." Indeed, through the invention of DNA topoisomerases, nature has found the tools it needs to solve the entanglement problem of the genetic material.

REFERENCES

Atlung, T. 1984. Allele-specific suppression of *dnaA*(ts) mutations by *rpoB* mutations in *Escherichia coli. Mol. Gen. Genet.* **197:** 125–128.

Avemann, K., R. Knippers, T. Koller, and J.M. Sogo. 1988. Camptothecin, a specific inhibitor of type I DNA topoisomerase, induces DNA breakage at replication forks. *Mol. Cell. Biol.* **8:** 3026–3034.

Bagdasarian, M.M., M. Izakowska, and M. Bagdasarian. 1977. Suppression of the *DnaA* phenotype by mutations in the *rpoB* cistron of ribonucleic acid polymerase in *Salmonella typhimurium* and *Escherichia coli. J. Bacteriol.* **130:** 577–582.

Baker, T.A. and A. Kornberg. 1988. Transcriptional activation of initiation of replication from the *E. coli* chromosomal origin: An RNA-DNA hybrid near *oriC*. *Cell* **55:** 113–123.

Baker, T.A., B.E. Funnel, and A. Kornberg. 1987. Helicase action of *dnaB* protein during replication from the *Escherichia coli* chromosomal origin *in vitro*. *J. Biol. Chem.* **262:** 6877–6885.

Baker, T.A., K. Sekimizu, B.E. Funnell, and A. Kornberg. 1986. Extensive unwinding of the plasmid template during staged enzymatic initiation of DNA replication from the origin of the *Escherichia coli* chromosome. *Cell* **45:** 53–64.

Bauer, W. and J. Vinograd. 1970. Interaction of closed circular DNA with intercalative dyes. II. The free energy of superhelix formation in SV40 DNA. *J. Mol. Biol.* **47:** 419–435.

Bliska, J.B. and N.R. Cozzarelli. 1987. Use of site-specific recombination as a probe of DNA structure and metabolism *in vivo*. *J. Mol. Biol.* **205:** 205–218.

Bramhill, D. and A. Kornberg. 1988. Duplex opening by *dnaA* protein at novel sequences in initiation of replication at the origin of the *E.* chromosome. *Cell* **52:** 743–755.

Brill, S.J., S. DiNardo, K. Voelkel-Meiman, and R. Sternglanz. 1987. Need for DNA topoisomerase activity as a swivel for DNA replication for transcription of ribosomal RNA. *Nature* **326:** 414–416.

Brown, P.O. and N.R. Cozzarelli. 1981. Catenation and knotting of duplex DNA by type I topoisomerase: A mechanism parallel with type 2 topoisomerases. *Proc. Natl. Acad. Sci.* **78:** 843–847.

Cairns, J. 1963. The bacterial chromosome and its manner of replication as seen by autoradiography. *J. Mol. Biol.* **6:** 208–213.

———. 1964. The chromosome of *Escherichia coli. Cold Spring Harbor Symp. Quant. Biol.* **28:** 43–45.

Cairns, J. and C.I. Davern. 1967. The mechanism of DNA replication in bacteria. *J. Cell. Physiol.* (suppl. 1) **70:** 65.

Champoux, J.J. and M.D. Been. 1980. Topoisomerases and the swivel problem. *ICN-UCLA Symp. Mol. Cell. Biol.* **19:** 809–815.

Cozzarelli, N.R. 1980. DNA gyrase and the supercoiling of DNA. *Science* **207:** 953–960.

Davidson, N. 1972. Effect of DNA length on the free energy of binding of an unwinding ligand to a superhelical DNA. *J. Mol. Biol.* **66:** 307.

Dean, F.B., J.A. Borowiec, Y. Ishimi, S. Deb, P. Tegtmyer, and J. Hurwitz. 1987a. Simian virus 40 large tumor antigen requires three core replication origin domains for DNA unwinding and replication in vitro. *Proc. Natl. Acad. Sci.* **84:** 8267–8271.

Dean, F.B., P. Bullock, Y. Murakami, C.R. Wobbe, L. Weissback, and J. Hurwitz. 1987b. Simian virus 40 (SV40) DNA replication: SV40 large T antigen unwinds DNA containing the SV40 origin of replication. *Proc. Natl. Acad. Sci.* **84:** 16–20.

Dean, F.B., M.A. Krasnow, R. Otter, M.M. Matzuk, S.J. Spengler, and N.R. Cozzarelli. 1983. *Escherichia coli* type I topoisomerases: Identification, mechanism, and role in recombination. *Cold Spring Harbor Symp. Quant. Biol.* **47:** 769–778.

Delbruck, M. 1954. On the replication of deoxyribonucleic acid (DNA). *Proc. Natl. Acad. Sci.* **40:** 783–788.

Delbruck, M. and G.S. Stent. 1957. On the mechanism of DNA replication. In *The chemical basis of heredity* (ed. W.D. McElroy and B. Glass), pp. 699–736. The Johns Hopkins Press, Baltimore.

DiGate, R.J. and K.J. Marians. 1988. Identification of a potent decatenating enzyme from *Escherichia coli. J. Biol. Chem.* **263:** 13366–13373.

DiNardo, S., K. Voelkel, and R. Sternglanz. 1984. DNA topoisomerase II mutant of *Saccharomyces cerevisiae:* Topoisomerase II is required for segregation of daughter molecules at the termination of DNA replication. *Proc. Natl. Acad. Sci.* **81:** 2616–2620.

Dodson, M., J. Roberts, R. McMacken, and H. Echols. 1985. Specialized nucleoprotein structures at the origin of replication of bacteriophage λ: Complexes with λ O protein and with λ O, lambda P, and *Escherichia coli dnaB* proteins. *Proc. Natl. Acad. Sci.* **82:** 4678–4682.

Dodson, M., F.B. Dean, P. Bullock, H. Echols, and J. Hurwitz. 1987. Unwinding of duplex DNA from the SV40 origin of replication by T antigen. *Science* **238:** 964–967.

Dodson, M., H. Echols, S. Wickner, C. Alfano, K. Mensa-Wilmot, B. Gomes, J. LeBowitz, J.D. Roberts, and R. McMacken. 1986. Specialized nucleoprotein structures at the origin of replication of bacteriophage λ: Localized unwinding of duplex DNA by a six-protein reaction. *Proc. Natl. Acad. Sci.* **83:** 7638–7642.

Dove, W.F., E. Hargrove, M. Ohashi, F. Haugli, and A. Guha. 1969. Replicator activation in λ. *Jpn. J. Genet.* (suppl.) **44:** 11–12.

Filutowicz, M. and T. Jonczk. 1983. The *gyrB* gene product functions in both initiation and chain elongation of *E. coli* chromosome replication; suppression of the initiation deficiency in *gyrB* ts mutants by a class of *rpoB* mutations. *Mol. Gen. Genet.* **191:** 282–287.

Fuller, R.S. and A. Kornberg. 1983. Purified *dnaA* protein in initiation of replication at the *Escherichia coli* chromosomal origin of replication. *Proc. Natl. Acad. Sci.* **80:** 5817–5821.

Furth, M., W. Dove, and B. Meyer. 1982. Specificity determinants for bacteriophage λ DNA replication. III. Activation of replication in λ ri^c mutants by transcription outside of *ori. J. Mol. Biol.* **154:** 65–83.

Gefter, M.L. 1975. DNA replication. *Annu. Rev. Biochem.* **44:** 45–78.

Gellert, M. 1981. DNA topoisomerases. *Annu. Rev. Biochem.* **50:** 879–910.

Gellert, M. and H. Nash. 1987. Communication between segments of DNA during site-specific recombination. *Nature* **325:** 401–404.

Giaever, G.N. and J.C. Wang. 1988. Supercoiling of intracellular DNA can occur in eukaryotic cells. *Cell* **55:** 849–856.

Gross, D.S. and W.T. Garrard. 1988. Nuclease hypersensitive sites in chromatin. *Annu. Rev. Biochem.* **57:** 159–197.

Hsieh, T.S. and J.C. Wang. 1975. Thermodynamic properties of superhelical DNAs. *Biochemistry* **14:** 527–535.

Itoh, H. and J. Tomizawa. 1980. Formation of an RNA primer for initiation of replication of ColE1 DNA by ribonuclease H. *Proc. Natl. Acad. Sci.* **77:** 2450–2454.

Kaguni, J.M. and A. Kornberg. 1984. Topoisomerase I confers specificity in enzymatic replication of the *Escherichia coli* chromosomal origin. *J. Biol. Chem.* **259:** 8578–8583.

Kelly, T.J. 1988. SV40 DNA replication. *J. Biol. Chem.* **263:** 17889–17892.

Kim, R.A. and J.C. Wang. 1989. Function of DNA topoisomerases as replication swivels in *Saccharomyces cerevisiae. J. Mol. Biol.* **208:** 257–267.

Kirkegaard, K. and J.C. Wang. 1985. Bacterial DNA topoisomerase I can relax positively supercoiled DNA containing a single-stranded loop. *J. Mol. Biol.* **185:** 625–637.

Kirkegaard, K., H. Buc, A. Spassky, and J. Wang. 1983. Mapping of single-stranded regions in duplex DNA at the sequence level: Single-strand specific cytosine methylation in RNA polymerase-promoter complexes. *Proc. Natl. Acad. Sci.* **80:** 2544–2548.

Kornberg, A. 1980. *DNA replication.* W.H. Freeman, San Francisco.

———. 1982. *DNA replication*, supplement. W.H. Freeman, San Francisco.

———. 1988a. DNA replication. *Biochim. Biophys. Acta* **951:** 235–239.

———. 1988b. DNA replication. *J. Biol. Chem.* **263:** 1–4.

Kornberg, A. and T.A. Baker. 1990. *DNA replication.* W.H. Freeman, San Francisco. (In press.)

Kreuzer, K.N. and N.R. Cozzarelli. 1980. Formation and resolution of catenanes by DNA gyrase. *Cell* **20:** 245–254.

Lark, K.D. 1972. Evidence for the direct involvement of RNA in the initiation of DNA replication in *E. coli* 15T⁻. *J. Mol. Biol.* **64:** 47–60.

Liu, L.F. and J.C. Wang. 1978. *Micrococcus luteus* gyrase: Active components and a model for its supercoiling of DNA. *Proc. Natl. Acad. Sci.* **75:** 2098–2102.

———. 1987. Supercoiling of the DNA template during transcription. *Proc. Natl. Acad. Sci.* **84:** 7024–7027.

Liu, L.F., C.C. Liu, and B.M. Alberts. 1980. Type II DNA topoisomerases. Enzymes that can unknot a topologically knotted DNA molecule via a reversible double-strand break. *Cell* **19:** 697–707.

Low, R.L., K. Arai, and A. Kornberg. 1981. Conservation of the primosome in successive stages of ϕX174 DNA replication. *Proc. Natl. Acad. Sci.* **78:** 1436–1440.

Low, R.L., J.M. Kaguni, and A. Kornberg. 1984. Potent catenation of supercoiled and gapped DNA circles by topoisomerase I in the presence of a hydrophilic polymer. *J. Biol. Chem.* **259:** 4576–4581.

McKinney, M.D. and J.A. Wechsler. 1983. RNA polymerase interaction with *dnaB* protein and λ P protein during λ replication. *J. Virol.* **48:** 551–554.

McMacken, R. and T.J. Kelly. 1986. DNA replication and recombination. *UCLA Symp. Mol. Cell. Biol. New Ser.* **47.**

Mensa-Wilmot, K., R. Seaby, C. Alfano, M.S. Wold, B. Gomes, and R. McMacken.

1989. Reconstitution of a nine-protein system that initiates bacteriophage lambda DNA replication. *J. Biol. Chem.* **264:** 2853–2861.

Minden, J.S. and K.J. Marians. 1985. Replication of pBR322 DNA *in vitro* with purified proteins: Requirement for topoisomerase I in the maintenance of template specificity. *J. Biol. Chem.* **260:** 9316–9325.

———. 1986. *Escherichia coli* topoisomerase I can segregate replicating pBR322 daughter DNA molecules *in vitro*. *J. Biol. Chem.* **261:** 11906–11917.

Mizuuchi, K., M.H. O'Dea, and M. Gellert. 1978. DNA gyrase: Subunit structure and ATPase activity of the purified enzyme. *Proc. Natl. Acad. Sci.* **75:** 5960–5963.

Noguchi, H., G.R. veer Reddy, and A.B. Pardee. 1983. Rapid incorporation of label from ribonucleoside diphosphates into DNA by a cell-free high molecular weight fraction from animal cell nuclei. *Cell* **32:** 443–451.

Nossal, N. 1983. Prokaryotic DNA replication systems. *Annu. Rev. Biochem.* **52:** 581–615.

Ogawa, T., T.A. Baker, A. van der Ende, and A. Kornberg. 1985. Initiation of enzymatic replication at the origin of the *Escherichia coli* chromosome: Contributions of RNA polymerase and primase. *Proc. Natl. Acad. Sci.* **82:** 3562–3566.

Richter, A., U. Strausfeld, and Knippers, R. 1987. Effects of VM-26 (teniposide), a specific inhibitor of type II DNA topoisomerase, on SV40 replication *in vivo*. *Nucleic Acids Res.* **15:** 3455–3478.

Saavedra, R.A. and J.A. Huberman. 1986. Both DNA topoisomerases and II relax 2 mm plasmid DNA in living yeast cells. *Cell* **15:** 773–783.

Schnos, M., K. Zahn, R.B. Inman, and F.R. Blattner. 1988. Initiation protein induced helix destabilization at the λ origin: A prepriming step in DNA replication. *Cell* **52:** 385–395.

Sogo, J.M., M. Greenstein, and A. Skalka. 1976. The circle mode of replication of bacteriophage lambda: The role of covalently closed templates and the formation of mixed catenated dimers. *J. Mol. Biol.* **103:** 537–562.

Stahl, H. and R. Knippers. 1987. The simian virus 40 large tumor antigen. *Biochim. Biophys. Acta* **910:** 1–10.

Sugino, A., N.P. Higgins, P.O. Brown, C.L. Peebles, and N.R. Cozzarelli. 1978. Energy coupling in DNA gyrase and the mechanism of action of novobiocin. *Proc. Natl. Acad. Sci.* **75:** 4838–4842.

Sundin, O. and A. Varshavsky. 1980. Terminal stages of SV40 DNA replication proceed via multiply intertwined catenated dimers. *Cell* **21:** 103–114.

———. 1981. Arrest of segregation leads to accumulation of highly intertwined catenated dimers: Dissection of the final stages of SV40 DNA replication. *Cell* **25:** 659–669.

Tse, Y.C. and J.C. Wang. 1980. *E. coli* and *M. luteus* DNA topoisomerase I can catalyze catenation or decatenation of double-stranded DNA rings. *Cell* **22:** 269–276.

Uemura, T. and M. Yanagida. 1986. Mitotic spindle pulls but fails to separate chromosomes in type II DNA topoisomerase mutants, uncoordinated mitosis. *EMBO J.* **5:** 1003–1010.

Umek, R.M. and D. Kowalski. 1987. Yeast regulatory sequences preferentially adopt a non-B conformation in supercoiled DNA. *Nucleic Acids Res.* **15:** 4467–4480.

———. 1988. The ease of DNA unwinding as a determinant of initiation at yeast replication origins. *Cell* **52:** 559–567.

Varshavsky, A., L. Levinger, O. Sundin, J. Barsoum, E. Ozkaynak, P. Swerdlow, and D. Finley. 1983. Cellular and SV40 chromatin: Replication, segregation, ubiquitination, nuclease-hypersensitive sites, HMG-containing nucleosomes, and heterochromatin-specific protein. *Cold Spring Harbor Symp. Quant. Biol.* **47:** 511–528.

Vinograd, J., J. Lebowitz, and R. Watson. 1968. Early and late helix-coil transitions in closed circular DNA: The number of superhelical turns in polyoma DNA. *J. Mol. Biol.* **33:** 173–197.

Vosberg, H.P. 1985. DNA topoisomerases: Enzymes that control DNA conformation. *Curr. Top. Microbiol. Immunol.* **114:** 19–102.

Wallis, J.W., G. Chrebet, G. Brodsky, M. Rolfe, and R. Rothstein. 1989. A hyper-recombination mutation in *S. cerevisiae* identifies a novel eukaryotic topoisomerase. *Cell* **58:** 409–419.

Wang, J.C. 1971. Interaction between DNA and *Escherichia coli* protein ω. *J. Mol. Biol.* **55:** 523–533.

———. 1983. DNA supercoiling: Structural effects and biological consequences. In *Genetic rearrangement: Biological consequences of DNA structure and genome arrangement* (ed. K.F. Chater et al.), pp. 1–26. Croom Helm, London.

———. 1985. DNA topoisomerases. *Annu. Rev. Biochem.* **54:** 665–697.

———. 1987. Recent studies of DNA topoisomerases. *Biochim. Biophys. Acta* **909:** 1–9.

———. 1988. DNA topoisomerases: Nature's solution to the topological ramifications of the double-helix structure of DNA. *Harvey Lect.* **81:** 93–110.

Wang, J.C. and G.N. Giaever. 1988. Action at a distance along a DNA. *Science* **240:** 300–304.

Wang, J.C., J.H. Jacobsen, and J.-M. Saucier. 1977. Physicochemical studies on interactions between DNA and RNA polymerase. Unwinding of the DNA helix by *Escherichia coli* RNA polymerase. *Nucleic Acids Res.* **4:** 1225–1241.

Wang, J.C., L.J. Peck, and K. Becherer. 1983. DNA supercoiling and its effects on DNA structure and function. *Cold Spring Harbor Symp. Quant. Biol.* **47:** 85–91.

Wasserman, S.A. and N.R. Cozzarelli. 1986. Biochemical topology: Application to DNA recombination and replication. *Science* **232:** 951–960.

Watson, J.D. and F.H.C. Crick. 1953. Genetic implications of the structure of deoxyribonucleic acid. *Nature* **171:** 964–967.

Weaver, D.T., S.C. Fields-Berry, and M.L. DePamphilis. 1985. The termination region for SV40 DNA replication directs the mode of separation for the two sibling molecules. *Cell* **41:** 565–575.

Wold, M.S., J.J. Li, and T.J. Kelly. 1987. Initiation of simian virus 40 DNA replication in vitro: Large-tumor-antigen- and origin-dependent unwinding of the template. *Proc. Natl. Acad. Sci.* **84:** 3643–3647.

Wu, H.-Y., S.H. Shyy, J.C. Wang, and L.F. Liu. 1988. RNA transcription generates negatively and positively supercoiled domains in the template. *Cell* **53:** 433–440.

Yanagida, M. and J.C. Wang. 1987. Yeast DNA topoisomerase and their structural genes. *Nucleic Acids Mol. Biol.* **1:** 196–209.

Yang, L., B.C. Jessee, K. Lau, H. Zhang, and L.F. Liu. 1989. Template supercoiling during ATP-dependent DNA helix tracking: Studies with simian virus 40 large tumor antigen. *Proc. Natl. Acad. Sci.* **86:** 6121–6125.

Yang, L., M.S. Wold, J.J. Li, T.J. Kelly, and L.F. Liu. 1987. Roles of DNA topoisomerases in simian virus SV40 DNA replication in vitro. *Proc. Natl. Acad. Sci.* **84:** 950–954.

Yuan, R., D.L. Hamilton, and J. Burckhardt. 1980. DNA translocation by the restriction enzyme *E. coli* K. *Cell* **20:** 237–244.

12

DNA Topoisomerase-mediated Illegitimate Recombination

Hideo Ikeda
The Institute of Medical Science
The University of Tokyo
P.O. Takanawa, Tokyo 108, Japan

Illegitimate recombination can be defined as a DNA rearrangement between nonhomologous and nonspecific sequences (Campbell 1962; Campbell 1971; Franklin 1971; Weisberg and Adhya 1977). It can generate deletions, duplications, insertions, substitutions, inversions, and transducing phage formation, and generally yields a novel junction at the site of crossover. This type of recombination is often found in bacteriophages, plasmids, and bacterial cells but is particularly frequent in mammalian somatic cells.

I. EXAMPLES OF ILLEGITIMATE RECOMBINATION

Deletions are the loss of a DNA segment from a chromosome. They are often found among spontaneous and induced mutations (Benzer 1961; Tessman 1962; Schwartz and Beckwith 1969). Deletion formation is usually independent of the key gene for homologous recombination, *recA*, in *Escherichia coli* (Franklin 1967; Inselburg 1967). A number of other *E. coli* genes involved in recombination and DNA repair—*recB*, *recC*, *uvrA*, *uvrB*, *uvrC*, *lig*, and *endA*—also do not seem to play a role in forming deletions (Anderson 1970; Spudich et al. 1970; see also Franklin 1971). Mutations in the DNA polymerase-I structural gene (*polA*) in-

DNA Topology and Its Biological Effects
 0-87969-348-7/90 $1.00 + 00

crease the frequency of deletions about 30-fold in the *tonB-trp* region of *E. coli* chromosome (Coukell and Yanofsky 1970).

Specialized transducing phages contain a portion of the bacterial genome that is contiguous to the lysogenic phage (Campbell 1962). For example, λdg transducing phages contain genes for galactose metabolism and are produced by abnormal excision of the λ prophage from the *E. coli* chromosome. This type of excision is independent of the phage *int* and *xis* site-specific recombination functions as well as host *recA* function (Gottesman and Yarmolinsky 1969). The crossovers occur at random sites in the phage and bacterial genomes and represent one kind of illegitimate recombination (Campbell 1971; Franklin 1971; Enquist and Weisberg 1977).

Various types of duplication mutants are formed in phage and bacteria (Anderson and Roth 1977). The simplest type, tandem duplication, is thought to be produced by an illegitimate recombination event between various sites of phage genomes. The tandem direct repeats are produced independently of host *recA* and λ *red* homologous recombination functions (Emmons et al. 1975).

There are many examples of illegitimate recombination in mammalian systems. SV40 and polyoma DNA are integrated into or excised from host chromosomes by illegitimate recombination (Gutai and Nathans 1978; Botchan et al. 1980; Stringer 1982; Ruley and Fried 1983; Smith and Berg 1984; Lin et al. 1985). DNA molecules transferred into mammalian cells undergo a variety of rearrangements such as deletion or insertion (Wilson et al. 1982; Calos et al. 1983; Razzaque et al. 1983; Subramani and Berg 1983). Secondary or more complicated rearrangements are often observed near these rearranged regions (Botchan et al. 1980; Stringer 1982). These features resemble those observed in the illegitimate recombinations of bacterial and phage systems. The functions involved in most examples of illegitimate recombination are not yet understood. It has recently been shown that some types of illegitimate recombination are mediated by DNA topoisomerases. In this chapter, the current status of in vivo and in vitro studies of illegitimate recombination mediated by DNA topoisomerases will be reviewed.

DNA topoisomerases are classified into two groups. In brief, type-I topoisomerases relax supercoiling by changing the linking number in steps of one. They have been found in bacteria (Wang 1971), the nuclei and mitochondria of eukaryotic cells (Champoux and Dulbecco 1972; Fairfield et al. 1979), and vaccinia virus particles (Bauer et al. 1977). Type-II topoisomerases mediate interconversion of the supercoiled and relaxed forms of DNA by changing the linking number in steps of two. They include bacterial DNA gyrase (Gellert et al. 1976), T4 DNA topoisomerase (Liu et al. 1979; Stetler et al. 1979), and eukaryotic enzymes (Miller et al. 1981; Goto and Wang 1982; Sander and Hsieh 1983; Shelton et al. 1983; Halligan et al. 1985).

II. DNA STRAND-JOINING MEDIATED BY TYPE-I DNA TOPOISOMERASES

It is known that some of type-I topoisomerases are able to promote site-specific recombination efficiently. Int protein of bacteriophage λ, a type-I topoisomerase, is involved in the integration of the phage genome into a specific site in the *E. coli* chromosome (Kikuchi and Nash 1979; Nash et al. 1981). The resolvase of the Tn*3* family of transposons also has a type-I topoisomerase activity and catalyzes recombination between two copies of the specific recombination site, *res,* of the transposon (Reed and Grindley 1981; Krasnow and Cozzarelli 1983).

Illegitimate recombination mediated by topoisomerase I can also occur near a replication origin of several filamentous ssDNA phages (M13, f1, fd) (Horiuchi et al. 1979; Shaller 1979). Michel and Ehrlich (1986a,b) have studied the formation of deletions in *E. coli* with hybrids composed of phage M13 and plasmid pHV33. Most of deletion endpoints were at the position of the nicks introduced into the M13 replication origin by the phage gene-II protein, a type-I topoisomerase. A similar recombination takes place at a preferred site in the transfer origin of the F factor, *oriT* (Horowitz and Deonier 1985). Illegitimate recombination between pBR322 and the vegetative replication origin region of the F factor also takes place independently of *recA* function (O'Conner et al. 1986).

The mammalian DNA topoisomerase I catalyzes the in vitro ligation of nonhomologous DNA fragments lacking any sequence homology or complementarity (Been and Champoux 1981; Halligan et al. 1982). This enzyme can cleave single-stranded DNA (ssDNA) and remains covalently bound to the 3′-phosphoryl end. Ordinarily, the break is resealed exactly, but if the broken ends are not held together by a complementary region of DNA, the covalently bound DNA can be joined to the 5′-OH end of ssDNA or double-stranded DNA (dsDNA) molecules with the release of the enzyme (Fig. 1). It is important to determine directly whether the topoisomerase I can promote recombination between two dsDNAs or can produce a biologically active recombinant DNA molecule.

Bullock et al. (1984) showed in an in vivo experiment that excision of a SV40 or polyoma provirus genome from the host chromosome takes place by crossover between two different sites of host and/or viral sequences that have 2- to 3-bp homologies. They pointed out that the sequences of the crossover sites are similar to the topoisomerase I con-

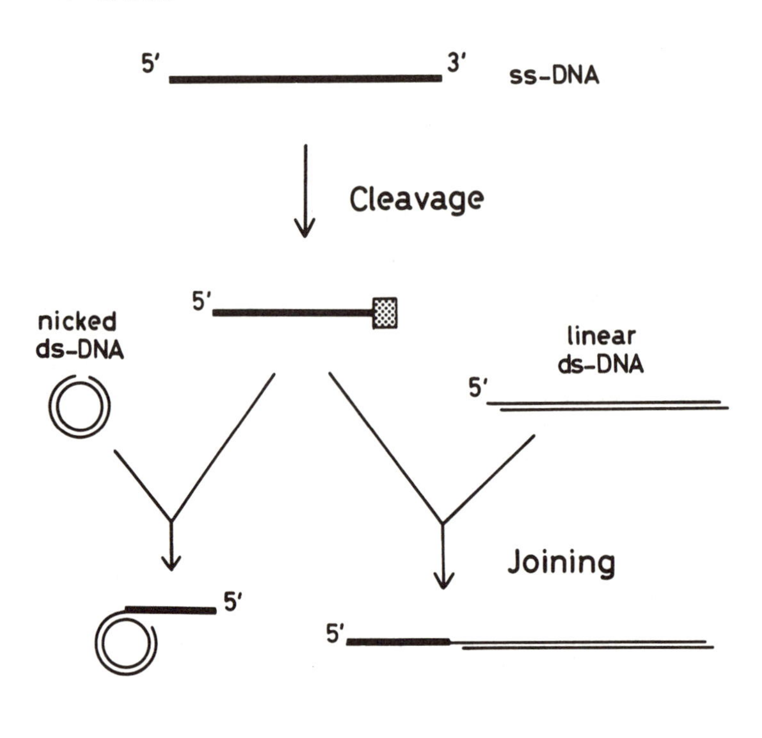

Figure 1 Model for strand-joining mediated by DNA topoisomerase I. DNA topoisomerase I cleaves ssDNA, resulting in the formation of a protein/DNA complex with the active protein linked to the 3′-end of the broken strand. The complex then reacts with either linear dsDNA or nicked circular DNA possessing a 5′-OH end, leading to the intermolecular strand transfer. (Reprinted, with permission, from Halligan et al. 1982.)

sensus cleavage site, 5′-PyrT↓T-3′ (Edwards et al. 1982). Direct comparison with the preferred topoisomerase I cleavage site in vitro showed a high correlation with the crossover-site sequences (Bullock et al. 1985). These data imply that excision of viral DNA is mediated by DNA topoisomerase I. Konopka (1988) compiled DNA sequences of illegitimate crossover regions in mammalian somatic cells and found that many crossover regions contain trinucleotides that are preferentially cleaved by mammalian topoisomerase I in vitro. Direct evidence for the involvement of eukaryotic topoisomerase I in illegitimate recombination was provided by the finding that vaccinia virus topoisomerase I expressed in *E. coli* cells promotes the *int*-independent prophage excision (Shuman 1989).

III. ILLEGITIMATE RECOMBINATION MEDIATED BY *E. COLI* DNA GYRASE

DNA gyrase is a type-II topoisomerase that introduces or removes negative supercoils, forms or resolves catenanes, and knots or unknots DNA (Gellert et al. 1976; Kreuzer and Cozzarelli 1980; Liu et al. 1980). The ability of DNA gyrase to make a double-stranded break plays an essential role in these topological alterations of the DNA (Gellert et al. 1977; Sugino et al. 1977; Brown and Cozzarelli 1979).

We have developed an in vitro system for the analysis of illegitimate recombination (Ikeda et al. 1981). It consists of an extract of induced λ lysogens of *E. coli* that were initially designed for the in vitro packaging of phage λ DNA. This packaging mixture contains a large amount of concatemeric λ DNA that is a good substrate for packaging. When a plasmid DNA that does not contain any sequence homologous to λ DNA, but contains the ampicillin-resistant (Ap^r) gene, is incubated with this mixture, the plasmid recombined with the endogeneous λ DNA. The resulting product was subsequently packaged into a phage particle and yielded ampicillin-resistant transducing phages at a frequency of approximately 3×10^{-7} per total plaque-forming units (pfu). These hybrid phages were formed by an insertion of the plasmid into λ DNA or by a substitution of a λ DNA segment by plasmid DNA. The sites of the crossovers seemed to be random, implying that the crossovers take place by illegitimate recombination.

The reaction takes place independently of both bacterial *recA* function as well as phage *int* and *red* functions. Oxolinic acid (45 μg/ml), an inhibitor of DNA gyrase that stabilizes an enzyme/DNA complex (Gellert et al. 1977; Sugino et al. 1977), stimulated recombination 13-fold. This is a specific effect on DNA gyrase because it had no effect on recombination by a lysate from a strain carrying an oxolinic-acid-resistant mutation in the *gyrA* gene. Coumermycin, a different type of gyrase inhibitor that blocks ATP access to the enzyme (Mizuuchi et al. 1978; Sugino et al. 1978), blocks oxolinic-acid-induced recombination. It does not affect recombination in a lysate from a mutant carrying a coumermycin-resistant mutation in the *gyrB* gene. If recombination merely required gyrase inhibition, then both drugs should have increased recombination. Hence, coumermycin should have, if anything, enhanced rather than blocked the induction of recombination by oxolinic acid. These results lead to the conclusion that oxolinic-acid-induced recombination is mediated directly by DNA gyrase.

Additional evidence for the involvement of DNA gyrase in illegitimate recombination comes from a modified in vitro system in which recombination is separated from the packaging step (Ikeda and

Shiozaki 1984). In the first step, purified λ DNA is incubated in an extract from *E. coli* cells possessing plasmid pKC16, which contains replication protein genes, the replication origin of phage λ, and Ap^r. The recombinant DNA produced in the first reaction is then packaged in a second step, and the frequency of the resulting Ap^r-transducing phages is measured. The frequency of recombination was increased from 3.7×10^{-8} to 1.9×10^{-6} Ap^r phages/total pfu when purified DNA gyrase A and B proteins were added to the recombination extract. It was increased further by 40-fold when oxolinic acid in addition to DNA gyrase was added. Thus, DNA gyrase promotes recombination in the first step even in the absence of oxolinic acid, and no λ-encoded function is required.

The structure of recombinants formed in the presence of oxolinic acid from a recombination between phage λ and pBR322 DNA was analyzed by heteroduplex mapping and DNA sequencing (Ikeda et al. 1982; Naito et al. 1984). Among nine isolates tested, two recombinants were formed by the simple insertion of the plasmid into the λ genome. In other cases, insertion of pBR322 was accompanied by a deletion in the λ or plasmid DNA. One end of the deletion coincided with one end of the pBR322 insertion in all cases. Recombination sites in these isolates seem to be distributed randomly on the λ and pBR322 genomes. On the basis of the comparison of the nucleotide sequences of recombination junctions of λ-pBR322 recombinants with those of parental λ and pBR322 DNAs, we have determined recombination sites for illegitimate events and found that the recombination sites of λ and pBR322 parental DNAs do not have a homologous sequence longer than 4 bp. There was absolutely no homology in one case. These results lead to the conclusion that homology is certainly not required for the DNA gyrase-mediated recombination and may play no role at all in our in vitro system.

Marvo et al. (1983) noted that some of the sites, where illegitimate recombination occurred, closely resembled known gyrase sites and matched the consensus sequence for gyrase-induced cleavage, 5′ Y-R-T↓G-N-Y-N-N-Y. Direct experiments have been performed to test whether DNA gyrase participates in illegitimate recombination in *E. coli* cells, and we found that *gyrA* mutations reduce the excision of pBR322 from the λ*plac*-pBR322 hybrid by 1/10 or 1/100. In contrast, oxolinic acid increased the excision of pBR322 approximately sixfold (Miura-Masuda and Ikeda 1990).

The *gyrA* mutation was also shown to affect the deletion of a cellular slime mold plasmid from a hybrid with another vector in *E. coli* (Saing et al. 1988). Barring some other indirect effects of these mutations, we conclude that DNA gyrase does indeed promote illegitimate recombination in vivo.

IV. ILLEGITIMATE RECOMBINATION MEDIATED BY T4 DNA TOPOISOMERASE

T4 DNA topoisomerase is also a type-II enzyme (Liu et al. 1979; Stetler et al. 1979). *E. coli* DNA gyrase and T4 DNA topoisomerase share many common properties although the T4 enzyme is not able to introduce supercoils into DNA in vitro. It has the capacity to make transient double-stranded breaks in DNA (Liu et al. 1980).

Ikeda (1986a,b) demonstrated that purified T4 DNA topoisomerase can mediate recombination between two dsDNA molecules in an in vitro system, producing a biologically active recombinant DNA. In the cross between two λ DNAs, the frequency of recombination was 1×10^{-3} recombinants/total pfu. Furthermore, oxolinic acid stimulated the recombination about tenfold as it did for DNA gyrase. At similar concentrations, oxolinic acid induces cleavage of dsDNA by T4 topoisomerase (Kreuzer and Alberts 1984) and stimulates recombination (Ikeda 1986a). Neutral sucrose gradient centrifugation showed that incubation of parental linear monomer λ DNAs with T4 DNA topoisomerase produced linear monomer recombinant DNA molecules that are packageable in vitro. The enzyme alone seems capable of completing recombination although some repair functions may have a role in the packaging extract.

All the recombinants between two λ DNA molecules examined contain a deletion or duplication at different sites on the phage genome. No two of the crossover sites were identical in nine isolates (Ikeda 1986a). This implies that the sites of the crossover are not specific as well as nonhomologous. This was confirmed by sequence analyses of the recombination junctions of the recombinant phage DNAs (Chiba et al. 1989). Therefore, recombination mediated by T4 DNA topoisomerase is a true illegitimate recombination.

Illegitimate recombination has been detected in vivo in the following phage T4 system. An *rII* deletion mutant, *r1589*, blocks only the function of *rII A* cistron although it extends into the *B* cistron (Champe and Benzer 1962). Another *rII* deletion mutation *r1236* is mostly in the *B* cistron and overlaps the *r1589*. When a cross is made between *r1589* and *r1236*, true rII^+ progeny cannot be formed. Instead, anomalous phages carrying an *rII* region from each parent have been detected that are phenotypically rII^+ (Weil et al. 1965). This in vivo system has been used to test whether or not T4 DNA topoisomerase participates in illegitimate recombination in T4-infected cells. The mutations in the structural genes for the topoisomerase, genes *39*, *52*, and *60*, all reduced illegitimate recombination between two T4 phage DNAs by about one-tenth (M. Honda and H. Ikeda, unpubl.). We therefore concluded that T4 DNA topoisomerase promotes illegitimate recombination in vivo.

V. ILLEGITIMATE RECOMBINATION MEDIATED BY MAMMALIAN DNA TOPOISOMERASE II

Illegitimate recombination is frequently observed in mammalian somatic cells. DNA molecules transferred into mammalian cells undergo a variety of rearrangements including deletion, insertion, and circularization (Calos et al. 1983; Razzaque et al. 1983; Roth and Wilson 1986). They also integrate at various chromosomal sites by illegitimate recombination (Smith and Berg 1984; Lin et al. 1985). It is therefore important to understand the mechanism by which illegitimate recombination occurs in these cells. The properties of mammalian DNA topoisomerase II are similar to those of T4 DNA topoisomerase (Miller et al. 1981; Shelton et al. 1983; Halligan et al. 1985; Heller et al. 1986). In particular, it has the capacity to make a transient double-stranded break in DNA. Thus, it is expected that mammalian DNA topoisomerase II can catalyze illegitimate recombination in vivo as well as in vitro.

Bae et al. (1988) have shown that purified calf thymus DNA topoisomerase II promotes recombination between two phage λ DNA molecules in an in vitro system, which is essentially the same recombination system as that used for the assay of T4 DNA topoisomerase (Ikeda 1986a). This recombination does not require other factors. The recombinant molecules contain duplications or deletions, and most crossovers take place between nonhomologous sequences of λ DNA as judged by the sequences of recombination junctions. Thus, recombination mediated by calf thymus DNA topoisomerase II is an illegitimate recombination that is similar to recombination mediated by phage T4 DNA topoisomerase or *E. coli* DNA gyrase.

Novobiocin, an inhibitor of *E. coli* DNA gyrase, inhibits this topoisomerase-mediated recombination in vitro at a concentration higher than that used for the DNA gyrase (Bae et al. 1988). This drug shows also an inhibitory effect on relaxation and unknotting activities of mammalian topoisomerase II at a concentration similar to that used for the inhibition of topoisomerase-mediated recombination (Miller et al. 1981). It is not yet known whether or not recombination takes place in vivo by DNA topoisomerase II in mammalian cells.

VI. MODEL FOR DNA TOPOISOMERASE-II-MEDIATED ILLEGITIMATE RECOMBINATION

The ability of DNA gyrase and other type-II topoisomerases to make a double-stranded break has been thought to play an essential role in the introduction or removal of supercoils, formation and resolution of catenanes, and knotting or unknotting of circular DNA (Gellert et al.

1977; Sugino et al. 1977; Kreuzer and Cozzarelli 1980; Liu et al. 1980; Mizuuchi et al. 1980). The cleavage activity may also be important in DNA gyrase-mediated illegitimate recombination. The critical question is whether there are any common sites on the DNA for the recombination and the cleavage reactions. *E. coli* DNA gyrase is known to cleave pBR322 DNA preferentially at two sites in vitro (Gellert et al. 1981) and at several sites in vivo (Lockshon and Morris 1985). However, no recombination site was found initially at these known gyrase-cleavage sites (Ikeda et al. 1982; Naito et al. 1984). Ikeda et al. (1984) produced a more sensitive assay of gyrase-cleavage sites near known recombination sites; the specificity of gyrase was not very high. Three of five recombination sites examined were found to be cleaved by DNA gyrase. These data suggest that the cleavage of DNA by gyrase has an important role in recombination.

The physicochemical properties of DNA gyrase and other type-II topoisomerases may be relevant to recombination. *E. coli* DNA gyrase consists of two subunit proteins of 105 and 95 kD coded by the *gyrA* and *gyrB* genes, respectively (Sugino et al. 1977; Higgins et al. 1978; Mizuuchi et al. 1978; Hansen and von Meyenburg 1979). The native enzyme contains an equimolar amount of the two subunits and exists in solution as an A_2B_2 complex (Klevan and Wang 1980; Mizuuchi et al. 1980; Sugino et al. 1980).

T4 topoisomerase cleaves dsDNA in vitro in the presence of oxolinic acid (Kreuzer and Alberts 1984). Some of the sites of recombination mediated by T4 DNA topoisomerase are also cleavage sites in vitro, implying that the cleavage activity plays a role in illegitimate recombination (Chiba et al. 1989). The purified T4 DNA topoisomerase consists of three subunit proteins of 63K, 52K, and 16K, known to be coded by genes *39*, *52*, and *60*, respectively (Liu et al. 1979; Stetler et al. 1979). The native form of T4 topoisomerase seems to be a dimer of each (Kreuzer and Jongeneel 1983; Seasholz and Greenberg 1983), as is DNA gyrase.

Mammalian DNA topoisomerase II is a homodimer of a single polypeptide of 170–180-kD subunits. Purified calf thymus DNA topoisomerase II is a dimer of 125-kD and/or 140-kD subunits, probably because of proteolysis during purification (Halligan et al. 1985). It also creates double-stranded breaks in vitro (Liu et al. 1983; Nelson et al. 1984).

Ikeda et al. (1982) proposed a DNA gyrase subunit exchange model for illegitimate recombination, based on the cleavage activity and physicochemical properties of DNA gyrase (Fig. 2). DNA gyrase binds to DNA as an A_2B_2 complex and transiently cleaves double strands of DNA, resulting in an intermediate in which each *gyrA* subunit covalently

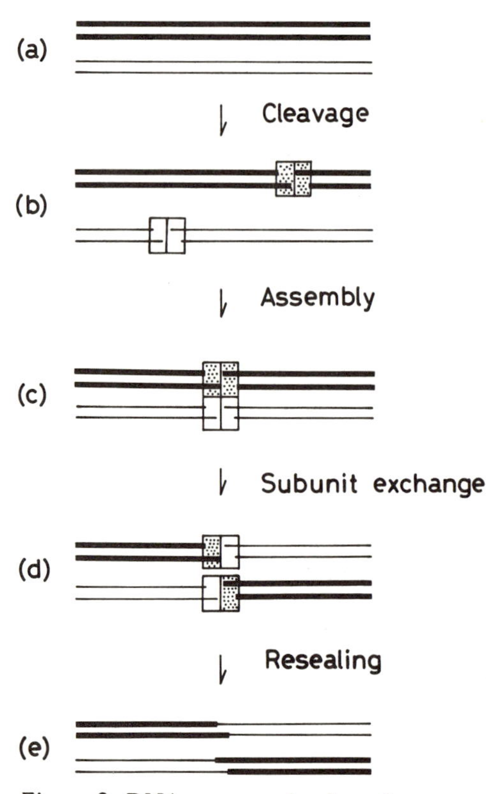

Figure 2 DNA gyrase subunit exchange model for illegitimate recombination. (*a*) Thin and thick lines represent two nonhomologous segments of DNA. (*b*) DNA gyrase binds with dsDNA and simultaneously cleaves it, leading to the formation of a protein/DNA complex with the protein linked to the 5′ end of the broken DNA. (*c*) Each box represents one AB subunit of DNA gyrase. The gyrase/DNA complex joins with another gyrase/DNA complex forming the tetrameric structure A_4B_4. (*d*) Dissociation of the tetrameric form to the dimeric form can be associated with subunit exchange that leads to the exchange of DNA strands. (*e*) Formation of recombinants. (Reprinted, with permission, from Ikeda et al. 1982.)

binds to each of the 5′ termini of the DNA at the cleavage site (Fig. 2b). Next, two gyrase/DNA complexes join to form a tetrameric complex (A_4B_4 form) (Fig. 2c). Here, we assume that the interactions between the four subunits are isologous so that the dissociation of the tetrameric complex to dimeric complexes occasionally results in subunit exchange with the accompanying exchange of DNA duplexes and resealing of the break in DNA strands (Fig. 2d,e). Alternatively, subunit exchange of the gyrase/DNA complex can be achieved by the dissociation of dimeric complex (A_2B_2 form) to a monomeric form (AB form) and the reassociation with another monomeric complex. However, the dissociation of the dimeric complex has not been detectable unless a denaturing agent is used (Gellert et al. 1977; Sugino et al. 1977). This model is also applicable to illegitimate recombination mediated by T4 DNA topoisomerase and mammalian DNA topoisomerase II.

DNA gyrase cleaves DNA with a 4-bp stagger and a protruding 5′ end (Morrison and Cozzarelli 1979). Because of the low specificity of cleavage, it is very unlikely that two protruding 5′ ends would be exactly complementary. Indeed, in most recombination junctions that have been formed in the in-vitro gyrase-mediated recombination, we found mismatched base pairs in putative intermediates of recombination junctions (Ikeda et al. 1982; Naito et al. 1984). These mismatched base pairs may be repaired, or the 5′ protruding ends may be joined covalently to the 3′ ends by the action of DNA gyrase even with the mismatches.

The gyrase subunit exchange model is further supported by data showing that secondary rearrangements are often observed near the recombination regions in vitro as well as in vivo (Ikeda et al. 1981, 1982; Marvo et al. 1983). Secondary rearrangements have been also found in mammalian systems (Botchan et al. 1980; Stringer 1982). In the former cases, for example, recombination between phage λ and plasmid DNA was accompanied by another deletion or duplication on one of the genomes at the same site of recombination. The model can explain these phenomena if the gyrase/DNA complex does not dissociate from DNA immediately after the first subunit exchange, and it often undergoes additional rounds of subunit exchange reactions with other gyrase/DNA complexes (Fig. 3). Other models for illegitimate recombination, described below, can hardly explain this feature.

VII. OTHER MODELS FOR ILLEGITIMATE RECOMBINATIONS

There are many examples of illegitimate recombination that take place between short homologous sequences (Farabaugh et al. 1978; Collins 1981; Jones et al. 1982) in addition to those between nonhomologous sequences. Albertini et al. (1982) have studied the nucleotide sequences of the endpoints of deletion mutations in the *E. coli lacI* gene, and they concluded that deletions are often formed by recombination between short homologous sequences. As originally proposed for frameshift mutations (Streisinger et al. 1967), they suggested a "slipped mispairing model" in which one of two homologous sequences unwinds during replication and mispairs with another sequence, resulting in deletion in the daughter

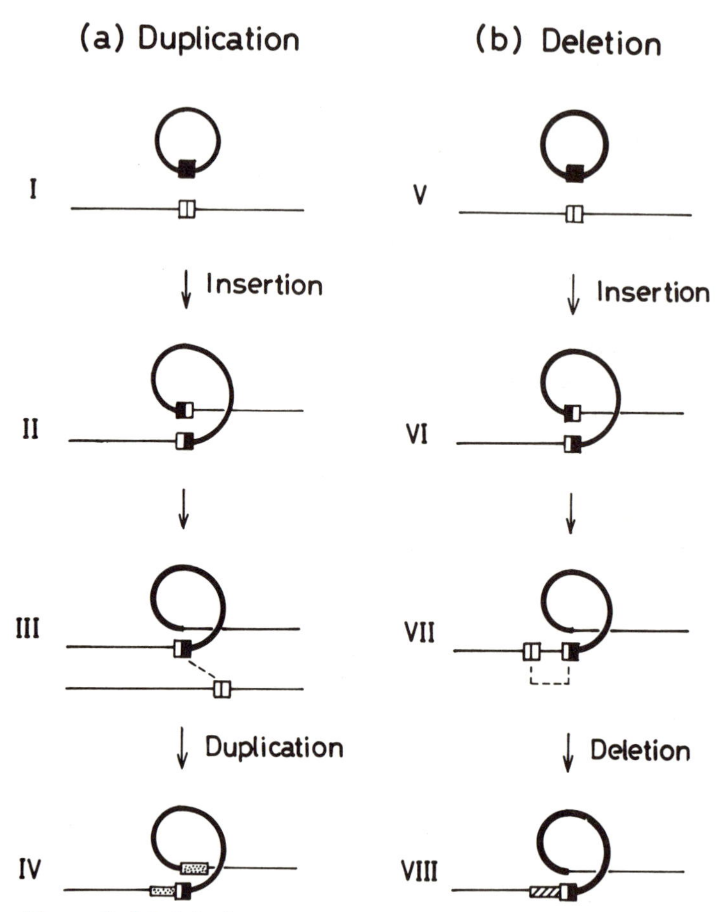

Figure 3 Models for secondary rearrangement mediated by DNA gyrase. (*a*) Circles and lines in *I* represent plasmid DNA and λ DNA, respectively. Rectangles attached to plasmid and λ DNA represent the AB form of DNA gyrase composed of one gyrase A protomer and one gyrase B protomer. Two AB forms constitute an A_2B_2 form that is a putative gyrase molecule. According to the DNA gyrase-subunit-exchange model, the subunit exchange between two intermediate complexes results in the exchange of DNA strands (*II*). One of the gyrase/DNA complexes (*III*) is still recombinogenic after the exchange. A second intermolecular recombination with a gyrase/DNA complex on another λ DNA molecule gives rise to the formation of a duplication (*IV*). (*b*) When recombination intermediate described as in *III* carries another DNA gyrase in another site of the DNA molelcule, it can undergo intramolecular recombination between the two gyrase/DNA complexes, leading to the formation of a deletion (*VIII*). (Reprinted, with permission, from Naito et al. 1984.)

strand. However, it has also been observed that illegitimate recombination mediated by *E. coli* DNA gyrase and phage T4 DNA topoisomerase in vivo may take place preferably between short homologous sequences (Saing et al. 1988; Miura-Masuda and Ikeda 1990; M. Honda and H. Ikeda, unpubl.).

Inverted repeats on phage DNA and plasmids are unstable in *E. coli* cells and result in deletions (Collins 1981; Lilley 1981; Hagan and Warren 1982; Goodchild et al. 1985). It has been reported also that these inverted repeats are stable in *recBC sbcB sbcC* mutants (Leach and Stahl 1983; Nader et al. 1986; Wyman et al. 1986) and that the *sbcC* genotype is the critical factor for this instability of the inverted repeats (Chalker et al. 1988). A model has been presented for this type of deletion formation in which stem-loop or cruciform structures serve as a substrate for excision or as a shortened template for DNA replication (Glickman and Ripley 1984). There is, however, evidence that the deletion formation in the long inverted repeats of a cellular slime mold plasmid is promoted by DNA gyrase in *E. coli* cells (Saing et al. 1988). Which model is correct for these gene rearrangements is an important question for further study.

VIII. POSSIBLE ROLE OF DNA TOPOISOMERASES IN GENE AMPLIFICATION

Gene amplification has often been observed in prokaryotic and in eukaryotic cells (Shimke 1982; Stark and Wahl 1984). The *E. coli lacZ* gene is often found to be amplified 40- to 200-fold at a frequency from 10^{-6} to 10^{-7} in the population (Tlsty et al. 1984). Integration of pBR322 into the *E. coli* chromosome leads to an amplification of up to 30-fold (Gutterson and Koshland 1983). Many cell lines transformed by either polyomavirus or SV40 contain tandem duplications of viral DNA integrated into chromosomal DNA (Tooze 1980). Novel joints have been identified in amplified DNAs by molecular cloning (Edlund et al. 1980; Ardeshir et al. 1983), implying that illegitimate recombination may have a role in gene amplification.

Botchan et al. (1979) proposed the "onion skin model" for the excision of the provirus genome in mammalian cells. This model involves unscheduled DNA replication followed by recombination between replicated daughter DNA strands. Roberts et al. (1983) extended this model to explain gene amplification. Unscheduled replication in a localized region of DNA may produce a "multi-eye structure" that may resolve into tandem repeats by multiple rounds of homologous or illegitimate recombination. In some cases of gene amplification, novel joints are formed by recombination between nonhomologous DNA sites that include trinucleotides that are DNA topoisomerase I cleavage sites in vitro (Hyrien et al. 1987; see also Konopka 1988). Ikeda and Shiozaki (1984) have pointed out a possibility that DNA gyrase may be involved in the formation of novel junctions in replicating DNA molecules. In these experiments, strict evidence for coupling between recombination and replication has not been presented yet. The mechanism responsible for the formation of novel joints during gene amplification is also in question.

ACKNOWLEDGMENTS

I thank A. Miura and Y. Tanaka for stimulating discussions, and I. Kobayashi, Y.-S. Bae, and N.R. Cozzarelli for critical readings of the manuscript. This work was supported by grants from the Ministry of Education of the Government of Japan and from the Nissan Science Foundation.

REFERENCES

Albertini, A.M., M. Hofer, M.P. Calos, and J.H. Miller. 1982. On the formation of spontaneous deletion: The importance of short sequence homologies in the generation of large deletions. *Cell* **29:** 319–328.

Anderson, C.W. 1970. Spontaneous deletion formation in several classes of *Escherichia coli* mutants deficient in recombination ability. *Mutat. Res.* **9:** 155–165.

Anderson, R.P. and J.R. Roth. 1977. Tandem genetic duplication in phage and fungi. *Annu. Rev. Microbiol.* **31:** 473–503.

Ardeshir, F., E. Giulotto, J. Zieg, O. Brison, W.S.L. Liao, and G.R. Stark. 1983. Structure of amplified DNA in different Syrian hamster cell lines resistant to N-(phosphonacetyl)-L-asparate. *Mol. Cell. Biol.* **3:** 2076–2088.

Bae, Y.-S., I. Kawasaki, H. Ikeda, and L.F. Liu. 1988. Illegitimate recombination mediated by calf thymus DNA topoisomerase II *in vitro*. *Proc. Natl. Acad. Sci.* **85:** 2076–2080.

Bauer, W.R., E.C. Ressner, J. Kates, and J.V. Patzke. 1977. A DNA nicking-closing enzyme encapsidated in vaccinia virus: Partial purification and properties. *Proc. Natl. Acad. Sci.* **74:** 1841–1845.

Been, M.D. and J.J. Champoux. 1981. DNA breakage and closure by rat liver type I topoisomerase: Separation of the half-reactions by using a single-stranded DNA substrate. *Proc. Natl. Acad. Sci.* **78:** 2883–2887.

Benzer, S. 1961. On the topology of the genetic fine structure. *Proc. Natl. Acad. Sci.* **47:** 403–415.

Botchan, M., W. Topp, and J. Sambrook. 1979. Studies on simian virus 40 excision from cellular chromosomes. *Cold Spring Harbor Symp. Quant. Biol.* **43:** 709–719.

Botchan, M., J. Stringer, T. Mitchison, and J. Sambrook. 1980. Integration and excision of SV40 DNA from the chromosome of a transformed cell. *Cell* **20:** 143–152.

Brown, P.O. and N.R. Cozzarelli. 1979. A sign inversion mechanism for enzymatic supercoiling of DNA. *Science* **206:** 1081–1083.

Bullock, P., J.J. Champoux, and M. Botchan. 1985. Association of crossover points with topoisomerase I cleavage sites: A model for nonhomologous recombination. *Science* **230:** 954–958.

Bullock, P., W. Forrester, and M. Botchan. 1984. DNA sequence studies of simian virus 40 chromosomal excision and integration in rat cells. *J. Mol. Biol.* **174:** 55–84.

Calos, M.P., J.S. Lebkowski, and M.R. Botchan. 1983. High mutation frequency in DNA transfected into mammalian cells. *Proc. Natl. Acad. Sci.* **80:** 3015–3019.

Campbell, A. 1962. The episomes. *Adv. Genet.* **11:** 101–146.

———. 1971. Genetic structure. In *The bacteriophage* λ (ed. A.D. Hershey), p. 13–44. Cold Spring Harbor Laboratory, Cold Spring Harbor, New York.

Chalker, A.F., D.R.F. Leach, and R.G. Lloyd. 1988. *Escherichia coli sbcC* mutants permit stable propagation of DNA replicons containing a long palindrome. *Gene* **71:** 201–205.

Champe, S.P. and S. Benzer. 1962. An active cistron fragment. *J. Mol. Biol.* **4:** 288–292.

Champoux, J.J. and R. Dulbecco. 1972. An activity from mammalian cells that untwists superhelical DNA-A possible swivel for DNA replication. *Proc. Natl. Acad. Sci.* **69:** 143–146.

Chiba, M., H. Shimizu, A. Fujimoto, H. Nashimoto, and H. Ikeda. 1989. Common sites for recombination and cleavage mediated by bacteriophage T4 DNA topoisomerase *in vitro*. *J. Biol. Chem.* **264:** 12785–12790.

Collins, J. 1981. Instability of palindrome DNA in *Escherichia coli*. *Cold Spring Harbor Symp. Quant. Biol.* **45:** 409–416.

Coukell, M.B. and C. Yanofsky 1970. Increased frequency of deletions in DNA polymerase mutants in *Escherichia coli*. *Nature* **228:** 633–635.

Edlund, T., T. Gründstom, G.R. Björk, and S. Normark. 1980. Tandem duplication induced by an unusual ampA1-, ampC-transducing lambda phage: A probe to initiate gene amplification. *Mol. Gen. Genet.* **180:** 249–257.

Edwards, K.A., B.D. Halligan, J.L. Davis, N.L. Nivera, and L.F. Liu. 1982. Recognition sites of eukaryotic DNA topoisomerase I: DNA nucleotide sequencing analysis of topo I cleavage sites on SV40 DNA. *Nucleic Acids Res.* **10:** 2565–2576.

Emmons, S.W., V. MacCosham, and R.L. Baldwin. 1975. Tandem genetic duplications in phage lambda. III. The frequency of duplication mutants in two derivatives of phage lambda is independent of known recombination systems. *J. Mol. Biol.* **91:** 133–146.

Enquist, L.W. and R.A. Weisberg. 1977. A genetic analysis of the *att-int-xis* region of coliphage lambda. *J. Mol. Biol.* **111:** 97–120.

Fairfield, F.R., W.R. Bauer, and M.V. Simpson. 1979. Mitochondria contain a distinct DNA topoisomerase. *J. Biol. Chem.* **254:** 9352–9354.

Farabaugh, P.J., U. Schmeissner, M. Hofer, and J.H. Miller. 1978. Genetic studies of the *lac* repressor. VII. On the molecular nature of spontaneous hotspots in the *lacI* gene of *Escherichia coli*. *J. Mol. Biol.* **126:** 847–863.

Franklin, N. 1967. Extraordinary recombinational events in *Escherichia coli*. Their independence of the rec^+ function. *Genetics* **55:** 699–707.

———. 1971. Illegitimate recombination. In *The bacteriophage* λ (ed. A.D. Hershey), p. 175–194. Cold Spring Harbor Laboratory, Cold Spring Harbor, New York.

Gellert, M., K. Mizuuchi, M.H. O'Dea, and H.A. Nash. 1976. DNA gyrase: An enzyme that introduces superhelical turns into DNA. *Proc. Natl. Acad. Sci.* **73:** 3872–3876.

Gellert, M., L.M. Fisher, H. Ohmori, M.H. O'Dea, and K. Mizuuchi. 1981. DNA gyrase: Site-specific interactions and transient double-strand breakage of DNA. *Cold Spring Harbor Symp. Quant. Biol.* **45:** 391–398.

Gellert, M., K. Mizuuchi, M.H., O'Dea, T. Itoh, and J. Tomizawa. 1977. Novobiocin and coumermycin inhibit DNA supercoiling catalyzed by DNA gyrase. *Proc. Natl. Acad. Sci.* **74:** 4772–4478.

Glickman, B.W. and L.S. Ripley. 1984. Structural intermediates of deletion mutagenesis: A role for palindromic DNA. *Proc. Natl. Acad. Sci.* **81:** 512–516.

Goodchild, J., J. Michniewicz, D. Seto-Young, and S. Narang. 1985. A novel deletion found during cloning of a synthetic palindromic DNA. *Gene* **33:** 367–371.

Goto, T. and J. Wang. 1982. Yeast DNA topoisomerase II. *J. Biol. Chem.* **257:** 5866–5872.

Gottesman, M.E. and M.B. Yarmolinsky. 1969. The integration and excision of the bacteriophage lambda genome. *Cold Spring Harbor Symp. Quant. Biol.* **33:** 735–747.

Gutai, M.W. and D. Nathans. 1978. Evolutionary variants of simian virus 40: Cellular DNA sequences and sequences at recombinant joints of substituted variants. *J. Mol. Biol.* **126:** 275–288.

Gutterson, N.I. and D.E. Koshland, Jr. 1983. Replacement and amplification of bacterial genes with sequences altered *in vitro*. *Proc. Natl. Acad. Sci.* **80:** 4894–4898.

Hagan, C.E. and G.J. Warren. 1982. Lethality of palindromic DNA and its use in selection of recombinant plasmids. *Gene* **19:** 147–151.

Halligan, B.D., K.A. Edwards, and L.F. Liu. 1985. Purification and characterization of a type II DNA topoisomerase I. *J. Biol. Chem.* **260:** 2475–2482.

Halligan, B.D., J.L. Davis, K.A. Edwards, and L.F. Liu. 1982. Intra- and intermolecular strand transfer by HeLa DNA topoisomerase. *J. Biol. Chem.* **257:** 3995–4000.

Hansen, F.G. and K. von Meyenburg. 1979. Characterization of the *dnaA*, *gyrB*, and other genes in the *dnaA* region of the *Escherichia coli* chromosome on specialized transducing phage λ*tna*. *Mol. Gen. Genet.* **175:** 135–144.

Heller, R.A., E.R. Shelton, V. Dietrich, S.C.R. Elgin, and D.L. Brutlag. 1986. Multiple forms and cellular localization of *Drosophila* DNA topoisomerase II. *J. Biol. Chem.* **261:** 8063–8069.

Higgins, N.P., C.L. Peebles, A. Sugino, and N.R. Cozzarelli. 1978. Purification of subunits of *Escherichia coli* DNA gyrase and reconstitution of enzymatic activity. *Proc. Natl. Acad. Sci.* **75:** 1773–1777.

Horiuchi, K., J.V. Ravetch, and N.D. Zinder. 1979. DNA replication of bacteriophage f1 in vivo. *Cold Spring Harbor Symp. Quant. Biol.* **43:** 389–399.

Horowitz, B. and R.C. Deonier. 1985. Formation of Δtra F′ plasmids: Specific recombination at *oriT*. *J. Mol. Biol.* **186:** 267–274.

Hyrien, O., M. Debatisse, G. Buttin, and B. Robert de Saint Vincent. 1987. A hotspot of novel amplification joints in a mosaic of *Alu*-like repeats and palindromic A+T-rich DNA. *EMBO J.* **6:** 2401–2408.

Ikeda, H. 1986a. Bacteriophage T4 DNA topoisomerase mediates illegitimate recombination *in vitro*. *Proc. Natl. Acad. Sci.* **83:** 922–926.

———. 1986b. Illegitimate recombination by T4 DNA topoisomerase *in vitro*. Recombinants between phage and plasmid DNA molecules. *Mol. Gen. Genet.* **202:** 518–520.

Ikeda, H. and M. Shiozaki. 1984. Nonhomologous recombination mediated by *Escherichia coli* DNA gyrase: Possible involvement of DNA replication. *Cold Spring Harbor Symp. Quant. Biol.* **49:** 401–409.

Ikeda, H., K. Aoki, and A. Naito. 1982. Illegitimate recombination mediated *in vitro* by DNA gyrase of *Escherichia coli*: Structure of recombinant DNA molecules. *Proc. Natl. Acad. Sci.* **79:** 3724–3728.

Ikeda, H., I. Kawasaki, and M. Gellert. 1984. Mechanism of illegitimate recombination: Common sites for recombinant and cleavage mediated by *E. coli* DNA gyrase. *Mol. Gen. Genet.* **196:** 546–549.

Ikeda, H., K. Moriya, and T. Matsumoto. 1981. In vitro study of illegitimate recombination: Involvement of DNA gyrase. *Cold Spring Harbor Symp. Quant. Biol.* **45:** 399–408.

Inselburg, J. 1967. Formation of deletion mutations in recombination-deficient mutants of *Escherichia coli. J. Bacteriol.* **94:** 1266–1267.

Jones, I.M., S.B. Primrose, and S.D. Ehrlich. 1982. Recombination between short direct repeats in a *recA* host. *Mol. Gen. Genet.* **188:** 486–489.

Kikuchi, Y. and H.A. Nash. 1979. Nicking-closing activity associated with bacteriophage λ*int* gene product. *Proc. Natl. Acad. Sci.* **76:** 3760–3764.

Klevan, L. and J.C. Wang. 1980. Deoxyribonucleic acid gyrase-deoxyribonucleic acid complex containing 140 base pairs of deoxyribonucleic acid and an $\alpha_2\beta_2$ protein core. *Biochemistry* **19:** 5229–5234.

Konopka, A.K. 1988. Compilation of DNA strand exchange sites for non-homologous recombination in somatic cells. *Nucleic Acids Res.* **16:** 1739–1758.

Krasnow, M.A. and N.R. Cozzarelli. 1983. Site-specific relaxation and recombination by the Tn3 resolvase: Recognition of the DNA path between oriented *res* sites. *Cell* **32:** 1313–1324.

Kreuzer, K.N. and B.M. Alberts. 1984. Site-specific recognition of bacteriophage T4 DNA by T4 type II DNA topoisomerase and *Escherichia coli* DNA gyrase. *J. Biol. Chem.* **259:** 5339–5346.

Kreuzer, K.N. and N.R. Cozzarelli. 1980. Formation and resolution of DNA catenanes by DNA gyrase. *Cell* **20:** 245–254.

Kreuzer, K.N. and C.V. Jongeneel. 1983. *Escherichia coli* phage T4 topoisomerase. *Methods Enzymol.* **100:** 144–160.

Leach, D.R.F. and F.W. Stahl. 1983. Viability of λ phages carrying a perfect palindrome in the absence of recombination nucleases. *Nature* **305:** 448–451.

Lilly, D.M.J. 1981. *In vivo* consequences of plasmid topology. *Nature* **292:** 380–382.

Lin, F.-L., K. Sperle, and N. Sternberg. 1985. Recombination in mouse L cells between DNA introduced into cells and homologous chromosomal sequences. *Proc. Natl. Acad. Sci.* **82:** 1391–1395.

Liu, L.F., C.-C. Liu, and B.M. Alberts. 1979. T4 DNA topoisomerase: A new ATP-dependent enzyme essential for initiation of T4 bacteriophage DNA replication. *Nature* **281:** 456–461.

———. 1980. Type II DNA topoisomerases: Enzymes that can unknot a topologically knotted DNA molecule via a reversible double-strand break. *Cell* **19:** 697–707.

Liu, L.F., T.C. Rowe, L. Yang, K.M. Tewey, and G.L. Chen. 1983. Cleavage of DNA by mammalian DNA topoisomerase II. *J. Biol. Chem.* **258:** 15365–15370.

Lockshon, D. and D.R. Morris. 1985. Sites of reaction of *Escherichia coli* DNA gyrase on pBR322 *in vivo* as revealed by oxolinic acid-induced plasmid linearization. *J. Mol. Biol.* **181:** 63–74.

Marvo, S.L., S.R. King, and S.R. Jaskunas. 1983. Role of short regions of homology in intermolecular illegitimate recombination events. *Proc. Natl. Acad. Sci.* **80:** 2452–2456.

Michel, B. and S.D. Ehrlich. 1986a. Illegitimate recombination at the replication origin of bacteriophage M13. Proc. Natl. Acad. Sci. **83:** 3386–3390.

———. 1986b. Illegitimate recombination occurs between the replication origin of the plasmid pC194 and a progressing replication fork. *EMBO J.* **5:** 3691–3696.

Miller, K.G., L.F. Liu, and P.T. Englund. 1981. A homologous type II DNA topoisomerase from HeLa cell nuclei. *J. Biol. Chem.* **256:** 9334–9339.

Miura-Masuda, A. and H. Ikeda. 1990. The gyrase of *Escherichia coli* participates in the formation of a spontaneous deletion by *recA*-independent recombination *in vivo*. *Mol. Gen. Genet*. (in press).

Mizuuchi, K., M.H. O'Dea, and M. Gellert. 1978. DNA gyrase: Subunit structure and ATPase activity of the purified enzyme. *Proc. Natl. Acad. Sci.* **75:** 5960–5963.

Mizuuchi, K., L.M. Fisher, M.H. O'Dea, and M. Gellert. 1980. DNA gyrase action involves the introduction of transient double-strand breaks into DNA. *Proc. Natl. Acad. Sci.* **77:** 1847–1851.

Morrison, A. and N.R. Cozzarelli. 1979. Site-specific cleavage of DNA by *E. coli* DNA gyrase. *Cell* **17:** 175–184.

Nader, W.F., G. Isenberg, and H.W. Sauer. 1986. Structure of *Physarum* actin gene locus *ardA*: A nonpalindromic sequence causes inviability of phage lambda and *recA*-independent deletions. *Gene* **48:** 133–144.

Naito, A., S. Naito, and H. Ikeda. 1984. Homology is not required for recombination mediated by DNA gyrase of *Escherichia coli*. *Mol. Gen. Genet*. **193:** 238–243.

Nash, H.A., K. Mizuuchi, L.W. Enquist, and R.A. Weisberg. 1981. Strand exchange in λ integrative recombination: Genetics, biochemistry, and models. *Cold Spring Harbor Symp. Quant. Biol.* **45:** 417–428.

Nelson, E.M., K.M. Tewey, and L.F. Liu. 1984. Mechanism of antitumor drug action: Poisoning of mammalian DNA topoisomerase II on DNA by 4′-(9-acridinylamino)-methanesulfon-m-anisidide. *Proc. Natl. Acad. Sci.* **81:** 1361–1365.

O'Conner, M.B., J.J. Kilbane, and M.H. Malamy. 1986. Site-specific and illegitimate recombination in the *oriV1* region of the F factor. *J. Mol. Biol.* **189:** 85–102.

Razzaque, A., H. Mizusawa, and M.M. Seidman. 1983. Rearrangement and mutagenesis of a shuttle vector plasmid after passage in mammalian cells. *Proc. Natl. Acad. Sci.* **80:** 3010–3014.

Reed, R.R. and N.D.F. Grindley. 1981. Transposon-mediated site-specific recombination in vitro: DNA cleavage and protein-DNA linkage at the recombination site. *Cell* **25:** 721–728.

Roberts, J.M., L.B. Buck, and R. Axel. 1983. A structure for amplified DNA. *Cell* **33:** 53–63.

Roth, D.B. and J.H. Wilson. 1986. Nonhomologous recombination in mammalian cells: Role for short sequence homologies in the joining reaction. *Mol. Cell. Biol.* **6:** 4295–4304.

Ruley, H.E. and M. Fried. 1983. Cultured illegitimate recombination events in mammalian cells involving very short sequence homologies. *Nature* **304:** 181–184.

Saing, K.M., H. Orii, Y. Tanaka, K. Yanagisawa, A. Miura, and H. Ikeda. 1988. Formation of deletion in *Escherichia coli* between direct repeats located in the long inverted repeats of a cellular slime mold plasmid: Participation of DNA gyrase. *Mol. Gen. Genet.* **214:** 1–5.

Sander, M. and T. Hsieh. 1983. Double strand DNA cleavage by type II DNA topoisomerase form *Drosophila melanogaster*. *J. Biol. Chem*. **258:** 8421–8428.

Schaller, H. 1979. The intergenic region and the origins for filamentous phage DNA replication. *Cold Spring Harbor Symp. Quant. Biol.* **43:** 401–408.

Schwartz, D.O. and J.R. Beckwith. 1969. Mutagens which cause deletions in *Escherichia coli*. *Genetics* **61:** 371–376.

Seasholz, A.F. and G.R. Greenberg. 1983. Purification of bacteriophage T4 gene 60 product and a role for this protein in DNA topoisomerase. *J. Biol. Chem.* **258:** 1221–1226.

Shelton, E.R., N. Osheroff, and D.L. Brutlag. 1983. DNA topoisomerase II from *Drosophila melanogaster*. Purification and physical characterization. *J. Biol. Chem.* **258:** 9530–9535.

Shimke, R.T., ed. 1982. *Gene amplification.* Cold Spring Harbor Laboratory, Cold Spring Harbor, New York.

Shuman, S. 1989. Vaccinia DNA topoisomerase I promotes illegitimate recombination in *Escherichia coli. Proc. Natl. Acad. Sci.* **86:** 3489–3493.

Smith, A.J.H. and P. Berg. 1984. Homologous recombination between defective *neo* genes in mouse 3T6 cells. *Cold Spring Harbor Symp. Quant. Biol.* **49:** 171–181.

Spudich, J.A., V. Horn, and C. Yanofsky. 1970. On the production of deletions in the chromosome of *Escherichia coli. J. Mol. Biol.* **53:** 49–67.

Stark, G.R. and G.M. Wahl. 1984. Gene amplification. *Annu. Rev. Biochem.* **53:** 447–491.

Stetler, G.L., G.J. King, and W.M. Huang. 1979. T4 DNA-delay proteins, required for specific DNA replication, form a complex that has ATP-dependent DNA topoisomerase activity. *Proc. Natl. Acad. Sci.* **76:** 3737–3741.

Streisinger, G., Y. Okada, J. Emrich, J. Newton, A. Tsugita, E. Terzaghi, and M. Inouye. 1967. Frameshift mutations and the genetic code. *Cold Spring Harbor Symp. Quant. Biol.* **31:** 77–84.

Stringer, J.R. 1982. DNA sequence homology and chromosomal deletion at a site of SV40 DNA integration. *Nature* **296:** 363–366.

Subramani, S. and P. Berg. 1983. Homologous and nonhomologous recombination in monkey cells. *Mol. Cell. Biol.* **3:** 1040–1052.

Sugino, A., N.P. Higgins, and N.R. Cozzarelli. 1980. DNA gyrase subunit stoichiometry and the covalent attachment of subunit A to DNA during DNA cleavage. *Nucleic Acids Res.* **8:** 3865–3874.

Sugino, A., C.L. Peebles, K.N. Kreuzer, and N.R. Cozzarelli. 1977. Mechanism of action of nalidixic acid: Purification of *Escherichia coli nalA* gene product and its relationship to DNA gyrase and a novel nicking-closing enzyme. *Proc. Natl. Acad. Sci.* **74:** 4767–4771.

Sugino, A., N.P. Higgins, P.O. Brown, C.L. Peebles, and N.R. Cozzarelli. 1978. Energy coupling in DNA gyrase and the mechanism of action of novobiocin. *Proc. Natl. Acad. Sci.* **75:** 4838–4842.

Tessman, I. 1962. The induction of large deletions by nitrous acid. *J. Mol. Biol.* **5:** 442–445.

Tlsty, T.D., A.M. Albertini, and J.H. Miller. 1984. Gene amplification in the *lac* region of *E. coli. Cell* **37:** 217–224.

Tooze, J., ed. 1980. *Molecular biology of tumor viruses,* 2nd edition: *DNA tumor viruses.* Cold Spring Harbor Laboratory, Cold Spring Harbor, New York.

Wang, J.C. 1971. Interaction between DNA and an *Escherichia coli* protein ω. *J. Mol. Biol.* **55:** 523–533.

Weil, J., B. Terzaghi, and J. Craseman. 1965. Partial diploidy in phage T4. *Genetics* **52:** 683–693.

Weisberg, R.A. and S. Adhya. 1977. Illegitimate recombination in bacteria and bacteriophage. *Annu. Rev. Genet.* **11:** 451–473.

Wilson, J.H., P.B. Berget, and J.M. Pipas. 1982. Somatic cell efficienty join unrelated DNA segments end to end. *Mol. Cell. Biol.* **2:** 1258–1269.

Wyman, A.R., K.F. Wertman, D. Baker, C. Helms, and W.H. Petri. 1986. Factors which equalize the representation of genome segments in recombinant libraries. *Gene* **49:** 263–271.

13
DNA Topoisomerase Modification

N. Patrick Higgins
Department of Biochemistry
University of Alabama at Birmingham
Birmingham, Alabama 35294

Ari M. Ferro and Baldomero M. Olivera
Department of Biology
University of Utah
Salt Lake City, Utah 84112

I. INTRODUCTION

The first reports of topoisomerase modification were published in 1982 and 1983 (Mills et al. 1982; Durban et al. 1983; Ferro et al. 1983; Jongstra-Bilen et al. 1983). Although a wide variety of posttranslational modifications of DNA topoisomerases may occur, this chapter focuses only on phosphorylation and poly(ADP-ribosylation), which have been observed both in vitro and in vivo. The experimental data suggest a regulatory role, but the precise cellular functions of these DNA topoisomerase modifications remain undefined at the present time.

A. Phosphorylation of DNA Topoisomerase at Serine Residues

The initial reports of phosphorylation in topoisomerases involved topoisomerase I purified from Novikoff hepatoma cells. A phosphoprotein of approximately 110 kD with a pI of 8.4 purified from cells labeled with ^{32}P (Durban et al. 1981) was subsequently identified as DNA topoisomerase I (Durban et al. 1983). A form of topoisomerase I could also be labeled in vitro by incubation with serine-specific protein kinases and [^{32}P]ATP. A protein kinase from an African Burkitt's lymphoma cell line that appeared to exhibit high affinity for topoisomerase I was purified (Mills et al. 1982). This serine kinase had an apparent K_m of approximately 0.3 μM for topoisomerase I; phosphorylation stimulated the DNA-relaxing activity of the enzyme three- to fivefold. The discovery of phosphorylation of topoisomerase I was followed by reports of phosphorylation of topoisomerase II. Sander et al. (1984) found a protein kinase activity that remained associated with *Drosophila* topoisomerase II through four purification steps. Phosphoserine was the predominant species modified by the topoisomerase-associated protein kinase.

Because phosphorylation and dephosphorylation of proteins are recognized as major mechanisms for regulating cellular functions, the reports of DNA topoisomerase phosphorylation stimulated studies with purified protein kinases. Casein kinase II, a messenger-independent kinase that uses acidic proteins like casein as substrate (Edelman et al. 1987), phosphorylated *Drosophila* DNA topoisomerase II; two to three phosphates are incorporated into each topoisomerase II polypeptide chain (Ackerman et al. 1985, 1988). In this study, no autophosphorylation of topoisomerase occurred, and serine was the only amino acid residue modified by casein kinase II. The reaction was moderately specific, showing an apparent K_m of 0.4 μM for topoisomerase II. Phosphorylation of the enzyme stimulated DNA relaxation activity approximately two- to threefold and was reversed by treatment with alkaline phosphatase.

Even more compelling is the evidence that casein kinase II type enzymes phosphorylate DNA topoisomerase I. The phosphorylation of Novikoff topoisomerase I increases enzymatic activity (Durban et al. 1983); the serine residue, which is the primary site of phosphorylation in vivo, is phosphorylated by purified casein kinase II in vitro (Durban et al. 1985). In *Xenopus* ovaries, phosphorylation of DNA topoisomerase I may be obligatory for enzymatic activity (Kaiserman et al. 1988). DNA topoisomerase activity was lost when the *Xenopus* enzyme was treated with phosphatase and was regained after treatment with ATP and a casein-kinase-II-like activity (Fig. 1). Indeed, dephosphorylation blocked the topoisomerase I nicking of DNA induced by the drug camptothecin. Thus, phosphorylation of this enzyme appears to be required for the formation of the initial covalent enzyme/DNA complex. Whether this total dependence on phosphorylation for catalytic activity is a special property of DNA topoisomerase from *Xenopus* ovaries or whether it is characteristic of most DNA topoisomerase I enzymes remains to be established.

Protein kinase C and a calmodulin-dependent protein kinase also phosphorylated topoisomerase II in vitro (Sahyoun et al. 1986). Protein kinase C is a well-characterized calcium-dependent and phospholipid-activated protein kinase. This cellular enzyme is the receptor for phorbol esters and tumor promoters with structural similarities to diacylglycerol.

1

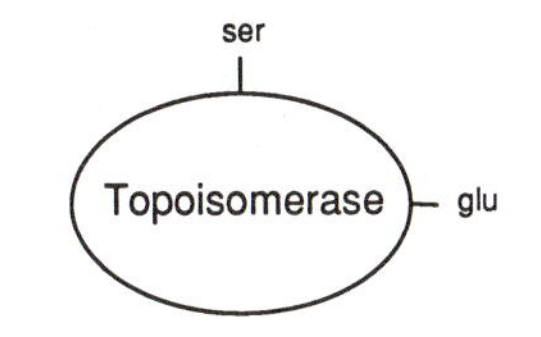

2

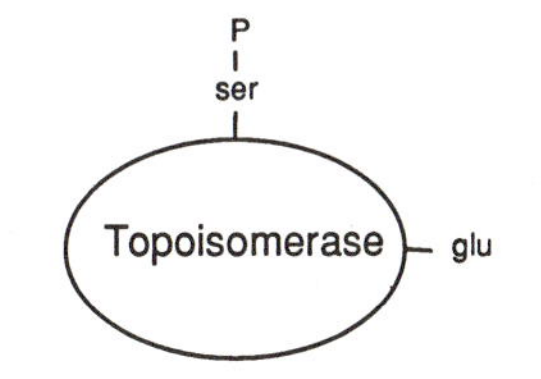

3

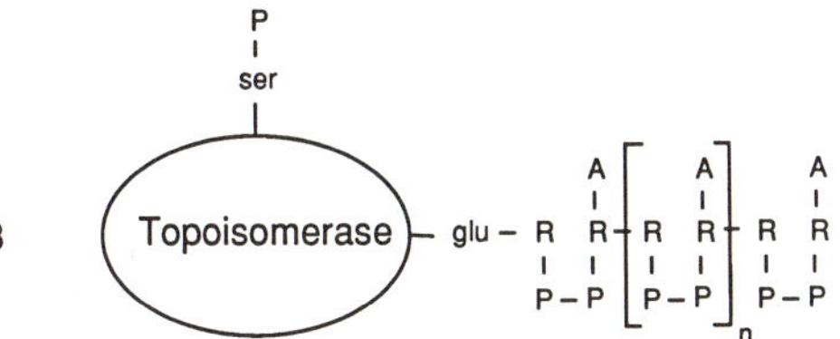

Figure 1 Some possible covalent modifications of DNA topoisomerases in higher eukaryotes. (*1*) Unmodified topoisomerase. (2) Topoisomerase phosphorylated at a serine residue. (*3*) Topoisomerase with both a phosphoserine and a poly(ADP-ribose) chain. A indicates adenine, R indicates ribose, P indicates phosphate. Type I and II topoisomerases can be modified by serine kinases, such as casein kinase II, to yield a phosphorylated enzyme as shown in structure *2*. In many cases, this modification stimulates activity and in *Xenopus* oocytes phosphorylation of serine may be required for activity. The reversibility of phosphorylation has been documented in vitro but not in vivo; this reaction is presumably carried out by a protein phosphatase. Topoisomerases are also substrates for poly(ADP-ribose) synthetase in higher eukaryotes (structure *3*). The modification site on the protein is probably a glutamate residue, and ADP-ribose chains of varying lengths are added to the enzyme. The DNA breaking-joining activity of the modified proteins becomes progressively more inhibited as the chains become longer. This process is reversed by a second nuclear enzyme, poly(ADP-ribose) glycohydrolase, which reactivates the modified enzyme.

Protein kinase C shows a high affinity for its topoisomerase target with an apparent K_m for *Drosophila* topoisomerase II of about 100 nM. The reaction requires calcium and phosphatidyl serine and is stimulated by phorbol esters. Approximately 1 mole of phosphate is incorporated per mole of topoisomerase II, suggesting that a single site on the enzyme might be phosphorylated. Like the studies described above with casein kinase II, topoisomerase phosphorylated by protein kinase C exhibited two- to threefold stimulation of DNA-relaxing activity. Studies of intact cells of the sponge *Geodia cydonium* are consistent with the involvement of protein kinase C in the phosphorylation of DNA topoisomerase II in vivo (Rottmann et al. 1987).

B. Other Phosphorylation Reactions

Other examples of topoisomerase phosphorylation have been described only in vitro. These include phosphorylation at tyrosine residues and the automodification of prokaryotic topoisomerases with the precise site of phosphorylation presently undefined.

Tyrosine protein kinase activity is intrinsic to numerous oncogene products and cell receptors that are implicated in modulation of cell growth. There are two reports of tyrosine protein kinases that can modify topoisomerases (Tse-Dinh et al. 1984, Goldberg et al. 1985). Rous sarcoma virus transforming gene product $pp60^{src}$ phosphorylates tyrosine moieties on *Escherichia coli* topoisomerase I, *Micrococcus luteus gyrA* subunit, calf thymus topoisomerase I, and calf thymus topoisomerase II. In the cases of the type-I topoisomerases of *E. coli* and calf thymus, phosphorylation of tyrosine residues is associated with loss of greater than 90% of the DNA-relaxing activity. This result indicates that the tyrosine involved in coupled DNA breakage and reunion activity may be a target of protein kinase modification. Similar results were obtained with a less well-defined 75-kD cellular protein kinase isolated from uninfected cells. However, since these kinases are normally membrane-bound proteins that may not travel to the nucleus, the significance of these results awaits further testing.

Prokaryotic DNA topoisomerases have autophosphorylation activity. Three subunits that make up the bacteriophage T4 DNA topoisomerase complex are encoded by products of genes *39*, *52*, and *60* (see Huang, this volume). Gene *39* has been cloned (Huang 1986a) and overproduced in uninfected *E. coli* cells. It has a strong ATP-binding site and intrinsic protein kinase activity (W.M. Huang, unpubl.). The gp39 subunit carries out an autophosphorylation reaction and also modifies gp52 and gp60 of the T4 enzyme. gp52 has significant homology with *gyrA* (Huang 1986b), and the A and B subunits of DNA gyrase are substrates for modification by the T4 DNA topoisomerase gp39 protein kinase activity.

The T4 gp39 and *E. coli gyrB* protein have regions of distinct protein homology, and the B subunit of gyrase also has a kinase activity that is weak by comparison with T4 gp39; *gyrB* protein undergoes an autophosphorylation reaction that is stimulated by Ca^{++} (W.M. Huang, unpubl.). The *gyrB* kinase activity also phosphorylates the *gyrA* subunit of *E. coli* gyrase. A major issue to be resolved is whether these protein kinase activities exhibited by DNA topoisomerase II subunits are a manifestation of the intrinsic catalytic mechanism in which ATP is cleaved as DNA strand passage occurs or, alternatively, serve some regulatory role.

C. Poly(ADP-ribosylation) of DNA Topoisomerases

The poly(ADP-ribosylation) of DNA topoisomerases has been shown to occur efficiently in vitro (Ferro et al. 1983; Jongstra-Bilen et al. 1983; Ferro and Olivera 1984; Darby et al. 1985). This modification is carried out by a nuclear enzyme, poly(ADP-ribose) synthetase, which has been extensively characterized in several eukaryotic systems (for reviews, see Hayaishi and Ueda 1982; Ferro and Olivera 1987; Ueda 1987). Nicotinamide-adenine dinucleotide (NAD) is the substrate for all ADP-ribosylation reactions. The glycosidic bond between the nicotinamide ring and ribose (a high-energy bond) is cleaved, and the ADP-ribose moiety is transferred to the DNA topoisomerase target (probably to a carboxyl group of a glutamate side chain although this has not been directly established for DNA topoisomerases). The synthetase then continues transferring ADP-ribose residues to form ribose-ribose glycosidic linkages so that a chain of ADP-ribose monomer units (poly[ADP-ribose]) is covalently attached to the DNA topoisomerase (see Fig. 1). An unusual property of the modifying enzyme in vivo is that the synthetase is quiescent without an activating DNA structure. Interruptions in the DNA double helix, including double-strand breaks and single-strand nicks, are absolutely required to activate poly(ADP-ribose) synthetase. In the absence of activating DNA, the modification is quickly removed by another nuclear enzyme, poly(ADP-ribose) glycohydrolase.

The human gene for poly(ADP-ribose) polymerase (huADPRP) has been cloned, sequenced, and mapped (Cherney et al. 1987; Kurosaki et al. 1987; Uchida et al. 1987). Human and mouse proteins are closely related evolutionarily, and domains for an NAD-binding site, an automodification segment, and a DNA-binding domain containing two "zinc fingers" have been identified (Mazen et al. 1989). In a mouse, this gene is on chromosome 1, closely linked to the autoimmune locus *gld* (*g*eneralized *l*ymphoproliferative *d*isorder) (Huppi et al. 1989).

In vitro, when DNA topoisomerase I from calf thymus is extensively modified by the homologous poly(ADPR-synthetase), DNA-relaxing activity is abolished (Ferro et al. 1983; Jongstra-Bilen et al.1983; Ferro and

Olivera 1984). DNA topoisomerase II can also be modified (Darby et al. 1985). It has been suggested that poly(ADP-ribosylation) inhibits the DNA topoisomerase activity because the poly(ADP-ribose) chains decrease the affinity of topoisomerases for DNA. Since each ADP-ribose monomer has two negative charges, strong electrostatic repulsions exist between these moieties and the DNA substrate for topoisomerases (Ferro et al. 1984). Consistent with this electrostatic repulsion model (see Ferro and Olivera 1982; Zahradka and Ebisuzaki 1982; Ferro et al. 1984), the extent of modification that can be carried out by poly(ADPR-synthetase) in vitro depends on ionic strength. At low ionic strength, relatively minimal levels of modification can be carried out before the enzyme can no longer bind DNA. As the ionic strength increases, much more extensive modification occurs (up to 40 ADP-ribose residues/protein).

DNA topoisomerase I is one of the best exogenous acceptor proteins for ADPR-synthetase. Only a small subset of targets, including DNA topoisomerases, are efficiently modified under physiological conditions in vitro. The modification has also been detected in vivo (Krupitza and Cerutti 1989). The biological consequences of poly(ADP-ribose)-synthetase-inhibition include a variety of effects on DNA repair and an elevation of the frequency of sister chromatid exchanges (Oikawa et al. 1980).

In the absence of any exogenous protein, the synthetase carries out an automodification reaction and becomes so extensively poly(ADP-ribosylated) that it no longer binds DNA and therefore becomes inactive. However, exogenous targets such as DNA topoisomerase I depress automodification by competing with the synthetase as a substrate. The poly(ADP-ribosylation) of the synthetase itself may serve as a "timing device" so each enzyme molecule catalyzes poly(ADP-ribosylation) at a DNA strand break for a limited period of time before the enzyme self-inactivates.

II. PERSPECTIVES

The experimental data described in this chapter suggest that the covalent modification of topoisomerases may play a regulatory role in vivo. DNA topoisomerases of higher eukaryotes are substrates for protein kinases and poly(ADPR-synthetase) in vitro and have been identified in the phosphorylated and ADP-ribosylated states in vivo.

The report that *Xenopus* DNA topoisomerase I requires phosphorylation for enzymatic activity indicates that in some vertebrate systems, the degree of phosphorylation may be the critical factor determining the amount of intracellular DNA topoisomerase activity. The question that needs to be addressed is how widespread is the requirement that DNA topoisomerases be phosphorylated in order to be active? Are the *Xenopus* results restricted to certain stages in the life cycle of vertebrates, or will this be a general feature of topoisomerases? The fact that yeast topoisomerase I can be cloned and expressed in *E. coli* in an active form (Bjornsti and Wang 1987) is indicative that this enzyme may be active without being phosphorylated.

In contrast with the stimulatory effects of serine phosphorylation, tyrosine phosphorylation and poly(ADP-ribosylation) are potential mechanisms for inactivating topoisomerases. However, the tyrosine kinases that have been studied to date are not nuclear proteins, and therefore the in vivo significance of tyrosine modification remains problematic. In contrast, poly(ADP-ribose) synthetase is a ubiquitous nuclear enzyme in higher eukaryotes; the evidence that this modification has physiological relevance is more compelling.

Phosphorylation and ADP-ribosylation are not specific to DNA topoisomerases; the best characterized enzymes relevant to DNA topoisomerase modification, casein kinase II and poly(ADP-ribose) synthetase, are known to modify a number of other proteins under physiological conditions. It seems likely that these enzymes are general triggers for changing the physiological state of the nucleus. The cellular function of these DNA topoisomerase modifications is best explored in the larger context of synthetase and casein kinase II physiology, and the entire spectrum of proteins modified needs to be considered.

Casein kinase II may be involved in cellular growth control. The enzyme appears to be activated when cells are proliferating. In contrast, poly(ADP-ribosylation) may be a general nuclear response to unscheduled interruptions in the DNA double helix such as strand breaks due to DNA damage. When the synthetase becomes activated, the enzyme modifies a number of proteins involved in the chromatin structure including histones and topoisomerases.

Thus, changes in the chromatin structure and metabolism may take place in the vicinity of a DNA strand break as a consequence of chromosomal protein ADP-ribosylation. Since sister chromatid exchange increases when synthetase is inhibited in vivo, it was previously suggested that poly(ADP-ribosylation) of topoisomerase may be part of a general mechanism of suppressing mitotic recombination events at DNA strand breaks. Such a mechanism would be important for long-lived multicellular organisms in order to decrease the probability that accumulated recessive mutations become homozygous. Clear evidence has been published recently that the activity of topoisomerases directly affect the frequency of crossing-over events (Christman et al. 1988; Kim and Wang 1989; Wallis et al. 1989).

ACKNOWLEDGMENTS

The research of the authors was supported by grants from the National Institutes of General Medical Sciences GM-25654 (to B.M.O.) and GM-33143 (to N.P.H.).

REFERENCES

Ackerman, P., C.V.C. Glover, and N. Osheroff. 1985. Phosphorylation of DNA topoisomerase II by casein kinase II: Modulation of eukaryotic topoisomerase II activity *in vitro*. *Proc. Natl. Acad. Sci.* **82:** 3164–3168.

———. 1988. Phosphorylation of DNA topoisomerase II *in vivo* and in total homogenates of *Drosophila* Kc cells. The role of casein kinase II. *J. Biol. Chem.* **263:** 12653–12660.

Bjornsti, M. and J.C. Wang. 1987. Expression of DNA topoisomerase I can complement a conditional-lethal topoisomerase I mutation in *E. coli*. *Proc. Natl. Acad. Sci.* **84:** 8971–8975.

Cherney, B.W., O.W. McBride, D. Chen, H. Alkhatib, K. Bhatia, P. Hensley, and M.E. Smulson. 1987. cDNA sequence, protein structure, and chromosomal location of the human gene for poly(ADP-ribose) polymerase. *Proc. Natl. Acad. Sci.* **84:** 8370–8374.

Christman, M.F., F.S. Deitrich, and G.R. Fink. 1988. Mitotic recombination in the rDNA of *S. cerevesiae* is suppressed by the combined action of topoisomerases I and II. *Cell* **55:** 413–425.

Darby, M.K., B. Schmitt, J. Jongstra-Bilen, and H.P. Vosberg. 1985. Inhibition of calf thymus type II DNA topoisomerase by poly ADP-ribosylation. *EMBO J.* **4:** 2129–2134.

Durban, E., M. Goodenough, J. Mills, and H. Busch. 1985. Topoisomerase I phosphorylation *in vitro* and in rapidly growing Novikoff hepatoma cells. *EMBO J.* **4:** 2921–2926.

Durban, E., J.S. Mills, D. Roll, and H. Busch. 1983. Phosphorylation of purified Novikoff hepatoma topoisomerase I. *Biochem. Biophys. Res. Commun.* **111:** 897–905.

Durban, E., D. Roll, G. Beckner, and H. Busch. 1981. Purification and characterization of a nuclear DNA binding phosphoprotein in fetal and tumor tissues. *Cancer Res.* **41:** 537–545.

Edelman, A.M., D.K. Blumenthal, and E.G. Krebs. 1987. Protein serine/threonine kinases. *Annu. Rev. Biochem.* **56:** 567–613.

Ferro, A.M. and B.M. Olivera. 1982. Poly (ADP-ribosylation) *in vitro*. Reaction parameters and enzyme mechanism. *J. Biol. Chem.* **257:** 7808–7813.

———. 1984. Poly (ADP-ribosylation) of DNA topoisomerase I from calf thymus. *J. Biol. Chem.* **259:** 547–559.

———. 1987. Intracellular pyridine nucleotide degradation and turnover. In *Pyridine nucleotide coenzymes* (ed. D. Dolphin et al.), p. 25–77. Wiley, New York.

Ferro, A.M., N.P. Higgins, and B.M. Olivera. 1983. Poly (ADP-ribosylation) of a DNA topoisomerase. *J. Biol. Chem.* **258:** 6000–6003.

Ferro, A.M., L.H. Thompson, and B.M. Olivera. 1984. Poly (ADP-ribosylation) and DNA topoisomerase I in different cell lines. In *Proteins involved in DNA replication* (ed. U. Hubscher and S. Spadari), p. 441–447. Plenum Press, New York.

Goldberg, A.R., T.W. Wong, and Y. Tse-Dinh. 1985. Properties of the major species of tyrosine protein kinase in rat liver: Effect on DNA topoisomerase activity. *Cancer Cells* **3:** 369–379.

Hayaishi, O. and K. Ueda, eds. 1982. *ADP-ribosylation reactions*, p. 674. Academic Press, New York.

Huang, W.M. 1986a. Nucleotide sequence of a type II DNA topoisomerase gene. Bacteriophage T4 gene 39. *Nucleic Acids. Res.* **14:** 7751–7765.

———. 1986b. The 52-protein subunit of T4 DNA topoisomerase is homologous to the *gyr*A protein of gyrase. *Nucleic Acids Res.* **14:** 7379–739O.

Huppi, K., K. Bhatia, D. Siwarski, D. Klinman, B. Cherney, and M. Smulson. 1989. Sequence and organization of the mouse poly (ADP-ribose) polymerase gene. *Nucleic Acids Res.* **9:** 3387–3401.

Jongstra-Bilen, J., M.-E. Ittel, C. Niedergang, H.-P. Vosberg, and P. Mandel. 1983. DNA topoisomerase from calf thymus is inhibited *in vitro* by poly (ADP-ribosylation). *Eur. J. Biochem.* **136:** 391–396.

Kaiserman, H.B., T.S. Ingebritsen, and R.M. Benbow. 1988. Regulation of *Xenopus laevis* DNA topoisomerase I activity by phosphorylation *in vitro*. *Biochemistry* **27:** 3216–3222.

Kim, R.A. and J.C. Wang. 1989. A subthreshold level of DNA topoisomerases leads to the excision of yeast rDNA as extrachromosomal rings. *Cell* **57:** 975–985.

Krupitza, G. and P. Cerutti. 1989. ADP-ribosylation of ADPR-transferase and topoisomerase I in intact mouse epidermal cells JB6. *Biochemistry* **28:** 2034–2040.

Kurosaki, T., H. Ushiro, N. Mitsuuchi, K. Kangawa, H. Matsuo, T. Hirose, S. Inayama, and Y. Shizuta. 1987. Primary structure of human poly (ADP-ribose) synthetase as deduced from cDNA sequence. *J. Biol. Chem.* **262:** 15990–15997.

Mazen, A., J. Menissier-deMurcia, M. Molinete, F. Simonin, G. Gradwohl, G. Poirier, and G. deMurcia. 1989. Poly (ADP-ribose) polymerase: A novel finger protein. *Nucleic Acids Res.* **17:** 4689–4698.

Mills, J.S., H. Busch, and E. Durban. 1982. Purification of a protein kinase from human Namalwa cells that phosphorylates topoisomerase I. *Biochem. Biophys. Res. Commun.* **109:** 1222–1227.

Oikawa, A., H. Tohda, M. Kanai, M. Miwa, and T. Sugimura. 1980. Inhibitors of poly (ADP-ribose) polymerase induce sister chromatid exchanges. *Biochem. Biophys. Res. Commun.* **97:** 1311–1316.

Rottmann, M., H.C. Schröder, M. Gramzow, K. Renneisen, B. Kurelec, A. Dorn, U. Friese, and W.E.G. Müller. 1987. Specific phosphorylation of proteins in pore complex-laminae from the sponge *Geodia cydonium* by the homologous aggregation factor and phorbol ester. Role of protein kinase C in the phosphorylation of DNA topoisomerase II. *EMBO J.* **6:** 3939–3944.

Sahyoun, N., M. Wolf, J. Besterman, T. Hsieh, M. Sander, H. Levine, K. Chang, and Cuatrecasas. 1986. Protein kinase C phosphorylates topoisomerase II: Topoisomerase activation and its possible role in phorbol ester-induced differentiation of HL-60 cells. *Proc. Natl. Acad. Sci.* **83:** 1603–1607.

Sander, M., J.M. Nolan, and T. Hsieh. 1984. A protein kinase activity tightly associated with *Drosophila* type II DNA topoisomerase. *Proc. Natl. Acad. Sci.* **81:** 6938–6942.

Tse-Dinh, Y., T. Wous, and A.R. Goldberg. 1984. Virus- and cell-encoded tyrosine protein kinases inactivate DNA topoisomerases *in vitro*. *Nature* **312:** 785–786.

Uchida, K., T. Morita, T. Sato, T. Ogura, R. Yamashita, S. Noguchi, H. Suzuki, H. Nyunaya, M. Miwa, and T. Sugimura. 1987. Nucleotide sequence of a full-length cDNA for human fibroblast poly (ADP-ribose) polymerase. *Biochem. Biophys. Res. Commun.* **148:** 617–622.

Ueda, K. 1987. ADP-ribosylation of proteins. Enzymology and biological significance. *Mol. Biol. Biochem. Biophys.* **137:** 1–237.

Wallis, J.W., G. Chrebet, G. Brodsky, M. Rolfe, and R. Rothstein. 1989. A hyper recombination mutation in *S. cerevesiae* identifies a novel eukaryotic topoisomerase. *Cell* **58:** 409–419.

Zahradka, P. and K. Ebisuzaki. 1982. A shuttle mechanism for DNA-protein interactions. *Eur. J. Biochem.* **127:** 579–585.

14

Anticancer Drugs That Convert DNA Topoisomerases into DNA Damaging Agents

Leroy F. Liu
Department of Biological Chemistry
Johns Hopkins School of Medicine
Baltimore, Maryland 21205

I. INTRODUCTION

The importance of DNA topoisomerases as molecular targets in clinical pharmacology has become increasingly evident (for review, see Ross 1985; Chen and Liu 1986; Bodley and Liu 1988; Drlica and Franco 1988; Liu 1989). Topoisomerases change the topology of DNA by two fundamentally different mechanisms. Type-I DNA topoisomerases transiently break one strand of the duplex DNA, whereas type-II DNA topoisomerases transiently break both DNA strands during each catalytic cycle. Although the catalytic activity of topoisomerases is important for many DNA functions, topoisomerase-targeting therapeutic agents exert their therapeutic action not through inhibition of the catalytic activity of topoisomerases but through a mechanism involving DNA damage. So far, all known topoisomerase-targeting therapeutics specifically and reversibly block the DNA rejoining step of topoisomerases, resulting in the trapping of a covalent enzyme/DNA intermediate termed the cleavable

DNA Topology and Its Biological Effects

complex (for review, see Liu 1989). The cleavable complex can be irreversibly converted to topoisomerase-linked DNA breaks upon addition of a strong protein denaturant. This type of "DNA lesion" differs from other types of covalent DNA modifications in its reversibility: The lesion disappears upon removal of the inhibitors.

According to this proposed mechanism of drug action, DNA can be viewed as the cotarget of this class of topoisomerase inhibitors. Because many of the cellular effects of these therapeutics are due to their "DNA damaging" effects rather than their inhibition of enzyme activities, this class of topoisomerase inhibitors can be viewed as "topoisomerase poisons" (Kreuzer and Cozzarelli 1979; Liu 1989).

The potential of topoisomerase poisons in clinical pharmacology has yet to be exploited fully. Those that have been in clinical use include the quinolone antibiotics that act on bacterial gyrase (DNA topoisomerase II) (Neu 1988) and anticancer agents that target mammalian DNA topoisomerases I and II (Liu 1989). Their potential use as antitrypanosomal (Shapiro et al. 1989) and antifungal agents (Figgitt et al. 1989) is being developed. Whereas the clinical value of topoisomerase poisons is well-established, the complicated cellular responses to these agents, including their cell-killing mechanism, has just begun to be clarified.

Topoisomerase poisons also serve as powerful investigative tools owing to the unique reversible damage that they introduce into chromosomal DNA. They have been used to probe the function and mechanism of topoisomerases, and they may be of great value in studies of cellular responses to DNA damage.

II. MECHANISM OF ENZYME INHIBITION

A. DNA Topoisomerase I Poisons

The only well-characterized type-I topoisomerase poison is camptothecin. Camptothecin (see Fig. 1 for the chemical structure) was originally isolated from the plant *Camptotheca acuminata* and had been used in Chinese herbal medicine for treating certain types of cancer, as well as ring worm infections (for review, see Horwitz 1975). The impressive antitumor activity in experimental animals had led to brief clinical trials in the United States in the early 1970s, but excessive toxic side effects prevented its further development. The recent identification of topoisomerase I as its primary target and the partial elucidation of its mechanism of action have revived interest in camptothecin and its analogs as potential therapeutic drugs. Camptothecin can exist either in the active lactone form or the much less active hydrolyzed, water soluble, sodium salt form (Hertzberg et al. 1989b). The 20 S-camptothecin, but not the 20 R-

topoisomerase I poison

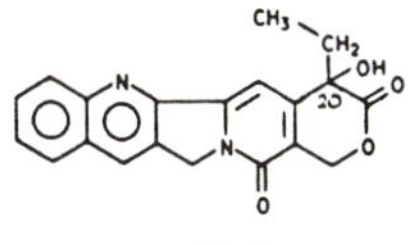

topoisomerase II poisons

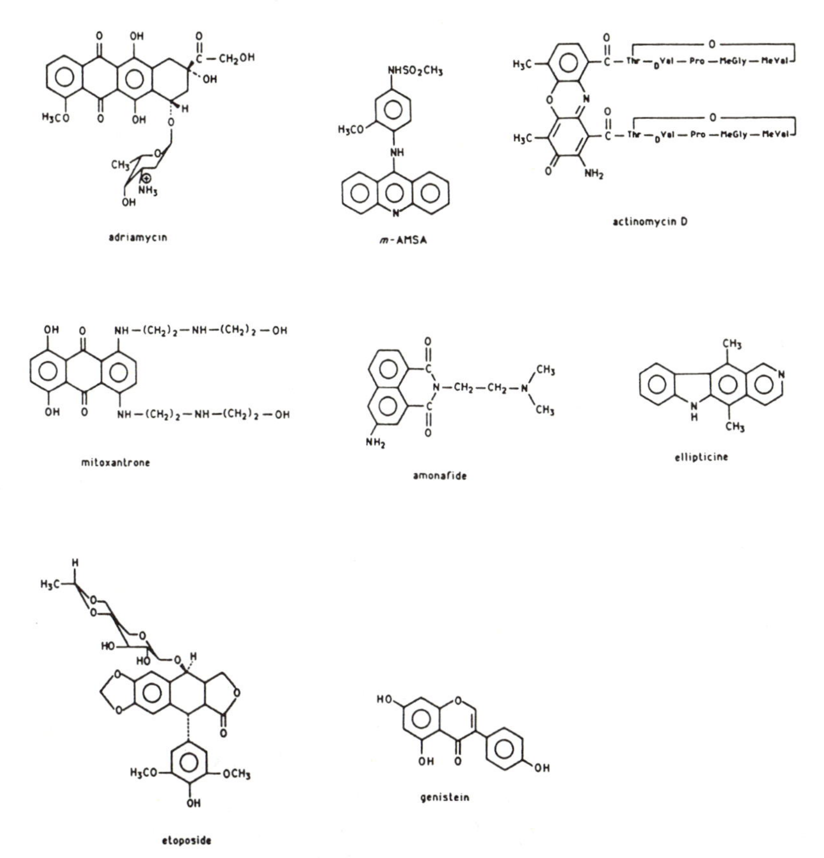

Figure 1 The chemical structures of mammalian topoisomerase poisons.

camptothecin enantiomer, is the biologically active optical isomer (Jaxel et al. 1989). A number of analogs of camptothecin with improved biological activity have been developed (Hsiang et al. 1989c; Jaxel et al. 1989). Two of them, 9-amino-camptothecin and 10,11-methylenedioxy-camptothecin, have been shown to produce long-term remissions in nude mice carrying human colon cancer xenografts (Giovanella et al. 1989).

A number of prominent effects of camptothecin have been noted in early studies in cultured mammalian cells. They include inhibition of DNA and RNA synthesis, chromosomal DNA fragmentation, S-phase-specific cytotoxicity, G_2 delay and/or arrest, elevation of sister chromatid exchanges, and induction of murine erythroleukemia cell (MELC) differentiation (for review, see Hsiang et al. 1985; E. Schneider and L.F. Liu, unpubl.). All these complex cellular effects may now be better explained by the interaction between camptothecin and DNA topoisomerase I (Hsiang et al. 1985, 1989b; Andoh et al. 1987; Hsiang and Liu 1988b; Nitiss and Wang 1988).

Camptothecin interferes with the breakage-rejoining reaction of mammalian DNA topoisomerase I by trapping the cleavable complex (Hsiang et al. 1985). The cleavable complex can be converted to a topoisomerase-I-linked, single-stranded DNA (ssDNA) break in the presence of a strong protein denaturant such as SDS or alkali. Topoisomerase I becomes covalently linked through a unique tyrosine, Tyr-723 in the human enzyme (Lynn et al. 1989), to the 3′-phosphoryl end of the broken DNA strand. The formation of the drug/topoisomerase I/DNA cleavable complex is rapid and highly reversible. A number of treatments, such as a large dilution of the reaction, addition of excess DNA or salts, or brief heating to temperatures above 55°C, are known to reverse rapidly the cleavable complexes (Hsiang and Liu 1988a, 1989). Figure 2 is a schematic diagram showing the mechanism of enzyme inhibition by camptothecin.

The relaxation of DNA by mammalian DNA topoisomerase I (M_r = 100K) is presumed to occur through the sequential formation of at least two types of complexes: the noncleavable complex (Fig. 2B) and the cleavable complex (Fig. 2C). The two complexes are normally at rapid equilibrium, and the formation of the cleavable complex (Fig. 2C) presumably precedes the "DNA strand passage" step (Fig. 2D), leading to linking number changes in closed circular DNAs. Camptothecin presumably binds reversibly to the enzyme/DNA binary complex to form the abortive drug/enzyme/DNA cleavable complex and thereby shifts the equilibrium toward the cleavable complex (Hertzberg et al. 1989a). The addition of a strong protein denaturant to the equilibrium reaction mixture converts a drug-trapped cleavable complex to a topoisomerase-I-linked, ssDNA break. The chemical structure of the drug/enzyme/DNA cleavable complex is not known. It is possible that the phosphodiester bond is cleaved within the cleavable complex and the broken strand containing the 5′-OH end and its complementary strand are not tightly bound by topoisomerase I at all times during the breakage-rejoining cycles (indicated by the horseshoe-shaped representation of topoisomer-

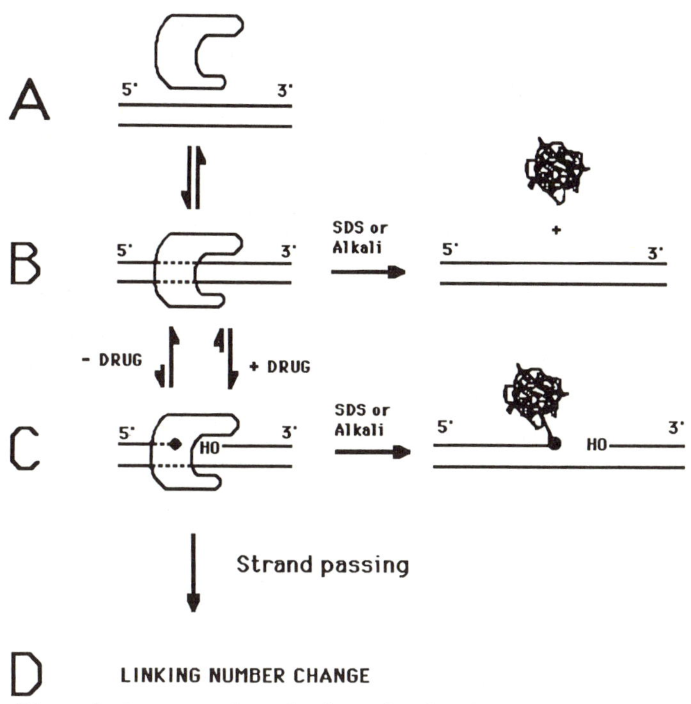

Figure 2 A proposed mechanism of action for mammalian DNA topoisomerase I poisons.

ase I in Fig. 2C). The protein/DNA contact(s) may be located primarily on the same side as the strand whose 3′-phosphoryl end is covalently linked to topoisomerase I. The strand complementary to the 5′-OH-end-containing strand presumably keeps the transiently broken strand containing the free 5′-OH end close to the protein/DNA complex. The lack of strong interaction between topoisomerase I and the cleaved 5′-OH-containing strand is supported by two observations: First, mammalian DNA topoisomerase I, unlike bacterial DNA topoisomerase I, can cleave ssDNA in the absence of a protein denaturant, and the cleaved ssDNA can rejoin covalently to another DNA molecule (Been and Champoux 1981; Halligan et al. 1982); and second, mammalian DNA topoisomerase I can cause spontaneous linearization (at the site of the nick) of a nicked circular DNA in the absence of a protein denaturant (McCoubrey and Champoux 1986).

B. DNA Topoisomerase II Poisons

Whereas bacterial DNA topoisomerase II (DNA gyrase) has been known for some time to be the target of quinolone antibiotics such as nalidixic acid and oxolinic acid (for review, see Cozzarelli 1980; Gellert 1981), mammalian DNA topoisomerase II has only recently been identified as the primary target of a number of structurally diverse anticancer drugs (for review, see Liu 1989). A partial list of these anticancer drugs is shown in Table 1. The chemical structures of some of them are shown in Figure 1.

Similar to bacterial DNA topoisomerase II poisons, mammalian DNA topoisomerase II poisons interfere with the breakage-rejoining reaction by trapping the cleavable complex (Liu et al. 1983; Nelson et al. 1984; Tewey et al. 1984a,b; Chen et al. 1984; Hsiang et al. 1989a; Markovits et al. 1989). The chemical structure of the cleavable complex is still unclear. However, some of its unusual properties have suggested the following mechanism of enzyme inhibition (Fig. 3). Similar to the interaction between mammalian DNA topoisomerase I and camptothecin, mammalian DNA topoisomerase II, a homodimeric protein (M_r = 170K), binds to DNA to form at least two different types of complexes in the absence of ATP (Liu et al. 1983; Nelson et al. 1984). These two complexes, termed the cleavable complex (Fig. 3C) and the noncleavable complex (Fig. 3B), are in rapid equilibrium. The enzyme-mediated strand-passing that follows cleavable complex formation may result from the collision between the cleavable complex and another duplex DNA (Fig. 3D). The requirement for ATP in the strand-passing reaction is unclear but is likely to be for "resetting" the enzyme back to the original conformation as has been suggested for other ATP-dependent type-II DNA topoisomerases (for review, see Cozzarelli 1980; Gellert 1981; Wang 1985).

Table 1 Mammalian DNA topoisomerase II poisons

Class	Examples
Intercalators	
anthracyclines	adriamycin, daunomycin
benzisoquinolinediones	amonafide, mitonafide
anthracenediones	mitoxantrone, bisantrene
acridines	*m*-AMSA
actinomycins	actinomycin D
ellipticines	2-methyl-9-hydroxy-ellipticinium acetate
Nonintercalators	
epipodophyllotoxins	etoposide (VP-16) teniposide (VM-26)
isoflavonoids	genestein

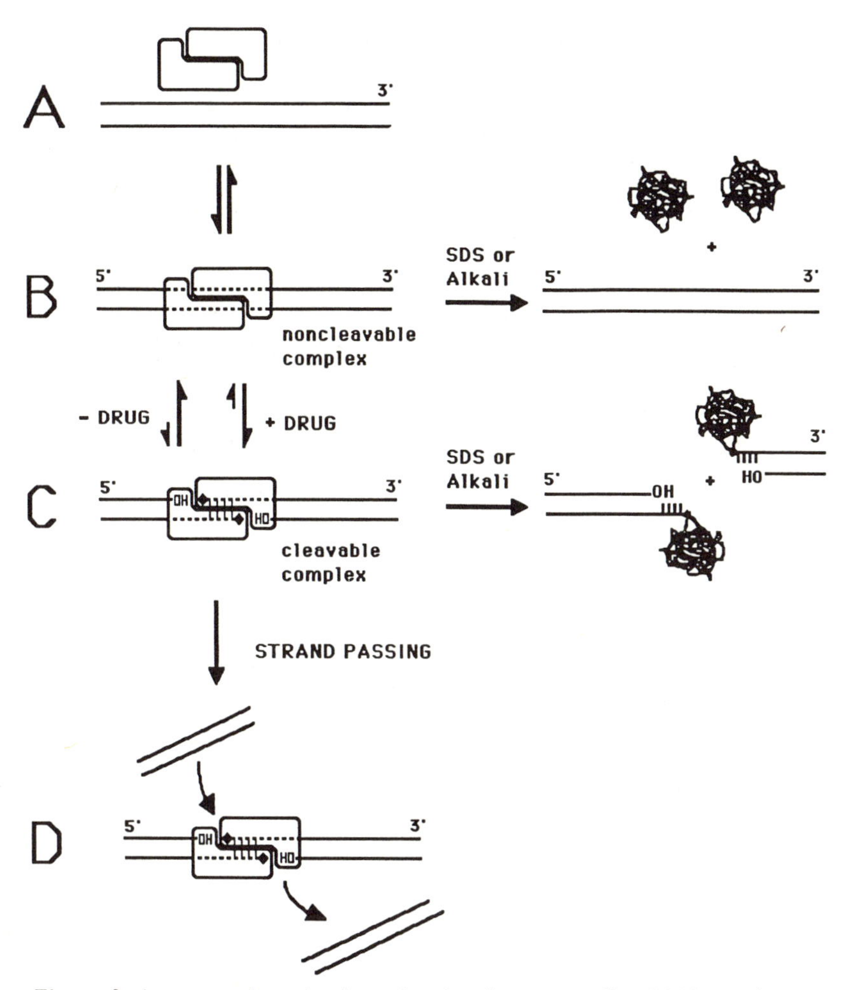

Figure 3 A proposed mechanism of action for mammalian DNA topoisomerase II poisons.

Studies of topoisomerase-II-linked DNA breaks have shown that a topoisomerase II polypeptide is covalently linked to the 5′-phosphoryl end of the broken DNA strand, and in the case of a double-strand break, the 3′-OH end is recessed by 4 bases (Liu et al. 1983; Sander and Hsieh 1983). The covalent linkage is between the 5′-phosphoryl end of DNA and Tyr-804 of the human enzyme (Rowe et al. 1986b; Tsai-Pflugfelder et al. 1988).

How topoisomerase II is poisoned by the structurally diverse drugs is unclear. One important clue is that, except for epipodophyllotoxins and isoflavonoids, all known mammalian topoisomerase II poisons are DNA intercalators that insert part of their planar structures between two adjacent base pairs in duplex DNA, causing a reduction in the rotation angle between adjacent base pairs by 10°–30° (see Table 1) (Waring 1981; Wilson and Jones 1981). A second is that, for intercalative poisons, there is an excellent correlation between the strength of intercalation and the efficiency of cleavable complex formation (Bodley et al. 1989). On the basis of these and other considerations, a misalignment model has been proposed to explain the inhibition mechanism by intercalative topoisomerase II poisons (Fig. 4).

In this model (see Fig. 4), topoisomerase II is proposed to contact DNA at at least two sites that flank the breakage point, thereby separating the DNA into two topological domains: the bound and unbound regions. In the absence of ATP, topoisomerase II may still carry out breakage-rejoining cycles that are uncoupled from strand-passing (Liu et al. 1983). The intercalation of drug within the bound DNA region effec-

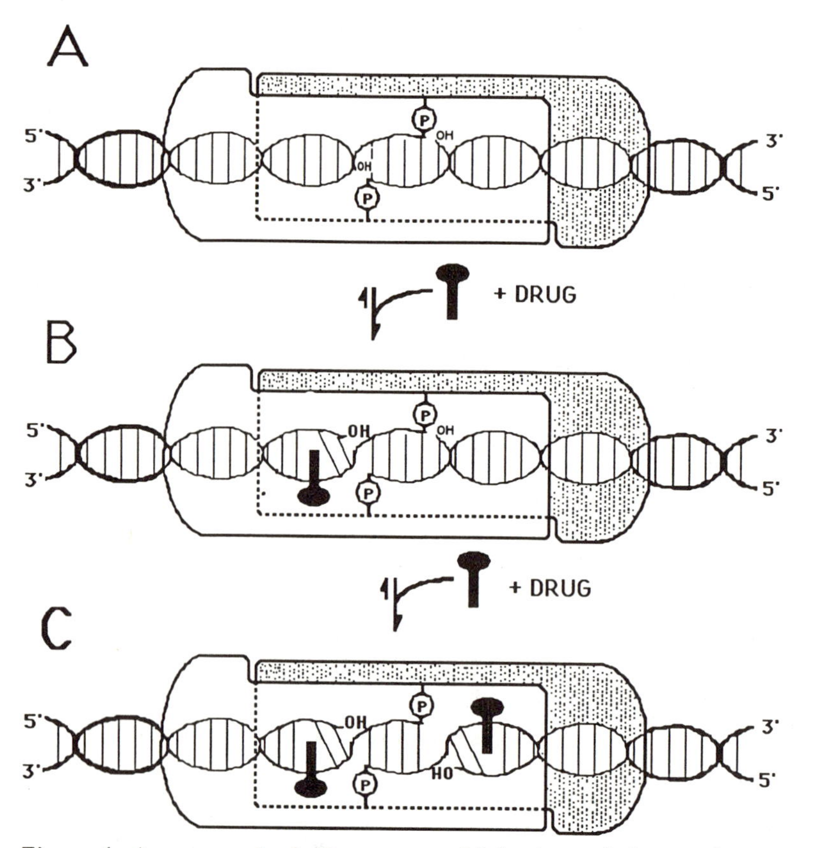

Figure 4 A proposed misalignment model for intercalative topoisomerase II poisons.

tively displaces the two transiently broken ends from their aligned position. DNA in the misaligned state may define the cleavable complex. The covalent interaction between the 5′-phosphoryl ends of the transiently broken DNA and the two topoisomerase subunits may provide an additional protein/DNA contact that further separates the bound DNA into two topological subdomains. The cleavable complex can form in the presence of either one (Fig. 4B) or two (Fig. 4C) bound drug molecules. The bulky side chain(s) of the drug is presumably important for steric interaction of the drug with topoisomerase II. According to this model, specific- and high-affinity interaction of the drug with the enzyme is not essential, which explains the diversity of drug structures of topoisomerase II poisons.

The misalignment model at the moment is still very speculative. The model cannot explain the action of nonintercalative topoisomerase II poisons such as epipodophyllotoxins and genestein. One possibility is that the nonintercalative epipodophyllotoxins may be very weak intercalators. However, experiments done so far have failed to demonstrate such an interaction (Chen et al. 1984; H.-Y. Wu and L.F. Liu, unpubl.). Interestingly, studies of quinolone antibiotics (bacterial DNA topoisomerase-II poisons) have also indicated that binding to DNA (and possibly intercalation) is a crucial parameter for trapping cleavable complexes (Shen and Pernet 1985).

III. CELLULAR RESPONSES TO TOPOISOMERASE POISONS

Topoisomerase poisons elicit rather diverse cellular responses. Many of these cellular responses are not directly related to inhibition of the enzymatic activities of the topoisomerases but are the consequences of the unusual cellular DNA damage induced by topoisomerase poisons. Some of the responses to topoisomerase poisons are discussed below with emphasis on mammalian cells.

A. Site-specific Breakage of Chromosomal DNA

One of the most prominent cellular effects of topoisomerase poisons is the apparent breakage of chromosomal DNA (for review, see Liu 1989). It is almost certain that the majority of the breaks that are rapidly induced in cells treated with topoisomerase poisons are due to the formation of the cleavable complexes. As expected for cleavable complexes, the breakage of chromosomal DNA in cells is reversible and is revealed only upon treatment with a protein denaturant. Each broken DNA strand is covalently linked to a topoisomerase polypeptide. One of the most convincing demonstrations for the presence of reversible cleavable complexes in cells treated with topoisomerase poisons is the heat reversal experiment. A brief heating (3 min) of cultured cells treated with topoisomerase poisons to temperatures above 55°C rapidly dissociated cleavable complexes. The reversal of the cleavable complex formation was evidenced by the religation of protein-linked DNA breaks and the release of bound topoisomerase molecules from chromosomal DNA (Hsiang and Liu 1988b; 1989).

The sites of breakage on chromosomal DNA induced by a number of topoisomerase poisons have been mapped. At a concentration of 25 μM, the topoisomerase I poison camptothecin can trap more than 80% of cellular topoisomerase I (about 10^5 to 10^6 copies/cell) on chromosomal DNA (Hsiang and Liu 1988b). These breakage sites have been localized within the transcribed region of a number of actively transcribed genes (Gilmour and Elgin 1987; Rowe et al. 1987; Stewart and Schutz 1987; Zhang et al. 1988). These mapping results are consistent with immunological studies that also show a strong association of topoisomerase I with actively transcribed genes (Fleischmann et al. 1984; Gilmour et al. 1986). This association may imply a functional involvement of topoisomerase I in RNA transcription.

Topoisomerase II poisons have also been used to localize topoisomerase II on chromatin (Yang et al. 1985; Rowe et al. 1986a). Topoisomerase II molecules are generally localized in nucleosome-free regions and are not enriched in actively transcribed regions (Rowe et al. 1986a). Active transcription, however, does alter the distribution of topoisomerase II cleavage sites on *Drosophila* hsp70 genes, especially at the 3′ end of the genes (Rowe et al. 1986a). However, the involvement of topoisomerase II in RNA transcription is not established.

Topoisomerase II cleavage sites on pBR322 DNA have also been determined in *Escherichia coli* cells treated with the gyrase poison, oxolinic acid (Lockshon and Morris 1985). The cleavage sites are widely distributed throughout the plasmid DNA. However, major gyrase cleavage sites are concentrated in between the 3′ ends of two divergently transcribed genes, *bla* and *tetA* (Lockshon and Morris 1985). More recent studies have shown that the enhanced gyrase cleavage between the two divergently transcribed genes requires both RNA transcription and protein synthesis. The use of either rifampicin or chloramphenicol reduced overall cleavage, especially in the region between the 3′ ends of *tetA* and *bla* genes (H.S. Koo et al., unpubl.). These mapping studies suggest that the local DNA conformation is altered significantly by the process of RNA transcription and support the "twin-supercoiled-domain" model of transcription (Liu and Wang 1987; Giaever and Wang 1988; Wu et al. 1988).

B. Inhibition of DNA Replication

All topoisomerase poisons strongly inhibit cellular DNA synthesis (for review, see Liu 1989). Inhibition of DNA synthesis may be critically involved in a number of cellular responses (e.g., cell death, induction of sister chromatid exchanges, and G_2 block). However, the mechanism of replication inhibition by topoisomerase poisons is poorly understood (Snapka 1986; Yang et al. 1987; Snapka et al. 1988). Recent studies both in cultured cells and in a cell-free SV40 replication system have suggested that the collision between the replication forks and drug-induced cleavable complexes results in replication fork arrest and possibly fork breakage (Hsiang et al. 1989b). One may speculate that such a collision could also be responsible for the strong induction of the SOS response in *E. coli* cells treated with gyrase poisons such as nalidixic and oxolinic acids (Drlica 1984). Studies in mammalian cells have suggested that the collision between the replication forks and cleavable complexes is responsible for G_2 arrest (P. D'Arpa et al., unpubl.) and cell death (Hsiang et al. 1989b). Whether such a collision also leads to other cellular responses awaits further studies.

C. Inhibition of RNA Synthesis

Topoisomerase poisons also inhibit RNA synthesis. However, the mechanism of inhibition is not clear. Most investigations have been done using the mammalian DNA topoisomerase I poison camptothecin. Camptothecin can reversibly inhibit RNA synthesis, especially the synthesis of high-molecular-weight RNA (for review, see Horwitz 1975). Nuclear run-on studies of human rRNA genes have indicated that the inhibitory effect is mostly on the elongation step of RNA synthesis (Zhang et al. 1988). The inhibitory effect on transcription elongation may be due to two different inhibitory mechanisms: inhibition of the enzymatic activity of topoisomerase I and the formation of drug/enzyme/DNA cleavable complexes on the template DNA. Although the contribution from these two different mechanisms for inhibition of RNA transcription is not clear, the formation of cleavable complexes on the DNA template has been shown to be responsible for the inhibition in vitro using purified bacterial RNA polymerases (H.Y. Hsiang and L.F. Liu, unpubl.).

The effects of topoisomerase II poisons on transcription are complex. The mammalian topoisomerase II poison VP-16 can either stimulate or inhibit uridine incorporation into RNA, depending on the doses (Grieder et al. 1974). However, nuclear run-on studies of human rRNA genes showed an initial decline in transcription by RNA polymerase I followed by rapid recovery and then elevated levels of RNA transcription (H. Zhang and L.F. Liu, unpubl.). These apparently conflicting results remain to be resolved.

D. Inhibition of Cell Division

Topoisomerase poisons, like other DNA damaging agents, also cause G_2 delay or arrest in cultured mammalian cells (Rao 1980). Very little is known about the mechanism of G_2 arrest in mammalian cells. In *E. coli,* cell division is coupled to DNA replication by at least three different mechanisms, all of which involve the essential cell division regulator *SfiB (SulB).* Inhibition of DNA replication results in cell-division inhibition (filamentation) because of the the activation of a number of cell-division inhibitors (e.g., *SulA* and *SfiC*) that inhibit the function of *SfiB (SulB)* (for review, see Liu 1989). Bacterial topoisomerase II poison nalidixic acid is a potent inhibitor of DNA synthesis and induces a number of cell-division inhibitors, one of which is the SOS protein *SulA (SfiA)* (Walker 1984). *SulA* has been shown to be responsible for acute cell killing in *E. coli lon*$^-$ strains (Walker 1984). In yeast, *RAD9* has been shown to be responsible for G_2 arrest upon treatment with DNA-damaging agents (Weinert and Hartwell 1988). Whether G_2 delay and/or arrest in mammalian cells in response to topoisomerase poisons is also controlled by a complicated network of cell-division inhibitors remains to be tested. The recent demonstration that the maturation promoting factor (MPF) is critically involved in the entry of G_2 cells into mitosis suggests that MPF might be involved in damage-induced G_2 delay and/or arrest in mammalian cells (for review, see Dunphy and Newport 1988).

E. Other Cellular Responses

Mammalian DNA topoisomerase poisons, like many other anticancer drugs that affect DNA metabolism, induce sister chromatid exchanges and chromosomal aberrations (Lim et al. 1986; Pommier et al. 1988), differentiation of MELC and HL60 cells (Marks and Rifkind 1978; A. Bodley and L.F. Liu, unpubl.), extensive degradation of chromosomal DNA into nucleosome-size DNA (Jaxel et al. 1988; Kaufmann 1989), and induction of the transcription of certain heat-shock genes (Rowe et al. 1986a; J.L. Hwang, pers. comm.). Although the biochemical mechanisms of these responses are not known, they are more likely to be related to

DNA damage induced by topoisomerase poisons than to the inhibition of the catalytic activity of topoisomerases.

IV. PHARMACOLOGICAL ASPECTS OF DNA TOPOISOMERASES

A. Mechanism of Cell Killing

Mammalian topoisomerase poisons are primarily used as anticancer drugs. A fundamental question concerns the mechanism of their tumor cell killing. Extensive studies in mammalian cells have suggested that tumor cell killing, like many other cellular responses to topoisomerase poisons, is related to the formation of a drug/enzyme/DNA cleavable complex rather than the inhibition of topoisomerase activity (for review, see Liu 1989).

The mechanisms of tumor cell killing by topoisomerase poisons are likely to be complex. The reversible nature of cleavable complexes induced by topoisomerase poisons has led to the suggestion that interaction of cleavable complexes with other cellular functions (e.g., replication machinery, helicases, proteases, or nucleases) is necessary to trigger cell death (Nelson et al. 1984). Recent studies in mammalian cells (Hsiang et al. 1989b) and in yeast (J. Nitiss and J.C. Wang, unpubl.) have suggested that the moving replication fork is one of the major cellular functions whose interaction with cleavable complexes leads to cell death and possibly other cellular responses (Fig. 5).

Studies in *E. coli lon*$^-$ strains have convincingly demonstrated that acute cell killing by topoisomerase poisons is not due to DNA degradation but involves the induction of an SOS protein *SulA (SfiA)* (Walker 1984). It is interesting to note that all antitumor agents that affect DNA metabolism are potential bacterial SOS inducers. The complex cellular responses to topoisomerase poisons in mammalian cells are also reminiscent of some of the SOS responses in bacteria (for review, see Liu 1989). It will be interesting to determine if cell division inhibition is at least in part responsible for tumor cell killing.

B. Topoisomerases in Tumor Cells

The cellular levels of human topoisomerases in tumor cells are important parameters in determining the efficacy of topoisomerase poisons in cancer chemotherapy (Potmesil et al. 1988). It is not surprising that both topoisomerase I and II are abundant enzymes in tumor cells (Chow and Ross 1987; Nelson et al. 1987; Heck et al. 1988; Hsiang et al. 1988; Potmesil et al. 1988). However, the regulatory mechanisms for topoisomerases I and II in tumor cells are quite different. The level of topoisomerase

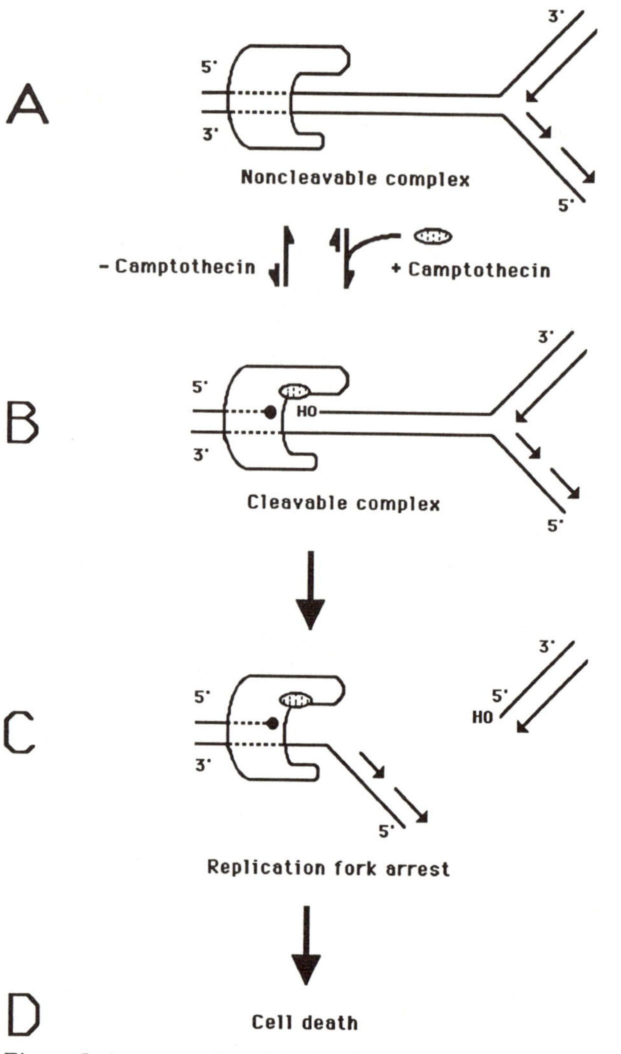

Figure 5 A proposed model of cell killing by topoisomerase I poisons.

II in cultured cells parallels the state of cell proliferation (Chow and Ross 1987; Hsiang et al. 1988). High levels of topoisomerase II are associated with proliferating cells, whereas very low levels of topoisomerase II are present in quiescent or differentiated cells (Bodley et al. 1987; Chow and Ross 1987; Hsiang et al. 1988). Growth parameters (e.g., cell density and serum growth factors) that affect cell proliferation show parallel effects on topoisomerase II levels. However, tumor cells maintain high levels of topoisomerase II probably because of their reduced sensitivity to these growth parameters. The high levels of topoisomerase II in tumor cells may be partly responsible for the high efficacy of topoisomerase II poisons in cancer chemotherapy.

Topoisomerase I, on the other hand, is present at a rather high level (approximately 10^5 copies/cell) in primary human skin fibroblasts regardless of the growth state of cells (Hsiang et al. 1988). Its level is also maintained relatively constant during MELC differentiation (Bodley et al. 1987). However, elevated levels of topoisomerase I have been detected in surgical specimens of human colon cancer. Furthermore, the level of topoisomerase I in human colon cancer cells parallels disease progression (Giovanella et al. 1989). These apparently conflicting results are further confronted with another surprising finding: Topoisomerase I mRNA, similar to c-*myc* mRNA, is transiently (2–10 hr poststimulation) elevated up to tenfold during mitogenic or TPA (a tumor promoter) stimulation of human skin fibroblasts without affecting the protein level significantly (Hwong et al. 1989). One speculation concerning the high levels of topoisomerase I in certain colon cancer cells is that altered regulation of topoisomerase I in tumor cells leads to constitutively high levels of topoisomerase I mRNA and protein. The recent demonstration of unprecedented antitumor activity of 9-amino-camptothecin against human colon cancer xenografts warrants extensive investigation into the possible altered regulation of topoisomerase I in tumor cells and encourages further development of antitumor drugs that are targeted to topoisomerase I (Giovanella et al. 1989).

ACKNOWLEDGMENTS

We are grateful to Drs. James C. Wang and Nicholas R. Cozzarelli, Peter D'Arpa, Hyeon-Sook Koo, Erasmus Schneider and Annette Bodley, and Mr. Kawai Lau for critical reading of the manuscript. This work was supported by an National Institutes of Health grant CA-39662.

REFERENCES

Andoh, T., K. Ishii, Y. Suzuki, Y. Ikegami, Y. Kusunoki, Y. Takemoto, and K. Okada. 1987. Characterization of a mammalian mutant with a camptothecin resistant DNA topoisomerase I. *Proc. Natl. Acad. Sci.* **84:** 5565–5569.

Been, M.D. and J.J. Champoux. 1981. DNA breakage and closure by rat liver type I topoisomerase: Separation of the half reactions by using a single-stranded DNA substrate. *Proc. Natl. Acad. Sci.* **78:** 5498–5502.

Bodley, A.L. and L.F. Liu. 1988. Topoisomerase as novel targets for cancer chemotherapy. *Biotechnology* **6:** 1315–1319.

Bodley, A.L., H.-Y. Wu, and L.F. Liu. 1987. Regulation of DNA topoisomerases during cellular differentiation. *NCI Monogr.* **4:** 31–35.

Bodley, A.L., L.F. Liu, M. Israel, F.C. Giuliani, R. Silber, S. Kirschenbaum, and M. Potmesil. 1989. DNA topoisomerase II-mediated Interaction of doxocrubicin and daunomycin congeners with DNA. *Cancer Res.* **49:** 5969–5978.

Chen, G.L. and L.F. Liu. 1986. DNA topoisomerases as therapeutic targets in cancer chemotherapy. *Annu. Rep. Med. Chem.* **21:** 257–262.

Chen, G.L., L. Yang, T.C. Rowe, B.D. Halligan, K.M. Tewey, and L.F. Liu. 1984. Nonintercalative antitumor drugs interfere with the breakage-reunion reaction of mammalian DNA topoisomerase II. *J. Biol. Chem.* **259:** 13560–13566.

Chow, K.C. and W.E. Ross. 1987. Topoisomerase-specific drug sensitivity in relation to cell cycle progression. *Mol. Cell. Biol.* **7:** 3119–3123.

Cozzarelli, N.R. 1980. DNA gyrase and the supercoiling of DNA. *Science* **207:** 953–960.

Drlica, K. 1984. Biology of bacterial DNA topoisomerases. *Microbiol. Rev.* **48:** 273–289.

Drlica, K. and R.J. Franco. 1988. Inhibitors of DNA topoisomerases. *Biochemistry* **27:** 2253–2259.

Dunphy, W.G. and J.W. Newport. 1988. Unraveling of mitotic control mechanisms. *Cell* **55:** 925–928.

Figgitt, D.P., S.P. Denyer, P.M. Dewick, D.E. Jackson, and P. Williams. 1989. Topoisomerase II: A potential target for novel antifungal agents. *Biochem. Biophys. Res. Commun.* **160:** 257–262.

Fleischmann, G., G. Pflugfelder, E.K. Steiner, K. Javaherian, G.C. Howard, J.C. Wang, and C.R. Elgin. 1984. *Drosophila* DNA topoisomerase I is associated with transcriptionally active regions of the genome. *Proc. Natl. Acad. Sci.* **81:** 6958–6962.

Gellert, M. 1981. DNA topoisomerases. *Annu. Rev. Biochem.* **50:** 879–910.

Giaever, G.N. and J.C. Wang. 1988. Supercoiling of intracellular DNA can occur in eukaryotic cells. *Cell* **55:** 849–856.

Gilmour, D.S. and C.R. Elgin. 1987. Localization of specific topoisomerae I interactions within the transcribed region of active heat shock genes by using the inhibitor camptothecin. *Mol. Cell. Biol.* **7:** 141–148.

Gilmour, D.S., G. Pflugfelder, J.C. Wang, and J.T. Lis. 1986. Topoisomerase I interactions within the transcribed region of active heat shock genes by using the inhibitor camptothecin. *Mol. Cell. Biol.* **7:** 141–148.

Giovanella, B.C., J.S. Stehlin, W.E. Wall, M.C. Wani, A.W. Nicholas, L.F. Liu, R. Silber, and M. Potmesil. 1989. DNA topoisomerase I targeted chemotherapy of human colon cancer in xenografts. *Science* **246:** 1046–1048.

Grieder, A., R. Maurer, and H. Stahelin. 1974. Effect of an epipodophyllotoxin derivative (VP-16-213) on macromolecular synthesis and mitosis in mastocytoma cells in vitro. *Cancer Res.* **34:** 1788–1793.

Halligan, B.D., J.L. Davis, K.A. Edwards, and L.F. Liu. 1982. Intra- and intermolecular strand transfer by HeLa DNA topoisomerase I. *J. Biol. Chem.* **257:** 3995–4000.

Heck, M.M., W.N. Hittelman, and W.C. Earnshaw. 1988. Differential expression of DNA topoisomerase I and II during the eukaryotic cell cycle. *Proc. Natl. Acad. Sci.* **85:** 1086–1090.

Hertzberg, R.P., M.J. Caranfa, and S.M. Hecht. 1989a. On the mechanism of topoisomerase I inhibition by camptothecin: Evidence for binding to an enzyme-DNA complex. *Biochemistry* **28:** 4629–4638.

Hertzberg, R.P., M.J. Caranfa, K.G. Holdern, D.R. Jakas, G. Gallagher, M.R. Mattern, S. Mong, J.O. Bartus, R.K. Johnson, and W.D. Kingsbury. 1989b. Modification of the hydroxy lactone ring of camptothecin: Inhibition of mammalian topoisomerase I and biological activity. *J. Med. Chem.* **32:** 715–720.

Horwitz, S.B. 1975. Camptothecin. In *Antibiotics: Mechanism of action of antimicrobial and antitumor agents* (ed. J.W. Corcoran and F.E. Hahn), vol. 3, p. 48–57. Springer-Verlag, New York.

Hsiang, Y. and L.F. Liu. 1988a. Identification of mammalian DNA topoisomerase I as an intracellular target of the anticancer drug camptothecin. *Cancer Res.* **48:** 1722–1726.

———. 1988b. Identification of mammalian DNA topoisomerase I as an intracellular target of the anticancer drug camptothecin. *Cancer Res.* **48:** 1722–1726.

———. 1989. Evidence for the reversibility of cellular DNA lesion induced by mammalian topoisomerase II poisons. *J. Biol. Chem.* **264:** 9713–9715.

Hsiang, Y., J. Jiang, and L.F. Liu. 1989a. Topoisomerase II-mediated DNA cleavage by amonafide and its structural analogs. *Mol. Pharmacol.* **36:** 371–376.

Hsiang, Y., M.G. Lihou, and L.F. Liu. 1989b. Arrest of replication forks by drug-stabilized topoisomerase I-DNA cleavable complexes as a mechanism of cell killing by camptothecin. *Cancer Res.* **49:** 5077–5082.

Hsiang, Y., H.-Y. Wu, and L.F. Liu. 1988. Proliferation-dependent regulation of DNA topoisomerase II in cultured human cells. *Cancer Res.* **48:** 3230–3235.

Hsiang, Y., R. Hertzberg, S. Hecht, and L.F. Liu. 1985. Camptothecin induces protein-linked DNA breaks via mammalian DNA topoisomerase I. *J. Biol. Chem.* **260:** 14873–14878.

Hsiang, Y., L.F. Liu, M.E. Wall, N.C. Wani, A.W. Nicholas, G. Manikumar, S. Kirschenbaum, R. Silber, and M. Potmesil. 1989c. DNA topoisomerase I mediated DNA cleavage and cytotoxicity of camptothecin analogs. *Cancer Res.* **49:** 4385–4389.

Hwong, C.L., M.S. Chen, and J. Hwang. 1989. Phorbol ester transiently increases topoisomerase I mRNA levels in human skin fibroblasts. *J. Biol. Chem.* **264:** 14923–14926.

Jaxel, C., K.W. Kohn, M.C. Wani, W.E. Wall, and Y. Pommier. 1989. Structure-activity study of the actions of camptothecin derivatives on mammalian topoisomerase I: Evidence for a specific receptor site and a relation to antitumor activity. *Cancer Res.* **49:** 1465–1469.

Jaxel, C., G. Taudou, C. Portemer, G. Mirambeau, J. Panijel, and M. Duguet. 1988. Topoisomerase inhibitors induce irreversible fragmentation of replicated DNA in concanavalin A stimulated splenocytes. *Biochemistry* **27:** 95–100.

Kaufmann, S.H. 1989. Induction of endonucleolytic DNA cleavage in human acute myelogenous leukemia cells by etoposide, camptothecin and other cytotoxic anticancer drugs. *Cancer Res.* **49:** 5870–5878.

Kreuzer, K.N. and N.R. Cozzarelli. 1979. *Escherichia coli* mutants thermosensitive for deoxyribonucleic acid gyrase subunit A: Effects on deoxyribonucleic acid replication, transcription, and bacteriophage growth. *J. Bacteriol.* **140:** 424–435.

Lim, M., L.F. Liu, D. Kacobson-Kram, and J.R. Williams. 1986. Induction of sister chromatid exchanges by inhibitors of topoisomerases. *Cell Biol. Toxicol.* **2:** 485–494.

Liu, L.F. 1989. DNA topoisomerase poisons as antitumor drugs. *Annu. Rev. Biochem.* **58:** 351–375.

Liu, L.F. and J.C. Wang. 1987. Supercoiling of the DNA template during transcription. *Proc. Natl. Acad. Sci.* **75:** 2098–2102.

Liu, L.F., T.C. Rowe, L. Yang, K.M. Tewey, and G.L. Chen. 1983. Cleavage of DNA by mammalian DNA topoisomerase II. *J. Biol. Chem.* **258:** 15365–15370.

Lockshon, D. and D.R. Morris. 1985. Sites of reaction of *Escherichia coli* DNA gyrase on pBR322 *in vivo* as revealed by oxolinic acid-induced plasmid linearization. *J. Mol. Biol.* **181:** 63–74.

Lynn, R.M., M.A. Bjornsti, P.R. Caron, and J.C. Wang. 1989. Peptide sequencing and site-directed mutagenesis identify tyrosine-727 as the active site tyrosine of *Saccharomyces cerevisiae* DNA topoisomerase I. *Proc. Natl. Acad. Sci.* **86:** 3559–3563.

Markovits, J., C. Linassier, P. Fosse, J. Couprie, J. Pierre, A. Jacquemin-Sablon, J.-M. Saucier, J.-B. Le Pecq, and A.K. Larsen. 1989. Inhibitory effects of the tyrosine kinase inhibitor genestein on mammalian DNA topoisomerase II. *Cancer Res.* **49:** 5111–5117.

Marks, P.A. and R.A. Rifkind. 1978. Erythroleukemic differentiation. *Annu. Rev. Biochem.* **47:** 419–448.

McCoubrey, W.K., Jr. and J.J. Champoux. 1986. The role of single-strand breaks in the catenation reaction catalyzed by the rat type I topoisomerase. *J. Biol. Chem.* **261:** 5130–5137.

Nelson, E.M., K.M. Tewey, and L.F. Liu. 1984. Mechanism of antitumor drugs. Poisoning of mammalian DNA topoisomerase II on DNA by an antitumor drug m-AMSA. *Proc. Natl. Acad. Sci.* **81:** 1361–1365.

Nelson, W.G., K.R. Cho, Y.-H. Hsiang, L.F. Liu, and D.S. Coffey. 1987. Growth-related elevations of DNA topoisomerase II levels found in Dunning R3327 rat prostatic adenocarcinomas. *Cancer Res.* **47:** 3246–3250.

Neu, H.C. 1988. Quinolones: A new class of antimicrobial agents with wide potential uses. *Med. Clin. North Am.* **72:** 623–636.

Nitiss, J. and J.C. Wang. 1988. DNA topoisomerase-targeting antitumor drugs can be studied in yeast. *Proc. Natl. Acad. Sci.* **85:** 7501–7505.

Pommier, Y., D. Kerrigan, J.M. Covey, C. Kao-Shan, and J. Whang-Peng. 1988. Sister chromatid exchanges, chromosomal aberrations, and cytotoxicity produced by antitumor topoisomerase II inhibitors in sensitive (DC3F) and resistant (DC3F/9-OHE) Chinese hamster cells. *Cancer Res.* **48:** 512–516.

Potmesil, M., Y. Hsiang, L.F. Liu, B. Bank, H. Grossberg, S. Kirschenbaum, A. Penziner, D. Kanganis, D. Knowles, F. Traganos, and R. Silber. 1988. Resistance of human leukemic and normal lymphocytes to DNA cleavage by drugs correlates with low levels of DNA topoisomerase II. *Cancer Res.* **48:** 3537–3543.

Rao, P.N. 1980. The molecular basis of drug-induced G2 arrest in mammalian cells. *Mol. Cell. Biochem.* **29:** 47–57.

Ross, W.E. 1985. DNA topoisomerases as targets for cancer therapy. *Biochem. Pharmacol.* **34:** 4191–4195.

Rowe, T.C., E. Couto, and D.J. Kroll. 1987. Camptothecin inhibits *Hsp* 70 heat-shock transcription and induces DNA strand breaks in *hsp* 70 genes in *Drosophila*. *NCI Monogr.* **4:** 49–53.

Rowe, T.C., J.C. Wang, and L.F. Liu. 1986a. *In vivo* localization of DNA topoisomerase II cleavage sites on *Drosophila* heat-shock chromatin. *Mol. Cell. Biol.* **6:** 985–992.

Rowe, T.C., G.L. Chen, Y.-H. Hsiang, and L.F. Liu. 1986b. DNA damage by antitumor acridines mediated by mammalian DNA topoisomerase II. *Cancer Res.* **46:** 2021–2026.

Sander, M. and T.S. Hsieh. 1983. Double strand DNA cleavage by type II DNA topoisomerase from *Drosophila melanogaster*. *J. Biol. Chem.* **258:** 8421–8428.

Shapiro, T.A., V.A. Klein, and P.T. Englund. 1989. Drug-promoted cleavage of kinetoplast DNA minicircles. *J. Biol. Chem.* **264:** 4173–4178.

Shen, L.L. and A.G. Pernet. 1985. Mechanism of inhibition of DNA gyrase by analogues of nalidixic acid: The target of the drugs is DNA. *Proc. Natl. Acad. Sci.* **82:** 307–311.

Snapka, R.M. 1986. Topoisomerase inhibitors can selectively interfere with different stages of simian virus 40 DNA replication. *Mol. Cell. Biol.* **6:** 4221–4227.

Snapka, R.M., M.A. Powelson, and J.M. Strayer. 1988. Swiveling and decatenation of replicating simian virus 40 genome in vivo. *Mol. Cell. Biol.* **8:** 515–521.

Stewart, A.F. and G. Schutz. 1987. Camptothecin-induced *in vivo* topoisomerase I cleavages in the transcriptionally active tyrosine aminotransferase gene. *Cell* **50:** 1109–1117.

Tewey, K.M., G.L. Chen, E.M. Nelson, and L.F. Liu. 1984a. Intercalative antitumor drugs interfere with the breakage-reunion reaction of mammalian DNA topoisomerase II. *J. Biol. Chem.* **259:** 9182–9187.

Tewey, K.M., T.C. Rowe, L. Yang, B.D. Halligan, and L.F. Liu. 1984b. Adriamycin-induced DNA damage mediated by mammalian DNA topoisomerase II. *Science* **226:** 466–468.

Tsai-Pflugfelder, M., L.F. Liu, A.A. Liu, K.M. Tewey, J. Whang-Peng, T. Knutsen, K. Huebner, C.M. Croce, and J.C. Wang. 1988. Cloning and sequencing of cDNA encoding human DNA topoisomerase II and localization of the gene to chromosome region 17q21-22. *Proc. Natl. Acad. Sci.* **85:** 7177–7181.

Walker, G.C. 1984. Mutagenesis and inducible responses to deoxyribonucleic acid damage in *Escherichia coli. Microbiol. Rev.* **48:** 60–93.

Wang, J.C. 1985. DNA topoisomerases. *Annu. Rev. Biochem.* **54:** 665–697.

Waring, M. 1981. DNA modification and cancer. *Annu. Rev. Biochem.* **50:** 159–192.

Weinert, T.A. and L.H. Hartwell. 1988. The RAD9 gene controls the cell cycle response to DNA damage in *Saccharomyces cerevisiae. Science* **241:** 317–322.

Wilson, W.D. and R.L. Jones. 1981. Intercalating drugs: DNA binding and molecular pharmacology. *Adv. Pharmacol. Chemother.* **18:** 177–222.

Wu, H.-Y., S. Shyy, J.C. Wang, and L.F. Liu. 1988. Transcription generates positively and negatively supercoiled domains in the template. *Cell* **53:** 433–440.

Yang, L., T.C. Rowe, E.M. Nelson, and L.F. Liu. 1985. *In vivo* mapping of DNA topoisomerase II-specific cleavage sites on SV40 chromatin. *Cell* **41:** 127–132.

Yang, L., M.S. Wold, J.J. Li, T.J. Kelly, and L.F. Liu. 1987. Roles of DNA topoisomerases in simian virus 40 DNA replication in vitro. *Proc. Natl. Acad. Sci.* **84:** 950–954.

Zhang, H., J.C. Wang, and L.F. Liu. 1988. Involvement of DNA topoisomerase I in the transcription of human ribosomal RNA genes. *Proc. Natl. Acad. Sci.* **85:** 1060–1064.

15

DNA Topoisomerases in Clinical Oncology

Milan Potmesil[1] and Robert Silber[2]

[1]Department of Radiology, Laboratory of Experimental Therapy
and [2]Department of Medicine, Division of Hematology
New York University School of Medicine
New York, New York 10016

I. INTRODUCTION

There is a compelling reason for studying human disease as the ultimate experimental system. This approach, highly practical and direct, can use already available information correlating clinical status, pathology, and pharmacology. This data base may help investigators to reevaluate current trends and to develop new directions in therapy. In this chapter, DNA topoisomerase research is discussed in the relation to biology and chemotherapy of selected malignancies.

It was recently recognized that some of the most valuable anticancer drugs interact with topoisomerase II in cell-free systems, as well as in intact cells (for review, see Chen and Liu 1986; Potmesil 1988). In addition, topoisomerase I was identified as a target for the plant alkaloid camptothecin (Hsiang et al. 1985; Hsiang and Liu 1988). The interactions between the drug and topoisomerase involve mostly natural products of microbial or plant origin and their analogs. This results in a complex of a topoisomerase molecule bound covalently to cleaved DNA (Chen and Liu 1986), which is, most likely, the initial event leading to cell death. Drug interaction with either topoisomerase I or II plays a crucial role in drug-exerted cytotoxicity.

II. ANTICANCER DRUGS INHIBITING TOPOISOMERASE I

In the early 1970s, camptothecin, a plant alkaloid isolated from *Camptotheca acuminata* of the Nyssaceae plant family (Wall et al. 1966), was briefly tested in phase-I and -II clinical trials (Gottlieb et al. 1970; Moertel et al. 1971; Muggia et al. 1972). Myelosuppression was the dose-limiting toxicity in cancer patients treated with camptothecin sodium salt, and cystitis was the most prominent nonhematopoietic complication. Because the purpose of phase-I studies is to establish drug toxicity, therapeutic responses can be evaluated only in some patients entered into the trials. However, it is interesting to note that responses were recorded among the patients with advanced gastrointestinal cancer, refractory to other treatments. Although the median duration of the response was slightly over 2 months, the performance status was improved in most of the patients treated on a weekly schedule (Gottlieb et al. 1970). In another study, two responses were seen among 15 patients with advanced cancer on this treatment, and no responses were seen among patients on a daily schedule. Daily treatment with camptothecin was associated with an unpredictable pattern of severe hematologic toxicity (Muggia et al. 1972).

The renewed interest in camptothecin analogs stems from two developments: (1) New, totally synthetic drugs have shown improved effectiveness against leukemia and cancer in experimental systems (Wani et al. 1980, 1987; Wall et al. 1986); and (2) topoisomerase I has been identified as a target not only for camptothecin, but also for all effective analogs tested (Hsiang and Liu 1988; Hsiang et al. 1989). As discussed later in this chapter, intermediate high levels of topoisomerase I are present in several types of human cancer and leukemia with a large fraction of cells confined in the G_0–G_1 phases (Hsiang et al. 1988; Potmesil et al. 1988a,b; Giovanella et al. 1989). It had been established earlier (Gallo et al. 1971) that camptothecin-treated G_0-phase cells cannot reenter the cell cycle. This observation is in general agreement with known effects of topoisomerase I inhibition by the drug.

A recent study of several new camptothecin analogs (Hsiang et al. 1989) has conclusively found a direct correlation between the effectiveness of the drugs against murine leukemia in vivo and their action on DNA topoisomerase I in vitro. These findings suggest that interaction between camptothecin analogs and the enzyme is an essential step leading to tumor cell death. The conclusion has further implications for drug developmental strategies: (1) It supports the contention that topoisomerase I is the principal target not only of camptothecin, but also of its analogs; and (2) although the mechanism of camptothecin cytotoxicity is still poorly understood, available biological and biochemical screens can now be applied to the selection of new, potentially useful anticancer drugs.

Numerous defense mechanisms exist in the biological world that are used by fungi, bacteria, and plants to protect themselves against their adversaries. The production of toxins, such as the alkaloid camptothecin or topoisomerase II inhibitor podophyllotoxin, could be part of a complex mechanism defending plants against competing organisms and may have been in place throughout evolution for millions of years. It seems reasonable to assume that new topoisomerase-targeting natural products, other than those presently known, will be identified and developed for clinical use.

III. ANTICANCER DRUGS INHIBITING TOPOISOMERASE II

Several investigations led to the identification of doxorubicin (adriamycin, DXR) and some other anthracyclines, acridine derivatives, and epipodophyllotoxins, as topoisomerase-II-targeted drugs (Nelson et al. 1984; Tewey et al. 1984; Yang et al. 1985a,b). The most extensively studied inhibitor of topoisomerase II has been DXR (Crooke 1981; Green et al. 1988; Potmesil 1988), an anthracycline with a broad utility in clinical oncology. An effective therapy of human cancers with topoisomerase-II-directed drugs, as much as any other cancer therapy, has been derived empirically by trial and error. DXR remains the drug of choice in soft-tissue sarcomas, and the drug is included in all effective protocols of combination chemotherapy or multimodal treatment. Combination chemotherapy with DXR regimens continues to be the best for primary drug treatment of metastatic breast cancer. There has been interest in intravesical and intraperitoneal application of DXR in patients with urinary bladder or ovarian cancer. DXR and more recently an epipodophyllotoxin VP-16 (etoposide) have found use in several protocols for patients with untreated or recurrent ovarian cancer. DXR is commonly given to patients with acute lymphocytic and nonlymphocytic leukemia. Treatment protocols also include another epipodophyllotoxin VM-26 (teniposide) together with DXR. DXR, VP-16, or their combination are used as part of effective regiments in relapsing patients with Hodgkin's disease and as a first-line treatment in aggressive non-Hodgkin's lymphomas. Some of the protocols are highly effective. A combination of DXR and six other drugs induces almost 100% complete remission in patients with advanced Hodgkin's disease and can achieve a disease-free long-term survival in 90% of the patients (Klimo and Connors 1988).

An acridine derivative, *m*-AMSA (4′-9-acridininylamino-methanesulfon-*m*-anisidide), is being used in the treatment of acute nonlymphocytic leukemia (Pinedo and Chabner 1986; Pinedo et al. 1988). VP-16 has been used as first-line treatment of acute nonlymphocytic leukemia, whereas its congener VM-26 appears to be an active agent in acute lymphocytic leukemia. The drugs are also active as a single agent or in combination treatment of patients with Hodgkin's disease. Combination therapy including DXR and VP-16 is the mainstay in the treatment of small-cell (oat-cell) lung carcinoma. VP-16 also has a substantial effect in terms of tumor reduction as a single agent and in patients who have relapsed. Protocols including VM-26 have also been applied in this cancer. A regimen including cisplatin, bleomycin, and VP-16 has been used in patients with germinal-cell testicular cancer who failed the initial course of chemotherapy. Most notably, 43% of patients thus treated achieve long-term complete remissions with some cures. VP-16 is also highly active in the Kaposi's sarcoma associated with AIDS (O'Dwyer et al. 1985; Pinedo et al. 1988).

Studies of DXR and daunorubicin analogs, the drugs in clinical use or preclinical testing, have shown that, in addition to direct topoisomerase-II-inhibitors, two other types of anthracyclines can be identified. Analogs of the first type need a metabolic activation before becoming topoisomerase-II-targeted. The second type of biologically active analogs apparently do not interact with topoisomerase II at all. This implies that the mechanism of cytotoxicity exerted by the latter group differs from the rest of anthracyclines (Bodley et al. 1989).

DNA topoisomerase levels were studied in three types of neoplasms, in chronic lymphocytic leukemia (CLL), non-Hodgkin's lymphomas, and colonic adenocarcinoma. The selected malignancies differ in their responses to the treatment with topoisomerase-II-directed drugs, and the results are discussed in following sections.

IV. CHRONIC LYMPHOCYTIC LEUKEMIA

CLL is characterized by lymphoid proliferation resulting in the accumulation of mature lymphocytes. In most instances, the disease develops from the malignant transformation of a B-cell lymphocyte and represents the most frequent leukemia in the United States and Europe (for review, see Gale and Foon 1987). Patients with lymphocytosis and no other symptoms (stage 0) or with additional lymph node enlargement (stage I) live in excess of 8 years and, during most of the time, may require no anticancer therapy. Approximately 75% of CLL patients have enlarged liver and/or spleen (stage II), anemia (stage III) or thrombocytopenia (stage IV), and are in intermediate-to-high risk of disease progression (Rai et al. 1975). Three classes of anticancer agents are commonly used: radiation therapy, alkylating agents, and corticosteroids.

DXR, a topoisomerase-II-targeted anthracycline, despite its wide application in the management of lymphoma, has little use in the combination treatment of CLL and no reported effect as a single agent. In an attempt to understand the unresponsiveness to DXR, topoisomerase II was assayed by immunoreaction with antibodies in CLL B lymphocytes obtained from 28 untreated patients (stages 0–IV) and in 7 normal donors (Potmesil et al. 1987, 1988b). The enzyme was neither detected in unfractionated nor in enriched CLL or normal B and T lymphocytes. The cells are mostly confined to the G_0–G_1 phases of the cell cycle. Exponentially growing tissue culture L1210 cells, used as a positive control, had approximately 7×10^5 topoisomerase II copies per cell. This suggests that the enzyme level is on the average 100-fold lower in CLL lymphocytes than in proliferating L1210 cells. Consistent with this conclusion, significant quantities of the enzyme were detected in L1210 cells by indirect immunofluorescence with topoisomerase-II antibodies, whereas no fluorescence was seen over the nuclei of CLL lymphocytes.

DXR produces protein-associated DNA breaks in numerous cell types, including L1210. The action is dose-dependent and results from the stabilization of the cleavable complex between DNA and topoisomerase II (Ross et al. 1978; Tewey et al. 1984). Since suboptimal levels of the substrate enzyme are present in CLL and normal lymphocytes, DXR (0.1–50.0 μg/ml) does not cause any detectable DNA breaks in these cells (Potmesil et al. 1987, 1988b). Exposure to two other topoisomerase-II-directed drugs, *m*-AMSA or VP-16, caused at least 50-fold less DNA cleavage in CLL than in L1210 cells. In either case, the differences in drug action cannot be accounted for by the differences in drug uptake (Potmesil et al. 1988b). It could be concluded that an extremely low level of topoisomerase II contributes to the ineffectiveness of DXR in the treatment of patients with aggressive CLL. The depletion of the topoisomerase II targets in G_0–G_1 phase cells may constitute the basis of a new mechanism of drug resistance. This type of resistance is very likely operational in leukemia and cancer with a large cohort of nonproliferating cells.

V. NON-HODGKIN'S LYMPHOMAS

The non-Hodgkin's lymphomas are classified by the cell of origin. In adults, most of the lymphomas are derived from a monoclonal population of B lymphocytes. The tumors may also be related to various developmental and functional cell subsets. It is well-recognized that the results of combination chemotherapy differ between patients with a high- or intermediate-grade (aggressive) lymphoma and those with a low-grade disease. With most treatment protocols, 60–70% of patients with diffuse histiocytic lymphoma (DHL) and with other diffuse aggressive lymphomas have a long-term disease-free remission. Many of these patients can be considered cured of the disease (Longo and DeVita 1983; DeVita et al. 1985, 1987, 1988; O'Connell et al. 1988). A large number of effective protocols include topoisomerase II inhibitors DXR and more recently VP-16. Several regimens listed in Table 1 include a variety of chemotherapeutic agents administered concomitantly or sequentially. Although single- or two-drug treatments of high-grade lymphomas are not commonly used, it is interesting to note that in terms of remission rates results almost equivalent to combination chemotherapy were achieved in untreated patients by VP-16 alone and by VP-16 combined with DXR (Jacobs et al. 1985). As a whole, these results are better than the treatment with any other single drug not belonging to the category of topoisomerase II inhibitors (DeVita et al. 1985). VP-16 is considerably less effective as a second-line regimen. All listed treatment protocols have been derived empirically, and no cohesive rational could be provided explaining why some patients fail the treatment and develop a recurrent disease.

In terms of remissions, the treatment of intermediate-to-low grade lymphomas with large-cell histology can be almost as beneficial as the

Table 1 First-line treatment of diffuse aggressive lymphoma

Regimen[a]	Number of patients	CR[b] (%)	Long-term remission (%)	Institute[c]
COMPLA	42	55	48	Chicago
B*A*COP	32	46	37	NCI
M-B*A*COD	101	72	48	S. Farber
ProM*ACE*-MOPP	79	74	65	NCI
	80	77	70	NCI
EA	32	54	18	Cape Town

Data from Longo and DeVita (1983); Jacobs et al. (1985); DeVita et al. (1987).

[a]COMPLA: cyclophosphamide + vincristine + methotrexate + leucovorin rescue + cytarabine; B*A*COP: bleomycin + *doxorubicin* + vincristine + prednisone; M-B*A*COD: methotrexate high dose + leucovorin rescue + bleomycin + *doxorubicin* + cyclophosphamide + vincristine + dexamethasone; ProM*ACE*-MOPP: cyclophosphamide + *doxorubicin* + *VP-16* + methotrexate high dose + chlormethin + vincristine + procarbazine + prednisone; *EA*: *VP-16* + *doxorubicin*.

[b]CR indicates complete remission

[c]NCI: National Cancer Institute.

treatment of DHL. The treatment of advanced nodular poorly differentiated lymphoma (NPDL) remains controversial. It is recognized that, in terms of long-term survival, intensive chemotherapy has not provided any definite advantages over less aggressive therapy or single-agent treatment, even if the remission induced by intensive regimens has been quicker (for review, see Steward et al. 1988). Complete remissions obtained by intensive intermittent maintenance therapy are achieved in 56% of patients with advanced disease. Yet, even with the best of protocols, only 30% of patients in large studies are disease-free 5 years later (Ezdinli et al. 1987; Steward et al. 1988). Thus, the final result compares unfavorably with the effects of chemotherapy in high-grade lymphoma such as DHL. If treated with intensive chemotherapy, patients with NPDL are at high risk of developing resistance to the drugs that would be needed when NPDL occasionally evolves to a high-rate aggressive lymphoma. Intensive chemotherapy could also lead to bone marrow aplasia or to a second malignancy. The overall benefit of an aggressive therapy of lower grade lymphoma over a single-drug treatment and no treatment at all is currently being reevaluated by an active clinical trial (Longo 1988). A preliminary report indicates that there is no survival difference between patients treated with the ProMACE-MOPP protocol (see Table 1 for abbreviations) plus radiation therapy and no treatment at all (Young et al. 1988).

On the basis of the above, several conclusions can be drawn: (1) Topoisomerase II inhibitors DXR and VP-16 are used in combination treatment, which cures many patients with high-to-intermediate grade lymphoma such as DHL. There also are effective protocols available that do not use any of topoisomerase II inhibitors. DXR and VP-16 alone also are of benefit in inducing but not in maintaining complete remissions in high-grade lymphoma patients. (2) In patients with DHL who have failed the first-line treatment, topoisomerase-II-directed drugs are less effective. (3) DXR or other topoisomerase II inhibitors are significantly less effective in the treatment of low-grade lymphoma, including NPDL.

A study has been undertaken with the intent to identify differences in the proliferation pattern and topoisomerase I and II levels between DHL and NPDL cells obtained from untreated patients (Table 2). It appears that topoisomerase II is growth regulated in lymphoid malignancies. We have shown in the preceding section that undetectable levels were found in G_0–G_1 phases of CLL cells. NPDL, with 1–2% of cells in S phase, has approximately 10^4 enzyme copies/cell, whereas in DHL, 23–36% of cells are in S phase, and the topoisomerase II level reaches on average 8×10^4 copies/cell. A noteworthy observation are intermediate-to-high levels of topoisomerase I in CLL, NPDL, and DHL (Potmesil et al.

Table 2 DNA topoisomerases in human leukemia and lymphoma

Type of cells	Number of enzyme copies/cell (x 10^5)	
	topoisomerase I	topoisomerase II[a]
Chronic lymphocytic B-cell leukemia (CLL)	5.4 (21)[b] ±2.7	n.d. (21)
Nodular poorly differentiated lymphoma (NPDL)	10.2 (6) ±3.6	0.1 (6) ±0.06
Diffuse histiocytic lymphona (DHL)	27.0 (7) ±9.0	0.8 (7) ±0.2
Normal blood B-cell lymphocytes	2.3 (6) ±1.6	n.d. (6)

Data from Potmesil et al. (1988a,b).
[a] n.d. indicates not detectable, i.e., < 7 x 10^3 copies/cell.
[b] Mean ±S.D. (no. of patients/donors).

1988a,b). Both enzymes were also titrated in malignant cells obtained from several patients with other lymphoid malignancies such as Burkitt's lymphoma, prolymphocytic leukemia, diffuse undifferentiated lymphoma, and nodular mixed lymphoma (Potmesil et al. 1988b).

VI. COLONIC CANCER

Colonic cancer remains one of the major problems facing the oncologist. Approximately 1 of 25 Americans will develop this disease during the course of their lifetime (Sugarbaker et al. 1985). Progress in the treatment of colorectal carcinoma has mainly come through improved operative and perioperative care, including the "no-touch" technique, better anesthesia, monitoring, and the use of antibiotics. The most commonly used classification of colonic cancer, the Astler-Coller modification of the Dukes staging system (Astler and Coller 1954) is shown in Table 3. It accounts for the degree of cancer growth through the colonic wall and the involvement of the lymph nodes and metastatic spread. Of patients with colonic cancer, 20–30% present with locally advanced or metastatic disease (see Table 3, D stage) and may have an extremely unfavorable prognosis. Other patients with lymph node involvement (C1–C4 stages) have a 40% recurrence rate at 5 years.

5-Fluorouracil (5FU), the major therapeutic drug used in adjuvant therapy and for the treatment of advanced disease, is only marginally effective. Used in multicenter studies as a single-agent treatment of ad-

Table 3 Modified Dukes classification of colorectal cancer

Stage	Depth of invasion of the colon/rectum	Lymph node involvement
A	muscularis mucosa	0
B1	to muscularis propria	0
B2	through muscularis propria or serosa	0
B3	through muscularis propria or serosa, invasion of contiguous organ	0
C1	to muscularis propria	+
C2	through muscularis propria or serosa	+
C3	through muscularis propria or serosa, invasion of contiguous organ	+
D	distant metastases	

Data from Astler and Coller (1954).

vanced disease, 5FU has shown response rates of only 20–25% (Sugarbaker et al. 1985). No other single agent has been convincingly shown to be more effective. In the adjuvant setting, 5FU has, when compared to no treatment at all, no statistically significant differences in survival. Studies of combination treatment include 5FU and methyl-*N*-(2-chloroethyl)-*N′*-cyclohexyl-*N*-nitrosourea, semustine, levamisole, leucovorin rescue, or bacillus Calmette-Guerin (for review, see Wittes 1988). All of the protocols have failed so far to show any convincing improvement in disease-free survival. The best results so far were obtained with adjuvant levamisole and 5FU given to patients with completely resected Dukes stage C colon cancer (Laurie et al. 1989). The multitude of approaches reflects the uncertainty surrounding this formidable task of improving the treatment of colonic cancer.

In the 1970s, numerous anticancer chemotherapeutic drugs, including topoisomerase II inhibitors DXR and actinomycin C, were tested as single agents or in combination treatments of colonic cancer (Moertel 1975; Macdonald and Neefe 1979). None of the protocols matched the meager effectiveness of 5FU. DXR alone or in combination had a disappointing response rate of less than 20% of short-term partial remissions. This has not been improved by the introduction of new DXR analogs, such as 4′epiDXR, into clinical oncology (Arcamone 1987). The drug induced short-term partial remission in only 3% of treated patients. As mentioned earlier, somewhat better results were obtained with camptothecin sodium salt, a topoisomerase-I-directed drug (Gottlieb et al. 1970; Moertel et al. 1971; Muggia et al. 1972). In view of the results indicating ineffectiveness of topoisomerase-II-directed drugs and the better results with a topoisomerase I inhibitor, both enzymes were assayed in surgical specimens of untreated patients with colonic cancer (Hsiang et al. 1988).

Enzyme titration in tumor tissue are shown in Table 4. In a group of untreated patients who underwent surgery for colonic cancer, enzyme levels were compared with histologic grades of the tumor and with the stage of the disease (Hsiang et al. 1988; Giovanella et al. 1989). Patients listed in the table are divided into three groups: (1) Clinical stages C1–C3 (tumor invading muscularis propria of the colonic wall to the invasion of contiguous organs with lymph node involvement and/or metastases); (2) clinical stages B1–B3 (tumors confined to the primary site with various depth of the wall penetration); and (3) clinical stages B1–B3 with undetectable levels of topoisomerase II. Significant differences were observed between the enzyme levels in most of the tumors and the colonic mucosa. Topoisomerase I was 13-fold higher in cancer stages C1–C3 (D) and 5-fold higher in tumor stages B1–B3 than in normal tissues. Topoisomerase II levels in the tumor stages C1–C3 (D) were 8-fold higher than in normal mucosa, and in the tumor stages B1–B3 (second group), they were 6-fold higher. In the third group of patients, topoisomerase II is substantially lower in tumor tissue than in normal mucosa.

In summary, relatively low levels of topoisomerase I and undetectable levels of topoisomerase II were observed in colonic adenocarcinoma confined to the colon, without lymph node involvement or metastases (patients nos. 18–23). The tumors of other patients (nos. 9–17) have low-to-elevated levels of both topoisomerases. With one exception, the rest of patients (nos. 1–8) with advanced disease propagated into the lymph nodes and/or with organ metastases have significantly elevated enzyme levels in the tumors.

VII. CONCLUSIONS AND PERSPECTIVES

Although combination and/or sequential chemotherapy improves or even cures some cancer patients, the mechanism of complex treatments is poorly understood. Mechanistic studies conducted in experimental systems have used single agents of various generic groups or in some cases a two-drug or two-modality combination treatment. A quantitative assessment of synergism, additivity, or antagonism between two drugs or modalities (for review, see Potmesil et al. 1986) could evaluate the effects. A study of potential clinical significance found enhanced DNA cleavage and cytotoxicity caused by *m*-AMSA in tissue-culture cells pretreated by *araC*, hydroxyurea, or 5-azacytidine (Zwelling et al. 1986). The three antimetabolites, also used in cancer chemotherapy, may en-

Table 4 DNA topoisomerase I and II levels in adenocarcinoma of the colon

Patient	Enzyme copies/cell ($x10^5$)[a,b]		Histology: grade of differentiation	Clinical stage[c]	Comment
	topoisomerase I	topoisomerase II			
1	36.3	17.4	moderate	C3 D	
2	32.0	8.1	moderate	C1 D	
3	21.3	14.3	poor	C2 D	
4	23.5	10.4	poor to moderate	C3	
5	78.0	10.2	poor to moderate	C3	lymphatic invasion
6	39.0	9.0	poor to moderate	C3	
7	47.0	6.2	moderate	C3	
8	3.0	3.1	moderate mucinous	C3	lymphatic invasion
9	21.3	20.3	moderate mucinous	B3	
10	18.0	6.2	moderate	B3	
11	17.0	6.2	moderate	B2	
12	16.0	5.0	moderate	B2	
13	12.0	8.7	moderate	B2	
14	12.0	7.9	moderate mucinous	B2	
15	3.0	3.6	moderate	B2	
16	3.6	3.0	moderate mucinous	B1	
17	32.0	1.6	moderate	B2	
18	19.0	n.d.	moderate	B2	
19	5.3	n.d.	moderate	B3	
20	3.2	n.d.	moderate	B2	
21	3.5	n.d.	moderate	B1	
22	2.8	n.d.	moderate	B1	
23	—	n.d.	moderate	B1	

Data from Hsiang et al. (1988).
[a]Normalized to DNA content and related to HeLa cell number.
[b]n.d. indicates not detectable, i.e., $<7 \times 10^3$ copies/cell; — indicates not done.
[c]Clinical stage, see Table 3 for explanation.

hance the effect of *m*-AMSA by the induction of cell proliferation accompanied by an increase of topoisomerase II levels. This could explain the schedule dependence and possible synergism between various topoisomerase-II-targeted drugs and *araC* in clinical settings. Other examples of drug synergism between VP-16 and cisplatin, DXR and cisplatin, and even between DXR and VP-16, were observed in treatment studies and remain mechanistically unexplained (Muggia and McVie 1987). Depletion of polyamines by an irreversible inhibitor of ornithine decarboxylase has been associated with altered drug effects on DNA. A similar mechanism was proposed for the enhancement of *m*-AMSA cytotoxicity by α-difluoromethylornithine (Zwelling et al. 1986). In addition to drugs, hormones have also been used in therapy of certain types of human cancer. It was shown that the treatment of estrogen-responsive human breast cancer cells by 17-β estradiol enhances their sensitivity to *m*-AMSA or to an anthracycline 4-demethoxydaunorubicin (Zwelling et al. 1983). There is evidence indicating that the selection of cells with the multi-drug-resistance phenotype actually increases their sensitivity to deoxycortisone and 1-dehydrotestosterone (Ling and Thompson 1973).

Acquired or de novo drug resistance to chemotherapy represents a final stumbling block in the management of patients with many types of cancer or leukemia. The specifics of biochemical mechanisms operating in clinical drug resistance are largely unknown although several extrapolations from experimental systems have been suggested. The potentially most damaging is the multi-drug-resistance phenotype. The phenotype is associated with the overexpression of P-glycoprotein, which plays a role in the decreased drug retention in a cell. A gene coding for the glycoprotein has been cloned, and its overexpression has been detected in colorectal cancer among other malignancies (for review, see Kramer et al. 1988; Moscow and Cowan 1988). The resistance involves not only the drugs interacting with topoisomerase II, but also other functionally dissimilar anticancer compounds. Another related type of resistance is caused by the alteration of the glutathion redox cycle and may interfere with DXR treatments (Kramer et al. 1988). The "atypical" multi-drug resistance, which operates against topoisomerase-II-directed drugs, is not connected with decreased drug accumulation (Beck et al. 1987). A suggestion has been made that this type of resistance is caused by an altered interaction between the drug and a cellular target, presumably topoisomerase II. It remains to be documented whether regulatory or structural changes of the enzyme are involved. Finally, extremely low level of topoisomerase II in nonproliferating cells, cells that may comprise a major part of a tumor, can be an important factor in the resistance involving drugs directed against the enzyme.

Several types of de novo drug resistance may be present in cancers that have minimal responses to chemotherapy. Since high levels of topoisomerase II were detected in colonic cancer, the total failure of DXR treatments can be due to the regulatory and/or structural changes of the enzyme, along with the overexpression of the *mdr1* gene and ensuing drug resistance. Unlike colorectal cancer, high-grade lymphomas in most patients are responsive to the treatment with topoisomerase-II-directed drugs. In some, however, the first-line regimen does not destroy all tumor cells. In a patient with a recurrent disease, responsive tumor cells may have been replaced by cells with the multi-drug-resistance phenotype. Consequently, a remission of the disease is rarely achieved by second-line treatments. In low-grade lymphoma, the resistance to chemotherapy by topoisomerase-II-directed drugs is likely due to a large number of cells with extremely low levels of the enzyme. Tumors depleted by the treatment are repopulated by cells recruited from the nonproliferating compartment. Presumably, these cells are equipped by complex biochemical and genetic mechanisms preventing their eradication.

The importance of DNA topoisomerases in clinical oncology is increasingly apparent. Further insights may be gained through the unknotting of several scientific skeins bearing on their action. A great deal of pharmacological and biochemical research will be required to define their role in combination therapy, which includes topoisomerase-II-targeted drugs.

It may well be that the development of topoisomerase-II-directed drugs has reached its maximal potential, and new directions have to be sought. Research of new drugs targeted at topoisomerase I seems to be a logical approach. An unprecedented efficacy of captothecin analogs in the treatment of cancer xenografts was reported recently (Giovanella et al. 1989). Another direction can be the development of drugs that stabilize topoisomerase II/DNA complexes with high efficiency. The synthesis of "hybrid" anthracycline nitrosoureas (Israel et al. 1988) can be a step in this direction.

ACKNOWLEDGMENTS

Parts of the reviewed research were done in collaboration with Drs. Bruce Bank, Annette Bodley, Howard Grossberg, Yaw-Huei Hsiang, Daniel Knowles, Leroy F. Liu, Frank Traganos, Monroe E. Wall, and Mansukhlal C. Wani. This work was supported in part by American Cancer Society grants CH-348C and RD-300, by Public Health Service grant CA-11655, and by grants from Farmitalia Carlo Erba and the Marcia Slater Society for Research in Leukemia.

REFERENCES

Arcamone, F. 1987. Clinically useful doxorubicin analogues. *Cancer Treat. Rep.* **14:** 159–161.

Astler, V.B. and F.A. Coller. 1954. The prognostic significance of direct extension of carcinoma of the colon and rectum. *Ann. Surg.* **139:** 846–852.

Beck, W.T., M.C. Cirtain, M.K. Danks, R.L. Felsted, A.R. Safa, J.S. Wolverton, D. Parker Suttle, and J.M. Trent. 1987. Pharmacological, molecular, and cytogenetic analysis of "atypical" multi-drug resistant human leukemic cell line. *Cancer Res.* **47:** 5455–5460.

Bodley, A., L.F. Liu, M. Israel, R. Seshadri, Y. Koseki, F.C. Giuliani, S. Kirschenbaum, R. Silber, and M. Potmesil. 1989. DNA topoisomerase II-mediated interaction of doxorubicin and daunorubicin congeners with DNA. *Cancer Res.* **49:** 5969–5978.

Chen, G.L. and L.F. Liu. 1986. DNA topoisomerases as targets in cancer chemotherapy. *Annu. Rep. Med. Chem.* **21:** 257–262.

Crooke, S.T. 1981. The anthracyclines. In *Cancer and chemotherapy* (ed. S.T. Crooke and A.W. Prestayko), vol. III, pp. 111–132. Academic Press, New York.

DeVita, V.T., Jr., S.M. Hubbard, and D.I. Longo. 1987. The chemotherapy of lymphoma: Looking back, moving forward—The Richard and Hilda Rosenthal Foundation Award Lecture. *Cancer Res.* **47:** 5810–5824.

DeVita, V.T., Jr., E.S. Jaffe, and S. Helleman. 1985. Hodgkin's disease and non-Hodgkin's lymphoma. In *Cancer* (ed. V.T. DeVita et al.), p. 1623–1709. Lippincott, Philadelphia.

DeVita, V.T., Jr., S.M. Hubbard, R.C. Young, and D.L. Longo. 1988. The role of chemotherapy in diffuse aggressive lymphoma. *Semin. Hematol.* (suppl. 2) **25:** 2–10.

Ezdinli, E.Z., D.P. Harrington, O. Kucuk, M.W. Silverstein, J. Anderson, and M.J. O'Connell. 1987. The effect of intensive intermittent maintenance therapy in advanced low-grade non-Hodgkin's lymphoma. *Cancer* **60:** 156–160.

Gale, R.P. and K.A. Foon. 1987. Biology of chronic lymphocytic leukemia. *Semin. Hematol.* **24:** 209–229.

Gallo, R.C., J. Whang-Peng, and R.H. Adamson. 1971. Studies on the anti-tumor activity, mechanism of action, and cell cycle effects of camptothecin. *J. Natl. Cancer Inst.* **46:** 789–795.

Giovanella, B.C., J.S. Stehlin, M.E. Wall, M.C. Wani, A.W. Nicholas, L.F. Liu, R.S. Silber, and M. Potmesil. 1989. DNA topoisomerase I-targeted chemotherapy of human colon cancer in xenografts. *Science* **246:** 1046–1048.

Gottlieb, J.A., A.M. Guarino, J.B. Call, V.T. Oliverio, and J.B. Block. 1970. Preliminary pharmacological and clinical evaluation of camptothecin sodium (NSC-100880). *Cancer Chemother. Rep.* **54:** 461–470.

Green M.D., F.M. Muggia, and R.H. Blum. 1988. A perspective from recent clinical trials. In *Anthracyclines and anthracenedione based anticancer agents* (ed. J.W. Lown), p. 667–714. Elsevier, Amsterdam.

Hsiang, Y.-H. and L.F. Liu. 1988. Identification of mammalian topoisomerase I as an intracellular target of the anticancer drug camptothecin. *Cancer Res.* **48:** 1722–1726.

Hsiang, Y.-H., R. Hertzberg, S. Hecht, and L.F. Liu. 1985. Camptothecin induced protein-linked DNA breaks via mammalian DNA topoisomerase I. *J. Biol. Chem.* **260:** 14873–14878.

Hsiang, Y.-H., L.F. Liu, H. Hochster, and M. Potmesil. 1988. Levels of topoisomerase I (topo-I) and topoisomerase II (topo-II) in human colorectal carcinoma and normal colonic mucosa. *Proc. Am. Assoc. Cancer Res.* **29:** 172. (Abstr.)

Hsiang, Y.-H., L.F. Liu, M.E. Wall, M.C. Wani, A.W. Nicholas, G. Manikumar, S. Kirschenbaum, R. Silber, and M. Potmesil. 1989. DNA topoisomerase I-mediated DNA cleavage and cytotoxicity of camptothecin analogues. *Cancer Res.* **49:** 4385–4389.

Israel, M., R. Seshadri, Y. Koseki, A. Bodley, L.F. Liu, S. Kirschenbaum, R. Silber, and M. Potmesil. 1988. Novel topoisomerase II-directed nitrosoureidoanthracyclines. In *The 2nd Conference on DNA Topoisomerases in Cancer Chemotherapy: Program and abstracts*, poster 29, p. 32. N.Y.U. Medical Center, New York.

Jacobs, P., H.S. King, D.M. Dent, and M. Hayes. 1985. VP16–213 in the treatment of stage III and IV diffuse large cell lymphoma. *Cancer* **56:** 1006–1013.

Klimo, P. and J.M. Connors. 1988. An update on the Vancouver experience in the management of advanced Hodgkin's disease treated with the MOPP/ABV hybrid program. *Semin. Hematol.* (suppl. 2) **25:** 34–40.

Kramer, R.A., J. Zakher, and G. Kim. 1988. Role of the glutathione redox cycle in acquired and de novo multidrug resistance. *Science* **241:** 694–687.

Laurie, J.A., C.G. Moertel, T.R. Fleming, H.S. Wieand, J.E. Leigh, J. Rubin, G.W. McCormack, J.B. Gerstner, J.E. Krook, J. Malliard, D.I. Twita, R.F. Morton, and J.F. Barlow. 1989. Surgical adjuvant therapy of large bowel carcinoma: An evaluation of levamisole and the combination of levamisole and 5-fluoro-uracil. *J. Clin. Oncol.* **7:** 1447–1456.

Longo, D.L. 1988. Physician data query—NCI computerized database for physicians. Active protocols, accession #000088.8804. National Cancer Institute, Bethesda, Maryland.

Longo, D.L. and V.T. DeVita. 1983. Lymphomas. In *Cancer chemotherapy*, annual 3 (ed. H.M. Pinedo and B.A. Chabner), p. 248–281. Elsevier, Amsterdam.

Ling, V. and L.H. Thompson. 1973. Reduced permeability in CHO cells as a mechanism of resistance to colchicine. *J. Cell Physiol.* **83:** 103–116.

Macdonald, J.S. and J. Neefe. 1979. Chemotherapy in the management of gastrointestinal cancer. *Abdom. Surg.* **21:** 126–133.

Moertel, C.G. 1975. Clinical management of advanced gastrointestinal cancer. *Cancer* **36:** 675–682.

Moertel, C.G., R.J. Reitemeier, A.J. Schutt, and R.G. Hahn. 1971. A phase II study of camptothecin (NSC 100880) in gastrointestinal cancer. *Proc. Am. Assoc. Cancer Res.* **12:** 18. (Abstr.).

Moscow, J.A. and K.H. Cowan. 1988. Multidrug resistance. *J. Natl. Cancer Inst.* **80:** 14–20.

Muggia, F.M. and G. McVie. 1987. Treatment strategies in relation to drug action. *Natl. Cancer Inst. Monogr.* **4:** 129–133.

Muggia, F.M., P.J. Creaven, H.H. Hansen, M.H. Cohen, and O.S. Selawry. 1972. Phase I clinical trials of weekly and daily treatment with camptothecin (NSC-100880): Correlation with preclinical studies. *Cancer Chemother. Rep.* **56:** 515–521.

Nelson, E.M., K.M. Tewey, and L.F. Liu. 1984. Mechanism of antitumor drug action: Poisoning of mammalian DNA topoisomerase II on DNA by 4′-(9-acridinylamino)-methanesulfon-m-anisidine. *Proc. Natl. Acad. Sci.* **81:** 1361–1365.

O'Connell, M.J., D.P. Harrington, J.D. Earle, G.J. Johnson, J.H. Glick, R.S. Neiman, and M.N. Silverstein. 1988. Chemotherapy followed by consolidation radiotherapy for the treatment of clinical stage II aggressive histologic type non-Hodgkin's lymphoma. *Cancer* **61:** 1754–1758.

O'Dwyer, P.J., B. Leyland-Jones, M.T. Alonso, S. Marsoni, and R.E. Wittes. 1985. Etoposide (VP-16-213). Current status of an active anticancer drug. *N. Engl. J. Med.* **312:** 692–700.

Pinedo, H.M. and B.A. Chabner, eds. 1986. *Cancer chemotherapy*, Annual 7. Elsevier,

Amsterdam.

Pinedo, H.M., B.A. Chabner, and D.L. Longo, eds. 1988. *Cancer chemotherapy and biological response modifications*, Annual 9. Elsevier, Amsterdam.

Potmesil, M. 1988. DNA topoisomerase II as intracellular target in anthracycline treatment of cancer. In *Anthracyclines and anthracenedione based anticancer agents* (ed. J.W. Lown), p. 447–474. Elsevier, Amsterdam.

Potmesil, M., M. Israel, S. Kirschenbaum, J. Bowen, and R. Silber. 1986. Effects of N-trifluoroacetyladriamycin-14-*O*-hemiadipate and radiation on L1210 cells. *Radiat. Res.* **105:** 147–157.

Potmesil, M., Y.-H. Hsiang, L.F. Liu, D. Knowles, S. Kirschenbaum, F. Traganos, and R. Silber. 1988a. Topoisomerase I (topo-I) and topoisomerase II (topo-II) levels in high and low grade lymphomas. *Proc. Am. Assoc. Cancer Res.* **29:** 176. (Abstr.).

Potmesil, M., Y.-H. Hsiang, L.F. Liu, H.-Y. Wu, F. Traganos, B. Bank, and B. Silber. 1987. DNA topoisomerase II as a potential factor in drug resistance of human malignancies. *Natl. Cancer Inst. Monogr.* **4:** 105–109.

Potmesil, M., Y.-H. Hsiang, L.F. Liu, D. Knowles, S. Kirschenbaum, T.J. Forlenza, A. Penziner, D. Kanganis, D. Knowles, F. Traganos, and R. Silber. 1988b. Resistance of human leukemic and normal lymphocytes to drug-induced DNA cleavage and low levels of DNA topoisomerase II. *Cancer Res.* **48:** 3537–3543.

Rai, K.R., A. Sawitsky, E.P. Cronkite, A. Chanana, R.N. Levy, and B.S. Pasternak. 1975. Clinical staging of chronic lymphocytic leukemia. *Blood* **46:** 219–234.

Ross, W.E., D.L. Glaubiger, and K.W. Kohn. 1978. Protein-associated DNA breaks in cells treated with adriamycin or ellipticine. *Biochim. Biophys. Acta* **519:** 23–30.

Steward, W.P., D. Crowther, L.J. McWilliam, J.M. Jones, D.P. Deakin, D.H. Todd, G. Blackledge, J. Wagstaff, J.H. Scarffe, and M. Harris. 1988. Maintenance chlorambucil after CVP in the management of advanced stage, low-grade histologic type non-Hodgkin's lymphoma. *Cancer* **61:** 441–447.

Sugarbaker, P.H., L.L. Gunderson, and R.E. Wittes. 1985. Colorectal cancer. In *Cancer*, 2nd edition (ed. S. Hellman et al.), p. 795–884. Lippincott, Philadelphia.

Tewey, K.M., T.C. Rowe, L. Yang, B.D. Halligan, and L.F. Liu. 1984. Adriamycin induced DNA damage by mammalian DNA topoisomerase II. *Science* **226:** 466–468.

Wall, M.E., M.C. Wani, S.M. Natschke, and A.W. Nicholas. 1986. Plant antitumor agents. 22. Isolation of 11-hydroxycamptothecin from Camptotheca acuminata Decne: Total synthesis and biological activity. *J. Med. Chem.* **29:** 1553–1555.

Wall, M.E., M.C. Wani, C.E. Cooke, K.H. Palmer, A.T. McPhail, and G.A. Slim. 1966. Plant antitumor agents. I. The isolation and structure of camptothecin, a novel alkaloidal leukemia and tumor inhibitor from Camptotheca acuminata. *J. Am. Chem. Soc.* **88:** 3888–3890.

Wani, M.C., A.W. Nicholas, G. Manikumar, and M.E. Wall. 1987. Plant antitumor agents. 25. Total synthesis and antileukemic activity of ring A substituted camptothecin analogues. Structure-activity correlations. *J. Med. Chem.* **30:** 1774–1779.

Wani, M.C., P.E. Ronman, J.T. Lindley, and M.E. Wall. 1980. Plant antitumor agents. 18. Synthesis and biological activity of camptothecin analogues. *J. Med. Chem.* **23:** 554–560.

Wittes, R.E. 1988. Progress in a desert: Oasis or mirage? *J. Natl. Cancer Inst.* **80:** 5–7.

Yang, L., T.C. Rowe, and L.F. Liu. 1985a. Identification of DNA topoisomerase II as an intracellular target of antitumor epipodophyllotoxins in simian virus 40-infected monkey cells. *Cancer Res.* **45:** 5872–5876.

Yang, L., T.C. Rowe, E.M. Nelson, and L.F. Liu. 1985b. In vivo mapping of DNA topoisomerase II-specific cleavage sites on SV40 chromatin. *Cell* **41:** 127–132.

Young, R.C., D.L. Longo, E. Glatstein, D.C. Ihde, E.S. Jaffe, and V.T. DeVita, Jr. 1988. The treatment of indolent lymphomas: Watchful waiting v. aggressive combined modality treatment. *Semin. Hematol.* (suppl. 2) **25:** 11–16.

Zwelling, L.A., D. Kerrigan, and M.E. Lippman. 1983. Protein-associated intercalator-induced DNA scission is enhanced by estrogen stimulation in human breast cancer cells. *Proc. Natl. Acad. Sci.* **80:** 6182–6186.

Zwelling, L.A., L. Silberman, and E. Estey. 1986. Intercalator-induced, topoisomerase II-mediated DNA cleavage and its modification by antineoplastic antimetabolites. *Int. J. Radiat. Oncol. Biol. Phys.* **12:** 1041–1047.

APPENDIX

Nucleotide Sequences and the Encoded Amino Acids of DNA Topoisomerase Genes

Wai Mun Huang
Department of Cellular Viral and Molecular Biology
University of Utah Medical Center
Salt Lake City, Utah 84132

The nucleotide sequences of DNA topoisomerase genes from many diverse sources have been determined recently. Although the nucleotide sequences of these genes are dissimilar, reflecting differences in codon use, some generalizations are possible regarding the encoded amino acid sequence for each class of topoisomerases. Sequence homology is recognized among all eukaryotic type-I DNA topoisomerases. They are distinct from their prokaryotic counterparts except for the homology of the *Saccharomyces cerevisiae TOP3*-encoded protein with the prokaryotic type-I proteins. The amino acid sequence homology among type-II enzymes is more extensive and includes all sequenced examples, regardless of source. Detailed discussions of homology can be found in the relevant chapters of this book and in the references of this list. Because these sequences are scattered in the data bases, sometimes under obscure headings, a current list (dated March, 1990) of 7 type-I genes and 15 type-II genes is compiled here. This list, and the corresponding sequences, will be furnished by the author, on request, on floppy diskettes compatible with microcomputers. The compilation will be updated as more sequence information becomes available. Members of the DNA topoisomerase community who would like their topoisomerase gene sequences included in updated compilations should contact the author at the above address.

The compiled DNA sequences provide only the coding information of these genes. Introns and the terminating codon of the genes are not included. Readers should refer to the original publications for flanking sequences and other characteristics of the genes. The size of the genes is given in number of coding amino acids. The accession number is from the original GenBank data base. Because of space limitations, only the literature reference in which the completed sequence first appeared is cited.

DNA topoisomerase genes whose sequences have been determined

Source	Genes	Size (aa)	Accession number	References
Type I				
Escherichia coli	*topA*	865	X04475	17[a]
	topB	653	J05076	4
Saccharomyces cerevisiae	*TOP1*	769	K03077	15
	TOP3	656	M24939	20
Schizosaccharomyces pombe	*TOP1*	812	X06201	19
Homo sapiens	*TOP1*	765	J03250	3
vaccina virus	*vvTOP*	314	M13209	12
Type II				
Escherichia coli	*gyrA*	875	M15631	14
	gyrB	804	M15548	1, 22
Klebsiella pneumoniae	*gyrA*	876	X16817	5
Bacillus subtilis	*gyrA*	821	X02369	11
	gyrB	638	X02369	11
Mycoplasma pneumoniae	*gyrB*	650	M30153	2
bacteriophage T4	*39*	519	X06220	9[a]
	60	160	M19728	10
	52	441	X04376	8
bacteriophage T2	*39*	605		7
Saccharomyces cerevisiae	*TOP2*	1429	M13814	6
Schizosaccharomyces pombe	*TOP2*	1431	X04326	18
Trypanosoma brucei	*TOP2*	1221	M26803	13
Drosophila melanogaster	*TOP2*	1447		21
Homo sapiens	*TOP2*	1530	J04088	16

[a]Sequence has been updated from this published version.

REFERENCES

1. Adachi, T., M. Mizuuchi, E. Robinson, E. Appella, M. O'dea, M. Gellert, and K. Mizuuchi. 1987. DNA sequence of the *E. coli gyr*B gene: Application of a new sequencing strategy. *Nucleic Acids Res.* **15:** 771–783.
2. Colman, S., P. Hu, and K. Bott. In preparation.

3. D'Arpa, P., P. Machlin, H. Ratrie, R. Rothfield, D. Cleveland, and W. Earnshaw. 1988. cDNA cloning of human DNA topoisomerase I: Catalytic activity of a 67.7 kDa carboxyl-terminal fragment. *Proc. Natl. Acad. Sci.* **85:** 2543–2547.
4. DiGate, R. and K. Marians. 1989. Molecular cloning and DNA sequence analysis of *Escherichia coli topB*, the gene encoding topoisomerase III. *J. Biol. Chem.* **264:** 17924–17930.
5. Dimri, G. and H.K. Das. 1990. Cloning and sequence analysis of *gyrA* gene of *Klebsiella pneumoniae*. *Nucleic Acids Res.* **18:** 151–156.
6. Giaever, G., R. Lynn, T. Goto, and J. Wang. 1986. The complete nucleotide sequence of the structural gene TOP2 of yeast DNA topoisomerase II. *J. Biol. Chem.* **261:** 12448–12454.
7. Gibson, A., M. Fang, and W.M. Huang. In preparation.
8. Huang, W.M. 1986a. The 52-protein subunit of T4 DNA topoisomerase is homologous to the gyrA-protein of gyrase. *Nucleic Acids Res.* **14:** 7379–7389.
9. Huang, W.M. 1986b. Nucleotide sequence of a type II DNA topoisomerase gene: Bacteriophage T4 gene 39. *Nucleic Acids Res.* **14:** 7751–7765.
10. Huang, W.M., S. Ao, S. Casjens, R. Orlandi, R. Zeikus, R. Weiss, D.M. Winge, and M. Fang. 1988. A persistent untranslated sequence within bacteriophage T4 DNA topoisomerase gene *60*. *Science* **239:** 1005–1012.
11. Moriya, S., N. Ogasawara, and H. Yoshikawa. 1985. Structure and function of the region of the replication origin of the *Bacillus subtilis* chromosome. III. Nucleotide sequence of some 10,000 base pairs in the origin region. *Nucleic Acids Res.* **13:** 2251–2265.
12. Shuman, S. and B. Moss. 1987. Identification of a vaccinia virus gene encoding a type I DNA topoisomerase. *Proc. Natl. Acad. Sci.* **84:** 7478–7482.
13. Strauss, P. and J. Wang. 1990. The *TOP2* gene of *Trypanosoma brucei:* A single-copy gene that shares extensive homology with other *TOP2* genes encoding eukaryotic DNA topoisomerase II. *Mol. Biochem. Parasitol.* **38:** 141–150.
14. Swanberg, S. and J. Wang. 1987. Cloning and sequencing of the *E. coli gyr*A gene encoding for the A subunit of DNA gyrases. *J. Mol. Biol.* **197:** 729–736.
15. Thrash, C., A. Bankier, B. Barrell, and R. Sternglanz. 1985. Cloning characterization, and sequence of the yeast DNA topoisomerase I gene. *Proc. Natl. Acad. Sci.* **82:** 4374–4378.
16. Tsai-Pflugfelder, M., L. Liu, A. Liu, K. Tewey, J. Whang-Peng, T. Knutsen, K. Huebner, C. Croce, and J. Wang. 1988. Cloning and sequencing of cDNA encoding human DNA topoisomerase II and localization of the gene to chromosome region 17q21-22. *Proc. Natl. Acad. Sci.* **85:** 7177–7181.
17. Tse-Dinh, Y. and J. Wang. 1986. Complete nucleotide sequence of the *topA* gene encoding *Escherichia coli* DNA topoisomerase I. *J. Mol. Biol.* **191:** 321–331.
18. Uemura, T., K. Morikawa, and M. Yanagida. 1986. The nucleotide sequence of the fission yeast DNA topoisomerase II gene: Structural and functional relationships to other DNA topoisomerases. *EMBO J.* **5:** 2355–2361.
19. Uemura, T., K. Morino, K. Uzawa, and M. Yanagida. 1987. Cloning and sequencing of *Schizosaccharomyces pombe* DNA topoisomerase I gene, and effect of gene disruption. *Nucleic Acids Res.* **15:** 9727–9739.
20. Wallis, J., G. Chrebet, G. Brodsky, M. Rolfe, and R. Rothstein. 1989. A hyper-recombination mutation in *S. cerevisiae* identifies a novel eukaryotic topoisomerase. *Cell* **58:** 409–419.
21. Wyckoff, E., D. Natalie, J. Nolan, M. Lee, and T. Hsieh. 1989. Structure of the *Drosophila* DNA topoisomerase II gene: Nucleotide sequence and homology among topoisomerases II. *J. Mol. Biol.* **205:** 1–13.
22. Yamagishi, J., H. Yoshida, M. Yamayoshi, and S. Nakamura. 1987. Nalidixic acid-resistant mutations of the *gyr*B gene of *Escherichia coli. Mol. Gen. Genet.* **204:** 367–373.

Escherichia coli topA

Met Gly Lys Ala Leu Val Ile Val Glu Ser Pro Ala Lys Ala Lys Thr Ile Asn Lys Tyr 20
ATG GGT AAA GCT CTT GTC ATC GTT GAG TCC CCG GCA AAA GCC AAA ACG ATC AAC AAG TAT 60

Leu Gly Ser Asp Tyr Val Val Lys Ser Ser Val Gly His Ile Arg Asp Leu Pro Thr Ser 40
CTG GGT AGT GAC TAC GTG GTG AAA TCC AGC GTC GGT CAC ATT CGC GAT TTG CCG ACC AGT 120

Gly Ser Ala Ala Lys Lys Ser Ala Asp Ser Thr Ser Thr Lys Thr Ala Lys Lys Pro Lys 60
GGC TCA GCT GCC AAA AAG AGT GCC GAC TCT ACC TCC ACC AAG ACG GCT AAA AAG CCT AAA 180

Lys Asp Glu Arg Gly Ala Leu Val Asn Arg Met Gly Val Asp Pro Trp His Asn Trp Glu 80
AAG GAT GAA CGT GGC GCT CTC GTC AAC CGT ATG GGG GTT GAC CCG TGG CAC AAT TGG GAG 240

Ala His Tyr Glu Val Leu Pro Gly Lys Glu Lys Val Val Ser Glu Leu Lys Gln Leu Ala 100
GCG CAC TAT GAA GTG TTG CCT GGT AAA GAG AAG GTC GTC TCT GAA CTG AAA CAA CTG GCT 300

Glu Lys Ala Asp His Ile Tyr Leu Ala Thr Asp Leu Asp Arg Glu Gly Glu Ala Ile Ala 120
GAA AAA GCC GAC CAC ATC TAT CTC GCA ACC GAC CTT GAC CGC GAA GGG GAA GCC ATT GCA 360

Trp Arg Leu Arg Glu Val Ile Gly Gly Asp Asp Ala Arg Tyr Ser Arg Val Val Phe Asn 140
TGG CGC CTG CGG GAA GTG ATT GGG GGT GAT GAT GCG CGC TAT AGC CGA GTG GTG TTT AAC 420

Glu Ile Ala Lys Asn Ala Ile Arg Gln Ala Phe Asn Lys Pro Gly Glu Leu Asn Ile Asp 160
GAA ATT GCT AAA AAC GCG ATC CGC CAA GCA TTT AAC AAA CCG GGT GAG CTG AAT ATT GAT 480

Arg Val Asn Ala Gln Arg Ala Arg Arg Phe Met Asp Arg Val Val Gly Tyr Met Val Ser 180
CGT GTT AAT GCC CAG CGG GCG CGT CGC TTT ATG GAC CGC GTG GTG GGG TAT ATG GTT TCG 540

Pro Leu Leu Trp Lys Lys Ile Ala Arg Gly Leu Ser Ala Gly Arg Val Gln Ser Val Ala 200
CCG CTG CTA TGG AAA AAG ATC GCT CGT GGC CTG TCT GCC GGT CGT GTG CAG TCG GTG GCG 600

Val Arg Leu Ala Val Glu Arg Glu Arg Glu Ile Lys Ala Phe Val Pro Glu Glu Phe Trp 220
GTT CGC CTG GCG GTC GAG CGT GAG CGT GAA ATT AAA GCG TTC GTG CCG GAA GAG TTC TGG 660

Glu Val Asp Ala Ser Thr Thr Thr Pro Ser Gly Glu Ala Leu Ala Leu Gln Val Thr His 240
GAA GTC GAT GCC AGC ACG ACC ACG CCA TCT GGT GAA GCG TTG GCG TTA CAG GTG ACT CAT 720

Gln Asn Asp Lys Pro Phe Arg Pro Val Asn Lys Glu Gln Thr Gln Ala Ala Val Ser Leu 260
CAG AAC GAC AAA CCG TTC CGT CCG GTC AAC AAA GAA CAA ACT CAG GCT GCG GTA AGT CTG 780

Leu Glu Lys Ala Arg Tyr Ser Val Leu Glu Arg Glu Asp Lys Pro Thr Thr Ser Lys Pro 280
CTG GAA AAA GCG CGC TAC AGC GTG CTG GAA CGT GAA GAC AAA CCG ACA ACC AGT AAA CCT 840

Ser Ala Pro Leu Ile Thr Ser Thr Leu Gln Gln Ala Ala Ser Thr Arg Leu Gly Phe Gly 300
AGC GCT CCT TTA ATT ACC TCT ACG CTG CAA CAA GCT GCC AGC ACC CGT CTT GGA TTT GGC 900

Val Lys Lys Thr Met Met Met Ala Gln Arg Leu Tyr Glu Ala Gly Tyr Ile Thr Tyr Met 320
GTG AAA AAA ACC ATG ATG ATG GCG CAG CGT TTG TAT GAA GCA GGC TAT ATC ACT TAC ATG 960

Arg Thr Asp Ser Thr Asn Leu Ser Gln Asp Ala Val Asn Met Val Arg Gly Tyr Ile Ser 340
CGT ACC GAC TCC ACT AAC CTG AGT CAG GAC GCG GTA AAT ATG GTT CGC GGT TAT ATC AGC 1020

Asp Asn Phe Gly Lys Lys Tyr Leu Pro Glu Ser Pro Asn Arg Tyr Ala Ser Lys Glu Asn 360
GAT AAT TTT GGT AAG AAA TAT CTG CCG GAA AGT CCG AAT CGG TAC GCC AGC AAA GAA AAC 1080

Ser Gln Glu Ala His Glu Ala Ile Arg Pro Ser Asp Val Asn Val Met Ala Glu Ser Leu 380
TCA CAG GAA GCG CAC GAA GCG ATT CGT CCT TCT GAC GTC AAT GTG ATG GCG GAA TCG CTG 1140

Lys Asp Met Glu Ala Asp Ala Gln Lys Leu Tyr Gln Leu Ile Trp Arg Gln Phe Val Ala 400
AAG GAT ATG GAA GCA GAT GCG CAG AAA CTG TAC CAG TTA ATC TGG CGT CAG TTC GTT GCC 1200

Cys Gln Met Thr Pro Ala Lys Tyr Asp Ser Thr Thr Leu Thr Val Gly Ala Gly Asp Phe 420
TGC CAG ATG ACC CCA GCG AAA TAT GAC TCC ACG ACG CTG ACC GTT GGT GCG GGC GAT TTC 1260

```
Arg Leu Lys Ala Arg Gly Arg Ile Leu Arg Phe Asp Gly Trp Thr Lys Val Met Pro Ala    440
CGC CTG AAA GCA CGC GGT CGT ATT TTG CGT TTT GAT GGC TGG ACA AAA GTG ATG CCT GCG   1320

Leu Arg Lys Gly Asp Glu Asp Arg Ile Leu Pro Ala Val Asn Lys Gly Asp Ala Leu Thr    460
TTG CGT AAA GGC GAT GAA GAT CGC ATC TTA CCA GCA GTT AAT AAA GGC GAT GCT CTG ACG   1380

Leu Val Glu Leu Thr Pro Ala Gln His Phe Thr Lys Pro Pro Ala Arg Phe Ser Glu Ala    480
CTC GTT GAA CTT ACA CCA GCC CAG CAC TTT ACC AAG CCG CCA GCC CGT TTC AGT GAA GCA   1440

Ser Leu Val Lys Glu Leu Glu Lys Arg Gly Ile Gly Arg Pro Ser Thr Tyr Ala Ser Ile    500
TCG CTG GTT AAA GAG CTG GAA AAA CGC GGT ATC GGT CGT CCG TCT ACC TAT GCG TCG ATC   1500

Ile Ser Thr Ile Gln Asp Arg Gly Tyr Val Arg Val Glu Asn Arg Arg Phe Tyr Ala Glu    520
ATT TCG ACC ATT CAG GAT CGT GGC TAC GTG CGA GTA GAA AAT CGT CGT TTC TAT GCG GAA   1560

Lys Met Gly Glu Ile Val Thr Asp Arg Leu Glu Glu Asn Phe Arg Glu Leu Met Asn Tyr    540
AAA ATG GGC GAA ATC GTC ACC GAT CGC CTT GAA GAA AAT TTC CGC GAG TTA ATG AAC TAC   1620

Asp Phe Thr Ala Gln Met Glu Asn Ser Leu Asp Gln Val Ala Asn His Glu Ala Glu Trp    560
GAC TTT ACC GCG CAG ATG GAA AAC AGC CTC GAC CAA GTG GCA AAT CAC GAA GCA GAG TGG   1680

Lys Ala Val Leu Asp His Phe Phe Ser Asp Phe Thr Gln Gln Leu Asp Lys Ala Glu Lys    580
AAA GCT GTA CTG GAT CAC TTC TTC TCG GAT TTC ACC CAG CAG TTA GAT AAA GCT GAA AAA   1740

Asp Pro Glu Glu Gly Gly Met Arg Pro Asn Gln Met Val Leu Thr Ser Ile Asp Cys Pro    600
GAT CCG GAA GAG GGT GGT ATG CGC CCG AAC CAG ATG GTT CTG ACC AGC ATT GAC TGC CCG   1800

Thr Cys Gly Arg Lys Met Gly Ile Arg Thr Ala Ser Thr Gly Val Phe Leu Gly Cys Ser    620
ACT TGT GGT CGC AAA ATG GGG ATT CGC ACA GCG AGC ACC GGG GTA TTC CTT GGC TGT TCT   1860

Gly Tyr Ala Leu Pro Pro Lys Glu Arg Cys Lys Thr Thr Ile Asn Val Val Pro Glu Asn    640
GGC TAT GCG CTG CCG CCG AAA GAG CGT TGC AAA ACC ACC ATT AAC GTG GTG CCG GAA AAC   1920

Glu Val Leu Asn Val Leu Glu Gly Glu Asp Ala Glu Thr Lys Pro Leu Arg Ala Lys Arg    660
GAA GTG CTG AAC GTG CTG GAA GGC GAA GAT GCT GAA ACC AAG CCG CTG CGC GCA AAA CGT   1980

Arg Cys Pro Lys Cys Gly Thr Ala Met Asp Ser Tyr Leu Ile Asp Pro Lys Arg Lys Leu    680
CGT TGC CCG AAA TGC GGC ACG GCG ATG GAC AGC TAT CTC ATC GAT CCG AAA CGT AAG TTG   2040

His Val Cys Gly Asn Asn Pro Thr Cys Asp Gly Tyr Glu Ile Glu Glu Gly Glu Phe Arg    700
CAT GTC TGT GGT AAT AAC CCA ACC TGC GAC GGT TAC GAG ATC GAA GAG GGC GAA TTC CGC   2100

Ile Lys Gly Tyr Asp Gly Pro Ile Val Glu Cys Glu Lys Cys Gly Ser Glu Met His Leu    720
ATT AAA GGT TAT GAC GGC CCG ATC GTT GAG TGT GAA AAA TGT GGC TCT GAA ATG CAC CTG   2160

Lys Met Gly Arg Phe Gly Lys Tyr Met Ala Cys Thr Asn Glu Glu Cys Lys Asn Thr Arg    740
AAA ATG GGG CGT TTC GGT AAA TAC ATG GCC TGC ACC AAC GAA GAG TGT AAA AAC ACA CGT   2220

Lys Ile Leu Arg Asn Gly Glu Val Ala Pro Pro Lys Glu Asp Pro Val Pro Leu Pro Glu    760
AAG ATT TTA CGT AAC GGC GAA GTG GCA CCA CCG AAA GAA GAT CCG GTG CCA TTA CCT GAG   2280

Leu Pro Cys Glu Lys Ser Asp Ala Tyr Phe Val Leu Arg Asp Gly Ala Ala Gly Val Phe    780
CTG CCG TGC GAA AAA TCA GAT GCT TAT TTC GTG CTG CGT GAC GGT GCT GCC GGT GTG TTC   2340

Leu Ala Ala Asn Thr Phe Pro Lys Ser Arg Glu Thr Arg Ala Pro Leu Val Glu Glu Leu    800
CTG GCT GCC AAC ACT TTC CCG AAA TCG CGT GAA ACG CGT GCG CCA CTG GTG GAA GAG CTT   2400

Tyr Arg Phe Arg Asp Arg Leu Pro Glu Lys Leu Arg Tyr Leu Ala Asp Ala Pro Gln Gln    820
TAT CGC TTC CGC GAC CGT CTG CCG GAA AAA CTG CGT TAT CTG GCC GAT GCG CCA CAG CAG   2460

Asp Pro Glu Gly Asn Lys Thr Met Val Arg Phe Ser Arg Lys Thr Lys Gln Gln Tyr Val    840
GAT CCG GAA GGT AAT AAG ACC ATG GTT CGC TTT AGC CGT AAA ACC AAA CAG CAA TAT GTC   2520
```

```
Ser Ser Glu Lys Asp Gly Lys Ala Thr Gly Trp Ser Ala Phe Tyr Val Asp Gly Lys Trp   860
TCT TCG GAA AAA GAC GGA AAG GCG ACT GGC TGG TCA GCA TTT TAT GTT GAT GGC AAA TGG  2580

Val Glu Gly Lys Lys   865
GTT GAA GGA AAA AAA  2595
```

Escherichia coli topB

```
Met Arg Leu Phe Ile Ala Glu Lys Pro Ser Leu Ala Arg Ala Ile Ala Asp Val Leu Pro    20
ATG CGG TTG TTT ATT GCC GAA AAA CCG AGT CTG GCG CGC GCC ATT GCT GAT GTC CTG CCC    60

Lys Pro His Arg Lys Gly Asp Gly Phe Ile Glu Cys Gly Asn Gly Gln Val Val Thr Trp    40
AAA CCG CAC CGG AAA GGC GAT GGC TTT ATC GAG TGC GGT AAT GGT CAG GTG GTG ACC TGG   120

Cys Ile Gly His Leu Leu Glu Gln Ala Gln Pro Asp Ala Tyr Asp Ser Arg Tyr Ala Arg    60
TGT ATC GGT CAC CTG CTT GAG CAG GCG CAG CCA GAC GCC TAC GAC AGC CGC TAT GCG CGC   180

Trp Asn Leu Ala Asp Leu Pro Ile Val Pro Glu Lys Trp Gln Leu Gln Pro Arg Pro Ser    80
TGG AAT CTT GCG GAT TTG CCG ATT GTC CCG GAA AAG TGG CAA TTA CAG CCC CGA CCC TCC   240

Val Thr Lys Gln Leu Asn Val Ile Lys Arg Phe Leu His Glu Ala Ser Glu Ile Val His   100
GTG ACC AAA CAA CTT AAC GTC ATC AAA CGG TTC CTG CAT GAA GCC AGC GAA ATC GTT CAC   300

Ala Gly Asp Pro Asp Arg Glu Gly Gln Leu Leu Val Asp Glu Val Leu Asp Tyr Leu Gln   120
GCC GGG GAC CCG GAT CGT GAA GGG CAA TTG CTG GTG GAT GAA GTG CTG GAC TAT CTG CAA   360

Leu Ala Pro Glu Lys Arg Gln Gln Val Gln Arg Cys Leu Ile Asn Asp Leu Asn Pro Gln   140
CTG GCA CCG GAA AAG CGC CAG CAG GTA CAG CGT TGC TTG ATA AAC GAC CTG AAC CCG CAG   420

Ala Val Glu Arg Ala Ile Asp Arg Leu Arg Ser Asn Ser Glu Phe Val Pro Leu Cys Val   160
GCG GTT GAG CGG GCG ATC GAC CGT CTT CGT TCC AAC AGT GAG TTT GTA CCG CTG TGC GTT   480

Ser Ala Leu Ala Arg Ala Arg Ala Asp Trp Leu Tyr Gly Ile Asn Met Thr Arg Ala Tyr   180
TCT GCG CTG GCG CGA GCG CGT GCC GAC TGG CTG TAC GGC ATC AAT ATG ACC CGT GCG TAT   540

Thr Ile Leu Gly Arg Asn Ala Gly Tyr Gln Gly Val Leu Ser Val Gly Arg Val Gln Thr   200
ACC ATT CTC GGT CGC AAT GCC GGT TAT CAG GGC GTA CTT TCC GTG GGA CGC GTG CAG ACG   600

Pro Val Leu Gly Leu Val Val Arg Arg Asp Glu Glu Ile Glu Asn Phe Val Ala Lys Asp   220
CCC GTG CTT GGG CTG GTG GTG CGC CGC GAT GAA GAG ATT GAA AAC TTC GTG GCG AAA GAC   660

Phe Phe Glu Val Lys Ala His Ile Val Thr Pro Ala Asp Glu Arg Phe Thr Ala Ile Trp   240
TTC TTT GAA GTC AAA GCA CAT ATC GTG ACA CCT GCC GAT GAG CGG TTT ACC GCT ATC TGG   720

Gln Pro Ser Glu Ala Cys Glu Pro Tyr Gln Asp Glu Glu Gly Arg Leu Leu His Arg Pro   260
CAA CCG AGC GAA GCG TGT GAA CCG TAC CAG GAT GAA GAA GGG CGC TTG TTA CAT CGT CCA   780

Leu Ala Glu His Val Val Asn Arg Ile Ser Gly Gln Pro Ala Ile Val Thr Ser Tyr Asn   280
CTG GCG GAG CAT GTG GTT AAC CGC ATT AGT GGT CAA CCG GCT ATT GTC ACC AGC TAT AAC   840

Asp Lys Arg Glu Ser Glu Ser Ala Pro Leu Pro Phe Ser Leu Ser Ala Leu Gln Ile Glu   300
GAT AAA CGG GAA TCA GAA TCC GCG CCG CTG CCT TTT TCG CTT TCA GCG TTG CAG ATT GAA   900

Ala Ala Lys Arg Phe Gly Leu Ser Ala Gln Asn Val Leu Asp Ile Cys Gln Lys Leu Tyr   320
GCG GCA AAA CGT TTT GGT CTG AGT GCG CAG AAC GTG CTT GAT ATC TGC CAG AAA CTG TAC   960

Glu Thr His Lys Leu Ile Thr Tyr Pro Arg Ser Asp Cys Arg Tyr Leu Pro Glu Glu His   340
GAA ACG CAC AAG CTA ATC ACT TAT CCG CGT TCT GAT TGT CGC TAT TTG CCA GAA GAA CAT  1020

Phe Ala Gly Arg His Ala Val Met Asn Ala Ile Ser Val His Ala Pro Asp Leu Leu Pro   360
TTT GCC GGA CGC CAC GCG GTG ATG AAT GCC ATC AGT GTT CAT GCA CCG GAT CTG TTG CCG  1080
```

```
Gln Pro Val Val Asp Pro Asp Ile Arg Asn Arg Cys Trp Asp Asp Lys Lys Val Asp Ala   380
CAG CCA GTG GTA GAT CCA GAT ATA CGC AAC CGC TGT TGG GAT GAC AAA AAG GTC GAT GCG  1140

His His Ala Ile Ile Pro Thr Ala Arg Ser Ser Ala Ile Asn Leu Thr Glu Asn Glu Ala   400
CAC CAC GCC ATC ATT CCG ACC GCA CGG AGT TCT GCG ATC AAC CTG ACG GAG AAC GAA GCG  1200

Lys Val Tyr Asn Leu Ile Ala Arg Gln Tyr Leu Met Gln Phe Cys Pro Asp Ala Val Phe   420
AAG GTC TAT AAC CTG ATT GCC CGT CAG TAT CTG ATG CAA TTC TGC CCG GAT GCG GTG TTC  1260

Arg Lys Cys Val Ile Glu Leu Asp Ile Ala Lys Gly Lys Phe Val Ala Lys Ala Arg Phe   440
CGC AAG TGT GTT ATC GAA CTG GAC ATT GCC AAA GGC AAA TTT GTC GCT AAA GCG CGT TTT  1320

Leu Ala Glu Ala Gly Trp Arg Thr Leu Leu Gly Ser Lys Glu Arg Asp Glu Glu Asn Asp   460
CTT GCT GAA GCA GGC TGG CGC ACG CTG TTA GGC AGC AAA GAG CGC GAT GAA GAA AAC GAC  1380

Gly Thr Pro Leu Pro Val Val Ala Lys Gly Asp Glu Leu Leu Cys Glu Lys Gly Glu Val   480
GGC ACG CCA CTG CCT GTG GTG GCG AAA GGC GAT GAG TTG CTG TGT GAA AAA GGT GAA GTG  1440

Val Glu Arg Gln Thr Gln Pro Pro Arg His Phe Thr Asp Ala Thr Leu Leu Ser Ala Met   500
GTA GAG CGG CAA ACC CAG CCG CCG CGC CAT TTT ACC GAT GCA ACA CTG CTT TCG GCG ATG  1500

Thr Gly Ile Ala Arg Phe Val Gln Asp Lys Asp Leu Lys Lys Ile Leu Arg Ala Thr Asp   520
ACC GGG ATC GCG CGC TTT GTG CAG GAT AAA GAT CTG AAA AAG ATC CTC CGT GCG ACC GAT  1560

Gly Leu Gly Thr Glu Ala Thr Arg Ala Gly Ile Ile Glu Leu Leu Phe Lys Arg Gly Phe   540
GGT CTG GGG ACA GAG GCA ACG CGT GCC GGG ATT ATT GAA CTG TTG TTC AAG CGT GGT TTC  1620

Leu Thr Lys Lys Gly Arg Tyr Ile His Ser Thr Asp Ala Gly Lys Ala Leu Phe His Ser   560
CTG ACC AAA AAA GGG CGC TAT ATC CAC TCC ACC GAC GCC GGA AAA GCG CTA TTC CAT TCG  1680

Leu Pro Glu Met Ala Thr Arg Pro Asp Met Thr Ala His Trp Glu Ser Val Leu Thr Gln   580
CTG CCG GAG ATG GCG ACG CGA CCG GAC ATG ACC GCG CAC TGG GAA TCG GTG CTG ACG CAA  1740

Ile Ser Glu Lys Gln Cys Arg Tyr Gln Asp Phe Met Gln Pro Leu Val Gly Thr Leu Tyr   600
ATC AGC GAA AAG CAG TGT CGC TAT CAG GAC TTT ATG CAG CCG CTG GTG GGG ACG CTA TAT  1800

Gln Leu Ile Asp Gln Ala Lys Arg Thr Pro Val Arg Gln Phe Arg Gly Ile Val Ala Pro   620
CAG CTT ATT GAT CAA GCC AAA CGT ACG CCG GTG CGG CAG TTT CGC GGC ATT GTG GCT CCG  1860

Gly Ser Gly Gly Ser Ala Asp Lys Lys Lys Ala Ala Pro Arg Lys Arg Ser Ala Lys Lys   640
GGC AGT GGT GGC AGT GCT GAT AAG AAA AAG GCT GCA CCG CGT AAA CGT AGT GCG AAA AAA  1920

Ser Pro Pro Ala Asp Glu Val Gly Ser Gly Ala Ile Ala   653
AGT CCG CCA GCA GAT GAA GTC GGA AGC GGG GCG ATA GCG  1959
```

Saccharomyces cerevisiae Top1

```
Met Thr Ile Ala Asp Ala Ser Lys Val Asn His Glu Leu Ser Ser Asp Asp Asp Asp Asp    20
ATG ACT ATT GCT GAT GCT TCC AAA GTT AAT CAT GAG TTG TCT TCT GAT GAC GAT GAC GAT    60

Val Pro Leu Ser Gln Thr Leu Lys Lys Arg Lys Val Ala Ser Met Asn Ser Ala Ser Leu    40
GTG CCA TTA TCT CAA ACT TTA AAA AAA AGA AAG GTG GCG TCC ATG AAC TCT GCC TCT CTT   120

Gln Asp Glu Ala Glu Pro Tyr Asp Ser Asp Glu Ala Ile Ser Lys Ile Ser Lys Lys Lys    60
CAA GAC GAA GCG GAA CCT TAT GAT AGT GAT GAG GCA ATC TCT AAG ATT TCC AAG AAA AAG   180

Thr Lys Lys Ile Lys Thr Glu Pro Val Gln Ser Ser Ser Leu Pro Ser Pro Pro Ala Lys    80
ACT AAG AAA ATA AAG ACC GAA CCA GTG CAA TCG TCG TCA TTA CCA TCG CCT CCA GCA AAG   240
```

```
Lys Ser Ala Thr Ser Lys Pro Lys Lys Ile Lys Lys Glu Asp Gly Asp Val Lys Val Lys   100
AAA AGC GCG ACA TCA AAG CCT AAA AAA ATC AAG AAA GAA GAT GGT GAT GTA AAG GTA AAA   300

Thr Thr Lys Lys Glu Glu Gln Glu Asn Glu Lys Lys Lys Arg Glu Glu Glu Glu Glu Glu   120
ACA ACT AAA AAG GAA GAA CAG GAG AAC GAA AAA AAG AAA CGA GAA GAA GAA GAA GAG GAG   360

Asp Lys Lys Ala Lys Glu Glu Glu Glu Glu Tyr Lys Trp Trp Glu Lys Glu Asn Glu Asp   140
GAC AAG AAA GCG AAG GAG GAG GAG GAA GAA TAT AAA TGG TGG GAA AAA GAA AAC GAA GAT   420

Asp Thr Ile Lys Trp Val Thr Leu Lys His Asn Gly Val Ile Phe Pro Pro Pro Tyr Gln   160
GAC ACC ATA AAA TGG GTC ACA CTG AAG CAT AAC GGT GTT ATA TTC CCT CCA CCA TAC CAG   480

Pro Leu Pro Ser His Ile Lys Leu Tyr Tyr Asp Gly Lys Pro Val Asp Leu Pro Pro Gln   180
CCC TTA CCA TCT CAC ATC AAA TTA TAT TAC GAT GGG AAG CCA GTA GAT TTA CCT CCG CAA   540

Ala Glu Glu Val Ala Gly Phe Phe Ala Ala Leu Leu Glu Ser Asp His Ala Lys Asn Pro   200
GCT GAA GAA GTA GCC GGG TTC TTT GCT GCC CTA TTA GAG AGT GAT CAT GCC AAA AAT CCT   600

Val Phe Gln Lys Asn Phe Phe Asn Asp Phe Leu Gln Val Leu Lys Glu Ser Gly Gly Pro   220
GTT TTC CAA AAG AAC TTC TTC AAT GAT TTC TTG CAA GTA CTG AAA GAA AGT GGT GGT CCC   660

Leu Asn Gly Ile Glu Ile Lys Glu Phe Ser Arg Cys Asp Phe Thr Lys Met Phe Asp Tyr   240
CTC AAT GGA ATT GAG ATA AAG GAA TTT TCT CGT TGC GAT TTC ACC AAA ATG TTT GAT TAC   720

Phe Gln Leu Gln Lys Glu Gln Lys Lys Gln Leu Thr Ser Gln Glu Lys Lys Gln Ile Arg   260
TTC CAG TTA CAA AAA GAA CAG AAA AAG CAA CTG ACT TCC CAA GAA AAG AAA CAG ATT CGT   780

Leu Glu Arg Glu Lys Phe Glu Glu Asp Tyr Lys Phe Cys Glu Leu Asp Gly Arg Arg Glu   280
TTG GAA AGA GAA AAA TTC GAG GAA GAT TAT AAA TTC TGT GAA TTA GAT GGC AGA AGG GAA   840

Gln Val Gly Asn Phe Lys Val Glu Pro Pro Asp Leu Phe Arg Gly Arg Gly Ala His Pro   300
CAA GTA GGG AAT TTC AAG GTT GAA CCT CCT GAT CTA TTT AGA GGT CGT GGC GCT CAC CCA   900

Lys Thr Gly Lys Leu Lys Arg Arg Val Asn Pro Glu Asp Ile Val Leu Asn Leu Ser Lys   320
AAA ACA GGC AAA TTG AAG AGA AGA GTG AAT CCT GAG GAT ATC GTT TTA AAT CTA AGT AAA   960

Asp Ala Pro Val Pro Pro Ala Pro Glu Gly His Lys Trp Gly Glu Ile Arg His Asp Asn   340
GAC GCA CCC GTT CCG CCA GCC CCA GAA GGG CAC AAG TGG GGT GAA ATC AGA CAC GAC AAT  1020

Thr Val Gln Trp Leu Ala Met Trp Arg Glu Asn Ile Phe Asn Ser Phe Lys Tyr Val Arg   360
ACC GTT CAA TGG TTA GCC ATG TGG AGA GAG AAT ATT TTC AAC TCA TTC AAA TAC GTC AGA  1080

Leu Ala Ala Asn Ser Ser Leu Lys Gly Gln Ser Asp Tyr Lys Lys Phe Glu Lys Ala Arg   380
TTG GCA GCG AAC TCT TCA TTG AAG GGT CAA AGT GAC TAC AAG AAG TTT GAA AAG GCG AGA  1140

Gln Leu Lys Ser Tyr Ile Asp Ala Ile Arg Arg Asp Tyr Thr Arg Asn Leu Lys Ser Lys   400
CAA TTG AAA TCC TAT ATC GAT GCC ATC AGA AGG GAT TAC ACG AGA AAT TTG AAA AGC AAA  1200

Val Met Leu Glu Arg Gln Lys Ala Val Ala Ile Tyr Leu Ile Asp Val Phe Ala Leu Arg   420
GTT ATG CTA GAG CGC CAA AAG GCC GTA GCC ATT TAT TTG ATC GAT GTA TTC GCT TTA AGA  1260

Ala Gly Gly Glu Lys Ser Glu Asp Glu Ala Asp Thr Val Gly Cys Cys Ser Leu Arg Tyr   440
GCC GGT GGT GAA AAA TCC GAA GAT GAA GCC GAT ACT GTG GGT TGT TGT TCA TTG CGA TAT  1320

Glu His Val Thr Leu Lys Pro Pro Asn Thr Val Ile Phe Asp Phe Leu Gly Lys Asp Ser   460
GAG CAT GTT ACT TTG AAA CCT CCG AAT ACT GTT ATC TTT GAT TTC TTA GGT AAG GAT TCT  1380

Ile Arg Phe Tyr Gln Glu Val Glu Val Asp Lys Gln Val Phe Lys Asn Leu Thr Ile Phe   480
ATT AGA TTT TAT CAA GAG GTA GAA GTT GAC AAA CAA GTT TTC AAA AAT TTG ACA ATT TTT  1440

Lys Arg Pro Pro Lys Gln Pro Gly His Gln Leu Phe Asp Arg Leu Asp Pro Ser Ile Leu   500
AAA AGG CCG CCC AAA CAG CCA GGA CAT CAA CTG TTT GAT CGT CTA GAT CCA TCT ATA CTG  1500
```

```
Asn Lys Tyr Leu Gln Asn Tyr Met Pro Gly Leu Thr Ala Lys Val Phe Arg Thr Tyr Asn   520
AAC AAA TAT CTA CAA AAC TAC ATG CCG GGA TTG ACT GCT AAA GTT TTC CGT ACA TAT AAT  1560

Ala Ser Lys Thr Met Gln Asp Gln Leu Asp Leu Ile Pro Asn Lys Gly Ser Val Ala Glu   540
GCT TCC AAA ACA ATG CAA GAT CAA CTG GAT CTA ATT CCA AAT AAA GGA TCT GTC GCA GAG  1620

Lys Ile Leu Lys Tyr Asn Ala Ala Asn Arg Thr Val Ala Ile Leu Cys Asn His Gln Arg   560
AAA ATA TTG AAG TAC AAC GCA GCA AAT AGA ACT GTA GCC ATC CTA TGT AAC CAT CAA AGG  1680

Thr Val Thr Lys Gly His Ala Gln Thr Val Glu Lys Ala Asn Asn Arg Ile Gln Glu Leu   580
ACT GTC ACG AAG GGG CAT GCA CAA ACA GTG GAA AAG GCC AAT AAT AGA ATA CAA GAG TTG  1740

Glu Trp Gln Lys Ile Arg Cys Lys Arg Ala Ile Leu Gln Leu Asp Lys Asp Leu Leu Lys   600
GAA TGG CAA AAG ATT CGT TGC AAG AGG GCC ATT TTA CAA TTG GAT AAG GAT CTT TTA AAG  1800

Lys Glu Pro Lys Tyr Phe Glu Glu Ile Asp Asp Leu Thr Lys Glu Asp Glu Ala Thr Ile   620
AAA GAG CCA AAA TAT TTC GAA GAA ATC GAC GAT TTG ACG AAA GAA GAT GAA GCC ACC ATT  1860

His Lys Arg Ile Ile Asp Arg Glu Ile Glu Lys Tyr Gln Arg Lys Phe Val Arg Glu Asn   640
CAC AAG AGA ATT ATT GAT AGA GAA ATT GAA AAA TAT CAG CGA AAA TTT GTT AGG GAG AAC  1920

Asp Lys Arg Lys Phe Glu Lys Glu Glu Leu Leu Pro Glu Ser Gln Leu Lys Glu Trp Leu   660
GAT AAG AGG AAA TTT GAA AAG GAA GAA TTA TTG CCG GAA AGT CAA TTG AAG GAA TGG TTG  1980

Glu Lys Val Asp Glu Lys Lys Gln Glu Phe Glu Lys Glu Leu Lys Thr Gly Glu Val Glu   680
GAG AAA GTC GAC GAA AAG AAA CAA GAA TTC GAA AAG GAA TTG AAA ACC GGT GAA GTG GAA  2040

Leu Lys Ser Ser Trp Asn Ser Val Glu Lys Ile Lys Ala Gln Val Glu Lys Leu Glu Gln   700
CTG AAA TCA AGT TGG AAT TCA GTC GAA AAA ATA AAA GCA CAA GTA GAG AAA TTA GAA CAG  2100

Arg Ile Gln Thr Ser Ser Ile Gln Leu Lys Asp Lys Glu Glu Asn Ser Gln Val Ser Leu   720
CGT ATC CAA ACT AGT TCC ATT CAG TTG AAA GAT AAA GAG GAA AAC TCC CAG GTT TCA CTG  2160

Gly Thr Ser Lys Ile Asn Tyr Ile Asp Pro Arg Leu Ser Val Val Phe Cys Lys Lys Tyr   740
GGC ACT TCC AAA ATC AAT TAT ATA GAC CCT AGA CTT TCT GTG GTA TTT TGC AAA AAG TAT  2220

Asp Val Pro Ile Glu Lys Ile Phe Thr Lys Thr Leu Arg Glu Lys Phe Lys Trp Ala Ile   760
GAT GTT CCG ATT GAA AAG ATT TTT ACA AAA ACC CTA AGA GAA AAA TTC AAA TGG GCC ATA  2280

Glu Ser Val Asp Glu Asn Trp Arg Phe   769
GAA TCG GTA GAT GAA AAT TGG AGG TTT  2307
```

Saccharomyces cerevisiae Top3

```
Met Lys Val Leu Cys Val Ala Glu Lys Asn Ser Ile Ala Lys Ala Val Ser Gln Ile Leu    20
ATG AAA GTG CTA TGT GTC GCA GAG AAA AAT TCT ATA GCG AAG GCA GTT TCA CAG ATC CTA    60

Gly Gly Gly Arg Ser Thr Ser Arg Asp Ser Gly Tyr Met Tyr Val Lys Asn Tyr Asp Phe    40
GGA GGA GGC AGA TCA ACT TCA AGG GAT TCC GGC TAC ATG TAT GTA AAG AAC TAT GAT TTC   120

Met Phe Ser Gly Phe Pro Phe Ala Arg Asn Gly Ala Asn Cys Glu Val Thr Met Thr Ser    60
ATG TTT AGT GGG TTC CCG TTT GCC AGA AAT GGG GCT AAC TGC GAA GTT ACC ATG ACT AGT   180

Val Ala Gly His Leu Thr Gly Ile Asp Phe Ser His Asp Ser His Gly Trp Gly Lys Cys    80
GTT GCA GGG CAC CTA ACA GGC ATT GAT TTC AGC CAT GAT TCG CAT GGG TGG GGA AAA TGC   240

Ala Ile Gln Glu Leu Phe Asp Ala Pro Leu Asn Glu Ile Met Asn Asn Asn Gln Lys Lys   100
GCC ATC CAA GAG TTA TTT GAT GCG CCA CTG AAC GAG ATT ATG AAT AAC AAC CAA AAA AAG   300

Ile Ala Ser Asn Ile Lys Arg Glu Ala Arg Asn Ala Asp Tyr Leu Met Ile Trp Thr Asp   120
ATA GCA AGC AAC ATC AAG CGA GAA GCG AGG AAT GCA GAC TAT CTG ATG ATA TGG ACA GAT   360
```

```
Cys Asp Arg Glu Gly Glu Tyr Ile Gly Trp Glu Ile Trp Gln Glu Ala Lys Arg Gly Asn   140
TGC GAC CGG GAA GGA GAG TAC ATC GGT TGG GAG ATA TGG CAG GAG GCC AAG AGA GGC AAC   420

Arg Leu Ile Gln Asn Asp Gln Val Tyr Arg Ala Val Phe Ser His Leu Glu Arg Gln His   160
AGG CTC ATA CAA AAT GAT CAA GTA TAC CGG GCA GTC TTT TCG CAT CTC GAA AGA CAA CAC   480

Ile Leu Asn Ala Ala Arg Asn Pro Ser Arg Leu Asp Met Lys Ser Val His Ala Val Gly   180
ATA TTA AAT GCA GCA CGA AAC CCA AGT CGA TTG GAT ATG AAG AGT GTG CAC GCT GTA GGC   540

Thr Arg Ile Glu Ile Asp Leu Arg Ala Gly Val Thr Phe Thr Arg Leu Leu Thr Glu Thr   200
ACG CGG ATT GAA ATC GAT CTT CGA GCA GGT GTT ACA TTC ACC AGA CTC TTA ACA GAA ACG   600

Leu Arg Asn Lys Leu Arg Asn Gln Ala Thr Met Thr Lys Asp Gly Ala Lys His Arg Gly   220
CTA CGA AAT AAA CTG AGA AAC CAA GCC ACC ATG ACC AAG GAT GGT GCA AAA CAC CGC GGT   660

Gly Asn Lys Asn Asp Ser Gln Val Val Ser Tyr Gly Thr Cys Gln Phe Pro Thr Leu Gly   240
GGT AAC AAG AAC GAC TCA CAA GTC GTA TCG TAT GGT ACA TGC CAG TTT CCA ACG CTC GGC   720

Phe Val Val Asp Arg Phe Glu Arg Ile Arg Asn Phe Val Pro Glu Glu Phe Trp Tyr Ile   260
TTT GTA GTA GAC AGG TTT GAA AGA ATA CGA AAT TTT GTT CCC GAA GAG TTC TGG TAT ATC   780

Gln Leu Val Val Glu Asn Lys Asp Asn Gly Gly Thr Thr Thr Phe Gln Trp Asp Arg Gly   280
CAA TTG GTA GTC GAA AAC AAA GAC AAC GGC GGA ACA ACA ACG TTC CAG TGG GAC AGG GGC   840

His Leu Phe Asp Arg Leu Ser Val Leu Thr Phe Tyr Glu Thr Cys Ile Glu Thr Ala Gly   300
CAC TTG TTC GAC CGG CTG AGC GTG TTA ACG TTT TAC GAG ACA TGC ATC GAA ACC GCC GGC   900

Asn Val Ala Gln Val Val Asp Leu Lys Ser Lys Pro Thr Thr Lys Tyr Arg Pro Leu Pro   320
AAT GTT GCT CAA GTA GTA GAC TTG AAA TCA AAG CCA ACA ACG AAA TAC AGA CCT TTA CCT   960

Leu Thr Thr Val Glu Leu Gln Lys Asn Cys Ala Arg Tyr Leu Arg Leu Asn Ala Lys Gln   340
CTG ACC ACA GTG GAG CTA CAA AAA AAC TGC GCC CGG TAC CTG CGT CTG AAC GCC AAA CAA  1020

Ser Leu Asp Ala Ala Glu Lys Leu Tyr Gln Lys Gly Phe Ile Ser Tyr Pro Arg Thr Glu   360
TCA CTA GAC GCA GCA GAA AAG CTA TAC CAA AAG GGG TTC ATA TCG TAT CCA AGA ACA GAG  1080

Thr Asp Thr Phe Pro His Ala Met Asp Leu Lys Ser Leu Val Glu Lys Gln Ala Gln Leu   380
ACT GAT ACT TTC CCA CAC GCA ATG GAC CTA AAA TCC TTG GTC GAA AAG CAA GCT CAA TTG  1140

Asp Gln Leu Ala Ala Gly Gly Arg Thr Ala Trp Ala Ser Tyr Ala Ala Ser Leu Leu Gln   400
GAC CAA CTC GCT GCA GGC GGC AGA ACC GCC TGG GCA TCG TAC GCG GCA TCG CTG CTC CAA  1200

Pro Glu Asn Thr Ser Asn Asn Asn Lys Phe Lys Phe Pro Arg Ser Gly Ser His Asp Asp   420
CCC GAA AAC ACA AGT AAC AAT AAC AAG TTC AAG TTT CCA CGA AGC GGC TCC CAT GAC GAC  1260

Lys Ala His Pro Pro Ile His Pro Ile Val Ser Leu Gly Pro Glu Ala Asn Val Ser Pro   440
AAA GCG CAT CCA CCA ATC CAC CCC ATC GTA AGT CTG GGG CCT GAA GCA AAT GTT TCG CCA  1320

Val Glu Arg Arg Val Tyr Glu Tyr Val Ala Arg His Phe Leu Ala Cys Cys Ser Glu Asp   460
GTG GAA AGA AGA GTA TAC GAG TAC GTG GCC AGG CAC TTT TTG GCA TGC TGC TCA GAG GAC  1380

Ala Lys Gly Gln Ser Met Thr Leu Val Leu Asp Trp Ala Val Glu Arg Phe Ser Ala Ser   480
GCC AAG GGC CAA TCG ATG ACC CTT GTG TTG GAC TGG GCC GTT GAA CGT TTC TCA GCT TCA  1440

Gly Leu Val Val Leu Glu Arg Asn Phe Leu Asp Val Tyr Pro Trp Ala Arg Trp Glu Thr   500
GGT CTC GTA GTC CTA GAG AGA AAT TTC CTC GAT GTT TAC CCT TGG GCC CGA TGG GAA ACC  1500

Thr Lys Gln Leu Pro Arg Leu Glu Met Asn Ala Leu Val Asp Ile Ala Lys Ala Glu Met   520
ACC AAG CAG TTA CCG CGG CTT GAA ATG AAT GCC CTC GTA GAC ATC GCG AAG GCC GAA ATG  1560

Lys Ala Gly Thr Thr Ala Pro Pro Lys Pro Met Thr Glu Ser Glu Leu Ile Leu Leu Met   540
AAG GCG GGC ACT ACG GCG CCG CCC AAG CCG ATG ACT GAG AGT GAA CTC ATT CTC CTC ATG  1620
```

```
Asp Thr Asn Gly Ile Gly Thr Asp Ala Thr Ile Ala Glu His Ile Asp Lys Ile Gln Val   560
GAT ACA AAC GGC ATT GGC ACA GAC GCC ACC ATT GCG GAG CAC ATA GAC AAG ATC CAA GTA  1680

Arg Asn Tyr Val Arg Ser Glu Lys Val Gly Lys Glu Thr Tyr Leu Gln Pro Thr Thr Leu   580
CGT AAT TAC GTT AGG AGC GAG AAA GTA GGC AAG GAA ACC TAC TTA CAA CCC ACG ACC CTG  1740

Gly Val Ser Leu Val His Gly Phe Glu Ala Ile Gly Leu Glu Asp Ser Phe Ala Lys Pro   600
GGT GTC TCA CTA GTG CAC GGC TTC GAG GCC ATC GGC CTC GAA GAC TCC TTT GCA AAG CCC  1800

Phe Gln Arg Arg Glu Met Glu Gln Asp Leu Lys Lys Ile Cys Glu Gly His Ala Ser Lys   620
TTC CAG CGC AGA GAA ATG GAG CAA GAC CTC AAG AAA ATC TGC GAA GGT CAT GCC TCC AAG  1860

Thr Asp Val Val Lys Asp Ile Val Glu Lys Tyr Arg Lys Tyr Trp His Lys Thr Asn Ala   640
ACT GAT GTT GTA AAG GAC ATA GTC GAG AAG TAT AGG AAG TAC TGG CAC AAG ACG AAT GCC  1920

Cys Lys Asn Thr Leu Leu Gln Val Tyr Asp Arg Val Lys Ala Ser Met   656
TGC AAG AAT ACT CTC TTG CAA GTT TAT GAC CGT GTC AAG GCA TCC ATG  1968
```

Schizosaccharomyces pombe Top1

```
Met Ser Ser Ser Asp Ser Val Ser Leu Ser Ile Arg Arg Arg Gln Arg Arg Gly Ser Ser    20
ATG TCT TCG TCT GAT TCC GTG TCT TTA TCC ATT AGA AGG AGG CAA AGA CGC GGT TCA TCT    60

Lys Arg Ile Ser Met Lys Glu Ser Asp Glu Glu Ser Asp Ser Ser Glu Asn His Pro Leu    40
AAA CGG ATT AGT ATG AAG GAG TCG GAC GAA GAG AGT GAT AGT TCT GAA AAT CAT CCT CTT   120

Ser Glu Ser Leu Asn Lys Lys Ser Lys Ser Glu Ser Asp Glu Asp Asp Ile Pro Ile Arg    60
AGT GAA TCG TTG AAT AAG AAA TCC AAA TCC GAG AGT GAT GAG GAT GAT ATA CCG ATT AGA   180

Lys Arg Arg Ala Ser Ser Lys Lys Asn Met Ser Asn Ser Ser Ser Lys Lys Arg Ala Lys    80
AAG AGA CGT GCT TCT TCA AAG AAA AAT ATG TCC AAT TCA AGC TCT AAA AAA AGA GCT AAA   240

Val Met Gly Asn Gly Gly Leu Lys Asn Gly Lys Lys Thr Ala Val Val Lys Glu Glu Glu   100
GTA ATG GGA AAT GGT GGT TTA AAA AAC GGA AAA AAA ACG GCG GTT GTA AAA GAG GAA GAG   300

Asp Phe Asn Glu Ile Ala Lys Pro Ser Pro Lys His Lys Arg Val Ser Lys Ala Asn Gly   120
GAT TTC AAT GAA ATT GCA AAG CCC TCA CCG AAG CAT AAA CGT GTT TCT AAA GCG AAT GGA   360

Ser Lys Asn Gly Ala Lys Ser Ala Val Lys Lys Glu Glu Ser Asp Thr Asp Asp Ser Val   140
AGT AAA AAT GGC GCG AAG TCA GCT GTT AAA AAA GAG GAA AGT GAC ACA GAT GAT TCA GTT   420

Pro Leu Arg Ala Val Ser Thr Val Ser Leu Thr Pro Tyr Lys Ser Glu Leu Pro Ser Gly   160
CCA TTG AGA GCT GTT TCG ACT GTA TCT TTA ACT CCT TAC AAA TCA GAA CTG CCT TCG GGA   480

Ala Ser Thr Thr Gln Asn Arg Ser Pro Asn Asp Glu Glu Asp Glu Asp Glu Asp Tyr Lys   180
GCT TCG ACC ACG CAA AAT CGC TCA CCA AAC GAT GAG GAA GAT GAA GAT GAG GAT TAC AAG   540

Trp Trp Thr Ser Glu Asn Ile Asp Asp Thr Gln Lys Trp Thr Thr Leu Glu His Asn Gly   200
TGG TGG ACT TCA GAA AAT ATC GAT GAT ACT CAA AAA TGG ACT ACA TTG GAG CAT AAT GGT   600

Val Ile Phe Ala Pro Pro Tyr Glu Pro Leu Pro Lys Asn Val Lys Leu Ile Tyr Asp Gly   220
GTA ATT TTT GCT CCA CCC TAT GAA CCT TTA CCA AAG AAC GTC AAG CTA ATT TAC GAT GGA   660

Asn Pro Val Asn Leu Pro Pro Glu Ala Glu Glu Val Ala Gly Phe Tyr Ala Ala Met Leu   240
AAC CCC GTA AAT CTT CCT CCC GAA GCA GAA GAA GTT GCT GGT TTT TAT GCT GCA ATG CTT   720

Glu Thr Asp His Ala Lys Asn Pro Val Phe Gln Asp Asn Phe Phe Arg Asp Phe Leu Lys   260
GAA ACC GAT CAT GCC AAA AAT CCT GTA TTC CAA GAT AAT TTT TTC CGT GAC TTC TTA AAG   780

Val Cys Asp Glu Cys Asn Phe Asn His Asn Ile Lys Glu Phe Ser Lys Cys Asp Phe Thr   280
GTC TGT GAT GAA TGT AAC TTT AAT CAC AAC ATT AAA GAG TTT TCT AAA TGT GAT TTT ACC   840
```

```
Gln Met Phe His His Phe Glu Gln Lys Arg Glu Glu Lys Lys Ser Met Pro Lys Glu Gln   300
CAA ATG TTT CAC CAT TTT GAG CAA AAG AGG GAA GAG AAG AAG AGT ATG CCG AAG GAA CAG   900

Lys Lys Ala Ile Lys Glu Lys Lys Asp Glu Glu Glu Glu Lys Tyr Lys Trp Cys Ile Leu   320
AAG AAG GCA ATA AAG GAG AAA AAA GAT GAG GAA GAG GAA AAA TAT AAA TGG TGC ATA CTT   960

Asp Gly Arg Lys Glu Lys Val Gly Asn Phe Arg Ile Glu Pro Pro Gly Leu Phe Arg Gly   340
GAT GGG AGA AAG GAG AAG GTT GGT AAC TTT CGT ATT GAA CCT CCA GGG TTA TTT CGT GGT  1020

Arg Gly Ser His Pro Lys Thr Gly Ser Leu Lys Arg Arg Val Tyr Pro Glu Gln Ile Thr   360
CGA GGT AGT CAT CCT AAA ACT GGT TCT TTA AAG CGT CGA GTA TAT CCT GAA CAA ATT ACC  1080

Ile Asn Ile Gly Glu Gly Val Pro Val Pro Glu Pro Leu Pro Gly His Gln Trp Ala Glu   380
ATT AAT ATT GGT GAA GGT GTA CCC GTT CCA GAA CCA CTC CCT GGG CAT CAA TGG GCG GAG  1140

Val Lys His Asp Asn Thr Val Thr Trp Leu Ala Thr Trp His Glu Asn Ile Asn Asn Asn   400
GTA AAG CAT GAT AAT ACA GTG ACC TGG TTG GCA ACC TGG CAT GAA AAT ATA AAT AAT AAT  1200

Val Lys Tyr Val Phe Leu Ala Ala Gly Ser Ser Leu Lys Gly Gln Ser Asp Leu Lys Lys   420
GTC AAA TAT GTC TTT TTA GCT GCA GGA AGT TCT CTA AAA GGA CAG AGT GAC TTA AAA AAA  1260

Tyr Glu Lys Ser Arg Lys Leu Lys Asp Tyr Ile Asp Asp Ile Arg Lys Gly Tyr Arg Lys   440
TAC GAA AAG TCA AGA AAG CTT AAG GAT TAT ATT GAT GAT ATC CGT AAA GGC TAC CGG AAA  1320

Asp Leu Lys Asn Glu Leu Thr Val Glu Arg Gln Arg Gly Thr Ala Met Tyr Leu Ile Asp   460
GAT TTG AAA AAT GAG TTA ACG GTT GAG CGT CAA AGA GGA ACT GCC ATG TAT TTA ATT GAT  1380

Val Phe Ala Leu Arg Ala Gly Asn Glu Lys Gly Glu Asp Glu Ala Asp Thr Val Gly Cys   480
GTT TTT GCT TTA AGA GCA GGA AAT GAA AAG GGT GAA GAC GAG GCG GAT ACT GTT GGT TGT  1440

Cys Ser Leu Arg Tyr Glu His Val Thr Leu Lys Pro Pro Arg Thr Val Val Phe Asp Phe   500
TGT TCA CTG CGA TAT GAA CAT GTT ACA CTG AAG CCA CCA CGA ACA GTC GTT TTC GAT TTT  1500

Leu Gly Lys Asp Ser Ile Arg Tyr Tyr Asn Glu Val Glu Val Asp Pro Gln Val Phe Lys   520
CTT GGT AAA GAT TCT ATT CGT TAC TAC AAC GAG GTT GAA GTT GAT CCC CAA GTT TTT AAA  1560

Asn Leu Lys Ile Phe Lys Arg Pro Pro Lys Lys Glu Gly Asp Leu Ile Phe Asp Arg Leu   540
AAT CTA AAG ATC TTT AAA CGT CCT CCC AAA AAA GAG GGT GAT TTA ATT TTC GAC CGT CTC  1620

Ser Thr Asn Ser Leu Asn Lys Tyr Leu Thr Ser Leu Met Asp Gly Leu Ser Ala Lys Val   560
AGT ACA AAC AGT CTT AAC AAA TAT CTG ACT AGC CTT ATG GAT GGA CTT TCA GCT AAA GTA  1680

Phe Arg Thr Tyr Asn Ala Ser Tyr Thr Met Ala Glu Glu Leu Lys Lys Met Pro Lys Asn   580
TTT CGT ACC TAC AAT GCT TCA TAC ACC ATG GCC GAG GAA CTT AAG AAA ATG CCT AAG AAC  1740

Leu Thr Leu Ala Asp Lys Ile Leu Phe Tyr Asn Arg Ala Asn Arg Thr Val Ala Ile Leu   600
CTC ACC CTT GCA GAC AAA ATA CTA TTT TAT AAT AGG GCA AAT AGG ACT GTT GCA ATT TTA  1800

Cys Asn His Gln Arg Ser Val Thr Lys Asn His Asp Val Gln Met Glu Arg Phe Ala Glu   620
TGT AAT CAT CAA CGT TCC GTA ACC AAA AAT CAC GAT GTT CAA ATG GAA CGG TTT GCC GAA  1860

Arg Ile Lys Ala Leu Gln Tyr Gln Arg Met Arg Leu Arg Lys Met Met Leu Asn Leu Glu   640
AGG ATT AAG GCA TTA CAA TAC CAG CGG ATG AGA CTG CGA AAA ATG ATG CTG AAT TTA GAG  1920

Pro Lys Leu Ala Lys Ser Lys Pro Glu Leu Leu Ala Lys Glu Glu Gly Ile Thr Asp Ser   660
CCC AAG CTT GCT AAA AGC AAG CCC GAA TTG CTG GCT AAA GAA GAA GGC ATT ACC GAT TCA  1980

Trp Ile Val Lys His His Glu Thr Leu Tyr Glu Leu Glu Lys Glu Lys Ile Lys Lys Lys   680
TGG ATC GTA AAA CAT CAC GAG ACA CTT TAC GAA CTA GAA AAA GAG AAA ATA AAA AAG AAA  2040

Phe Asp Arg Glu Asn Glu Lys Leu Ala Ala Glu Asp Pro Lys Ser Met Leu Pro Glu Ser   700
TTC GAT CGT GAG AAC GAA AAA TTA GCT GCT GAG GAT CCC AAA TCA ATG CTT CCG GAA TCT  2100
```

```
Glu Leu Glu Val Arg Leu Lys Ala Ala Asp Glu Leu Lys Lys Ala Leu Asp Ala Glu Leu   720
GAA TTG GAA GTT CGA TTG AAA GCG GCT GAT GAG TTG AAG AAA GCG CTG GAC GCT GAA CTT  2160

Lys Ser Lys Lys Val Asp Pro Gly Arg Ser Ser Met Glu Gln Leu Glu Lys Arg Leu Asn   740
AAA AGC AAA AAA GTC GAT CCA GGT CGT TCT TCG ATG GAA CAA CTT GAG AAA AGA TTA AAC  2220

Lys Leu Asn Glu Arg Ile Asn Val Met Arg Thr Gln Met Ile Asp Lys Asp Glu Asn Lys   760
AAA CTC AAT GAA CGA ATT AAT GTT ATG CGT ACT CAG ATG ATC GAT AAA GAC GAG AAT AAA  2280

Thr Thr Ala Leu Gly Thr Ser Lys Ile Asn Tyr Ile Asp Pro Arg Leu Thr Tyr Ser Phe   780
ACT ACT GCT TTG GGT ACA AGT AAG ATT AAC TAC ATA GAC CCG AGG CTT ACT TAT TCG TTC  2340

Ser Lys Arg Glu Asp Val Pro Ile Glu Lys Leu Phe Ser Lys Thr Ile Arg Asp Lys Phe   800
AGC AAG CGA GAA GAC GTT CCT ATT GAG AAG CTG TTT AGT AAG ACG ATT CGT GAC AAG TTC  2400

Asn Trp Ala Ala Asp Thr Pro Pro Asp Trp Lys Trp   812
AAT TGG GCT GCT GAT ACA CCT CCG GAT TGG AAG TGG  2436
```

Homo sapiens Top1

```
Met Ser Gly Asp His Leu His Asn Asp Ser Gln Ile Glu Ala Asp Phe Arg Leu Asn Asp    20
ATG AGT GGG GAC CAC CTC CAC AAC GAT TCC CAG ATC GAA GCG GAT TTC CGA TTG AAT GAT    60

Ser His Lys His Lys Asp Lys His Lys Asp Arg Glu His Arg His Lys Glu His Lys Lys    40
TCT CAT AAA CAC AAA GAT AAA CAC AAA GAT CGA GAA CAC CGG CAC AAA GAA CAC AAG AAG   120

Glu Lys Asp Arg Glu Lys Ser Lys His Ser Asn Ser Glu His Lys Asp Ser Glu Lys Lys    60
GAG AAG GAC CGG GAA AAG TCC AAG CAT AGC AAC AGT GAA CAT AAA GAT TCT GAA AAG AAA   180

His Lys Glu Lys Glu Lys Thr Lys His Lys Asp Gly Ser Ser Glu Lys His Lys Asp Lys    80
CAC AAA GAG AAG GAG AAG ACC AAA CAC AAA GAT GGA AGC TCA GAA AAG CAT AAA GAC AAA   240

His Lys Asp Arg Asp Lys Glu Lys Arg Lys Glu Glu Lys Val Arg Ala Ser Gly Asp Ala   100
CAT AAA GAC AGA GAC AAG GAA AAA CGA AAA GAG GAA AAG GTT CGA GCC TCT GGG GAT GCA   300

Lys Ile Lys Lys Glu Lys Glu Asn Gly Phe Ser Ser Pro Pro Gln Ile Lys Asp Glu Pro   120
AAA ATA AAG AAG GAG AAG GAA AAT GGC TTC TCT AGT CCA CCA CAA ATT AAA GAT GAA CCT   360

Glu Asp Asp Gly Tyr Phe Val Pro Pro Lys Glu Asp Ile Lys Pro Leu Lys Arg Pro Arg   140
GAA GAT GAT GGC TAT TTT GTT CCT CCT AAA GAG GAT ATA AAG CCA TTA AAG AGA CCT CGA   420

Asp Glu Asp Asp Val Asp Tyr Lys Pro Lys Lys Ile Lys Thr Glu Asp Thr Lys Lys Glu   160
GAT GAG GAT GAT GTT GAT TAT AAA CCT AAG AAA ATT AAA ACA GAA GAT ACC AAG AAG GAG   480

Lys Lys Arg Lys Leu Glu Glu Glu Glu Asp Gly Lys Leu Lys Lys Pro Lys Asn Lys Asp   180
AAG AAA AGA AAA CTA GAA GAA GAA GAG GAT GGT AAA TTG AAA AAA CCC AAG AAT AAA GAT   540

Lys Asp Lys Lys Val Pro Glu Pro Asp Asn Lys Lys Lys Lys Pro Lys Lys Glu Glu Glu   200
AAA GAT AAA AAA GTT CCT GAG CCA GAT AAC AAG AAA AAG AAG CCG AAG AAA GAA GAG GAA   600

Gln Lys Trp Lys Trp Trp Glu Glu Glu Arg Tyr Pro Glu Gly Ile Lys Trp Lys Phe Leu   220
CAG AAG TGG AAA TGG TGG GAA GAA GAG CGC TAT CCT GAA GGC ATC AAG TGG AAA TTC CTA   660

Glu His Lys Gly Pro Val Phe Ala Pro Pro Tyr Glu Pro Leu Pro Glu Asn Val Lys Phe   240
GAA CAT AAA GGT CCA GTA TTT GCC CCA CCA TAT GAG CCT CTT CCA GAG AAT GTC AAG TTT   720

Tyr Tyr Asp Gly Lys Val Met Lys Leu Ser Pro Lys Ala Glu Glu Val Ala Thr Phe Phe   260
TAT TAT GAT GGT AAA GTC ATG AAG CTG AGC CCC AAA GCA GAG GAA GTA GCT ACG TTC TTT   780

Ala Lys Met Leu Asp His Glu Tyr Thr Thr Lys Glu Ile Phe Arg Lys Asn Phe Phe Lys   280
GCA AAA ATG CTC GAC CAT GAA TAT ACT ACC AAG GAA ATA TTT AGG AAA AAT TTC TTT AAA   840
```

```
Asp Trp Arg Lys Glu Met Thr Asn Glu Glu Lys Asn Ile Ile Thr Asn Leu Ser Lys Cys   300
GAC TGG AGA AAG GAA ATG ACT AAT GAA GAG AAG AAT ATT ATC ACC AAC CTA AGC AAA TGT   900

Asp Phe Thr Gln Met Ser Gln Tyr Phe Lys Ala Gln Thr Glu Ala Arg Lys Gln Met Ser   320
GAT TTT ACC CAG ATG AGC CAG TAT TTC AAA GCC CAG ACG GAA GCT CGG AAA CAG ATG AGC   960

Lys Glu Glu Lys Leu Lys Ile Lys Glu Glu Asn Glu Lys Leu Leu Lys Glu Tyr Gly Phe   340
AAG GAA GAG AAA CTG AAA ATC AAA GAG GAG AAT GAA AAA TTA CTG AAA GAA TAT GGA TTC  1020

Cys Ile Met Asp Asn His Lys Glu Arg Ile Ala Asn Phe Lys Ile Glu Pro Pro Gly Leu   360
TGT ATT ATG GAT AAC CAC AAA GAG AGG ATT GCT AAC TTC AAG ATA GAG CCT CCT GGA CTT  1080

Phe Arg Gly Arg Gly Asn His Pro Lys Met Gly Met Leu Lys Arg Arg Ile Met Pro Glu   380
TTC CGT GGC CGC GGC AAC CAC CCC AAG ATG GGC ATG CTG AAG AGA CGA ATC ATG CCC GAG  1140

Asp Ile Ile Ile Asn Cys Ser Lys Asp Ala Lys Val Pro Ser Pro Pro Pro Gly His Lys   400
GAT ATA ATC ATC AAC TGT AGC AAA GAT GCC AAG GTT CCT TCT CCT CCT CCA GGA CAT AAG  1200

Trp Lys Glu Val Arg His Asp Asn Lys Val Thr Trp Leu Val Ser Trp Thr Glu Asn Ile   420
TGG AAA GAA GTC CGG CAT GAT AAC AAG GTT ACT TGG CTG GTT TCC TGG ACA GAG AAC ATC  1260

Gln Gly Ser Ile Lys Tyr Ile Met Leu Asn Pro Ser Ser Arg Ile Lys Gly Glu Lys Asp   440
CAA GGT TCC ATT AAA TAC ATC ATG CTT AAC CCT AGT TCA CGA ATC AAG GGT GAG AAG GAC  1320

Trp Gln Lys Tyr Glu Thr Ala Arg Arg Leu Lys Lys Cys Val Asp Lys Ile Arg Asn Gln   460
TGG CAG AAA TAC GAG ACT GCT CGG CGG CTG AAA AAA TGT GTG GAC AAG ATC CGG AAC CAG  1380

Tyr Arg Glu Asp Trp Lys Ser Lys Glu Met Lys Val Arg Gln Arg Ala Val Ala Leu Tyr   480
TAT CGA GAA GAC TGG AAG TCC AAA GAG ATG AAA GTC CGG CAG AGA GCT GTA GCC CTG TAC  1440

Phe Ile Asp Lys Leu Ala Leu Arg Ala Gly Asn Glu Lys Glu Glu Gly Glu Thr Ala Asp   500
TTC ATC GAC AAG CTT GCT CTG AGA GCA GGC AAT GAA AAG GAG GAA GGA GAA ACA GCG GAC  1500

Thr Val Gly Cys Cys Ser Leu Arg Val Glu His Ile Asn Leu His Pro Glu Leu Asp Gly   520
ACT GTG GGC TGC TGC TCA CTT CGT GTG GAG CAC ATC AAT CTA CAC CCA GAG TTG GAT GGT  1560

Gln Glu Tyr Val Val Glu Phe Asp Phe Leu Gly Lys Asp Ser Ile Arg Tyr Tyr Asn Lys   540
CAG GAA TAT GTG GTA GAG TTT GAC TTC CTC GGG AAG GAC TCC ATC AGA TAC TAT AAC AAG  1620

Val Pro Val Glu Lys Arg Val Phe Lys Asn Leu Gln Leu Phe Met Glu Asn Lys Gln Pro   560
GTC CCT GTT GAG AAA CGA GTT TTT AAG AAC CTA CAA CTA TTT ATG GAG AAC AAG CAG CCC  1680

Glu Asp Asp Leu Phe Asp Arg Leu Asn Thr Gly Ile Leu Asn Lys His Leu Gln Asp Leu   580
GAG GAT GAT CTT TTT GAT AGA CTC AAT ACT GGT ATT CTG AAT AAG CAT CTT CAG GAT CTC  1740

Met Glu Gly Leu Thr Ala Lys Val Phe Arg Thr Tyr Asn Ala Ser Ile Thr Leu Gln Gln   600
ATG GAG GGC TTG ACA GCC AAG GTA TTC CGT ACG TAC AAT GCC TCC ATC ACG CTA CAG CAG  1800

Gln Leu Lys Glu Leu Thr Ala Pro Asp Glu Asn Ile Pro Ala Lys Ile Leu Ser Tyr Asn   620
CAG CTA AAA GAA CTG ACA GCC CCG GAT GAG AAC ATC CCA GCG AAG ATC CTT TCT TAT AAC  1860

Arg Ala Asn Arg Ala Val Ala Ile Leu Cys Asn His Gln Arg Ala Pro Pro Lys Thr Phe   640
CGT GCC AAT CGA GCT GTT GCA ATT CTT TGT AAC CAT CAG AGG GCA CCA CCA AAA ACT TTT  1920

Glu Lys Ser Met Met Asn Leu Gln Thr Lys Ile Asp Ala Lys Lys Glu Gln Leu Ala Asp   660
GAG AAG TCT ATG ATG AAC TTG CAA ACT AAG ATT GAT GCC AAG AAG GAA CAG CTA GCA GAT  1980

Ala Arg Arg Asp Leu Lys Ser Ala Lys Ala Asp Ala Lys Val Met Lys Asp Ala Lys Thr   680
GCC CGG AGA GAC CTG AAA AGT GCT AAG GCT GAT GCC AAG GTC ATG AAG GAT GCA AAG ACG  2040

Lys Lys Val Val Glu Ser Lys Lys Lys Ala Val Gln Arg Leu Glu Glu Gln Leu Met Lys   700
AAG AAG GTA GTA GAG TCA AAG AAG AAG GCT GTT CAG AGA CTG GAG GAA CAG TTG ATG AAG  2100
```

```
Leu Glu Val Gln Ala Thr Asp Arg Glu Glu Asn Lys Gln Ile Ala Leu Gly Thr Ser Lys   720
CTG GAA GTT CAA GCC ACA GAC CGA GAG GAA AAT AAA CAG ATT GCC CTG GGA ACC TCC AAA  2160

Leu Asn Tyr Leu Asp Pro Arg Ile Thr Val Ala Trp Cys Lys Lys Trp Gly Val Pro Ile   740
CTC AAT TAT CTG GAC CCT AGG ATC ACA GTG GCT TGG TGC AAG AAG TGG GGT GTC CCA ATT  2220

Glu Lys Ile Tyr Asn Lys Thr Gln Arg Glu Lys Phe Ala Trp Ala Ile Asp Met Ala Asp   760
GAG AAG ATT TAC AAC AAA ACC CAG CGG GAG AAG TTT GCC TGG GCC ATT GAC ATG GCT GAT  2280

Glu Asp Tyr Glu Phe   765
GAA GAC TAT GAG TTT  2295
```

Vaccinia viral topoisomerase gene

```
Met Arg Ala Leu Phe Tyr Lys Asp Gly Lys Leu Phe Thr Asp Asn Asn Phe Leu Asn Pro    20
ATG CGT GCA CTT TTT TAT AAA GAT GGT AAA CTC TTT ACC GAT AAT AAT TTT TTA AAT CCT    60

Val Ser Asp Asp Asn Pro Ala Tyr Glu Val Leu Gln His Val Lys Ile Pro Thr His Leu    40
GTA TCA GAC GAT AAT CCA GCG TAT GAG GTT TTG CAA CAT GTT AAA ATT CCT ACT CAT TTA   120

Thr Asp Val Val Val Tyr Glu Gln Thr Trp Glu Glu Ala Leu Thr Arg Leu Ile Phe Val    60
ACA GAT GTA GTA GTA TAT GAA CAA ACG TGG GAA GAG GCA TTA ACT AGA TTA ATT TTT GTG   180

Gly Ser Asp Ser Lys Gly Arg Arg Gln Tyr Phe Tyr Gly Lys Met His Val Gln Asn Arg    80
GGA AGC GAT TCA AAA GGA CGT AGA CAA TAC TTT TAC GGA AAA ATG CAT GTA CAG AAT CGC   240

Asn Ala Lys Arg Asp Arg Ile Phe Val Arg Val Tyr Asn Val Met Lys Arg Ile Asn Cys   100
AAC GCT AAA AGA GAT CGT ATT TTT GTT AGA GTA TAT AAC GTT ATG AAA CGA ATT AAT TGT   300

Phe Ile Asn Lys Asn Ile Lys Lys Ser Ser Thr Asp Ser Asn Tyr Gln Leu Ala Val Phe   120
TTT ATA AAC AAA AAT ATA AAG AAA TCG TCC ACA GAT TCC AAT TAT CAG TTG GCG GTT TTT   360

Met Leu Met Glu Thr Met Phe Phe Ile Arg Phe Gly Lys Met Lys Tyr Leu Lys Glu Asn   140
ATG TTA ATG GAA ACT ATG TTT TTT ATT AGA TTT GGT AAA ATG AAA TAT CTT AAG GAG AAT   420

Glu Thr Val Gly Leu Leu Thr Leu Lys Asn Lys His Ile Glu Ile Ser Pro Asp Glu Ile   160
GAA ACA GTA GGG TTA TTA ACA CTA AAA AAT AAA CAC ATA GAA ATA AGT CCC GAT GAA ATA   480

Val Ile Lys Phe Val Gly Lys Asp Lys Val Ser His Glu Phe Val Val His Lys Ser Asn   180
GTT ATC AAG TTT GTA GGA AAG GAC AAA GTT TCA CAT GAA TTT GTT GTT CAT AAG TCT AAT   540

Arg Leu Tyr Lys Pro Leu Leu Lys Leu Thr Asp Asp Ser Ser Pro Glu Glu Phe Leu Phe   200
AGA CTA TAT AAG CCG CTA TTG AAA CTG ACG GAT GAT TCT AGT CCC GAA GAA TTT CTG TTC   600

Asn Lys Leu Ser Glu Arg Lys Val Tyr Glu Cys Ile Lys Gln Phe Gly Ile Arg Ile Lys   220
AAC AAA CTA AGT GAA CGA AAG GTA TAT GAA TGT ATC AAA CAG TTT GGT ATT AGA ATC AAG   660

Asp Leu Arg Thr Tyr Gly Val Asn Tyr Thr Phe Leu Tyr Asn Phe Trp Thr Asn Val Lys   240
GAT CTC CGA ACG TAT GGA GTC AAT TAT ACG TTT TTA TAT AAT TTT TGG ACA AAT GTA AAG   720

Ser Ile Ser Pro Leu Pro Ser Pro Lys Lys Leu Ile Ala Leu Thr Ile Lys Gln Thr Ala   260
TCC ATA TCT CCT CTT CCA TCA CCA AAA AAG TTA ATA GCG TTA ACT ATC AAA CAA ACT GCT   780

Glu Val Val Gly His Thr Pro Ser Ile Ser Lys Arg Ala Tyr Met Ala Thr Thr Ile Leu   280
GAA GTG GTA GGT CAT ACT CCA TCA ATT TCA AAA AGA GCT TAT ATG GCA ACG ACT ATT TTA   840

Glu Met Val Lys Asp Lys Asn Phe Leu Asp Val Val Ser Lys Thr Thr Phe Asp Glu Phe   300
GAA ATG GTA AAG GAT AAA AAT TTT TTA GAT GTA GTA TCT AAA ACT ACG TTC GAT GAA TTC   900

Leu Ser Ile Val Val Asp His Val Lys Ser Ser Thr Asp Gly   314
CTA TCT ATA GTC GTA GAT CAC GTT AAA TCA TCT ACG GAT GGA   942
```

Escherichia coli gyrA

```
Met Ser Asp Leu Ala Arg Glu Ile Thr Pro Val Asn Ile Glu Glu Glu Leu Lys Ser Ser    20
ATG AGC GAC CTT GCG AGA GAA ATT ACA CCG GTC AAC ATT GAG GAA GAG CTG AAG AGC TCC    60

Tyr Leu Asp Tyr Ala Met Ser Val Ile Val Gly Arg Ala Leu Pro Asp Val Arg Asp Gly    40
TAT CTG GAT TAT GCG ATG TCG GTC ATT GTT GGC CGT GCG CTG CCA GAT GTC CGA GAT GGC   120

Leu Lys Pro Val His Arg Arg Val Leu Tyr Ala Met Asn Val Leu Gly Asn Asp Trp Asn    60
CTG AAG CCG GTA CAC CGT CGC GTA CTT TAC GCC ATG AAC GTA CTA GGC AAT GAC TGG AAC   180

Lys Ala Tyr Lys Lys Ser Ala Arg Val Val Gly Asp Val Ile Gly Lys Tyr His Pro His    80
AAA GCC TAT AAA AAA TCT GCC CGT GTC GTT GGT GAC GTA ATC GGT AAA TAC CAT CCC CAT   240

Gly Asp Ser Ala Val Tyr Asp Thr Ile Val Arg Met Ala Gln Pro Phe Ser Leu Arg Tyr   100
GGT GAC TCG GCG GTC TAT GAC ACG ATT GTC CGC ATG GCG CAG CCA TTC TCG CTG CGT TAT   300

Met Leu Val Asp Gly Gln Gly Asn Phe Gly Ser Ile Asp Gly Asp Ser Ala Ala Ala Met   120
ATG CTG GTA GAC GGT CAG GGT AAC TTC GGT TCT ATC GAC GGC GAC TCT GCG GCG GCA ATG   360

Arg Tyr Thr Glu Ile Arg Leu Ala Lys Ile Ala His Glu Leu Met Ala Asp Leu Glu Lys   140
CGT TAT ACG GAA ATC CGT CTG GCG AAA ATT GCC CAT GAA CTG ATG GCC GAT CTC GAA AAA   420

Glu Thr Val Asp Phe Val Asp Asn Tyr Asp Gly Thr Glu Lys Ile Pro Asp Val Met Pro   160
GAG ACG GTC GAT TTC GTT GAT AAC TAT GAC GGC ACG GAA AAA ATT CCG GAC GTC ATG CCA   480

Thr Lys Ile Pro Asn Leu Leu Val Asn Gly Ser Ser Gly Ile Ala Val Gly Met Ala Thr   180
ACC AAA ATT CCT AAC CTG CTG GTG AAC GGT TCT TCC GGT ATC GCC GTA GGT ATG GCA ACC   540

Asn Ile Pro Pro His Asn Leu Thr Glu Val Ile Asn Gly Cys Leu Ala Tyr Ile Asp Asp   200
AAC ATC CCG CCG CAC AAC CTG ACG GAA GTC ATC AAC GGT TGT CTG GCG TAT ATT GAT GAT   600

Glu Asp Ile Ser Ile Glu Gly Leu Met Glu His Ile Pro Gly Pro Asp Phe Pro Thr Ala   220
GAA GAC ATC AGC ATT GAA GGG CTG ATG GAA CAC ATC CCG GGG CCG GAC TTC CCG ACG GCG   660

Ala Ile Ile Asn Gly Arg Arg Gly Ile Glu Glu Ala Tyr Arg Thr Gly Arg Gly Lys Val   240
GCA ATC ATT AAC GGT CGT CGC GGT ATT GAA GAA GCT TAC CGT ACC GGT CGC GGC AAG GTG   720

Tyr Ile Arg Ala Arg Ala Glu Val Glu Val Asp Ala Lys Thr Gly Arg Glu Thr Ile Ile   260
TAT ATC CGC GCT CGC GCA GAA GTG GAA GTT GAC GCC AAA ACC GGT CGT GAA ACC ATT ATC   780

Val His Glu Ile Pro Tyr Gln Val Asn Lys Ala Arg Leu Ile Glu Lys Ile Ala Glu Leu   280
GTC CAC GAA ATT CCG TAT CAG GTA AAC AAA GCG CGC CTG ATC GAG AAG ATT GCG GAA CTG   840

Val Lys Glu Lys Arg Val Glu Gly Ile Ser Ala Leu Arg Asp Glu Ser Asp Lys Asp Gly   300
GTA AAA GAA AAA CGC GTG GAA GGC ATC AGC GCG CTG CGT GAC GAG TCT GAC AAA GAC GGT   900

Met Arg Ile Val Ile Glu Val Lys Arg Asp Ala Val Gly Glu Val Val Leu Asn Asn Leu   320
ATG CGC ATC GTG ATT GAA GTG AAA CGC GAT GCG GTC GGT GAA GTT GTG CTC AAC AAC CTC   960

Tyr Ser Gln Thr Gln Leu Gln Val Ser Phe Gly Ile Asn Met Val Ala Leu His His Gly   340
TAC TCC CAG ACC CAG TTG CAG GTT TCT TTC GGT ATC AAC ATG GTG GCA TTG CAC CAT GGT  1020

Gln Pro Lys Ile Met Asn Leu Lys Asp Ile Ile Ala Ala Phe Val Arg His Arg Arg Glu   360
CAG CCG AAG ATC ATG AAC CTG AAA GAC ATC ATC GCG GCG TTT GTT CGT CAC CGC CGT GAA  1080

Val Val Thr Arg Arg Thr Ile Phe Glu Leu Arg Lys Ala Arg Asp Arg Ala His Ile Pro   380
GTG GTG ACC CGT CGT ACT ATT TTC GAA CTG CGT AAA GCT CGC GAT CGT GCT CAT ATC CCT  1140

Glu Ala Leu Ala Ala Ala Leu Ala Asn Ile Asp Pro Ile Ile Glu Leu Ile Arg His Ala   400
GAA GCA TTA GCC GCG GCG CTG GCG AAC ATC GAC CCG ATC ATC GAA CTG ATC CGT CAT GCG  1200
```

Pro Thr Pro Ala Glu Ala Lys Thr Ala Leu Val Ala Asn Pro Trp Gln Leu Gly Asn Val 420
CCG ACG CCT GCA GAA GCG AAA ACT GCG CTG GTT GCT AAT CCG TGG CAG CTG GGC AAC GTT 1260

Ala Ala Met Leu Glu Arg Ala Gly Asp Asp Ala Ala Arg Pro Glu Trp Leu Glu Pro Glu 440
GCC GCG ATG CTC GAA CGT GCT GGC GAC GAT GCT GCG CGT CCG GAA TGG CTG GAG CCA GAG 1320

Phe Gly Val Arg Asp Gly Leu Tyr Tyr Leu Thr Glu Gln Gln Ala Gln Ala Ile Leu Asp 460
TTC GGC GTG CGT GAT GGT CTG TAC TAC CTG ACC GAA CAG CAA GCT CAG GCG ATT CTG GAT 1380

Leu Arg Leu Gln Lys Leu Thr Gly Leu Glu His Glu Lys Leu Leu Asp Glu Tyr Lys Glu 480
CTG CGT TTG CAG AAA CTG ACC GGT CTT GAG CAC GAA AAA CTG CTC GAC GAA TAC AAA GAG 1440

Leu Leu Asp Gln Ile Ala Glu Leu Leu Arg Ile Leu Gly Ser Ala Asp Arg Leu Met Glu 500
CTG CTG GAT CAG ATC GCG GAA CTG TTG CGT ATT CTT GGT AGC GCC GAT CGT CTG ATG GAA 1500

Val Ile Arg Glu Glu Leu Glu Leu Val Arg Glu Gln Phe Gly Asp Lys Arg Arg Thr Glu 520
GTG ATC CGT GAA GAG CTG GAG CTG GTT CGT GAA CAG TTC GGT GAC AAA CGT CGT ACT GAA 1560

Ile Thr Ala Asn Ser Ala Asp Ile Asn Leu Glu Asp Leu Ile Thr Gln Glu Asp Val Val 540
ATC ACC GCC AAC AGC GCA GAC ATC AAC CTC GAA GAT CTG ATC ACC CAG GAA GAT GTG GTC 1620

Val Thr Leu Ser His Gln Gly Tyr Val Lys Tyr Gln Pro Leu Ser Glu Tyr Glu Ala Gln 560
GTG ACG CTC TCT CAC CAG GGC TAC GTT AAG TAT CAG CCG CTT TCT GAA TAC GAA GCG CAG 1680

Arg Arg Gly Gly Lys Gly Lys Ser Ala Ala Arg Ile Lys Glu Glu Asp Phe Ile Asp Arg 580
CGT CGT GGC GGG AAA GGT AAA TCT GCC GCA CGT ATT AAA GAA GAA GAC TTT ATC GAC CGA 1740

Leu Leu Val Ala Asn Thr His Asp His Ile Leu Cys Phe Ser Ser Arg Gly Arg Val Tyr 600
CTG CTG GTG GCG AAC ACT CAC GAC CAT ATT CTG TGC TTC TCC AGC CGT GGT CGC GTC TAT 1800

Ser Met Lys Val Tyr Gln Leu Pro Glu Ala Thr Arg Gly Ala Arg Gly Arg Pro Ile Val 620
TCG ATG AAA GTT TAT CAG TTG CCG GAA GCC ACT CGT GGC GCG CGC GGT CGT CCG ATC GTC 1860

Asn Leu Leu Pro Leu Glu Gln Asp Glu Arg Ile Thr Ala Ile Leu Pro Val Thr Glu Phe 640
AAC CTG CTG CCG CTG GAG CAG GAC GAA CGT ATC ACT GCG ATC CTG CCA GTG ACC GAG TTT 1920

Glu Glu Gly Val Lys Val Phe Met Ala Thr Ala Asn Gly Thr Val Lys Lys Thr Val Leu 660
GAA GAA GGC GTG AAA GTC TTC ATG GCG ACC GCT AAC GGT ACC GTG AAG AAA ACT GTC CTC 1980

Thr Glu Phe Asn Arg Leu Arg Thr Ala Gly Lys Val Ala Ile Lys Leu Val Asp Gly Asp 680
ACC GAG TTC AAC CGT CTG CGT ACC GCC GGT AAA GTG GCG ATC AAA CTG GTT GAC GGC GAT 2040

Glu Leu Ile Gly Val Asp Leu Thr Ser Gly Glu Asp Glu Val Met Leu Phe Ser Ala Glu 700
GAG CTG ATC GGC GTT GAC CTG ACC AGC GGC GAA GAC GAA GTA ATG CTG TTC TCC GCT GAA 2100

Gly Lys Val Val Arg Phe Arg Glu Ser Ser Val Arg Ala Met Gly Cys Asn Thr Thr Gly 720
GGT AAA GTG GTG CGC TTT AGA GAG TCT TCT GTC CGT GCG ATG GGC TGC AAC ACC ACC GGT 2160

Val Arg Gly Ile Arg Leu Gly Glu Gly Asp Lys Val Val Ser Leu Ile Val Pro Arg Gly 740
GTT CGC GGT ATT CGC TTA GGT GAA GGC GAT AAA GTC GTC TCT CTG ATC GTG CCT CGT GGC 2220

Asp Gly Ala Ile Leu Thr Ala Thr Gln Asn Gly Tyr Gly Lys Arg Thr Ala Val Ala Glu 760
GAT GGC GCA ATC CTC ACC GCA ACG CAA AAC GGT TAC GGT AAA CGT ACC GCA GTG GCG GAA 2280

Tyr Pro Thr Lys Ser Arg Ala Thr Lys Gly Val Ile Ser Ile Lys Val Thr Glu Arg Asn 780
TAC CCA ACC AAG TCG CGT GCG ACG AAA GGG GTT ATC TCC ATC AAG GTT ACC GAA CGT AAC 2340

Gly Leu Val Val Gly Ala Val Gln Val Asp Asp Cys Asp Gln Ile Met Met Ile Thr Asp 800
GGT TTA GTT GTT GGC GCG GTA CAG GTA GAT GAC TGC GAC CAG ATC ATG ATG ATC ACC GAT 2400

Ala Gly Thr Leu Val Arg Thr Arg Val Ser Glu Ile Ser Ile Val Gly Arg Asn Thr Gln 820
GCC GGT ACG CTG GTA CGT ACT CGC GTT TCG GAA ATC AGC ATC GTG GGC CGT AAC ACC CAG 2460

Gly Val Ile Leu Ile Arg Thr Ala Glu Asp Glu Asn Val Val Gly Leu Gln Arg Val Ala 840
GGC GTG ATC CTC ATC CGT ACT GCG GAA GAT GAA AAC GTA GTG GGT CTG CAA CGT GTT GCT 2520

Glu Pro Val Asp Glu Glu Asp Leu Asp Thr Ile Asp Gly Ser Ala Ala Glu Gly Asp Asp 860
GAA CCG GTT GAC GAG GAA GAT CTG GAT ACC ATC GAC GGC AGT GCC GCG GAA GGG GAC GAT 2580

Glu Ile Ala Pro Glu Val Asp Val Asp Asp Glu Pro Glu Glu Glu 875
GAA ATC GCT CCG GAA GTG GAC GTT GAC GAC GAG CCA GAA GAA GAA 2625

Escherichia coli gyrB

Met Ser Asn Ser Tyr Asp Ser Ser Ser Ile Lys Val Leu Lys Gly Leu Asp Ala Val Arg 20
ATG TCG AAT TCT TAT GAC TCC TCC AGT ATC AAA GTC CTG AAA GGG CTG GAT GCG GTG CGT 60

Lys Arg Pro Gly Met Tyr Ile Gly Asp Thr Asp Asp Gly Thr Gly Leu His His Met Val 40
AAG CGC CCG GGT ATG TAT ATC GGC GAC ACG GAT GAC GGC ACC GGT CTG CAC CAC ATG GTA 120

Phe Glu Val Val Asp Asn Ala Ile Asp Glu Ala Leu Ala Gly His Cys Lys Glu Ile Ile 60
TTC GAG GTG GTA GAT AAC GCT ATC GAC GAA GCG CTC GCG GGT CAC TGT AAA GAA ATT ATC 180

Val Thr Ile His Ala Asp Asn Ser Val Ser Val Gln Asp Asp Gly Arg Gly Ile Pro Thr 80
GTC ACC ATT CAC GCC GAT AAC TCT GTC TCT GTA CAG GAT GAC GGG CGC GGC ATT CCG ACC 240

Gly Ile His Pro Glu Glu Gly Val Ser Ala Ala Glu Val Ile Met Thr Val Leu His Ala 100
GGT ATT CAC CCG GAA GAG GGC GTA TCG GCG GCG GAA GTG ATC ATG ACC GTT CTG CAC GCA 300

Gly Gly Lys Phe Asp Asp Asn Ser Tyr Lys Val Ser Gly Gly Leu His Gly Val Gly Val 120
GGC GGT AAA TTT GAC GAT AAC TCC TAT AAA GTG TCC GGC GGT CTG CAC GGC GTT GGT GTT 360

Ser Val Val Asn Ala Leu Ser Gln Lys Leu Glu Leu Val Ile Gln Arg Glu Gly Lys Ile 140
TCG GTA GTA AAC GCC CTG TCG CAA AAA CTG GAG CTG GTT ATC CAG CGC GAG GGT AAA ATT 420

His Arg Gln Ile Tyr Glu His Gly Val Pro Gln Ala Pro Leu Ala Val Thr Gly Glu Thr 160
CAC CGT CAG ATC TAC GAA CAC GGT GTA CCG CAG GCC CCG CTG GCG GTT ACC GGC GAG ACT 480

Glu Lys Thr Gly Thr Met Val Arg Phe Trp Pro Ser Leu Glu Thr Phe Thr Asn Val Thr 180
GAA AAA ACC GGC ACC ATG GTG CGT TTC TGG CCC AGC CTC GAA ACC TTC ACC AAT GTG ACC 540

Glu Phe Glu Tyr Glu Ile Leu Ala Lys Arg Leu Arg Glu Leu Ser Phe Leu Asn Ser Gly 200
GAG TTC GAA TAT GAA ATT CTG GCG AAA CGT CTG CGT GAG TTG TCG TTC CTC AAC TCC GGC 600

Val Ser Ile Arg Leu Arg Asp Lys Arg Asp Gly Lys Glu Asp His Phe His Tyr Glu Gly 220
GTT TCC ATT CGT CTG CGC GAC AAG CGC GAC GGC AAA GAA GAC CAC TTC CAC TAT GAA GGC 660

Gly Ile Lys Ala Phe Val Glu Tyr Leu Asn Lys Asn Lys Thr Pro Ile His Pro Asn Ile 240
GGC ATC AAG GCG TTC GTT GAA TAT CTG AAC AAG AAC AAA ACG CCG ATC CAC CCG AAT ATC 720

Phe Tyr Phe Ser Thr Glu Lys Asp Gly Ile Gly Val Glu Val Ala Leu Gln Trp Asn Asp 260
TTC TAC TTC TCC ACT GAA AAA GAC GGT ATT GGC GTC GAA GTG GCG TTG CAG TGG AAC GAT 780

Gly Phe Gln Glu Asn Ile Tyr Cys Phe Thr Asn Asn Ile Pro Gln Arg Asp Gly Gly Thr 280
GGC TTC CAG GAA AAC ATC TAC TGC TTT ACC AAC AAC ATT CCG CAG CGT GAC GGC GGT ACT 840

His Leu Ala Gly Phe Arg Ala Ala Met Thr Arg Thr Leu Asn Ala Tyr Met Asp Lys Glu 300
CAC CTG GCA GGC TTC CGT GCG GCG ATG ACC CGT ACC CTG AAC GCC TAC ATG GAC AAA GAA 900

Gly Tyr Ser Lys Lys Ala Lys Val Ser Ala Thr Gly Asp Asp Ala Arg Glu Gly Leu Ile 320
GGC TAC AGC AAA AAA GCC AAA GTC AGC GCC ACC GGT GAC GAT GCG CGT GAA GGC CTG ATT 960

Ala Val Val Ser Val Lys Val Pro Asp Pro Lys Phe Ser Ser Gln Thr Lys Asp Lys Leu 340
GCG GTC GTT TCC GTG AAA GTG CCG GAC CCG AAA TTC TCC TCC CAG ACC AAA GAC AAA CTG 1020

```
Val Ser Ser Glu Val Lys Ser Ala Val Glu Gln Gln Met Asn Glu Leu Leu Ala Glu Tyr    360
GTT TCT TCT GAG GTG AAA TCG GCG GTT GAA CAG CAG ATG AAC GAA CTG CTG GCA GAA TAC   1080

Leu Leu Glu Asn Pro Thr Asp Ala Lys Ile Val Val Gly Lys Ile Ile Asp Ala Ala Arg    380
CTG CTG GAA AAC CCA ACC GAC GCG AAA ATC GTG GTT GGC AAA ATT ATC GAT GCT GCC CGT   1140

Ala Arg Glu Ala Ala Arg Arg Ala Arg Glu Met Thr Arg Arg Lys Gly Ala Leu Asp Leu    400
GCC CGT GAA GCG GCG CGT CGC GCG CGT GAA ATG ACC CGC CGT AAA GGT GCG CTC GAC TTA   1200

Ala Gly Leu Pro Gly Lys Leu Ala Asp Cys Gln Glu Arg Asp Pro Ala Leu Ser Glu Leu    420
GCG GGC CTG CCG GGC AAA CTG GCA GAC TGC CAG GAA CGC GAT CCG GCG CTT TCC GAA CTG   1260

Tyr Leu Val Glu Gly Asp Ser Ala Gly Gly Ser Ala Lys Gln Gly Arg Asn Arg Lys Asn    440
TAC CTG GTG GAA GGG GAC TCC GCG GGC GGC TCT GCG AAG CAG GGG CGT AAC CGC AAG AAC   1320

Gln Ala Ile Leu Pro Leu Lys Gly Lys Ile Leu Asn Val Glu Lys Ala Arg Phe Asp Lys    460
CAG GCG ATT CTG CCG CTG AAG GGT AAA ATC CTC AAC GTC GAG AAA GCG CGC TTC GAT AAG   1380

Met Leu Ser Ser Gln Glu Val Ala Thr Leu Ile Thr Ala Leu Gly Cys Gly Ile Gly Arg    480
ATG CTC TCT TCT CAG GAA GTG GCG ACG CTT ATC ACC GCG CTT GGC TGT GGT ATC GGT CGT   1440

Asp Glu Tyr Asn Pro Asp Lys Leu Arg Tyr His Ser Ile Ile Ile Met Thr Asp Ala Asp    500
GAC GAG TAC AAC CCG GAC AAA CTG CGT TAT CAC AGC ATC ATC ATC ATG ACC GAT GCG GAC   1500

Val Asp Gly Ser His Ile Arg Thr Leu Leu Leu Thr Phe Phe Tyr Arg Gln Met Pro Glu    520
GTC GAC GGC TCG CAC ATT CGT ACG CTG CTG TTG ACC TTC TTC TAT CGT CAG ATG CCG GAA   1560

Ile Val Glu Arg Gly His Val Tyr Ile Ala Gln Pro Pro Leu Tyr Lys Val Lys Lys Gly    540
ATC GTT GAA CGC GGT CAC GTC TAC ATC GCT CAG CCG CCG CTG TAC AAA GTG AAG AAA GGC   1620

Lys Gln Glu Gln Tyr Ile Lys Asp Asp Glu Ala Met Asp Gln Tyr Gln Ile Ser Ile Ala    560
AAG CAG GAA CAG TAC ATT AAA GAC GAC GAA GCG ATG GAT CAG TAC CAG ATC TCT ATC GCG   1680

Leu Asp Gly Ala Thr Leu His Thr Asn Ala Ser Ala Pro Ala Leu Ala Gly Glu Ala Leu    580
CTG GAC GGC GCA ACG CTG CAC ACC AAC GCC AGT GCA CCG GCA TTG GCT GGC GAA GCG TTA   1740

Glu Lys Leu Val Ser Glu Tyr Asn Ala Thr Gln Lys Met Ile Asn Arg Met Glu Arg Arg    600
GAG AAA CTG GTA TCT GAG TAC AAC GCG ACG CAG AAA ATG ATC AAT CGT ATG GAG CGT CGT   1800

Tyr Pro Lys Ala Met Leu Lys Glu Leu Ile Tyr Gln Pro Thr Leu Thr Glu Ala Asp Leu    620
TAT CCG AAA GCA ATG CTG AAA GAG CTT ATC TAT CAG CCG ACG TTG ACG GAA GCT GAC CTT   1860

Ser Asp Glu Gln Thr Val Thr Arg Trp Val Asn Ala Leu Val Ser Glu Leu Asn Asp Lys    640
TCT GAT GAG CAG ACC GTT ACC CGC TGG GTG AAC GCG CTG GTC AGC GAA CTG AAC GAC AAA   1920

Glu Gln His Gly Ser Gln Trp Lys Phe Asp Val His Thr Asn Ala Glu Gln Asn Leu Phe    660
GAA CAG CAC GGC AGC CAG TGG AAG TTT GAT GTT CAC ACC AAT GCT GAG CAA AAC CTG TTC   1980

Glu Pro Ile Val Arg Val Arg Thr His Gly Val Asp Thr Asp Tyr Pro Leu Asp His Glu    680
GAG CCG ATT GTT CGC GTG CGT ACC CAC GGT GTG GAT ACT GAC TAT CCG CTG GAT CAC GAG   2040

Phe Ile Thr Gly Gly Glu Tyr Arg Arg Ile Cys Thr Leu Gly Glu Lys Leu Arg Gly Leu    700
TTT ATC ACC GGT GGC GAA TAT CGT CGT ATC TGC ACG CTG GGT GAG AAA CTG CGT GGC TTG   2100

Leu Glu Glu Asp Ala Phe Ile Glu Arg Gly Glu Arg Arg Gln Pro Val Ala Ser Phe Glu    720
CTG GAA GAA GAT GCG TTT ATC GAA CGT GGC GAG CGT CGT CAG CCG GTA GCC AGC TTC GAG   2160

Gln Ala Leu Asp Trp Leu Val Lys Glu Ser Arg Arg Gly Leu Ser Ile Gln Arg Tyr Lys    740
CAG GCG CTG GAC TGG CTG GTG AAA GAG TCC CGT CGC GGC CTC TCC ATC CAG CGT TAT AAA   2220

Gly Leu Gly Glu Met Asn Pro Glu Gln Leu Trp Glu Thr Thr Met Asp Pro Glu Ser Arg    760
GGT CTG GGC GAG ATG AAC CCG GAA CAG CTG TGG GAA ACC ACT ATG GAC CCG GAA AGT CGT   2280
```

Arg Met Leu Arg Val Thr Val Lys Asp Ala Ile Ala Ala Asp Gln Leu Phe Thr Thr Leu 780
CGT ATG CTG CGC GTT ACC GTT AAA GAT GCG ATT GCT GCC GAC CAG TTG TTC ACC ACG CTG 2340

Met Gly Asp Ala Val Glu Pro Arg Arg Ala Phe Ile Glu Glu Asn Ala Leu Lys Ala Ala 800
ATG GGC GAC GCC GTT GAA CCG CGC CGT GCG TTT ATT GAA GAG AAC GCC CTG AAA GCG GCG 2400

Asn Ile Asp Ile 804
AAT ATC GAT ATT 2412

Klebsiella pneumoniae gyrA

Met Ser Asp Leu Ala Arg Glu Ile Thr Pro Val Asn Ile Glu Glu Glu Leu Lys Ser Ser 20
ATG AGC GAC CTT GCG AGA GAA ATT ACA CCG GTC AAC ATT GAG GAA GAG CTG AAG AGC TCG 60

Tyr Leu Asp Tyr Ala Met Ser Val Ile Val Gly Arg Ala Leu Pro Asp Val Arg Asp Gly 40
TAT CTG GAT TAC GCG ATG TCG GTC ATT GTT GGC CGT GCG CTG CCG GAT GTC CGA GAT GGC 120

Leu Lys Pro Val His Arg Arg Val Leu Tyr Ala Met Asn Val Leu Gly Asn Asp Trp Asn 60
CTG AAG CCG GTA CAC CGT CGC GTA CTA TAC GCC ATG AAC GTA TTG GGC AAT GAC TGG AAC 180

Lys Ala Tyr Lys Lys Ser Ala Arg Val Val Gly Asp Val Ile Gly Lys Tyr His Pro His 80
AAA GCC TAT AAA AAA TCT GCC CGT GTC GTT GGT GAC GTA ATC GGT AAA TAC CAC CCT CAT 240

Gly Asp Thr Ala Val Tyr Asp Thr Ile Val Arg Met Ala Gln Pro Phe Ser Leu Arg Tyr 100
GGT GAT ACT GCC GTG TAT GAC ACC ATT GTA CGT ATG GCG CAG CCA TTC TCC CTG CGT TAC 300

Met Leu Val Asp Gly Gln Gly Asn Phe Gly Ser Val Asp Gly Asp Ser Ala Ala Ala Met 120
ATG CTG GTA GAT GGC CAG GGT AAC TTC GGT TCT GTC GAC GGC GAC TCC GCC GCA GCG ATG 360

Arg Tyr Thr Glu Ile Arg Met Ser Lys Ile Ala His Glu Leu Met Ala Asp Leu Glu Lys 140
CGT TAT ACG GAA ATC CGT ATG TCG AAG ATC GCC CAT GAA CTG ATG GCC GAC CTG GAA AAA 420

Glu Thr Val Asp Phe Val Asp Asn Tyr Asp Gly Thr Glu Lys Ile Pro Asp Val Met Pro 160
GAG ACG GTC GAT TTC GTC GAT AAC TAT GAC GGC ACG GAA AAG ATC CCT GAC GTT ATG CCG 480

Thr Lys Ile Pro Asn Leu Leu Val Asn Gly Ser Phe Gly Ile Ala Val Gly Met Ala Thr 180
ACC AAA ATC CCG AAC CTG TTA GTC AAC GGT TCG TTC GGT ATC GCG GTA GGT ATG GCG ACC 540

Asn Ile Pro Pro His Asn Leu Thr Glu Val Ile Asn Gly Arg Leu Ala Tyr Val Glu Asp 200
AAC ATT CCG CCG CAC AAC CTG ACC GAA GTG ATC AAC GGT CGT CTG GCC TAC GTT GAA GAC 600

Glu Glu Ile Ser Ile Glu Gly Leu Met Glu His Ile Pro Gly Pro Asp Phe Pro Thr Ala 220
GAA GAG ATC AGC ATT GAA GGG CTG ATG GAA CAT ATT CCG GGC CCG GAC TTC CCG ACC GCC 660

Ala Ile Ile Asn Gly Arg Arg Gly Ile Glu Glu Ala Tyr Arg Thr Gly Arg Gly Lys Val 240
GCG ATC ATC AAC GGT CGC CGC GGC ATT GAA GAG GCC TAT CGT ACC GGT CGC GGT AAA GTG 720

Tyr Ile Cys Ala Arg Ala Glu Val Glu Ala Asp Ala Lys Thr Gly Arg Glu Thr Ile Ile 260
TAC ATT TGC GCC CGC GCG GAA GTG GAA GCT GAC GCG AAA ACC GGT CGC GAA ACC ATC ATC 780

Val His Glu Ile Pro Tyr Gln Val Asn Lys Ala Arg Leu Ile Glu Lys Ile Ala Glu Leu 280
GTG CAT GAA ATT CCG TAT CAG GTG AAC AAA GCG CGC CTG ATT GAG AAA ATC GCT GAG CTG 840

Val Lys Glu Lys Arg Val Glu Gly Ile Ser Ala Leu Arg Asp Glu Ser Asp Lys Asp Gly 300
GTC AAA GAA AAA CGC GTC GAA GGC ATC AGC GCG CTG CGT GAC GAG TCT GAT AAA GAC GGC 900

Met Arg Ile Val Ile Glu Val Lys Arg Asp Ala Val Gly Arg Val Val Leu Asn Asn Leu 320
ATG CGT ATA GTG ATT GAA GTG AAG CGC GAT GCG GTG GGT AGG GTT GTG CTC AAC AAC CTC 960

Tyr Ser Gln Thr Gln Leu Gln Val Ser Phe Gly Ile Asn Met Val Ala Leu His His Gly 340
TAC TCG CAG ACT CAG CTG CAG GTC TCC TTC GGT ATC AAC ATG GTT GCC CTG CAC CAT GGT 1020

```
Gln Pro Lys Ile Met Asn Leu Lys Glu Ile Ile Ala Ala Phe Val Arg His Arg Arg Glu   360
CAG CCG AAG ATC ATG AAT CTG AAA GAA ATT ATT GCC GCC TTC GTG CGC CAC CGC CGC GAA  1080

Val Val Thr Arg Arg Thr Ile Leu Ala Leu Arg Lys Ala Arg Asp Arg Ala Asp Ile Leu   380
GTG GTG ACC CGC CGT ACG ATT TTA GCA CTG CGT AAA GCC CGT GAT CGG GCG GAC ATC CTT  1140

Glu Ala Leu Ser Ile Ala Leu Ala Asn Ile Asp Pro Ile Ile Glu Leu Ile Arg Arg Ala   400
GAA GCG CTG TCG ATT GCC CTG GCC AAC ATC GAT CCG ATT ATT GAG CTG ATT CGC CGC GCG  1200

Pro Thr Pro Ala Glu Ala Lys Ala Gly Leu Ile Ala Arg Ser Trp Asp Leu Gly Asn Val   420
CCG ACG CCG GCG GAA GCG AAA GCT GGC TTA ATC GCC CGT TCA TGG GAT CTG GGC AAC GTT  1260

Ser Ala Met Leu Glu Ala Gly Asp Asp Ala Ala Arg Pro Glu Trp Leu Glu Pro Glu Phe   440
TCC GCG ATG CTG GAA GCG GGC GAT GAC GCC GCG CGT CCG GAA TGG CTG GAA CCT GAA TTC  1320

Gly Val Arg Asp Gly Gln Tyr Tyr Leu Thr Glu Gln Gln Ala Gln Ala Ile Leu Asp Leu   460
GGC GTG CGC GAC GGC CAG TAC TAC CTG ACC GAA CAG CAG GCA CAG GCG ATT CTG GAT CTG  1380

Arg Leu Gln Lys Leu Thr Gly Leu Glu His Glu Lys Leu Leu Asp Glu Tyr Lys Glu Leu   480
CGT TTG CAG AAA CTG ACC GGC CTT GAG CAT GAA AAA CTG CTC GAC GAA TAC AAA GAG CTG  1440

Leu Glu Gln Ile Ala Glu Leu Leu His Ile Leu Gly Ser Ala Asp Arg Leu Met Glu Val   500
CTG GAA CAG ATT GCG GAA CTG CTG CAT ATT CTG GGC AGC GCC GAT CGC CTG ATG GAA GTT  1500

Ile Arg Glu Glu Leu Glu Leu Val Arg Glu Gln Phe Gly Asp Ala Arg Arg Thr Asp Ile   520
ATC CGC GAA GAG CTG GAG CTG GTC CGT GAA CAG TTC GGC GAC GCG CGC CGT ACC GAC ATC  1560

Thr Ala Asn Ser Val Asp Ile Asn Ile Glu Asp Leu Ile Thr Gln Glu Asp Val Val Val   540
ACC GCT AAC AGC GTA GAC ATC AAC ATC GAA GAC CTG ATC ACC CAG GAA GAT GTT GTC GTG  1620

Thr Leu Ser His Glu Gly Tyr Val Lys Tyr Gln Pro Val Asn Asp Tyr Glu Ala Gln Arg   560
ACC CTT TCG CAT GAG GGC TAC GTG AAG TAT CAG CCG GTT AAC GAC TAC GAA GCG CAG CGT  1680

Arg Gly Gly Lys Gly Lys Ser Ala Pro Arg Ile Lys Glu Glu Asp Phe Ile Asp Arg Leu   580
CGC GGC GGT AAA GGC AAA TCT GCG CCG CGT ATC AAA GAA GAA GAC TTT ATC GAC CGT CTG  1740

Leu Val Ala Asn Thr His Asp Thr Ile Leu Cys Phe Ser Ser Arg Gly Arg Leu Tyr Trp   600
CTG GTG GCC AAC ACC CAC GAC ACG ATC CTC TGC TTC TCC AGC CGC GGT CGT CTC TAC TGG  1800

Met Lys Val Tyr Gln Val Pro Glu Ala Ser Arg Gly Ala Arg Gly Arg Pro Ile Val Asn   620
ATG AAA GTG TAT CAG GTG CCG GAA GCC AGC CGC GGC GCC CGC GGT CGT CCG ATC GTC AAC  1860

Leu Leu Pro Leu Glu Ala Asn Glu Arg Tyr Thr Ala Ile Leu Pro Val Arg Glu Tyr Glu   640
CTG CTG CCG CTG GAA GCC AAC GAG CGC TAC ACC GCG ATC CTG CCG GTC CGC GAA TAC GAA  1920

Glu Gly Val Asn Val Phe Met Ala Thr Ala Ser Gly Thr Val Lys Lys Thr Pro Ala Asp   660
GAG GGC GTG AAC GTC TTT ATG GCC ACC GCC AGC GGT ACC GTG AAG AAA ACG CCA GCT GAC  1980

Glu Phe Ser Arg Pro Arg Ser Ala Gly Ile Ile Ala Val Asn Leu Asn Glu Gly Asp Glu   680
GAA TTC AGC CGT CCG CGT TCC GCC GGT ATT ATC GCG GTT AAC CTT AAC GAA GGC GAC GAA  2040

Leu Ile Gly Val Asp Leu Thr Ser Gly Gln Asp Glu Val Met Leu Phe Ser Ala Ala Gly   700
CTA ATA GGC GTC GAT CTG ACC TCC GGT CAG GAC GAA GTC ATG CTG TTC TCC GCC GCC GGT  2100

Lys Val Val Arg Phe Lys Glu Asp Asp Val Arg Ala Met Gly Arg Thr Ala Thr Gly Val   720
AAA GTG GTG CGC TTT AAA GAG GAC GAC GTC CGC GCG ATG GGC CGT ACC GCC ACC GGC GTG  2160

Arg Gly Ile Lys Leu Ala Gly Glu Asp Lys Val Val Ser Leu Ile Val Pro Arg Gly Glu   740
CGC GGC ATC AAG CTG GCC GGC GAG GAC AAG GTG GTC TCG CTG ATC GTT CCG CGC GGC GAA  2220

Gly Arg Ile Leu Thr Ala Thr Glu Asn Gly Tyr Arg Lys Arg Thr Ala Val Ala Glu Tyr   760
GGC CGC ATC CTG ACC GCT ACG GAA AAC GGT TAC CGT AAA CGT ACC GCA GTG GCG GAA TAT  2280
```

```
Pro Thr Lys Ser Arg Ala Thr Gln Gly Val Ile Ser Ile Lys Val Thr Glu Arg Asn Gly   780
CCA ACC AAG TCG CGT GCG ACC CAG GGC GTT ATC TCT ATC AAG GTC ACC GAA CGC AAC GGT  2340

Ser Val Val Gly Ala Val Gln Val Asp Asp Cys Asp Gln Ile Met Met Ile Thr Asp Ala   800
TCC GTT GTT GGC GCG GTG CAG GTA GAT GAC TGC GAC CAG ATT ATG ATG ATC ACC GAT GCG  2400

Gly Thr Leu Val Arg Ile Arg Val Ser Glu Val Ser Ile Val Gly Arg Asn Thr Gln Gly   820
GGC ACG CTG GTA CGA ATC CGC GTT TCG GAA GTG AGC ATT GTC GGT CGT AAC ACT CAG GGC  2460

Val Ile Leu Ile Arg Thr Ala Glu Asp Glu Asn Val Val Ala Leu Gln Arg Val Ala Glu   840
GTG ATT CTG ATC CGC ACC GCG GAA GAT GAA AAT GTG GTT GCT CTG CAG CGC GTT GCT GAG  2520

Pro Val Asp Asp Glu Glu Leu Asp Ala Ile Asp Gly Ser Ala Ala Glu Gly Asp Glu Asp   860
CCG GTT GAT GAC GAA GAG CTG GAT GCC ATC GAC GGC AGC GCG GCG GAA GGC GAT GAG GAT  2580

Ile Ala Pro Glu Ala Asp Thr Asp Asp Asp Ile Ala Glu Asp Glu Glu   876
ATC GCG CCG GAA GCG GAT ACC GAC GAT GAT ATT GCC GAA GAC GAA GAG  2628
```

Bacillus subtilis gyrA

```
Met Ser Glu Gln Asn Thr Pro Gln Val Arg Glu Ile Asn Ile Ser Gln Glu Met Arg Thr    20
ATG AGT GAA CAA AAC ACA CCA CAA GTT CGT GAA ATA AAT ATC AGT CAG GAA ATG CGT ACG    60

Ser Phe Leu Asp Tyr Ala Met Ser Val Ile Val Ser Arg Ala Leu Pro Asp Val Arg Asp    40
TCC TTC TTG GAT TAT GCA ATG AGC GTT ATC GTG TCC CGT GCT CTT CCG GAT GTT CGT GAC   120

Gly Leu Lys Pro Val His Arg Arg Ile Leu Tyr Ala Met Asn Asp Leu Gly Met Thr Ser    60
GGT TTA AAA CCG GTT CAT AGA CGG ATT TTG TAT GCA ATG AAT GAT TTA GGC ATG ACA AGT   180

Asp Lys Pro Tyr Lys Lys Ser Ala Arg Ile Val Gly Glu Val Ile Gly Lys Tyr His Pro    80
GAC AAG CCT TAT AAA AAA TCC GCG CGT ATC GTT GGA GAA GTT ATC GGG AAA TAC CAC CCG   240

His Gly Asp Ser Ala Val Tyr Glu Ser Met Val Arg Met Ala Gln Asp Phe Asn Tyr Arg   100
CAC GGT GAT TCA GCG GTA TAT GAA TCC ATG GTC AGA ATG GCT CAG GAT TTC AAC TAC CGT   300

Tyr Met Leu Val Asp Gly His Gly Asn Phe Gly Ser Val Asp Gly Asp Ser Ala Ala Ala   120
TAT ATG CTC GTT GAC GGT CAC GGA AAC TTC GGT TCT GTT GAC GGA GAC TCA GCG GCG GCC   360

Met Arg Tyr Thr Glu Ala Arg Met Ser Lys Ile Ser Met Glu Ile Leu Arg Asp Ile Thr   140
ATG CGT TAT ACA GAA GCA CGA ATG TCT AAA ATC TCA ATG GAG ATT CTT CGC GAC ATC ACA   420

Lys Asp Thr Ile Asp Tyr Gln Asp Asn Tyr Asp Gly Ser Glu Arg Glu Pro Val Val Met   160
AAA GAC ACA ATC GAT TAC CAG GAT AAC TAT GAC GGG TCA GAA AGA GAA CCT GTC GTT ATG   480

Pro Ser Arg Phe Pro Asn Leu Leu Val Asn Gly Ala Ala Gly Ile Ala Val Gly Met Ala   180
CCT TCA AGG TTC CCG AAT CTG CTC GTG AAC GGT GCT GCC GGC ATT GCG GTA GGT ATG GCA   540

Thr Asn Ile Pro Pro His Gln Leu Gly Glu Ile Ile Asp Gly Val Leu Ala Val Ser Glu   200
ACA AAC ATT CCT CCG CAC CAG CTG GGA GAA ATC ATT GAC GGT GTA CTT GCT GTT AGT GAG   600

Asn Pro Asp Ile Thr Ile Pro Glu Leu Met Glu Val Ile Pro Gly Pro Asp Phe Pro Thr   220
AAT CCG GAC ATT ACA ATT CCA GAG CTT ATG GAA GTC ATT CCA GGG CCT GAT TTC CCG ACC   660

Ala Gly Gln Ile Leu Gly Arg Ser Gly Ile Arg Lys Ala Tyr Glu Ser Gly Arg Gly Ser   240
GCG GGT CAA ATC TTG GGA CGC AGC GGT ATC CGG AAA GCA TAC GAA TCA GGC CGA GGC TCT   720

Ile Thr Ile Arg Ala Lys Ala Glu Ile Glu Gln Thr Ser Ser Gly Lys Glu Arg Ile Ile   260
ATC ACG ATC CGG GCA AAA GCT GAG ATC GAA CAA ACA TCT TCG GGT AAA GAA AGA ATT ATC   780
```

Val Thr Glu Leu Pro Tyr Gln Val Asn Lys Ala Lys Leu Ile Glu Lys Ile Ala Asp Leu 280
GTT ACA GAG TTA CCT TAC CAA GTA AAT AAG GCG AAA TTA ATT GAG AAA ATT GCT GAT CTC 840

Val Arg Asp Lys Lys Ile Glu Gly Ile Thr Asp Leu Arg Asp Glu Ser Asp Arg Thr Gly 300
GTA AGG GAC AAA AAG ATA GAG GGT ATC ACA GAT CTG CGT GAT GAG TCA GAT CGT ACA GGT 900

Met Arg Ile Val Ile Glu Ile Arg Arg Asp Ala Asn Ala Asn Val Ile Leu Asn Asn Leu 320
ATG AGA ATT GTC ATT GAA ATC AGA CGC GAT GCC AAT GCG AAT GTT ATC TTA AAC AAT CTG 960

Tyr Lys Gln Thr Ala Leu Gln Thr Ser Phe Gly Ile Asn Leu Leu Ala Leu Val Asp Gly 340
TAC AAA CAA ACT GCT CTA CAA ACA TCT TTT GGC ATC AAC CTG CTT GCG CTT GTT GAT GGC 1020

Gln Pro Lys Val Leu Thr Leu Lys Gln Cys Leu Glu His Tyr Leu Asp His Gln Lys Val 360
CAG CCG AAA GTT TTA ACT CTT AAG CAA TGC CTG GAG CAT TAC CTT GAC CAT CAA AAA GTT 1080

Val Ile Arg Arg Arg Thr Ala Tyr Glu Leu Arg Lys Ala Glu Ala Arg Ala His Ile Leu 380
GTC ATT AGA CGC CGT ACT GCT TAT GAA TTG CGT AAA GCA GAA GCG AGA GCT CAT ATC TTG 1140

Glu Gly Leu Arg Val Ala Leu Asp His Leu Asp Ala Val Ile Ser Leu Ile Arg Asn Ser 400
GAA GGA TTG AGA GTT GCA CTG GAT CAT CTC GAT GCA GTT ATC TCC CTT ATC CGT AAT TCT 1200

Gln Thr Ala Glu Ile Ala Arg Thr Gly Leu Ile Glu Gln Phe Ser Leu Thr Glu Lys Gln 420
CAA ACG GCT GAA ATT GCG AGA ACA GGT TTA ATT GAA CAA TTC TCA CTG ACA GAG AAG CAA 1260

Ala Gln Ala Ile Leu Asp Met Arg Leu Gln Arg Leu Thr Gly Leu Glu Arg Glu Lys Ile 440
GCA CAA GCG ATC CTT GAC ATG AGG CTC CAG CGT TTA ACG GGA CTG GAA CGT GAA AAG ATC 1320

Glu Glu Glu Tyr Gln Ser Leu Val Lys Leu Ile Ala Glu Leu Lys Asp Ile Leu Ala Asn 460
GAA GAA GAA TAC CAG TCT CTT GTT AAA TTA ATT GCA GAG CTA AAA GAC ATC TTG GCA AAT 1380

Glu Tyr Lys Val Leu Glu Ile Ile Arg Glu Glu Leu Thr Glu Ile Lys Glu Arg Phe Asn 480
GAA TAT AAA GTG CTT GAG ATC ATT CGT GAA GAA CTC ACG GAA ATC AAA GAG CGT TTT AAC 1440

Asp Glu Arg Arg Thr Glu Ile Val Thr Ser Gly Leu Glu Thr Ile Glu Asp Glu Asp Leu 500
GAT GAA AGA CGT ACT GAG ATC GTC ACT TCT GGA CTG GAG ACA ATT GAA GAT GAA GAT CTC 1500

Ile Glu Arg Glu Asn Ile Val Val Thr Leu Thr His Asn Gly Tyr Val Lys Arg Leu Pro 520
ATC GAG AGA GAA AAT ATC GTA GTT ACT CTG ACG CAC AAC GGA TAC GTC AAA CGT CTT CCT 1560

Ala Ser Thr Tyr Arg Ser Gln Lys Arg Gly Gly Lys Gly Val Gln Gly Met Gly Thr Asn 540
GCA TCA ACT TAC CGC AGT CAA AAA CGG GGC GGA AAA GGT GTA CAA GGT ATG GGA ACA AAC 1620

Glu Asp Asp Phe Val Glu His Leu Ile Ser Thr Ser Thr His Asp Thr Ile Leu Phe Phe 560
GAA GAT GAT TTC GTT GAA CAT TTG ATC TCT ACG TCA ACT CAT GAC ACG ATT CTC TTC TTC 1680

Ser Asn Lys Gly Lys Val Tyr Arg Ala Lys Gly Tyr Glu Ile Pro Glu Tyr Gly Arg Thr 580
TCG AAC AAG GGG AAA GTG TAT CGT GCA AAA GGG TAT GAA ATC CCT GAA TAC GGC AGA ACG 1740

Ala Lys Gly Ile Pro Ile Ile Asn Leu Leu Glu Val Glu Lys Gly Glu Trp Ile Asn Ala 600
GCA AAA GGA ATC CCG ATT ATT AAC CTG CTG GAG GTA GAA AAG GGT GAG TGG ATC AAC GCG 1800

Ile Ile Pro Val Thr Glu Phe Asn Ala Glu Leu Tyr Leu Phe Phe Thr Thr Lys His Gly 620
ATT ATT CCA GTC ACG GAA TTC AAT GCG GAG CTT TAC CTC TTC TTC ACT ACA AAG CAT GGG 1860

Val Ser Lys Arg Thr Ser Leu Ser Gln Phe Ala Asn Ile Arg Asn Asn Gly Leu Ile Ala 640
GTT TCA AAA CGA ACA TCG CTA TCT CAA TTC GCT AAT ATC CGC AAC AAT GGT CTA ATT GCT 1920

Leu Ser Leu Arg Glu Asp Asp Glu Leu Met Gly Val Arg Leu Thr Asp Gly Thr Lys Gln 660
CTG AGT CTT CGT GAA GAT GAT GAA CTG ATG GGT GTA CGT CTG ACT GAC GGC ACA AAA CAA 1980

Ile Ile Ile Gly Thr Lys Asn Gly Leu Leu Ile Arg Phe Pro Glu Thr Asp Val Arg Glu 680
ATC ATC ATT GGA ACG AAA AAC GGT TTA CTG ATT CGT TTC CCT GAA ACA GAT GTC CGA GAG 2040

Met Gly Arg Thr Ala Ala Gly Val Lys Gly Ile Thr Leu Thr Asp Asp Asp Val Val Val 700
ATG GGA AGA ACT GCG GCG GGC GTA AAA GGC ATC ACC CTG ACG GAT GAC GAC GTT GTT GTC 2100

Gly Met Glu Ile Leu Glu Glu Glu Ser His Val Leu Ile Val Thr Glu Lys Gly Tyr Gly 720
GGC ATG GAG ATT TTA GAG GAA GAA TCA CAC GTC CTT ATC GTA ACT GAA AAA GGG TAC GGA 2160

Lys Arg Thr Pro Ala Glu Glu Tyr Arg Thr Gln Ser Arg Gly Gly Lys Gly Leu Lys Thr 740
AAA CGA ACT CCT GCT GAA GAG TAC AGA ACC CAA AGC CGG GGC GGA AAA GGA CTC AAA ACA 2220

Ala Lys Ile Thr Glu Asn Asn Gly Gln Leu Val Ala Val Lys Ala Thr Lys Gly Glu Glu 760
GCG AAA ATC ACC GAG AAC AAC GGC CAA CTA GTA GCA GTG AAA GCT ACT AAA GGT GAA GAG 2280

Asp Leu Met Ile Ile Thr Ala Ser Gly Val Leu Ile Arg Met Asp Ile Asn Asp Ile Ser 780
GAT CTA ATG ATT ATT ACA GCT AGC GGC GTA CTC ATC AGA ATG GAC ATC AAT GAT ATC TCC 2340

Ile Thr Gly Arg Val Thr Gln Gly Val Arg Leu Ile Arg Met Ala Glu Glu Glu His Val 800
ATC ACC GGA CGT GTC ACT CAA GGT GTG CGT CTC ATC AGA ATG GCA GAA GAA GAG CAT GTT 2400

Ala Thr Val Ala Leu Val Glu Lys Asn Glu Glu Asp Glu Asn Glu Glu Glu Gln Glu Glu 820
GCT ACA GTA GCT TTA GTT GAG AAA AAC GAA GAA GAT GAG AAT GAA GAA GAA CAA GAA GAA 2460

Val 821
GTG 2463

Bacillus subtilis gyrB

Met Glu Gln Gln Gln Asn Ser Tyr Asp Glu Asn Gln Ile Gln Val Leu Glu Gly Leu Glu 20
ATG GAA CAG CAG CAA AAC AGT TAT GAT GAA AAT CAG ATA CAG GTA CTA GAA GGA TTG GAA 60

Ala Val Arg Lys Arg Pro Gly Met Tyr Ile Gly Ser Thr Asn Ser Lys Gly Leu His His 40
GCT GTT CGT AAA AGA CCG GGG ATG TAT ATC GGT TCG ACA AAC AGC AAA GGC CTT CAC CAC 120

Leu Val Trp Glu Ile Val Asp Asn Ser Ile Asp Glu Ala Leu Ala Gly Tyr Cys Thr Asp 60
TTG GTA TGG GAA ATT GTC GAC AAT AGT ATT GAC GAA GCC CTC GCC GGT TAT TGT ACG GAT 180

Ile Asn Ile Gln Ile Glu Lys Asp Asn Ser Ile Thr Val Val Asp Asn Gly Arg Gly Ile 80
ATC AAT ATC CAA ATC GAA AAA GAC AAC AGT ATC ACG GTT GTA GAT AAT GGC CGC GGT ATT 240

Pro Val Gly Ile His Glu Lys Met Gly Arg Pro Ala Val Glu Val Ile Met Thr Val Leu 100
CCA GTC GGT ATT CAT GAA AAA ATG GGC CGT CCT GCG GTA GAA GTC ATT ATG ACG GTG CTT 300

His Ala Gly Gly Lys Phe Asp Gly Ser Gly Tyr Lys Val Ser Gly Gly Leu His Gly Val 120
CAT GCC GGA GGA AAA TTT GAC GGA AGC GGC TAT AAA GTA TCC GGA GGA TTA CAC GGT GTA 360

Gly Ala Ser Val Val Asn Ala Leu Ser Thr Glu Leu Asp Val Thr Val His Arg Asp Gly 140
GGT GCG TCG GTC GTA AAC GCA CTA TCA ACA GAG CTT GAT GTG ACG GTT CAC CGT GAC GGT 420

Lys Ile His Arg Gln Thr Tyr Lys Arg Gly Val Pro Val Thr Asp Leu Glu Ile Ile Gly 160
AAA ATT CAC CGC CAA ACC TAT AAA CGC GGA GTT CCG GTT ACA GAC CTT GAA ATC ATT GGC 480

Glu Thr Asp His Thr Gly Thr Thr Thr His Phe Val Pro Asp Pro Glu Ile Phe Ser Glu 180
GAA ACG GAT CAT ACA GGA ACG ACG ACA CAT TTT GTC CCG GAC CCT GAA ATT TTC TCA GAA 540

Thr Thr Glu Tyr Asp Tyr Asp Leu Leu Ala Asn Arg Val Arg Glu Leu Ala Phe Leu Thr 200
ACA ACC GAG TAT GAT TAC GAT CTG CTT GCC AAC CGC GTG CGT GAA TTA GCC TTT TTA ACA 600

Lys Gly Val Asn Ile Thr Ile Glu Asp Lys Arg Glu Gly Gln Glu Arg Lys Asn Glu Tyr 220
AAG GGC GTA AAC ATC ACG ATT GAA GAT AAA CGT GAA GGA CAA GAG CGC AAA AAT GAA TAC 660

His Tyr Glu Gly Gly Ile Lys Ser Tyr Val Glu Tyr Leu Asn Arg Ser Lys Glu Val Val 240
CAT TAC GAA GGC GGA ATT AAA AGT TAT GTA GAG TAT TTA AAC CGC TCT AAA GAG GTT GTC 720

```
His Glu Glu Pro Ile Tyr Ile Glu Gly Glu Lys Asp Gly Ile Thr Val Glu Val Ala Leu    260
CAT GAA GAG CCG ATT TAC ATT GAA GGC GAA AAG GAC GGC ATT ACG GTT GAA GTG GCT TTG    780

Gln Tyr Asn Asp Ser Tyr Thr Ser Asn Ile Tyr Ser Phe Thr Asn Asn Ile Asn Thr Tyr    280
CAA TAC AAT GAC AGC TAC ACA AGC AAC ATT TAC TCG TTT ACA AAC AAC ATT AAC ACG TAC    840

Glu Gly Gly Thr His Glu Ala Gly Phe Lys Thr Gly Leu Thr Arg Val Ile Asn Asp Tyr    300
GAA GGC GGT ACC CAT GAA GCT GGC TTC AAA ACG GGC CTG ACT CGT GTT ATC AAC GAT TAC    900

Ala Arg Lys Lys Gly Leu Ile Lys Glu Asn Asp Pro Asn Leu Ser Gly Asp Asp Val Arg    320
GCC AGA AAA AAA GGG CTT ATT AAA GAA AAT GAT CCA AAC CTA AGC GGA GAT GAC GTA AGG    960

Glu Gly Leu Thr Ala Ile Ile Ser Ile Lys His Pro Asp Pro Gln Phe Glu Gly Gln Thr    340
GAA GGG CTG ACA GCG ATT ATT TCA ATC AAA CAC CCT GAT CCG CAG TTT GAG GGC CAA ACA   1020

Lys Thr Lys Leu Gly Asn Ser Glu Ala Arg Thr Ile Thr Asp Thr Leu Phe Ser Thr Ala    360
AAA ACA AAG CTG GGC AAC TCA GAA GCA CGG ACG ATC ACC GAT ACG TTA TTT TCT ACG GCG   1080

Met Glu Thr Phe Met Leu Glu Asn Pro Asp Ala Ala Lys Lys Ile Val Asp Lys Gly Leu    380
ATG GAA ACA TTT ATG CTG GAA AAT CCA GAT GCA GCC AAA AAA ATT GTC GAT AAA GGT TTA   1140

Met Ala Ala Arg Ala Arg Met Ala Ala Lys Lys Ala Arg Glu Leu Thr Arg Arg Lys Ser    400
ATG GCG GCA AGA GCA AGA ATG GCT GCG AAA AAA GCG CGT GAA CTA ACA CGC CGT AAG AGT   1200

Ala Leu Glu Ile Ser Asn Leu Pro Gly Lys Leu Ala Asp Cys Ser Ser Lys Asp Pro Ser    420
GCT TTG GAA ATT TCA AAC CTG CCC GGT AAG TTA GCG GAC TGC TCT TCA AAA GAT CCG AGC   1260

Ile Ser Glu Leu Tyr Ile Val Glu Gly Asp Ser Ala Gly Gly Ser Ala Lys Gln Gly Arg    440
ATC TCC GAG TTA TAT ATC GTA GAG GGT GAC TCT GCC GGA GGA TCT GCT AAA CAA GGA CGC   1320

Asp Arg His Phe Gln Ala Ile Leu Pro Leu Arg Gly Lys Ile Leu Asn Val Glu Lys Ala    460
GAC AGA CAT TTC CAA GCC ATT TTG CCG CTT AGA GGT AAA ATC CTA AAC GTT GAA AAG GCC   1380

Arg Leu Asp Lys Ile Leu Ser Asn Asn Glu Val Arg Ser Met Ile Thr Ala Leu Gly Thr    480
AGA CTG GAT AAA ATC CTT TCT AAC AAC GAA GTT CGC TCT ATG ATC ACA GCG CTC GGC ACA   1440

Gly Ile Gly Glu Asp Phe Asn Leu Glu Lys Ala Arg Tyr His Lys Val Val Ile Met Thr    500
GGT ATC GGA GAA GAC TTC AAC CTT GAG AAA GCC CGT TAC CAC AAA GTT GTC ATT ATG ACA   1500

Asp Ala Asp Val Asp Gly Ala His Ile Arg Thr Leu Leu Leu Thr Phe Phe Tyr Arg Tyr    520
GAT GCC GAT GTT GAC GGC GCG CAC ATC AGA ACA CTG CTG TTA ACG TTC TTT TAC AGA TAT   1560

Met Arg Gln Ile Ile Glu Asn Gly Tyr Val Tyr Ile Ala Gln Pro Pro Leu Tyr Lys Val    540
ATG CGC CAA ATT ATC GAA AAT GGC TAC GTG TAC ATT GCG CAG CCG CCG CTC TAC AAG GTT   1620

Gln Gln Gly Lys Arg Val Glu Tyr Ala Tyr Asn Asp Lys Glu Leu Glu Glu Leu Leu Lys    560
CAA CAG GGG AAA CGC GTT GAA TAT GCA TAC AAT GAC AAG GAG CTT GAA GAG CTG TTA AAA   1680

Thr Leu Pro Gln Thr Pro Lys Pro Gly Leu Gln Arg Tyr Lys Gly Leu Gly Glu Met Asn    580
ACT CTT CCT CAA ACG CCT AAG CCT GGA CTG CAG CGT TAC AAA GGT CTT GGT GAA ATG AAT   1740

Ala Thr Gln Leu Trp Glu Thr Thr Met Asp Pro Ser Ser Arg Thr Leu Leu Gln Val Thr    600
GCC ACC CAG CTA TGG GAG ACA ACC ATG GAT CCT AGC TCC AGA ACA CTT CTT CAG GTA ACT   1800

Leu Glu Asp Ala Met Asp Ala Asp Glu Thr Phe Glu Met Leu Met Gly Asp Lys Val Glu    620
CTT GAA GAT GCA ATG GAT GCG GAT GAG ACT TTT GAA ATG CTT ATG GGC GAC AAG GTA GAA   1860

Pro Arg Arg Asn Phe Ile Glu Ala Asn Ala Arg Tyr Val Lys Asn Leu Asp Ile   638
CCG CGC CGA AAC TTC ATA GAA GCG AAT GCG AGA TAC GTT AAA AAT CTT GAC ATC  1914
```

Mycoplasma pneumoniae gyrB

Met Glu Asp Asn Asn Lys Thr Gln Ala Tyr Asp Ser Ser Ser Ile Lys Ile Leu Glu Gly 20
ATG GAA GAC AAT AAC AAA ACG CAA GCT TAC GAT TCC AGT AGC ATT AAG ATT CTT GAA GGC 60

Leu Glu Ala Val Arg Lys Arg Pro Gly Met Tyr Ile Gly Ser Thr Gly Glu Glu Gly Leu 40
TTA GAG GCC GTG CGT AAG CGC CCC GGC ATG TAC ATC GGT TCC ACC GGC GAA GAA GGG TTG 120

His His Met Ile Trp Glu Ile Ile Asp Asn Ser Ile Asp Glu Ala Met Gly Gly Phe Ala 60
CAC CAC ATG ATC TGG GAA ATC ATT GAT AAC TCA ATT GAC GAG GCC ATG GGT GGC TTT GCG 180

Ser Thr Val Lys Leu Thr Leu Lys Asp Asn Phe Val Thr Ile Val Glu Asp Asp Gly Arg 80
AGT ACT GTC AAA CTA ACC CTT AAA GAC AAC TTT GTC ACC ATT GTG GAG GAT GAT GGC CGG 240

Gly Ile Pro Val Asp Ile His Pro Lys Thr Asn Arg Ser Thr Val Glu Thr Val Phe Thr 100
GGC ATT CCC GTT GAC ATC CAT CCT AAG ACC AAC CGC TCC ACC GTT GAA ACA GTG TTT ACG 300

Val Leu His Ala Gly Gly Lys Phe Asp Asn Asp Ser Tyr Lys Val Ser Gly Gly Leu His 120
GTA CTC CAC GCT GGG GGT AAG TTT GAC AAC GAT AGC TAT AAG GTG TCA GGG GGA CTC CAC 360

Gly Val Gly Ala Ser Val Val Asn Ala Leu Ser Ser Ser Phe Lys Val Trp Val Ala Arg 140
GGT GTG GGT GCC TCG GTT GTT AAT GCT TTA AGT TCC TCC TTT AAG GTA TGG GTA GCC AGG 420

Glu His Gln Gln Tyr Phe Leu Ala Phe His Asn Gly Gly Glu Val Ile Gly Asp Leu Val 160
GAG CAC CAA CAG TAC TTC CTA GCG TTC CAC AAC GGT GGG GAA GTG ATT GGT GAT CTT GTC 480

Asn Glu Gly Lys Cys Asp Lys Glu His Gly Thr Lys Val Glu Phe Val Pro Asp Phe Thr 180
AAC GAA GGT AAG TGT GAC AAG GAA CAC GGT ACC AAG GTT GAG TTT GTC CCG GAC TTT ACC 540

Val Met Glu Lys Ser Asp Tyr Lys Gln Thr Val Ile Ala Ser Arg Leu Gln Gln Leu Ala 200
GTG ATG GAA AAG AGT GAT TAC AAA CAA ACG GTA ATT GCT AGT AGG TTG CAA CAA TTA GCC 600

Phe Leu Asn Lys Gly Ile Gln Ile Asp Phe Val Asp Glu Arg Arg Gln Asn Pro Gln Ser 220
TTT CTT AAC AAA GGG ATC CAA ATT GAC TTT GTT GAT GAA CGC CGT CAA AAT CCG CAA AGC 660

Phe Ser Trp Lys Tyr Asp Gly Gly Leu Val Gln Tyr Ile His His Leu Asn Asn Glu Lys 240
TTT TCT TGG AAG TAT GAT GGT GGC TTA GTC CAA TAC ATC CAC CAC CTC AAC AAC GAA AAG 720

Glu Pro Leu Phe Glu Asp Ile Ile Phe Gly Glu Lys Thr Asp Thr Val Lys Ser Val Ser 260
GAA CCT TTA TTT GAG GAC ATT ATC TTT GGT GAA AAA ACC GAT ACT GTT AAA TCA GTT AGC 780

Arg Asp Glu Ser Tyr Thr Ile Lys Val Glu Val Ala Phe Gln Tyr Asn Lys Thr Tyr Asn 280
CGT GAT GAG AGC TAC ACA ATT AAG GTG GAA GTG GCC TTT CAG TAC AAC AAG ACG TAT AAC 840

Gln Ser Ile Phe Ser Phe Cys Asn Asn Ile Asn Thr Thr Glu Gly Gly Thr His Val Glu 300
CAA TCA ATC TTT AGT TTT TGT AAC AAC ATT AAT ACC ACT GAA GGC GGT ACC CAT GTT GAA 900

Gly Phe Arg Asn Ala Leu Val Lys Ile Ile Asn Arg Phe Ala Val Glu Asn Lys Phe Leu 320
GGC TTT CGC AAT GCC TTA GTG AAG ATC ATT AAC CGC TTT GCT GTC GAA AAC AAG TTC TTA 960

Lys Glu Thr Asp Glu Lys Ile Thr Arg Asp Asp Ile Cys Glu Gly Leu Thr Ala Ile Ile 340
AAG GAA ACG GAT GAA AAG ATT ACC CGC GAT GAC ATC TGT GAA GGG TTA ACG GCC ATT ATC 1020

Ser Ile Lys His Pro Asn Pro Gln Tyr Glu Gly Gln Thr Lys Lys Lys Leu Gly Asn Thr 360
TCA ATT AAG CAC CCC AAC CCC CAG TAT GAG GGT CAA ACC AAA AAG AAA TTG GGT AAC ACC 1080

Glu Val Arg Pro Leu Val Asn Ser Ile Val Ser Glu Ile Phe Glu Arg Phe Met Leu Glu 380
GAA GTG CGT CCT CTA GTT AAC AGT ATT GTT AGT GAA ATC TTT GAG CGC TTT ATG TTG GAA 1140

Asn Pro Gln Glu Ala Asn Ala Ile Ile Arg Lys Thr Leu Leu Ala Gln Glu Ala Arg Arg 400
AAC CCC CAG GAA GCC AAC GCC ATT ATT CGC AAA ACG CTG TTA GCG CAA GAA GCA CGG CGC 1200

Arg Ser Gln Glu Ala Arg Glu Leu Thr Arg Arg Lys Ser Pro Phe Asp Ser Gly Ser Leu 420
CGT AGT CAG GAA GCA CGG GAA CTG ACC CGC AGG AAA TCT CCC TTT GAC AGT GGT TCA CTC 1260

```
Pro Gly Lys Leu Ala Asp Cys Thr Thr Arg Asp Pro Ser Ile Ser Glu Leu Tyr Ile Val   440
CCG GGT AAG TTA GCA GAC TGT ACC ACC CGT GAT CCC TCA ATT AGT GAA CTG TAC ATT GTC  1320

Glu Gly Asp Ser Ala Gly Gly Thr Ala Lys Thr Gly Arg Asp Arg Tyr Phe Gln Ala Ile   460
GAG GGA GAT AGC GCT GGT GGT ACG GCT AAA ACG GGG CGG GAC CGT TAC TTC CAA GCG ATC  1380

Leu Pro Leu Arg Gly Lys Ile Leu Asn Val Glu Lys Ser His Phe Glu Gln Ile Phe Asn   480
CTA CCA TTA AGG GGT AAG ATC CTC AAT GTG GAA AAG TCC CAC TTT GAA CAA ATC TTT AAT  1440

Asn Val Glu Ile Ser Ala Leu Val Met Ala Val Gly Cys Gly Ile Lys Pro Asp Phe Glu   500
AAT GTG GAA ATC TCG GCT TTA GTC ATG GCC GTT GGC TGT GGC ATT AAA CCA GAC TTC GAG  1500

Leu Glu Lys Leu Arg Tyr Asn Lys Ile Ile Ile Met Thr Asp Ala Asp Val Asp Gly Ala   520
TTG GAA AAA CTG CGG TAT AAC AAG ATC ATT ATC ATG ACC GAT GCT GAT GTC GAT GGG GCG  1560

His Ile Arg Thr Leu Leu Leu Thr Phe Phe Phe Arg Phe Met Tyr Pro Leu Val Glu Gln   540
CAC ATC CGT ACC CTC CTT TTA ACC TTC TTC TTT CGC TTT ATG TAT CCC TTG GTG GAA CAG  1620

Gly Asn Ile Tyr Ile Ala Gln Pro Pro Leu Tyr Lys Val Ser Tyr Ser Asn Lys Asp Leu   560
GGC AAC ATT TAC ATT GCC CAA CCC CCA CTG TAT AAG GTC TCT TAC TCT AAT AAG GAC TTA  1680

Tyr Met Gln Thr Asp Val Gln Leu Glu Glu Trp Lys Gln Gln His Pro Asn Leu Lys Tyr   580
TAC ATG CAA ACC GAT GTC CAA CTA GAG GAG TGG AAG CAA CAA CAC CCT AAT CTG AAA TAC  1740

Asn Leu Gln Arg Tyr Lys Gly Leu Gly Glu Met Asp Ala Ile Gln Leu Trp Glu Thr Thr   600
AAC TTA CAA CGC TAC AAG GGT TTA GGG GAA ATG GAT GCC ATC CAA CTG TGG GAA ACC ACG  1800

Met Asp Pro Lys Val Arg Thr Leu Leu Lys Val Thr Val Glu Asp Ala Ser Ile Ala Asp   620
ATG GAT CCC AAG GTG CGT ACC TTA CTA AAG GTA ACC GTT GAA GAT GCT TCA ATT GCT GAT  1860

Lys Ala Phe Ser Leu Leu Met Gly Asp Glu Val Pro Pro Arg Arg Glu Phe Ile Glu Gln   640
AAA GCC TTC TCC TTG TTA ATG GGC GAT GAA GTT CCC CCA AGA CGG GAA TTC ATC GAA CAA  1920

Asn Ala Arg Asn Val Lys Asn Ile Asp Ile   650
AAC GCA CGT AAC GTA AAG AAC ATT GAT ATC  1950
```

T4 gene 39

```
Met Ile Lys Asn Glu Ile Lys Ile Leu Ser Asp Ile Glu His Ile Lys Lys Arg Ser Gly    20
ATG ATT AAG AAT GAA ATT AAA ATT CTG AGC GAT ATT GAA CAT ATC AAA AAG CGT AGT GGC    60

Met Tyr Ile Gly Ser Ser Ala Asn Glu Thr His Glu Arg Phe Met Phe Gly Lys Trp Glu    40
ATG TAC ATT GGC TCT TCT GCT AAT GAA ACG CAT GAG CGC TTT ATG TTT GGT AAA TGG GAA   120

Ser Val Gln Tyr Val Pro Gly Leu Val Lys Leu Ile Asp Glu Ile Ile Asp Asn Ser Val    60
AGT GTT CAG TAT GTA CCT GGT CTT GTT AAG CTT ATT GAT GAA ATT ATC GAT AAC TCA GTA   180

Asp Glu Gly Ile Arg Thr Lys Phe Lys Phe Ala Asn Lys Ile Asn Val Thr Ile Lys Asn    80
GAT GAA GGT ATT CGT ACT AAG TTT AAA TTC GCG AAT AAA ATT AAT GTT ACT ATT AAA AAC   240

Asn Gln Val Thr Val Glu Asp Asn Gly Arg Gly Ile Pro Gln Ala Met Val Lys Thr Pro   100
AAT CAA GTA ACA GTT GAA GAT AAC GGT CGC GGT ATT CCA CAA GCG ATG GTT AAA ACA CCT   300

Thr Gly Glu Glu Ile Pro Gly Pro Val Ala Ala Trp Thr Ile Pro Lys Ala Gly Gly Asn   120
ACC GGT GAA GAA ATT CCT GGT CCT GTT GCC GCA TGG ACT ATT CCA AAA GCA GGT GGT AAC   360

Phe Gly Asp Asp Lys Glu Arg Val Thr Gly Gly Met Asn Gly Val Gly Ser Ser Leu Thr   140
TTT GGT GAT GAT AAA GAA CGC GTC ACC GGC GGT ATG AAC GGT GTT GGT TCT AGT TTG ACA   420
```

Asn Ile Phe Ser Val Met Phe Val Gly Glu Thr Gly Asp Gly Gln Asn Asn Ile Val Val 160
AAC ATT TTT TCT GTG ATG TTT GTC GGT GAA ACT GGC GAC GGT CAA AAT AAT ATT GTA GTT 480

Arg Cys Ser Asn Gly Met Glu Asn Lys Ser Trp Glu Asp Ile Pro Gly Lys Trp Lys Gly 180
CGT TGT TCA AAT GGC ATG GAA AAT AAA TCA TGG GAA GAT ATT CCT GGA AAA TGG AAA GGA 540

Thr Arg Val Thr Phe Ile Pro Asp Phe Met Ser Phe Glu Thr Asn Glu Leu Ser Gln Val 200
ACT CGT GTT ACT TTC ATT CCT GAT TTT ATG TCA TTT GAA ACT AAT GAG CTG TCC CAA GTT 600

Tyr Leu Asp Ile Thr Leu Asp Arg Leu Gln Thr Leu Ala Val Val Tyr Pro Asp Ile Gln 220
TAT CTT GAC ATT ACA CTG GAT CGT CTC CAG ACA CTT GCT GTA GTT TAT CCT GAT ATT CAA 660

Phe Thr Phe Asn Gly Lys Lys Val Gln Gly Asn Phe Lys Lys Tyr Ala Arg Gln Tyr Asp 240
TTT ACC TTT AAT GGT AAA AAG GTT CAG GGC AAT TTT AAG AAA TAT GCA CGG CAG TAT GAT 720

Glu His Ala Ile Val Gln Glu Gln Glu Asn Cys Ser Ile Ala Val Gly Arg Ser Pro Asp 260
GAA CAT GCT ATT GTT CAA GAG CAA GAA AAT TGT TCT ATT GCG GTT GGT CGT TCA CCG GAT 780

Gly Phe Arg Gln Leu Thr Tyr Val Asn Asn Ile His Thr Lys Asn Gly Gly His His Ile 280
GGT TTT CGT CAA TTA ACA TAC GTC AAT AAC ATT CAT ACT AAG AAT GGT GGC CAT CAC ATT 840

Asp Cys Ala Met Asp Asp Ile Cys Glu Asp Leu Ile Pro Gln Ile Lys Arg Lys Phe Lys 300
GAC TGC GCT ATG GAT GAT ATT TGT GAA GAC CTT ATT CCA CAA ATC AAA CGT AAG TTC AAA 900

Ile Asp Val Thr Lys Ala Arg Val Lys Glu Cys Leu Thr Ile Val Met Phe Val Arg Asp 320
ATT GAT GTG ACT AAA GCA CGT GTC AAA GAA TGT TTG ACT ATC GTT ATG TTT GTT CGT GAT 960

Met Lys Asn Met Arg Leu Ile Arg Gln Thr Lys Glu Arg Leu Thr Ser Pro Phe Gly Glu 340
ATG AAA AAC ATG CGA TTG ATT CGT CAA ACT AAA GAG CGT TTG ACT TCT CCA TTT GGC GAA 1020

Ile Arg Ser His Ile Gln Leu Asp Ala Lys Lys Ile Ser Arg Asp Ile Leu Asn Asn Glu 360
ATT CGT AGT CAT ATT CAA CTT GAT GCT AAA AAG ATT TCA CGT GAT ATT CTA AAT AAT GAA 1080

Ala Ile Leu Met Pro Ile Ile Glu Ala Ala Leu Ala Arg Lys Leu Ala Ala Glu Lys Ala 380
GCA ATT CTA ATG CCG ATT ATT GAA GCT GCT TTG GCT CGT AAA TTG GCG GCA GAA AAA GCA 1140

Ala Glu Thr Lys Ala Ala Lys Lys Ala Ser Lys Ala Lys Val His Lys His Ile Lys Ala 400
GCA GAA ACT AAA GCA GCT AAA AAG GCT TCT AAA GCT AAG GTT CAT AAA CAT ATC AAA GCG 1200

Asn Leu Cys Gly Lys Asp Ala Asp Thr Thr Leu Phe Leu Thr Glu Gly Asp Ser Ala Ile 420
AAT CTT TGC GGT AAA GAT GCT GAT ACT ACT CTT TTC TTG ACT GAG GGT GAT TCG GCT ATC 1260

Gly Tyr Leu Ile Asp Val Arg Asp Lys Glu Leu His Gly Gly Tyr Pro Leu Arg Gly Lys 440
GGA TAT CTT ATT GAT GTT CGT GAT AAA GAA CTT CAC GGT GGT TAT CCA TTG CGT GGT AAA 1320

Val Leu Asn Ser Trp Gly Met Ser Tyr Ala Asp Met Leu Lys Asn Lys Glu Leu Phe Asp 460
GTT CTC AAC AGT TGG GGT ATG TCT TAT GCA GAT ATG CTT AAA AAC AAA GAA CTA TTT GAT 1380

Ile Cys Ala Ile Thr Gly Leu Val Leu Gly Glu Lys Ala Phe Glu Glu Lys Glu Asp Gly 480
ATT TGC GCA ATC ACT GGT CTA GTT CTC GGT GAA AAA GCG TTT GAA GAA AAA GAA GAT GGC 1440

Glu Trp Phe Thr Phe Glu Leu Asn Gly Asp Thr Ile Ile Val Asn Glu Asn Asp Glu Val 500
GAG TGG TTT ACT TTC GAG CTA AAT GGC GAT ACA ATT ATC GTA AAT GAA AAT GAT GAA GTA 1500

Gln Ile Asn Gly Lys Trp Ile Thr Val Gly Glu Leu Arg Lys Lys Ser Ile Met Thr 519
CAG ATT AAT GGT AAA TGG ATA ACA GTA GGT GAA TTA CGC AAA AAA TCT ATA ATG ACT 1557

T4 gene 52

Met Gln Leu Asn Asn Arg Asp Leu Lys Ser Ile Ile Asp Asn Glu Ala Leu Ala Tyr Ala 20
ATG CAA CTG AAT AAT CGC GAT TTA AAA AGT ATC ATT GAT AAT GAA GCA TTG GCT TAT GCT 60

Met Tyr Thr Val Glu Asn Arg Ala Ile Pro Asn Met Ile Asp Gly Phe Lys Pro Val Gln 40
ATG TAC ACG GTT GAA AAT CGT GCT ATC CCA AAT ATG ATT GAT GGA TTT AAG CCA GTT CAA 120

Arg Phe Val Ile Ala Arg Ala Leu Asp Leu Ala Arg Gly Asn Lys Asp Lys Phe His Lys 60
CGA TTT GTT ATT GCT CGA GCT CTT GAT TTG GCA CGA GGA AAT AAA GAT AAG TTT CAC AAA 180

Leu Ala Ser Ile Ala Gly Gly Val Ala Asp Leu Gly Tyr His His Gly Glu Thr Leu His 80
CTC GCT TCT ATT GCA GGT GGT GTA GCG GAC CTT GGA TAT CAT CAT GGT GAA ACT CTG CAC 240

Lys Ser Gln Cys Leu Met Ala Asn Thr Trp Asn Asn Asn Phe Pro Leu Leu Asp Gly Gln 100
AAG AGC CAG TGC TTG ATG GCT AAT ACT TGG AAT AAT AAC TTT CCT CTG TTA GAT GGT CAA 300

Gly Asn Phe Gly Ser Arg Thr Val Gln Lys Ala Ala Ala Ser Arg Tyr Ile Phe Ala Arg 120
GGA AAC TTT GGT TCT CGT ACT GTC CAA AAG GCA GCG GCA AGT CGT TAT ATT TTT GCT CGT 360

Val Ser Lys Asn Phe Tyr Asn Val Tyr Lys Asp Thr Glu Tyr Ala Pro Val His Gln Asp 140
GTA AGT AAA AAT TTC TAT AAC GTA TAT AAA GAT ACT GAA TAT GCT CCG GTA CAT CAA GAT 420

Lys Glu His Ile Pro Pro Ala Phe Tyr Leu Pro Ile Ile Pro Thr Val Leu Leu Asn Gly 160
AAA GAA CAC ATT CCG CCT GCT TTC TAT TTG CCT ATT ATT CCT ACT GTT CTT CTT AAT GGC 480

Val Ser Gly Ile Ala Thr Gly Tyr Ala Thr Tyr Ile Leu Pro His Ser Val Ser Ser Val 180
GTT TCC GGT ATT GCA ACT GGT TAT GCA ACT TAC ATT CTT CCT CAT AGT GTT TCT TCT GTC 540

Lys Lys Ala Val Leu Gln Ala Leu Gln Gly Lys Lys Val Thr Lys Pro Lys Val Glu Phe 200
AAG AAA GCT GTA CTG CAA GCT CTT CAA GGA AAG AAA GTA ACT AAA CCG AAA GTA GAA TTC 600

Pro Glu Phe Arg Gly Glu Val Val Glu Ile Asp Gly Gln Tyr Glu Ile Arg Gly Thr Tyr 220
CCA GAA TTT CGT GGT GAA GTC GTT GAA ATT GAT GGG CAA TAT GAA ATT CGT GGA ACA TAT 660

Lys Phe Thr Ser Arg Thr Gln Met His Ile Thr Glu Ile Pro Tyr Lys Tyr Asp Arg Glu 240
AAG TTT ACT TCA CGA ACT CAA ATG CAT ATC ACT GAG ATT CCG TAT AAG TAT GAT CGT GAA 720

Thr Tyr Val Ser Lys Ile Leu Asp Pro Leu Glu Asn Lys Gly Phe Ile Thr Trp Asp Asp 260
ACT TAT GTG AGT AAA ATC TTA GAC CCA CTT GAA AAT AAA GGC TTC ATT ACA TGG GAT GAT 780

Ala Cys Gly Glu His Gly Phe Gly Phe Lys Val Lys Phe Arg Lys Glu Tyr Ser Leu Ser 280
GCT TGT GGT GAG CAT GGT TTT GGC TTC AAA GTT AAA TTC CGC AAA GAA TAT TCT TTG AGC 840

Asp Asn Glu Glu Glu Arg His Ala Lys Ile Met Lys Asp Phe Gly Leu Ile Glu Arg Arg 300
GAT AAC GAA GAA GAA CGC CAT GCA AAA ATT ATG AAA GAC TTC GGA CTG ATT GAG CGT CGT 900

Ser Gln Asn Ile Thr Val Ile Asn Glu Lys Gly Lys Leu Gln Val Tyr Asp Asn Val Val 320
TCC CAG AAT ATT ACG GTT ATT AAT GAG AAA GGA AAG CTG CAA GTT TAC GAT AAC GTA GTT 960

Asp Leu Ile Lys Asp Phe Val Glu Val Arg Lys Thr Tyr Val Gln Lys Arg Ile Asp Asn 340
GAT TTA ATT AAA GAC TTT GTT GAA GTT CGT AAA ACT TAT GTC CAA AAA CGA ATT GAT AAC 1020

Lys Ile Lys Glu Thr Glu Ser Ala Phe Arg Leu Ala Phe Ala Lys Ala His Phe Ile Lys 360
AAA ATT AAA GAA ACT GAG TCA GCT TTT CGT TTA GCC TTT GCC AAA GCA CAT TTC ATT AAG 1080

Lys Val Ile Ser Gly Glu Ile Val Val Gln Gly Lys Thr Arg Lys Glu Leu Thr Glu Glu 380
AAA GTA ATT TCA GGT GAA ATT GTT GTA CAG GGT AAA ACT CGT AAA GAA CTG ACC GAA GAA 1140

Leu Ser Lys Ile Asp Met Tyr Ser Ser Tyr Val Asp Lys Leu Val Gly Met Asn Ile Phe 400
CTT TCG AAA ATC GAT ATG TAT TCT TCT TAT GTT GAT AAA CTA GTT GGA ATG AAT ATT TTT 1200

His Met Thr Ser Asp Glu Ala Lys Lys Leu Ala Glu Glu Ala Lys Ala Lys Lys Glu Glu 420
CAT ATG ACT TCC GAT GAA GCA AAG AAA CTT GCT GAA GAA GCT AAA GCT AAA AAA GAA GAA 1260

Asn Glu Tyr Trp Lys Thr Thr Asp Val Val Thr Glu Tyr Thr Lys Asp Leu Glu Glu Ile 440
AAC GAA TAT TGG AAA ACT ACT GAT GTA GTT ACT GAA TAC ACC AAA GAT TTA GAG GAA ATC 1320

```
Lys   441
AAA  1323
```

T4 gene 60

```
Met Lys Phe Val Lys Ile Asp Ser Ser Ser Val Asp Met Lys Lys Tyr Lys Leu Gln Asn    20
ATG AAA TTT GTA AAA ATT GAT TCT TCT AGC GTT GAT ATG AAA AAA TAT AAA TTG CAG AAC    60

Asn Val Arg Arg Ser Ile Lys Ser Ser Ser Met Asn Tyr Ala Asn Val Ala Ile Met Thr    40
AAT GTT CGT CGT TCT ATT AAA TCC TCT TCA ATG AAC TAT GCG AAT GTC GCT ATT ATG ACA   120

Asp Ala Asp His Asp Gly                                                   Leu    47
GAC GCA GAT CAC GAT GGA TAGCCTTCGGGCTATCTATAGAAATACCTCATAATTAAGAGATTATTGGA TTA   191

Gly Ser Ile Tyr Pro Ser Leu Leu Gly Phe Phe Ser Asn Trp Pro Glu Leu Phe Glu Gln    67
GGT TCT ATT TAT CCT TCT CTG CTC GGA TTT TTT AGT AAT TGG CCA GAA TTG TTT GAG CAA   251

Gly Arg Ile Arg Phe Val Lys Thr Pro Val Ile Ile Ala Gln Val Gly Lys Lys Gln Glu    87
GGA CGA ATT CGC TTT GTC AAA ACT CCT GTA ATC ATC GCT CAG GTC GGT AAA AAA CAA GAA   311

Trp Phe Tyr Thr Val Ala Glu Tyr Glu Ser Ala Lys Asp Ala Leu Pro Lys His Ser Ile   107
TGG TTT TAT ACA GTC GCT GAA TAT GAG AGT GCC AAA GAT GCT CTA CCT AAA CAT AGC ATC   371

Arg Tyr Ile Lys Gly Leu Gly Ser Leu Glu Lys Ser Glu Tyr Arg Glu Met Ile Gln Asn   127
CGT TAT ATT AAG GGA CTT GGC TCT TTG GAA AAA TCT GAA TAT CGT GAG ATG ATT CAA AAC   431

Pro Val Tyr Asp Val Val Lys Leu Pro Glu Asn Trp Lys Glu Leu Phe Glu Met Leu Met   147
CCA GTA TAT GAT GTT GTT AAA CTT CCT GAG AAC TGG AAA GAG CTT TTT GAA ATG CTC ATG   490

Gly Asp Asn Ala Asp Leu Arg Lys Glu Trp Met Ser Gln   160
GGA GAT AAT GCT GAC CTT CGT AAA GAA TGG ATG AGC CAG   530
```

T2 gene 39

```
Met Ile Lys Asn Glu Ile Lys Ile Leu Ser Asp Ile Glu His Ile Lys Lys Arg Ser Gly    20
ATG ATT AAG AAT GAA ATT AAA ATT CTG AGC GAT ATT GAA CAC ATC AAA AAG CGT AGT GGC    60

Met Tyr Ile Gly Ser Ser Ala Asn Glu Met His Glu Arg Phe Leu Phe Gly Lys Trp Glu    40
ATG TAT ATT GGC TCT TCT GCT AAT GAA ATG CAT GAG CGC TTT CTG TTT GGT AAA TGG GAA   120

Ser Val Gln Tyr Val Pro Gly Leu Val Lys Leu Ile Asp Glu Ile Ile Asp Asn Ser Val    60
AGT GTT CAG TAT GTA CCT GGT CTT GTT AAG CTT ATT GAT GAA ATT ATC GAT AAC TCA GTA   180

Asp Glu Gly Ile Arg Thr Lys Phe Lys Leu Ala Asn Lys Ile Asn Val Thr Ile Lys Asn    80
GAT GAA GGT ATT CGT ACT AAG TTT AAA TTA GCA AAT AAA ATT AAT GTT ACT ATT AAA AAC   240

Asn Gln Val Thr Val Glu Asp Asn Gly Arg Gly Ile Pro Gln Ala Met Val Lys Thr Pro   100
AAT CAA GTA ACA GTT GAA GAT AAC GGT CGT GGT ATT CCA CAA GCG ATG GTT AAA ACA CCT   300

Thr Gly Glu Glu Ile Pro Gly Pro Val Ala Ala Trp Thr Ile Pro Lys Ala Gly Gly Asn   120
ACT GGT GAA GAA ATT CCT GGT CCA GTT GCT GCA TGG ACT ATT CCA AAA GCA GGT GGT AAC   360

Phe Gly Asp Asp Lys Glu Arg Val Thr Gly Gly Met Asn Gly Val Gly Ser Ser Leu Thr   140
TTT GGT GAT GAT AAA GAA CGC GTC ACC GGT GGT ATG AAT GGT GTT GGT TCT AGT TTG ACA   420

Asn Ile Phe Ser Val Met Phe Val Gly Glu Thr Gly Asp Gly Gln Asn Asn Ile Val Val   160
AAC ATT TTT TCT GTG ATG TTT GTC GGT GAA ACT GGC GAT GGT CAA AAT AAT ATT GTA GTT   480
```

```
Arg Cys Ser Asn Gly Met Glu Asn Lys Ser Trp Glu Asp Ile Pro Gly Lys Trp Lys Gly   180
CGT TGT TCA AAT GGC ATG GAA AAT AAA TCA TGG GAA GAT ATT CCT GGA AAA TGG AAA GGA   540

Thr Arg Val Thr Phe Ile Pro Asp Phe Met Ser Phe Glu Thr Asn Glu Leu Ser Gln Val   200
ACT CGT GTT ACT TTC ATT CCT GAT TTT ATG TCA TTT GAA ACT AAT GAG CTG TCC CAA GTT   600

Tyr Leu Asp Ile Thr Leu Asp Arg Leu Gln Thr Leu Ala Val Val Tyr Pro Asp Ile Gln   220
TAT CTT GAC ATT ACA CTT GAT CGT CTC CAG ACG CTT GCT GTA GTT TAT CCT GAT ATT CAA   660

Phe Thr Phe Asn Gly Lys Lys Val Gln Gly Asn Phe Lys Lys Tyr Ala Arg Gln Tyr Asp   240
TTT ACC TTT AAT GGT AAA AAG GTT CAG GGC AAT TTT AAG AAA TAT GCA CGA CAG TAT GAT   720

Glu His Ala Ile Val Gln Glu Gln Glu Asn Cys Ser Ile Ala Val Gly Arg Ser Pro Asp   260
GAA CAT GCT ATT GTT CAA GAA CAA GAA AAT TGT TCT ATT GCG GTT GGT CGT TCA CCG GAT   780

Gly Phe Arg Gln Leu Thr Tyr Val Asn Asn Ile His Thr Lys Asn Gly Gly His His Ile   280
GGT TTT CGT CAG TTG ACG TAC GTC AAT AAC ATT CAT ACT AAG AAT GGT GGC CAT CAT ATT   840

Asp Cys Val Met Asp Asp Ile Cys Glu Asp Leu Ile Pro Gln Ile Lys Arg Lys Phe Lys   300
GAC TGT GTT ATG GAT GAT ATT TGT GAA GAC CTT ATT CCA CAA ATC AAA CGT AAA TTC AAA   900

Ile Asp Val Thr Lys Ala Arg Val Lys Glu Cys Leu Thr Ile Val Met Phe Val Arg Asp   320
ATT GAT GTA ACT AAA GCA CGT GTT AAA GAA TGT TTG ACT ATC GTT ATG TTT GTT CGC GAT   960

Met Lys Asn Met Arg Phe Asp Ser Gln Thr Lys Glu Arg Leu Thr Ser Pro Phe Gly Glu   340
ATG AAA AAC ATG CGA TTT GAC TCT CAA ACT AAA GAA CGA CTT ACT TCT CCT TTT GGT GAA  1020

Ile Arg Ser His Ile Gln Leu Asp Ala Lys Lys Ile Ser Arg Ala Ile Leu Asn Asn Glu   360
ATT CGT AGT CAT ATT CAA CTT GAT GCT AAA AAG ATT TCA CGC GCT ATT CTA AAT AAT GAA  1080

Ala Ile Leu Met Pro Ile Ile Glu Ala Ala Leu Ala Arg Lys Leu Ala Ala Glu Lys Ala   380
GCA ATT TTA ATG CCA ATT ATT GAA GCA GCA TTA GCT CGT AAA TTG GCG GCG GAA AAA GCA  1140

Ala Glu Thr Lys Ala Ala Lys Lys Ala Ser Lys Ala Lys Val His Lys His Ile Lys Ala   400
GCA GAG ACA AAG GCA GCT AAA AAA GCT TCT AAA GCT AAG GTT CAT AAA CAT ATC AAA GCG  1200

Asn Leu Cys Gly Lys Asp Ala Asp Thr Thr Leu Phe Leu Thr Glu Gly Asp Ser Ala Ile   420
AAT CTT TGT GGT AAA GAT GCT GAT ACT ACT CTT TTC TTG ACT GAG GGT GAT TCT GCT ATC  1260

Gly Tyr Leu Ile Asp Val Arg Asp Lys Glu Leu His Gly Gly Tyr Pro Leu Arg Gly Lys   440
GGA TAT CTT ATT GAT GTT CGT GAT AAA GAA CTT CAT GGT GGT TAT CCA TTG CGT GGT AAA  1320

Val Leu Asn Ser Trp Gly Met Ser Tyr Ala Asp Met Leu Lys Asn Lys Glu Leu Phe Asp   460
GTT CTT AAT AGC TGG GGT ATG TCA TAT GCC GAT ATG CTT AAA AAC AAA GAA CTA TTT GAT  1380

Ile Cys Ala Ile Thr Gly Leu Val Leu Gly Glu Lys Ala Glu Asn Leu Asn Tyr His Asn   480
ATT TGC GCA ATC ACT GGT CTA GTT CTT GGT GAA AAA GCT GAA AAC TTG AAT TAT CAT AAT  1440

Ile Ala Ile Met Thr Asp Ala Asp His Asp Gly Leu Gly Ser Ile Tyr Pro Ser Leu Leu   500
ATT GCT ATT ATG ACT GAT GCT GAC CAT GAT GGT CTA GGA AGC ATT TAT CCT TCT CTG CTC  1500

Gly Phe Phe Ser Asn Trp Pro Glu Leu Phe Glu Gln Gly Arg Ile Arg Phe Val Lys Thr   520
GGA TTT TTT AGT AAT TGG CCA GAA TTG TTT GAG CAA GGA CGA ATT CGC TTT GTC AAA ACT  1560

Pro Val Ile Ile Ala Gln Val Gly Lys Lys Gln Glu Trp Phe Tyr Thr Val Ala Glu Tyr   540
CCT GTA ATC ATC GCT CAG GTC GGT AAA AAA CAA GAA TGG TTT TAT ACA GTC GCT GAA TAT  1620

Glu Ser Ala Lys Asp Ala Leu Pro Lys His Ser Ile Arg Tyr Ile Lys Gly Leu Gly Ser   560
GAG AGT GCC AAA GAT GCT CTA CCT AAA CAT AGC ATC CGT TAT ATT AAA GGA CTT GGC TCT  1680

Leu Glu Lys Ser Glu Tyr Arg Glu Met Ile Gln Asn Pro Val Tyr Asp Val Val Lys Leu   580
TTG GAA AAA TCT GAA TAT CGT GAA ATG ATT CAA AAT CCA GTA TAT GAT GTT GTT AAA CTT  1740
```

Pro Glu Asn Trp Lys Glu Leu Phe Glu Met Leu Met Gly Asp Asn Ala Asp Leu Arg Lys 600
CCT GAG AAC TGG AAA GAG CTT TTT GAA ATG CTC ATG GGA GAT AAT GCT GAC CTT CGT AAA 1800

Glu Trp Met Ser Gln 605
GAA TGG ATG AGC CAG 1815

Saccharomyces cerevisiae Top2

Met Ser Thr Glu Pro Val Ser Ala Ser Asp Lys Tyr Gln Lys Ile Ser Gln Leu Glu His 20
ATG TCA ACT GAA CCG GTA AGC GCC TCT GAT AAA TAT CAG AAA ATT TCT CAA CTG GAA CAT 60

Ile Leu Lys Arg Pro Asp Thr Tyr Ile Gly Ser Val Glu Thr Gln Glu Gln Leu Gln Trp 40
ATC TTA AAA AGA CCA GAC ACT TAT ATC GGT TCT GTT GAA ACT CAA GAG CAG CTG CAA TGG 120

Ile Tyr Asp Glu Glu Thr Asp Cys Met Ile Glu Lys Asn Val Thr Ile Val Pro Gly Leu 60
ATA TAC GAT GAA GAG ACC GAT TGC ATG ATT GAA AAA AAT GTC ACA ATT GTA CCA GGG TTG 180

Phe Lys Ile Phe Asp Glu Ile Leu Val Asn Ala Ala Asp Asn Asn Lys Val Arg Asp Pro 80
TTC AAA ATC TTT GAT GAA ATC TTA GTC AAT GCG GCA GAT AAT AAT AAA GTT CGT GAT CCA 240

Ser Met Lys Arg Ile Asp Val Asn Ile His Ala Glu Glu His Thr Ile Glu Val Lys Asn 100
TCG ATG AAA CGA ATC GAT GTA AAC ATA CAT GCT GAG GAA CAT ACT ATA GAA GTG AAA AAT 300

Asp Gly Lys Gly Ile Pro Ile Glu Ile His Asn Lys Glu Asn Ile Tyr Ile Pro Glu Met 120
GAT GGA AAA GGT ATT CCC ATA GAG ATT CAT AAC AAG GAG AAT ATT TAT ATT CCT GAA ATG 360

Ile Phe Gly His Leu Leu Thr Ser Ser Asn Tyr Asp Asp Asp Glu Lys Lys Val Thr Gly 140
ATA TTT GGT CAT TTG TTG ACA TCA TCC AAT TAT GAT GAT GAT GAG AAG AAA GTC ACT GGT 420

Gly Arg Asn Gly Tyr Gly Ala Lys Leu Cys Asn Ile Phe Ser Thr Glu Phe Ile Leu Glu 160
GGT AGA AAC GGT TAT GGT GCT AAG CTT TGT AAT ATA TTT TCC ACT GAA TTC ATA TTA GAA 480

Thr Ala Asp Leu Asn Val Gly Gln Lys Tyr Val Gln Lys Trp Glu Asn Asn Met Ser Ile 180
ACT GCA GAT CTA AAT GTT GGC CAG AAA TAT GTT CAA AAA TGG GAA AAT AAC ATG AGC ATT 540

Cys His Pro Pro Lys Ile Thr Ser Tyr Lys Lys Gly Pro Ser Tyr Thr Lys Val Thr Phe 200
TGC CAC CCC CCA AAA ATA ACA TCT TAC AAG AAG GGT CCA TCA TAT ACA AAG GTG ACA TTT 600

Lys Pro Asp Leu Thr Arg Phe Gly Met Lys Glu Leu Asp Asn Asp Ile Leu Gly Val Met 220
AAG CCG GAT TTA ACC AGA TTC GGA ATG AAA GAG CTA GAT AAT GAT ATC TTA GGA GTG ATG 660

Arg Arg Arg Val Tyr Asp Ile Asn Gly Ser Val Arg Asp Ile Asn Val Tyr Leu Asn Gly 240
CGA AGA AGA GTT TAT GAT ATC AAT GGT TCT GTT CGT GAC ATT AAT GTC TAT CTG AAT GGC 720

Lys Ser Leu Lys Ile Arg Asn Phe Lys Asn Tyr Val Glu Leu Tyr Leu Lys Ser Leu Glu 260
AAG TCC TTA AAG ATA AGA AAT TTC AAA AAT TAT GTT GAA CTC TAC TTG AAA TCA CTC GAA 780

Lys Lys Arg Gln Leu Asp Asn Gly Glu Asp Gly Ala Ala Lys Ser Asp Ile Pro Thr Ile 280
AAA AAA AGA CAA CTA GAT AAC GGT GAG GAC GGT GCC GCT AAG TCT GAT ATC CCG ACT ATT 840

Leu Tyr Glu Arg Ile Asn Asn Arg Trp Glu Val Ala Phe Ala Val Ser Asp Ile Ser Phe 300
CTT TAT GAG AGA ATA AAC AAC AGA TGG GAA GTT GCT TTT GCG GTT TCT GAT ATC TCT TTT 900

Gln Gln Ile Ser Phe Val Asn Ser Ile Ala Thr Thr Met Gly Gly Thr His Val Asn Tyr 320
CAA CAA ATT TCT TTT GTG AAT TCC ATT GCA ACT ACC ATG GGT GGT ACC CAT GTC AAT TAC 960

Ile Thr Asp Gln Ile Val Lys Lys Ile Ser Glu Ile Leu Lys Lys Lys Lys Lys Lys Ser 340
ATA ACA GAC CAA ATT GTA AAA AAA ATT TCA GAA ATT TTG AAG AAA AAG AAG AAA AAA AGT 1020

Val Lys Ser Phe Gln Ile Lys Asn Asn Met Phe Ile Phe Ile Asn Cys Leu Ile Glu Asn 360
GTG AAG TCT TTT CAG ATT AAA AAT AAT ATG TTC ATT TTC ATT AAT TGT TTG ATT GAG AAT 1080

```
Pro Ala Phe Thr Ser Gln Thr Lys Glu Gln Leu Thr Thr Arg Val Lys Asp Phe Gly Ser   380
CCT GCA TTT ACC TCA CAA ACA AAA GAG CAA CTG ACA ACA AGA GTC AAA GAT TTT GGG TCC  1140

Arg Cys Glu Ile Pro Leu Glu Tyr Ile Asn Lys Ile Met Lys Thr Asp Leu Ala Thr Arg   400
CGT TGT GAG ATT CCT CTT GAA TAT ATT AAT AAG ATT ATG AAA ACT GAT TTG GCT ACA AGA  1200

Met Phe Glu Ile Ala Asp Ala Asn Glu Glu Asn Ala Leu Lys Lys Ser Asp Gly Thr Arg   420
ATG TTT GAA ATT GCC GAC GCA AAT GAA GAA AAT GCG CTA AAG AAG TCT GAT GGT ACA AGG  1260

Lys Ser Arg Ile Thr Asn Tyr Pro Lys Leu Glu Asp Ala Asn Lys Ala Gly Thr Lys Glu   440
AAA AGC AGA ATT ACT AAT TAC CCT AAA CTG GAA GAT GCC AAC AAA GCC GGT ACA AAA GAA  1320

Gly Tyr Lys Cys Thr Leu Val Leu Thr Glu Gly Asp Ser Ala Leu Ser Leu Ala Val Ala   460
GGC TAT AAA TGT ACT TTA GTT CTG ACA GAA GGG GAT TCC GCC TTG TCA TTA GCT GTT GCA  1380

Gly Leu Ala Val Val Gly Arg Asp Tyr Tyr Gly Cys Tyr Pro Leu Arg Gly Lys Met Leu   480
GGT TTA GCT GTT GTT GGT AGA GAT TAT TAT GGT TGT TAT CCA CTT CGT GGT AAA ATG CTG  1440

Asn Val Arg Glu Ala Ser Ala Asp Gln Ile Leu Lys Asn Ala Glu Ile Gln Ala Ile Lys   500
AAT GTT AGA GAA GCT AGT GCT GAT CAG ATA CTA AAA AAC GCG GAA ATT CAA GCC ATT AAA  1500

Lys Ile Met Gly Leu Gln His Arg Lys Lys Tyr Glu Asp Thr Lys Ser Leu Arg Tyr Gly   520
AAA ATT ATG GGG TTA CAA CAT CGC AAG AAA TAT GAA GAT ACA AAA TCT TTA AGA TAT GGG  1560

His Leu Met Ile Met Thr Asp Gln Asp His Asp Gly Ser His Ile Lys Gly Leu Ile Ile   540
CAT CTT ATG ATC ATG ACC GAT CAA GAT CAT GAT GGT TCG CAT ATT AAA GGT TTA ATT ATA  1620

Asn Phe Leu Glu Ser Ser Phe Leu Gly Leu Leu Asp Ile Gln Gly Phe Leu Leu Glu Phe   560
AAC TTT TTA GAA AGC TCA TTT CTC GGT CTT TTG GAT ATC CAA GGT TTC TTA CTT GAA TTC  1680

Ile Thr Pro Ile Ile Lys Val Ser Ile Thr Lys Pro Thr Lys Asn Thr Ile Ala Phe Tyr   580
ATA ACT CCG ATC ATC AAA GTT TCC ATC ACT AAA CCA ACA AAA AAC ACT ATT GCA TTC TAC  1740

Asn Met Pro Asp Tyr Glu Lys Trp Arg Glu Glu Glu Ser His Lys Phe Thr Trp Lys Gln   600
AAT ATG CCG GAC TAT GAA AAA TGG AGA GAG GAA GAA TCG CAC AAA TTT ACT TGG AAG CAG  1800

Lys Tyr Tyr Lys Gly Leu Gly Thr Ser Leu Ala Gln Glu Val Arg Glu Tyr Phe Ser Asn   620
AAG TAT TAT AAA GGA TTA GGG ACT TCT CTA GCA CAA GAA GTC CGA GAA TAT TTT TCG AAC  1860

Leu Asp Arg His Leu Lys Ile Phe His Ser Leu Gln Gly Asn Asp Lys Asp Tyr Ile Asp   640
TTG GAC AGA CAT TTG AAA ATA TTC CAT TCT TTG CAG GGT AAT GAT AAA GAT TAC ATT GAT  1920

Leu Ala Phe Ser Lys Lys Lys Ala Asp Asp Arg Lys Glu Trp Leu Arg Gln Tyr Glu Pro   660
TTA GCT TTC TCC AAG AAA AAG GCA GAT GAC CGT AAA GAA TGG CTG AGA CAA TAC GAA CCT  1980

Gly Thr Val Leu Asp Pro Thr Leu Lys Glu Ile Pro Ile Ser Asp Phe Ile Asn Lys Glu   680
GGT ACT GTT TTA GAC CCT ACT TTA AAA GAG ATT CCA ATT AGC GAC TTC ATT AAT AAG GAA  2040

Leu Ile Leu Phe Ser Leu Ala Asp Asn Ile Arg Ser Ile Pro Asn Val Leu Asp Gly Phe   700
TTA ATC CTT TTT TCT TTG GCC GAT AAT ATA CGG TCG ATT CCC AAT GTT TTA GAT GGA TTT  2100

Lys Pro Gly Gln Arg Lys Val Leu Tyr Gly Cys Phe Lys Lys Asn Leu Lys Ser Glu Leu   720
AAA CCT GGC CAA AGA AAA GTT CTT TAT GGT TGT TTC AAA AAA AAT TTA AAG TCG GAA CTG  2160

Lys Val Ala Gln Leu Ala Pro Tyr Val Ser Glu Cys Thr Ala Tyr His His Gly Glu Gln   740
AAA GTA GCT CAA CTT GCA CCA TAC GTG AGC GAA TGT ACG GCA TAT CAC CAT GGT GAG CAG  2220

Ser Leu Ala Gln Thr Ile Ile Gly Leu Ala Gln Asn Phe Val Gly Ser Asn Asn Ile Tyr   760
TCA TTG GCA CAA ACT ATT ATT GGG CTA GCC CAA AAC TTT GTT GGG TCC AAC AAT ATT TAC  2280

Leu Leu Leu Pro Asn Gly Ala Phe Gly Thr Arg Ala Thr Gly Gly Lys Asp Ala Ala Ala   780
TTG CTA TTA CCT AAC GGT GCT TTC GGT ACA AGA GCC ACT GGT GGT AAA GAT GCA GCG GCA  2340
```

```
Ala Arg Tyr Ile Tyr Thr Glu Leu Asn Lys Leu Thr Arg Lys Ile Phe His Pro Ala Asp   800
GCG AGA TAT ATC TAC ACA GAA TTG AAC AAA TTA ACT CGT AAG ATA TTT CAC CCT GCT GAT  2400

Asp Pro Leu Tyr Lys Tyr Ile Gln Glu Asp Glu Lys Thr Val Glu Pro Glu Trp Tyr Leu   820
GAT CCA TTA TAC AAA TAT ATA CAA GAA GAT GAG AAA ACA GTG GAG CCA GAG TGG TAT TTA  2460

Pro Ile Leu Pro Met Ile Leu Val Asn Gly Ala Glu Gly Ile Gly Thr Gly Arg Ser Thr   840
CCA ATT CTT CCT ATG ATT CTT GTT AAC GGT GCT GAG GGT ATT GGC ACT GGC AGG AGT ACT  2520

Tyr Ile Pro Pro Phe Asn Pro Leu Glu Ile Ile Lys Asn Ile Arg His Leu Met Asn Asp   860
TAC ATT CCT CCA TTC AAC CCA TTG GAA ATT ATA AAG AAT ATA AGA CAT TTA ATG AAC GAC  2580

Glu Glu Leu Glu Gln Met His Pro Trp Phe Arg Gly Trp Thr Gly Thr Ile Glu Glu Ile   880
GAG GAG CTT GAG CAA ATG CAT CCG TGG TTT AGG GGA TGG ACC GGT ACT ATT GAA GAA ATT  2640

Glu Pro Leu Arg Tyr Arg Met Tyr Gly Arg Ile Glu Gln Ile Gly Asp Asn Val Leu Glu   900
GAG CCT CTG CGT TAC AGA ATG TAC GGT AGG ATT GAA CAA ATT GGA GAT AAC GTC TTA GAA  2700

Ile Thr Glu Leu Pro Ala Arg Thr Trp Thr Ser Thr Ile Lys Glu Tyr Leu Leu Leu Gly   920
ATA ACT GAG TTG CCA GCC AGA ACT TGG ACA TCG ACT ATA AAG GAG TAC TTA CTT TTA GGT  2760

Leu Ser Gly Asn Asp Lys Ile Lys Pro Trp Ile Lys Asp Met Glu Glu Gln His Asp Asp   940
TTA AGC GGT AAC GAT AAA ATA AAA CCC TGG ATC AAA GAT ATG GAG GAG CAG CAC GAT GAT  2820

Asn Ile Lys Phe Ile Ile Thr Leu Ser Pro Glu Glu Met Ala Lys Thr Arg Lys Ile Gly   960
AAC ATC AAA TTC ATA ATC ACG CTA TCA CCT GAG GAA ATG GCT AAA ACA AGG AAA ATA GGT  2880

Phe Tyr Glu Arg Phe Lys Leu Ile Ser Pro Ile Ser Leu Met Asn Met Val Ala Phe Asp   980
TTT TAT GAA AGA TTT AAA CTA ATT TCG CCT ATA AGT TTG ATG AAT ATG GTC GCA TTT GAT  2940

Pro His Gly Lys Ile Lys Lys Tyr Asn Ser Val Asn Glu Ile Leu Ser Glu Phe Tyr Tyr  1000
CCT CAC GGG AAA ATC AAG AAG TAC AAT TCC GTG AAT GAA ATA TTA AGC GAA TTT TAC TAC  3000

Val Arg Leu Glu Tyr Tyr Gln Lys Arg Lys Asp His Met Ser Glu Arg Leu Gln Trp Glu  1020
GTC AGA CTA GAA TAC TAT CAA AAA AGA AAA GAC CAT ATG AGC GAA AGG TTA CAG TGG GAG  3060

Val Glu Lys Tyr Ser Phe Gln Val Lys Phe Ile Lys Met Ile Ile Glu Lys Glu Leu Thr  1040
GTA GAG AAA TAC TCT TTC CAA GTA AAA TTT ATT AAA ATG ATT ATT GAA AAG GAG TTA ACA  3120

Val Thr Asn Lys Pro Arg Asn Ala Ile Ile Gln Glu Leu Glu Asn Leu Gly Phe Pro Arg  1060
GTC ACC AAT AAG CCT AGG AAC GCT ATT ATC CAA GAA CTT GAG AAT TTA GGG TTC CCC AGA  3180

Phe Asn Lys Glu Gly Lys Pro Tyr Tyr Gly Ser Pro Asn Asp Glu Ile Ala Glu Gln Ile  1080
TTT AAT AAG GAA GGT AAA CCA TAT TAT GGA AGT CCT AAC GAT GAG ATA GCT GAA CAA ATT  3240

Asn Asp Val Lys Gly Ala Thr Ser Asp Glu Glu Asp Glu Glu Ser Ser His Glu Asp Thr  1100
AAC GAC GTA AAA GGC GCA ACT TCT GAT GAA GAA GAT GAA GAA AGT TCA CAC GAA GAT ACT  3300

Glu Asn Val Ile Asn Gly Pro Glu Glu Leu Tyr Gly Thr Tyr Glu Tyr Leu Leu Gly Met  1120
GAA AAT GTT ATA AAT GGT CCT GAA GAA CTA TAT GGC ACA TAT GAA TAT TTA TTA GGA ATG  3360

Arg Ile Trp Ser Leu Thr Lys Glu Arg Tyr Gln Lys Leu Leu Lys Gln Lys Gln Glu Lys  1140
AGA ATA TGG TCA TTG ACC AAG GAA AGA TAT CAA AAG CTG TTG AAA CAA AAA CAA GAA AAG  3420

Glu Thr Glu Leu Glu Asn Leu Leu Lys Leu Ser Ala Lys Asp Ile Trp Asn Thr Asp Leu  1160
GAG ACA GAG TTG GAA AAC TTG TTA AAA CTT TCC GCG AAA GAT ATA TGG AAC ACT GAC TTG  3480

Lys Ala Phe Glu Val Gly Tyr Gln Glu Phe Leu Gln Arg Asp Ala Glu Ala Arg Gly Gly  1180
AAG GCT TTT GAG GTG GGA TAT CAA GAA TTT TTG CAA CGA GAT GCA GAA GCT CGC GGT GGT  3540

Asn Val Pro Asn Lys Gly Ser Lys Thr Lys Gly Lys Gly Lys Arg Lys Leu Val Asp Asp  1200
AAT GTT CCC AAT AAA GGG AGC AAA ACG AAA GGT AAA GGA AAA AGA AAG CTT GTT GAC GAC  3600
```

```
Glu Asp Tyr Asp Pro Ser Lys Lys Asn Lys Lys Ser Thr Ala Arg Lys Gly Lys Lys Ile  1220
GAA GAC TAC GAC CCA TCA AAA AAA AAC AAG AAA AGT ACT GCT AGA AAG GGC AAA AAA ATT  3660

Lys Leu Glu Asp Lys Asn Phe Glu Arg Ile Leu Leu Glu Gln Lys Leu Val Thr Lys Ser  1240
AAG TTA GAG GAT AAG AAT TTT GAA AGG ATT TTG TTA GAA CAA AAA CTA GTA ACC AAA AGC  3720

Lys Ala Pro Thr Lys Ile Lys Lys Glu Lys Thr Pro Ser Val Ser Glu Thr Lys Thr Glu  1260
AAG GCG CCT ACA AAG ATT AAA AAA GAG AAA ACG CCT TCT GTT TCA GAA ACA AAA ACA GAA  3780

Glu Glu Glu Asn Ala Pro Ser Ser Thr Ser Ser Ser Ser Ile Phe Asp Ile Lys Lys Glu  1280
GAA GAA GAG AAT GCT CCT TCT TCC ACG AGT TCT TCT TCT ATT TTC GAC ATA AAG AAA GAA  3840

Asp Lys Asp Glu Gly Glu Leu Ser Lys Ile Ser Asn Lys Phe Lys Lys Ile Ser Thr Ile  1300
GAT AAA GAT GAG GGC GAA CTG AGT AAG ATT TCG AAC AAG TTT AAA AAA ATT AGC ACG ATT  3900

Phe Asp Lys Met Gly Ser Thr Ser Ala Thr Ser Lys Glu Asn Thr Pro Glu Gln Asp Asp  1320
TTT GAC AAA ATG GGT TCT ACT TCC GCT ACA TCG AAG GAA AAT ACA CCA GAA CAG GAC GAT  3960

Val Ala Thr Lys Lys Asn Gln Thr Thr Ala Lys Lys Thr Ala Val Lys Pro Lys Leu Ala  1340
GTA GCC ACT AAA AAA AAT CAA ACA ACC GCT AAA AAA ACA GCT GTA AAA CCT AAA TTG GCC  4020

Lys Lys Pro Val Arg Lys Gln Gln Lys Val Val Glu Leu Ser Gly Glu Ser Asp Leu Glu  1360
AAG AAG CCA GTC AGG AAA CAA CAA AAA GTT GTG GAA CTA TCT GGT GAA AGC GAC CTA GAA  4080

Ile Leu Asp Ser Tyr Thr Asp Arg Glu Asp Ser Asn Lys Asp Glu Asp Asp Ala Ile Pro  1380
ATT TTA GAT TCA TAC ACT GAT CGG GAA GAT AGC AAT AAA GAT GAA GAT GAT GCT ATA CCA  4140

Gln Arg Ser Arg Arg Gln Arg Ser Ser Arg Ala Ala Ser Val Pro Lys Lys Ser Tyr Val  1400
CAA CGA TCA AGG AGA CAA AGA TCG TCG AGA GCT GCG TCG GTT CCT AAG AAA TCT TAC GTT  4200

Glu Thr Leu Glu Leu Ser Asp Asp Ser Phe Ile Glu Asp Asp Glu Glu Glu Asn Gln Gly  1420
GAA ACT TTA GAA TTA TCT GAC GAC AGT TTC ATC GAA GAT GAT GAA GAG GAA AAC CAA GGA  4260

Ser Asp Val Ser Phe Asn Glu Glu Asp  1429
TCA GAT GTT TCG TTC AAT GAA GAG GAT  4287
```

Schizosaccharomyces pombe Top2

```
Met Thr Ala Ser Glu Gln Ile Pro Leu Val Thr Asn Asn Gly Asn Gly Asn Ser Asn Val   20
ATG ACA GCT TCT GAA CAG ATA CCA CTA GTT ACC AAC AAC GGA AAT GGA AAT TCT AAT GTA   60

Ser Thr Gln Tyr Gln Arg Leu Thr Pro Arg Glu His Val Leu Arg Arg Pro Asp Thr Tyr   40
TCT ACA CAG TAC CAG CGT CTT ACA CCG AGA GAG CAT GTG CTA AGA CGT CCG GAT ACA TAC  120

Ile Gly Ser Ile Glu Pro Thr Thr Ser Glu Met Trp Val Phe Asp Ser Glu Lys Asn Lys   60
ATT GGC AGT ATT GAG CCA ACG ACT TCT GAA ATG TGG GTC TTT GAC TCT GAG AAG AAC AAG  180

Leu Asp Tyr Lys Ala Val Thr Tyr Val Pro Gly Leu Tyr Lys Ile Phe Asp Glu Ile Ile   80
CTG GAT TAC AAA GCA GTG ACC TAT GTT CCA GGC CTT TAC AAG ATA TTT GAC GAA ATA ATT  240

Val Asn Ala Ala Asp Asn Lys Val Arg Asp Pro Asn Met Asn Thr Leu Lys Val Thr Leu  100
GTC AAT GCT GCT GAC AAC AAA GTT CGC GAC CCA AAC ATG AAT ACC CTA AAG GTG ACT TTG  300

Asp Pro Glu Ala Asn Val Ile Ser Ile Tyr Asn Asn Gly Lys Gly Ile Pro Ile Glu Ile  120
GAT CCT GAA GCA AAT GTC ATA TCA ATT TAT AAC AAT GGC AAG GGC ATT CCT ATA GAG ATT  360

His Asp Lys Glu Lys Ile Tyr Ile Pro Glu Leu Ile Phe Gly Asn Leu Leu Thr Ser Ser  140
CAT GAT AAA GAG AAA ATT TAC ATC CCC GAG CTT ATT TTT GGT AAC TTG CTT ACT TCA AGC  420

Asn Tyr Asp Asp Asn Gln Lys Lys Val Thr Gly Gly Arg Asn Gly Tyr Gly Ala Lys Leu  160
AAT TAC GAC GAT AAT CAA AAG AAG GTA ACA GGA GGA AGA AAT GGT TAC GGC GCA AAA CTT  480
```

```
Cys Asn Ile Phe Ser Thr Glu Phe Val Val Glu Thr Ala Asp Lys Glu Arg Met Lys Lys   180
TGC AAT ATT TTT TCT ACT GAA TTT GTC GTT GAG ACA GCT GAT AAG GAA AGG ATG AAG AAA   540

Tyr Lys Gln Thr Trp Tyr Asp Asn Met Ser Arg Lys Ser Glu Pro Val Ile Thr Ser Leu   200
TAT AAA CAA ACT TGG TAT GAC AAT ATG TCC AGA AAG TCT GAG CCA GTG ATC ACT TCT CTC   600

Lys Lys Pro Asp Glu Tyr Thr Lys Ile Thr Phe Lys Pro Asp Leu Ala Lys Phe Gly Met   220
AAG AAA CCC GAT GAA TAC ACA AAA ATT ACA TTC AAA CCC GAT TTA GCA AAA TTT GGT ATG   660

Asp Lys Ile Asp Asp Asp Met Val Ser Ile Ile Lys Arg Arg Ile Tyr Asp Met Ala Gly   240
GAC AAG ATT GAT GAT GAT ATG GTA TCT ATT ATA AAA CGC CGT ATT TAT GAT ATG GCA GGT   720

Thr Val Arg Glu Thr Lys Val Tyr Leu Asn Asn Glu Arg Ile Ser Ile Ser Gly Phe Lys   260
ACC GTT CGT GAA ACT AAA GTT TAC CTG AAC AAT GAA CGC ATC AGT ATA AGT GGA TTT AAG   780

Lys Tyr Val Glu Met Tyr Leu Ala Ser Asp Thr Lys Pro Asp Glu Glu Pro Pro Arg Val   280
AAA TAT GTG GAG ATG TAT CTT GCT TCT GAC ACT AAA CCT GAT GAA GAG CCT CCT AGG GTA   840

Ile Tyr Glu His Val Asn Asp Arg Trp Asp Val Ala Phe Ala Val Ser Asp Gly Gln Phe   300
ATT TAT GAA CAT GTT AAT GAC CGT TGG GAC GTG GCC TTT GCT GTT TCA GAT GGA CAG TTT   900

Lys Gln Val Ser Phe Val Asn Asn Ile Ser Thr Ile Arg Gly Gly Thr His Val Asn Tyr   320
AAG CAG GTT TCG TTT GTT AAC AAT ATT TCT ACC ATT CGT GGA GGT ACA CAT GTT AAT TAT   960

Val Ala Asn Lys Ile Val Asp Ala Ile Asp Glu Val Val Lys Lys Glu Asn Lys Lys Ala   340
GTT GCC AAT AAA ATT GTC GAT GCT ATA GAT GAA GTC GTC AAG AAG GAG AAT AAA AAG GCA  1020

Pro Val Lys Ala Phe Gln Ile Lys Asn Tyr Val Gln Val Phe Val Asn Cys Gln Ile Glu   360
CCC GTT AAA GCG TTT CAA ATA AAA AAT TAC GTA CAG GTA TTT GTC AAT TGT CAA ATA GAG  1080

Asn Pro Ser Phe Asp Ser Gln Thr Lys Glu Thr Leu Thr Thr Lys Val Ser Ala Phe Gly   380
AAC CCA TCC TTT GAT TCC CAA ACA AAG GAG ACT CTC ACC ACG AAA GTT TCT GCT TTT GGC  1140

Ser Gln Cys Thr Leu Ser Asp Lys Phe Leu Lys Ala Ile Lys Lys Ser Ser Val Val Glu   400
TCT CAA TGC ACA TTA AGT GAT AAA TTT TTA AAA GCT ATC AAA AAA TCT TCT GTA GTG GAG  1200

Glu Val Leu Lys Phe Ala Thr Ala Lys Ala Asp Gln Gln Leu Ser Lys Gly Asp Gly Gly   420
GAG GTT TTA AAA TTT GCT ACC GCT AAG GCA GAT CAG CAA CTC AGT AAG GGC GAT GGG GGT  1260

Leu Arg Ser Arg Ile Thr Gly Leu Thr Lys Leu Glu Asp Ala Asn Lys Ala Gly Thr Lys   440
TTG CGT TCT CGT ATA ACG GGT TTA ACC AAA CTC GAA GAC GCA AAT AAA GCG GGT ACC AAA  1320

Glu Ser His Lys Cys Val Leu Ile Leu Thr Glu Gly Asp Ser Ala Lys Ser Leu Ala Val   460
GAA TCC CAT AAG TGT GTC CTT ATC CTC ACT GAG GGA GAT TCT GCT AAA TCT TTA GCT GTT  1380

Ser Gly Leu Ser Val Val Gly Arg Asp Tyr Tyr Gly Val Phe Pro Leu Arg Gly Lys Leu   480
TCG GGT CTC AGT GTT GTT GGT AGA GAC TAT TAT GGA GTT TTT CCG TTA AGA GGA AAA CTA  1440

Leu Asn Val Arg Glu Ala Ser His Ser Gln Ile Leu Asn Asn Lys Glu Ile Gln Ala Ile   500
CTG AAT GTA CGG GAG GCT TCT CAT TCC CAA ATC CTA AAT AAC AAA GAA ATT CAA GCT ATC  1500

Lys Lys Ile Met Gly Phe Thr His Lys Lys Thr Tyr Thr Asp Val Lys Gly Leu Arg Tyr   520
AAG AAA ATC ATG GGA TTC ACT CAT AAA AAA ACT TAT ACC GAT GTG AAA GGT CTT CGT TAT  1560

Gly His Leu Met Ile Met Thr Asp Gln Asp His Asp Gly Ser His Ile Lys Gly Leu Ile   540
GGT CAT TTA ATG ATT ATG ACA GAT CAA GAT CAT GAT GGA TCA CAT ATT AAA GGT TTA ATT  1620

Ile Asn Tyr Leu Glu Ser Ser Tyr Pro Ser Leu Leu Gln Ile Pro Gly Phe Leu Ile Gln   560
ATT AAT TAT TTA GAA TCC TCG TAT CCT TCT CTT TTG CAA ATT CCT GGA TTC TTG ATC CAA  1680

Phe Ile Thr Pro Ile Ile Lys Cys Thr Arg Gly Asn Gln Val Gln Ala Phe Tyr Thr Leu   580
TTT ATC ACA CCT ATC ATT AAG TGC ACT CGT GGC AAT CAG GTT CAA GCA TTT TAC ACT TTA  1740
```

```
Pro Glu Tyr Glu Tyr Trp Lys Glu Ala Asn Asn Asn Gly Arg Gly Trp Lys Ile Lys Tyr    600
CCC GAG TAC GAA TAC TGG AAG GAA GCT AAT AAC AAT GGA CGT GGT TGG AAA ATT AAA TAC   1800

Tyr Lys Gly Leu Gly Thr Ser Asp His Asp Asp Met Lys Ser Tyr Phe Ser Asp Leu Asp    620
TAC AAG GGT TTG GGA ACA AGT GAC CAT GAC GAT ATG AAA AGT TAT TTT TCA GAC CTT GAT   1860

Arg His Met Lys Tyr Phe His Ala Met Gln Glu Lys Asp Ala Glu Leu Ile Glu Met Ala    640
CGT CAC ATG AAG TAC TTT CAT GCC ATG CAG GAG AAG GAT GCT GAA TTA ATT GAA ATG GCA   1920

Phe Ala Lys Lys Lys Ala Asp Val Arg Lys Glu Trp Leu Arg Thr Tyr Arg Pro Gly Ile    660
TTT GCC AAG AAA AAG GCC GAT GTG CGT AAG GAA TGG CTT AGA ACG TAT CGA CCT GGT ATA   1980

Tyr Met Asp Tyr Thr Gln Pro Gln Ile Pro Ile Asp Asp Phe Ile Asn Arg Glu Leu Ile    680
TAT ATG GAT TAT ACT CAA CCT CAG ATT CCG ATT GAT GAC TTC ATT AAT CGA GAG CTT ATC   2040

Gln Phe Ser Met Ala Asp Asn Ile Arg Ser Ile Pro Ser Val Val Asp Gly Leu Lys Pro    700
CAA TTC AGT ATG GCT GAT AAT ATC CGT TCG ATT CCT TCA GTA GTA GAT GGC TTG AAG CCT   2100

Gly Gln Arg Lys Val Val Tyr Tyr Cys Phe Lys Arg Asn Leu Val His Glu Thr Lys Val    720
GGT CAG CGT AAA GTT GTC TAT TAT TGT TTT AAA CGC AAT CTC GTC CAT GAA ACT AAA GTC   2160

Ser Arg Leu Ala Gly Tyr Val Ala Ser Glu Thr Ala Tyr His His Gly Glu Val Ser Met    740
AGT AGA CTT GCC GGC TAT GTT GCT AGT GAA ACT GCA TAC CAC CAT GGC GAG GTT TCG ATG   2220

Glu Gln Thr Ile Val Asn Leu Ala Gln Asn Phe Val Gly Ser Asn Asn Ile Asn Leu Leu    760
GAG CAA ACT ATA GTT AAT CTT GCC CAA AAT TTT GTT GGC AGT AAC AAC ATT AAT TTG CTG   2280

Met Pro Asn Gly Gln Phe Gly Thr Arg Ser Glu Gly Gly Lys Asn Ala Ser Ala Ser Arg    780
ATG CCT AAT GGA CAA TTC GGT ACA CGA TCT GAG GGT GGA AAA AAT GCA TCG GCT TCA AGG   2340

Tyr Leu Asn Thr Ala Leu Ser Pro Leu Ala Arg Val Leu Phe Asn Ser Asn Asp Asp Gln    800
TAT CTA AAT ACA GCT TTA TCA CCC TTA GCA CGT GTT TTA TTT AAT TCC AAT GAT GAC CAG   2400

Leu Leu Asn Tyr Gln Asn Asp Glu Gly Gln Trp Ile Glu Pro Glu Tyr Tyr Val Pro Ile    820
CTT CTT AAT TAC CAA AAC GAC GAA GGC CAG TGG ATT GAA CCA GAG TAT TAT GTG CCA ATT   2460

Leu Pro Met Val Leu Val Asn Gly Ala Glu Gly Ile Gly Thr Gly Trp Ser Thr Phe Ile    840
CTT CCC ATG GTA CTT GTC AAT GGA GCC GAA GGT ATT GGT ACT GGC TGG TCT ACT TTT ATT   2520

Pro Asn Tyr Asn Pro Lys Asp Ile Thr Ala Asn Leu Arg His Met Leu Asn Gly Glu Pro    860
CCT AAT TAT AAC CCT AAG GAT ATA ACT GCT AAT TTA AGA CAT ATG CTT AAT GGC GAG CCT   2580

Leu Glu Ile Met Thr Pro Trp Tyr Arg Gly Phe Arg Gly Ser Ile Thr Lys Val Ala Pro    880
TTG GAA ATT ATG ACT CCC TGG TAC CGC GGC TTT CGT GGA AGT ATA ACG AAA GTT GCG CCT   2640

Asp Arg Tyr Lys Ile Ser Gly Ile Ile Asn Gln Ile Gly Glu Asn Lys Val Glu Ile Thr    900
GAC AGG TAC AAA ATA TCT GGT ATA ATT AAC CAA ATT GGT GAA AAC AAA GTA GAA ATT ACT   2700

Glu Leu Pro Ile Arg Phe Trp Thr Gln Asp Met Lys Glu Tyr Leu Glu Ala Gly Leu Val    920
GAA TTG CCT ATA CGA TTT TGG ACT CAG GAT ATG AAG GAG TAT CTG GAG GCT GGT CTT GTT   2760

Gly Thr Glu Lys Ile Arg Lys Phe Ile Val Asp Tyr Glu Ser His His Gly Glu Gly Ile    940
GGT ACG GAA AAA ATT CGT AAA TTC ATC GTG GAC TAT GAA AGC CAT CAC GGT GAA GGA ATT   2820

Val His Phe Asn Val Thr Leu Thr Glu Ala Gly Met Lys Glu Ala Leu Asn Glu Ser Leu    960
GTT CAC TTT AAT GTC ACG CTT ACC GAA GCT GGC ATG AAA GAA GCA TTA AAT GAA TCC TTG   2880

Glu Val Lys Phe Lys Leu Ser Arg Thr Gln Ala Thr Ser Asn Met Ile Ala Phe Asp Ala    980
GAA GTT AAG TTT AAA CTG TCG CGT ACT CAA GCA ACG AGT AAT ATG ATT GCT TTT GAT GCA   2940

Ser Gly Arg Ile Lys Lys Tyr Asp Ser Val Glu Asp Ile Leu Thr Glu Phe Tyr Glu Val   1000
TCT GGA CGG ATC AAG AAG TAT GAC AGT GTT GAA GAT ATT TTG ACC GAG TTT TAC GAA GTA   3000
```

```
Arg Leu Arg Thr Tyr Gln Arg Arg Lys Glu His Met Val Asn Glu Leu Glu Lys Arg Phe  1020
CGT TTA AGG ACA TAC CAG AGA CGT AAA GAG CAT ATG GTT AAT GAA TTG GAA AAA AGG TTT  3060

Asp Arg Phe Ser Asn Gln Ala Arg Phe Ile His Met Ile Ile Glu Gly Glu Leu Val Val  1040
GAT AGA TTT TCA AAT CAA GCG CGG TTT ATT CAT ATG ATT ATA GAG GGG GAG CTT GTA GTT  3120

Ser Lys Lys Lys Lys Lys Asp Leu Ile Val Glu Leu Lys Glu Lys Lys Phe Gln Pro Ile  1060
TCA AAA AAG AAG AAA AAG GAT CTC ATT GTG GAG CTG AAG GAA AAG AAG TTT CAA CCT ATC  3180

Ser Lys Pro Lys Lys Gly His Leu Val Asp Leu Glu Val Glu Asn Ala Leu Ala Glu Glu  1080
AGT AAA CCA AAG AAA GGG CAT TTG GTT GAT TTA GAA GTT GAA AAT GCT CTT GCT GAA GAA  3240

Glu Gln Ser Gly Asp Val Ser Gln Asp Glu Asp Ser Asp Ala Tyr Asn Tyr Leu Leu Ser  1100
GAA CAA TCT GGT GAT GTT TCG CAA GAC GAG GAT TCA GAT GCC TAC AAT TAC CTT CTT TCA  3300

Met Pro Leu Trp Ser Leu Thr Tyr Glu Arg Tyr Val Glu Leu Leu Lys Lys Lys Asp Glu  1120
ATG CCT TTA TGG TCT TTG ACC TAT GAA CGG TAT GTG GAA CTT CTC AAG AAA AAG GAC GAA  3360

Val Met Ala Glu Leu Asp Ala Leu Ile Lys Lys Thr Pro Lys Glu Leu Trp Leu His Asp  1140
GTA ATG GCC GAA CTG GAT GCT TTG ATT AAA AAA ACT CCT AAA GAA TTG TGG CTT CAT GAT  3420

Leu Asp Ala Phe Glu His Ala Trp Asn Lys Val Met Asp Asp Ile Gln Arg Glu Met Leu  1160
TTG GAT GCA TTT GAG CAT GCT TGG AAT AAG GTT ATG GAT GAT ATT CAG AGA GAA ATG TTA  3480

Glu Glu Glu Gln Ser Ser Arg Asp Phe Val Asn Arg Thr Lys Lys Lys Pro Arg Gly Lys  1180
GAA GAA GAG CAA TCC TCT AGA GAT TTT GTA AAT CGA ACG AAA AAG AAG CCT CGA GGT AAA  3540

Ser Thr Gly Thr Arg Lys Pro Arg Ala Ile Ala Gly Ser Ser Ser Ser Thr Ala Val Lys  1200
TCC ACT GGT ACT AGG AAA CCC AGG GCT ATT GCC GGA TCC TCC TCA AGT ACT GCT GTT AAA  3600

Lys Glu Ala Ser Ser Glu Ser Lys Pro Ser Thr Thr Asn Arg Lys Gln Gln Thr Leu Leu  1220
AAA GAA GCT AGC AGC GAA TCC AAA CCA TCT ACA ACA AAT CGA AAG CAA CAA ACT TTA CTT  3660

Glu Phe Ala Ala Ser Lys Glu Pro Glu Lys Ser Ser Asp Ile Asn Ile Val Lys Thr Glu  1240
GAG TTC GCA GCG TCG AAA GAA CCT GAG AAA TCA TCT GAC ATA AAC ATT GTC AAG ACA GAG  3720

Asp Asn Ser His Gly Leu Ser Val Glu Glu Asn Arg Ile Ser Lys Ser Pro Gly Leu Asp  1260
GAT AAC TCT CAT GGC TTA TCA GTT GAA GAA AAT CGC ATA TCT AAG AGC CCG GGA TTG GAT  3780

Ser Ser Asp Ser Gly Lys Ser Arg Lys Arg Ser Gln Ser Val Asp Ser Glu Asp Ala Gly  1280
AGC AGT GAT AGT GGA AAA TCA AGA AAA AGA AGT CAA TCT GTT GAC TCT GAA GAC GCA GGT  3840

Ser Lys Lys Pro Val Lys Lys Ile Ala Ala Ser Ala Ser Gly Arg Gly Arg Lys Thr Asn  1300
AGT AAA AAA CCC GTC AAA AAA ATA GCA GCC TCG GCG TCT GGA AGA GGA AGA AAG ACT AAT  3900

Lys Pro Val Ala Thr Thr Ile Phe Ser Ser Asp Asp Glu Asp Asp Leu Leu Pro Ser Ser  1320
AAA CCA GTT GCT ACT ACC ATA TTT TCT TCA GAC GAT GAA GAT GAT TTG TTG CCT AGC AGT  3960

Leu Lys Pro Ser Thr Ile Thr Ser Thr Lys Ala Ser Ala Lys Asn Lys Gly Lys Lys Ala  1340
TTG AAA CCT TCC ACC ATA ACA TCC ACG AAG GCT TCT GCG AAG AAT AAA GGG AAA AAA GCA  4020

Ser Ser Val Lys Lys Gln Ser Pro Glu Asp Asp Asp Asp Asp Phe Ile Ile Pro Gly Ser  1360
AGC TCG GTA AAG AAA CAA TCA CCT GAA GAT GAC GAT GAT GAT TTT ATT ATT CCA GGT AGT  4080

Ser Ser Thr Pro Lys Ala Ser Ser Thr Asn Ala Glu Pro Pro Glu Asp Ser Asp Ser Pro  1380
AGT TCA ACT CCA AAA GCT AGT TCG ACT AAC GCA GAG CCA CCA GAA GAT TCA GAT TCT CCG  4140

Ile Arg Lys Arg Pro Thr Arg Arg Ala Ala Ala Thr Val Lys Thr Pro Ile Tyr Val Asp  1400
ATT AGA AAG AGG CCA ACC AGA AGA GCA GCG GCT ACA GTG AAA ACA CCT ATT TAC GTT GAT  4200

Pro Ser Phe Asp Ser Met Asp Glu Pro Ser Met Gln Asp Asp Ser Phe Ile Val Asp Asn  1420
CCC TCG TTT GAT AGC ATG GAT GAA CCC AGC ATG CAA GAT GAT TCT TTT ATC GTC GAT AAC  4260
```

```
Asp Glu Asp Val Asp Asp Tyr Asp Glu Ser Asp  1431
GAT GAG GAT GTA GAC GAT TAT GAT GAG AGT GAT  4293
```

Trypanosome brucei Top2

```
Met Ala Glu Ala His Lys Tyr Lys Lys Leu Thr Pro Ile Glu His Val Leu Thr Arg Pro    20
ATG GCG GAG GCA CAC AAG TAT AAG AAG CTC ACA CCT ATT GAG CAT GTA CTC ACA CGA CCA    60

Glu Met Tyr Ile Gly Ser Leu Asp Thr Thr Ala Thr Pro Met Phe Ile Tyr Asp Glu Gln    40
GAG ATG TAC ATT GGT AGT CTC GAC ACA ACG GCA ACC CCC ATG TTT ATA TAC GAT GAA CAG   120

Lys Gly His Met Val Trp Glu Thr Val Lys Leu Asn His Gly Leu Leu Lys Ile Val Asp    60
AAG GGT CAC ATG GTG TGG GAG ACG GTG AAA CTG AAT CAC GGT TTG CTG AAA ATC GTG GAT   180

Glu Ile Leu Leu Asn Ala Ser Asp Asn Ile Ser Asn Arg Ser Ala Arg Met Thr Tyr Ile    80
GAA ATT CTG CTA AAT GCA TCT GAT AAC ATC TCC AAC AGA AGT GCG CGC ATG ACG TAT ATC   240

Arg Val Thr Ile Thr Asp Thr Gly Glu Ile Thr Ile Glu Asn Asp Gly Ala Gly Ile Pro   100
CGC GTG ACC ATC ACG GAC ACG GGT GAG ATT ACT ATA GAG AAC GAC GGT GCT GGG ATC CCC   300

Ile Val Arg Ser Arg Glu His Lys Leu Tyr Ile Pro Glu Met Val Phe Gly His Leu Leu   120
ATC GTA CGC AGT CGG GAG CAT AAA TTA TAT ATA CCA GAG ATG GTA TTC GGT CAC CTA CTT   360

Thr Ser Ser Asn Tyr Asp Asp Asp Asn Gln Asn Ala Val Ala Gly Arg His Gly Tyr Gly   140
ACC AGC TCT AAT TAT GAT GAC GAT AAC CAA AAT GCA GTT GCT GGT CGC CAC GGT TAC GGT   420

Ala Lys Leu Thr Asn Ile Leu Ser Leu Ser Phe Ser Val Cys Cys Arg Thr Asn Gly Arg   160
GCA AAG CTA ACC AAC ATT CTT TCC CTG AGC TTT TCC GTC TGC TGC CGC ACA AAT GGG AGG   480

Glu Phe His Met Ser Trp Gln Asp His Met Arg Lys Ala Thr Ala Pro Arg Val Ser Asn   180
GAG TTT CAC ATG AGT TGG CAG GAT CAC ATG AGG AAG GCA ACG GCT CCA CGC GTT TCA AAC   540

Val Gly Thr Lys Glu Lys Asn Val Thr Arg Val Lys Phe Leu Pro Asp Tyr Glu Arg Phe   200
GTC GGC ACA AAA GAG AAA AAT GTC ACG CGT GTG AAG TTT CTC CCC GAC TAC GAG CGA TTT   600

Gly Met Lys Glu Lys Lys Ile Ser Asn Asp Met Lys Arg Val Leu Tyr Lys Arg Ile Met   220
GGC ATG AAG GAG AAG AAA ATT TCA AAC GAC ATG AAG CGT GTG CTC TAC AAG CGC ATT ATG   660

Asp Leu Ser Ala Met Phe Pro Asn Ile Gln Ile Thr Leu Asn Gly Ser Ser Phe Gly Phe   240
GAC CTT TCC GCA ATG TTT CCG AAT ATT CAA ATA ACC CTG AAC GGC TCA TCC TTT GGC TTC   720

Lys Ser Phe Lys Asp Tyr Ala Thr Leu Tyr Ser Ala Met Thr Pro Lys Gly Glu Lys Pro   260
AAG TCC TTT AAG GAC TAT GCG ACT CTG TAC AGT GCC ATG ACC CCA AAG GGA GAG AAA CCA   780

Pro Pro Pro Tyr Val Tyr Glu Ser Lys Ser Gly Cys Val Ala Phe Ile Pro Ser Val Val   280
CCG CCA CCA TAC GTA TAC GAG AGT AAA AGC GGT TGC GTT GCC TTC ATT CCT TCA GTA GTC   840

Pro Gly Val Arg Arg Met Phe Gly Val Val Asn Gly Val Val Thr Tyr Asn Gly Gly Thr   300
CCC GGG GTG CGG CGG ATG TTT GGT GTG GTC AAC GGT GTG GTA ACG TAT AAT GGC GGT ACG   900

His Cys Asn Ala Ala Gln Asp Ile Leu Thr Gly Cys Leu Asp Gly Val Glu Arg Glu Leu   320
CAT TGC AAT GCT GCG CAG GAT ATA TTG ACC GGC TGC CTC GAT GGC GTG GAA CGG GAA TTA   960

Lys Lys Glu Asn Lys Val Met Asp Thr Asn Arg Val Leu Arg His Phe Thr Ile Leu Val   340
AAG AAG GAG AAC AAA GTG ATG GAC ACT AAT CGA GTG CTT CGT CAC TTC ACT ATT CTA GTT  1020

Phe Leu Val Gln Val Gln Pro Lys Phe Asp Ser Gln Asn Lys Ala Arg Leu Val Ser Thr   360
TTC CTC GTG CAG GTG CAG CCA AAG TTT GAT TCT CAG AAT AAA GCT CGA CTT GTT TCT ACC  1080

Pro Thr Met Pro Arg Val Pro Arg Gln Asp Val Met Lys Tyr Leu Leu Arg Met Pro Phe   380
CCC ACG ATG CCC CGT GTT CCT CGG CAA GAT GTG ATG AAA TAT CTT CTG CGC ATG CCT TTT  1140
```

```
Leu Glu Ala His Val Ser Thr Ile Thr Gly Gln Leu Ala Gln Glu Leu Asn Lys Glu Ile   400
CTC GAG GCT CAT GTG AGT ACT ATT ACG GGG CAG TTA GCG CAG GAA CTA AAT AAG GAG ATC  1200

Gly Thr Gly Arg Arg Met Ser Ser Lys Thr Leu Leu Thr Ser Ile Thr Lys Leu Val Asp   420
GGC ACC GGA CGC CGT ATG AGT AGC AAA ACC CTC CTG ACC TCC ATA ACG AAA CTG GTA GAT  1260

Ala Thr Ser Thr Arg Arg Asp Pro Lys His Thr Arg Thr Leu Ile Val Thr Glu Gly Asp   440
GCA ACT TCT ACA CGC CGT GAC CCA AAA CAT ACC CGC ACG TTA ATT GTT ACT GAG GGT GAC  1320

Ser Ala Lys Ala Leu Ala Gln Asn Ser Leu Ser Ser Asp Gln Lys Arg Tyr Thr Gly Val   460
TCC GCA AAG GCT CTC GCG CAG AAC TCT TTA TCG AGT GAC CAA AAG CGA TAT ACA GGC GTA  1380

Phe Pro Leu Arg Gly Lys Leu Leu Asn Val Arg Asn Lys Asn Leu Lys Arg Leu Arg Asn   480
TTT CCG CTT CGG GGT AAG CTG CTA AAC GTG CGT AAC AAG AAT CTT AAG CGA CTG AGG AAC  1440

Cys Lys Glu Leu Gln Glu Leu Phe Cys Ala Leu Gly Leu Glu Leu Asp Lys Asp Tyr Thr   500
TGC AAG GAG TTG CAG GAA CTG TTT TGT GCC CTG GGG CTT GAG CTA GAT AAA GAT TAC ACC  1500

Asp Ala Asp Glu Leu Arg Tyr Gln Arg Ile Leu Ile Met Thr Asp Gln Asp Ala Asp Gly   520
GAC GCC GAT GAA TTA CGG TAC CAA CGC ATA CTT ATC ATG ACA GAT CAG GAC GCA GAT GGC  1560

Ser His Ile Lys Gly Leu Val Ile Asn Ala Phe Glu Ser Leu Trp Pro Ser Leu Leu Val   540
TCA CAC ATT AAG GGT TTG GTT ATC AAC GCG TTC GAG TCT TTG TGG CCC TCG TTG CTG GTA  1620

Arg Asn Pro Gly Phe Ile Ser Ile Phe Ser Thr Pro Ile Val Lys Ala Arg Leu Arg Asp   560
CGC AAT CCT GGG TTC ATC TCT ATA TTC TCC ACA CCC ATC GTA AAG GCA CGA CTG CGC GAC  1680

Lys Ser Val Val Ser Phe Phe Ser Met Lys Glu Phe His Lys Trp Gln Arg Ser Asn Ala   580
AAG TCG GTG GTA TCC TTC TTC AGC ATG AAG GAG TTT CAC AAG TGG CAG CGC TCA AAT GCA  1740

Asn Thr Pro Tyr Thr Cys Lys Tyr Tyr Lys Gly Leu Gly Thr Ser Thr Thr Ala Glu Gly   600
AAT ACA CCA TAC ACA TGT AAG TAC TAT AAG GGT CTC GGT ACT TCT ACC ACT GCT GAG GGA  1800

Lys Glu Tyr Phe Lys Asp Met Glu Lys His Thr Met Arg Leu Leu Val Asp Arg Ser Asp   620
AAA GAG TAC TTC AAG GAT ATG GAG AAA CAC ACA ATG CGC TTA CTC GTG GAC CGC TCC GAT  1860

His Lys Leu Leu Asp Asn Val Phe Asp Ser Gln Glu Val Glu Trp Arg Lys Asp Trp Met   640
CAT AAG CTT CTT GAC AAT GTT TTC GAC TCA CAG GAG GTA GAA TGG CGA AAG GAC TGG ATG  1920

Thr Lys Ala Asn Ala Phe Thr Gly Glu Val Asp Ile Asp Arg Ser Lys Lys Met Leu Thr   660
ACC AAG GCG AAT GCT TTT ACC GGC GAG GTA GAT ATT GAT CGT AGC AAG AAA ATG CTA ACG  1980

Val Thr Asp Phe Val His Lys Glu Met Val His Phe Ala Leu Val Gly Asn Ala Arg Ala   680
GTC ACA GAT TTT GTG CAT AAG GAG ATG GTT CAT TTC GCC CTT GTT GGT AAT GCC CGT GCG  2040

Leu Ala His Ser Val Asp Gly Leu Lys Pro Ser Gln Arg Lys Ile Ile Trp Ala Leu Met   700
CTT GCG CAC TCT GTA GAC GGG CTT AAG CCT TCT CAG CGA AAG ATT ATT TGG GCT CTT ATG  2100

Arg Arg Ser Gly Asn Glu Ala Ala Lys Val Ala Gln Leu Ser Gly Tyr Ile Ser Glu Ala   720
CGG CGG TCC GGT AAT GAG GCG GCG AAG GTG GCA CAA CTA TCA GGT TAC ATA TCA GAA GCT  2160

Ser Ala Phe His His Gly Glu Thr Ser Leu Gln Glu Thr Met Ile Lys Met Ala Gln Ser   740
TCC GCT TTT CAT CAT GGT GAG ACT TCA TTG CAG GAG ACG ATG ATT AAG ATG GCG CAG AGC  2220

Phe Thr Gly Gly Asn Asn Val Asn Leu Leu Val Pro Glu Gly Gln Phe Gly Ser Arg Gln   760
TTC ACT GGT GGT AAC AAC GTC AAC CTT CTC GTC CCT GAG GGT CAG TTC GGT TCT CGT CAG  2280

Gln Leu Gly Asn Asp His Ala Ala Pro Arg Tyr Ile Phe Thr Lys Leu Ser Lys Val Ala   780
CAA CTC GGT AAT GAT CAT GCG GCG CCC CGT TAC ATT TTC ACA AAG CTT TCA AAA GTA GCC  2340

Arg Leu Leu Phe Pro Ser Glu Asp Asp Pro Leu Leu Asp Tyr Ile Val Glu Glu Gly Gln   800
CGC TTG CTT TTC CCT AGT GAA GAT GAC CCA TTG CTA GAC TAC ATT GTG GAA GAG GGT CAG  2400
```

```
Gln Val Glu Pro Asn His Tyr Val Pro Ile Leu Pro Leu Leu Leu Cys Asn Gly Ser Val   820
CAG GTG GAG CCG AAC CAT TAC GTT CCA ATC CTA CCG CTG CTC CTC TGC AAC GGA AGT GTG  2460

Gly Ile Gly Phe Gly Phe Ser Ser Asn Ile Pro Pro Phe His Arg Leu Asp Val Ser Ala   840
GGC ATC GGT TTC GGG TTT TCG TCG AAT ATT CCA CCA TTC CAC CGG TTG GAC GTA TCT GCA  2520

Ala Val Arg Ala Met Ile Ser Gly Glu Arg Ala Lys Ser Val Val Arg Arg Leu Val Pro   860
GCG GTA CGA GCG ATG ATT AGC GGC GAA CGT GCC AAG TCG GTT GTC CGT CGA CTT GTG CCG  2580

Trp Ala Val Gly Phe Gln Gly Glu Ile Arg Arg Gly Pro Glu Gly Glu Phe Ile Ala Val   880
TGG GCT GTA GGC TTT CAG GGT GAG ATA CGT CGT GGC CCC GAA GGA GAG TTT ATT GCT GTG  2640

Gly Thr Tyr Thr Tyr Cys Lys Gly Gly Arg Val His Val Thr Glu Leu Pro Trp Thr Cys   900
GGA ACG TAT ACT TAC TGT AAG GGT GGT CGT GTG CAT GTT ACG GAG CTT CCT TGG ACG TGT  2700

Ser Val Glu Ala Phe Arg Glu His Ile Ser Tyr Leu Ala Thr Lys Asp Ile Val Asn Arg   920
AGC GTT GAA GCA TTC CGT GAG CAC ATT TCT TAC CTC GCC ACA AAG GAT ATT GTT AAC CGC  2760

Ile Ala Asp Tyr Ser Gly Ala Asn His Val Asp Ile Asp Val Glu Val Ala Gln Gly Ala   940
ATT GCC GAC TAT TCC GGC GCC AAT CAC GTT GAC ATT GAT GTG GAA GTT GCT CAG GGT GCG  2820

Val Asn Thr Tyr Ala Glu Cys Glu Ser Glu Leu Gly Leu Thr Gln Arg Ile His Ile Asn   960
GTG AAC ACG TAT GCT GAG TGC GAG TCG GAA CTT GGC CTC ACG CAA CGT ATT CAC ATC AAC  2880

Gly Thr Val Phe Ser Pro Asn Gly Thr Leu Ser Pro Leu Glu Ser Asp Leu Thr Pro Val   980
GGT ACA GTC TTT TCA CCG AAT GGA ACT CTT TCA CCT CTG GAA AGT GAC CTC ACA CCC GTC  2940

Leu Gln Trp His Tyr Asp Arg Arg Leu Asp Leu Tyr Lys Lys Arg Arg Gln Arg Asn Leu  1000
CTC CAG TGG CAC TAC GAC CGC AGA CTT GAT TTA TAT AAA AAG AGG CGA CAA CGT AAT TTG  3000

Thr Leu Leu Glu Gln Glu Leu Ala Arg Glu Lys Ser Thr Leu Lys Phe Val Gln His Phe  1020
ACG CTG TTG GAG CAG GAA TTG GCC AGA GAG AAG TCG ACA CTC AAA TTT GTG CAA CAC TTC  3060

Gly Ala Gly His Ile Asp Phe Ala Asn Ala Thr Glu Ala Thr Leu Glu Lys Val Cys Ser  1040
GGT GCC GGC CAC ATT GAC TTT GCG AAT GCT ACG GAG GCA ACA CTT GAA AAG GTG TGT TCA  3120

Lys Leu Gly Leu Val Arg Val Asp Asp Ser Phe Asp Tyr Ile Leu Arg Lys Pro Ile Thr  1060
AAG TTA GGG TTA GTA CGT GTA GAT GAC TCG TTC GAC TAC ATT TTG CGT AAA CCC ATC ACG  3180

Phe Tyr Thr Lys Thr Ser Phe Glu Asn Leu Leu Lys Lys Ile Ala Glu Thr Glu Arg Arg  1080
TTC TAT ACC AAA ACA AGT TTT GAA AAT CTT CTC AAG AAG ATC GCG GAG ACG GAG CGG CGC  3240

Ile Glu Ala Leu Lys Lys Thr Thr Pro Val Gln Leu Trp Leu Gly Glu Leu Asp Gln Phe  1100
ATT GAA GCT CTC AAG AAG ACA ACC CCT GTG CAG TTG TGG TTG GGC GAA CTT GAT CAA TTT  3300

Asp Arg Phe Phe Gln Asp His Glu Lys Lys Met Val Glu Ala Ile Leu Lys Glu Arg Arg  1120
GAT CGC TTC TTT CAG GAC CAT GAG AAA AAG ATG GTG GAG GCT ATT TTG AAG GAA AGA AGG  3360

Gln Arg Ser Pro Pro Ser Asp Leu Leu Pro Gly Leu Gln Gln Pro Arg Leu Glu Val Glu  1140
CAG CGA TCA CCC CCG AGC GAC CTT CTC CCT GGT CTT CAA CAG CCG CGT CTG GAG GTG GAG  3420

Glu Ala Lys Gly Gly Lys Lys Phe Glu Met Arg Val Gln Val Arg Lys Tyr Val Pro Pro  1160
GAG GCA AAG GGT GGT AAA AAA TTT GAG ATG CGG GTT CAG GTG CGA AAG TAC GTC CCT CCA  3480

Pro Thr Lys Arg Gly Ala Gly Gly Arg Ser Asp Gly Asp Gly Gly Ala Thr Ala Ala Gly  1180
CCA ACC AAG CGG GGA GCA GGG GGC CGT AGT GAT GGT GAC GGA GGC GCA ACT GCG GCC GGT  3540

Ala Ala Ala Ala Val Gly Gly Arg Gly Glu Lys Lys Gly Pro Gly Arg Ala Gly Gly Val  1200
GCT GCT GCA GCC GTG GGG GGA AGA GGT GAA AAG AAG GGC CCG GGA CGA GCC GGT GGG GTT  3600

Arg Arg Met Val Leu Asp Ala Leu Ala Lys Arg Val Thr Arg Leu Leu Pro Arg Leu Leu  1220
CGA CGT ATG GTA CTC GAC GCC CTT GCG AAG CGT GTG ACA CGT TTG CTG CCG CGT CTG CTG  3660
```

```
Phe  1221
TTC  3663
```

Drosophila melanogaster Top2

```
Met Glu Asn Gly Asn Lys Ala Leu Ser Ile Glu Gln Met Tyr Gln Lys Lys Ser Gln Leu    20
ATG GAG AAC GGA AAC AAG GCC CTG TCC ATC GAA CAG ATG TAC CAG AAG AAG TCG CAG CTG    60

Glu His Ile Leu Leu Arg Pro Asp Ser Tyr Ile Gly Ser Val Glu Phe Thr Lys Glu Leu    40
GAG CAC ATC CTG CTG CGA CCC GAC TCG TAC ATC GGA TCC GTG GAG TTC ACC AAG GAG CTG   120

Met Trp Val Tyr Asp Asn Ser Gln Asn Arg Met Val Gln Lys Glu Ile Ser Phe Val Pro    60
ATG TGG GTG TAT GAC AAC TCC CAA AAC CGC ATG GTG CAG AAG GAG ATC TCT TTC GTG CCC   180

Gly Leu Tyr Lys Ile Phe Asp Glu Ile Leu Val Asn Ala Ala Asp Asn Lys Gln Arg Asp    80
GGC CTC TAC AAG ATA TTC GAC GAG ATT CTT GTG AAT GCG GCA GAT AAC AAG CAG CGA GAC   240

Lys Ser Met Asn Thr Ile Lys Ile Asp Ile Asp Pro Glu Arg Asn Met Val Ser Val Trp   100
AAG AGC ATG AAC ACC ATT AAG ATC GAC ATC GAT CCG GAG CGC AAT ATG GTG TCC GTG TGG   300

Asn Asn Gly Gln Gly Ile Pro Val Thr Met His Lys Glu Gln Lys Met Tyr Val Pro Thr   120
AAC AAC GGC CAG GGC ATT CCG GTG ACC ATG CAC AAG GAG CAG AAG ATG TAC GTT CCA ACG   360

Met Ile Phe Gly His Leu Leu Thr Ser Ser Asn Tyr Asn Asp Asp Glu Lys Lys Val Thr   140
ATG ATC TTT GGT CAT CTG CTG ACC TCG TCG AAC TAC AAC GAC GAT GAG AAG AAG GTC ACT   420

Gly Gly Arg Asn Gly Tyr Gly Ala Lys Leu Cys Asn Ile Phe Ser Thr Ser Phe Thr Val   160
GGC GGC AGG AAC GGA TAC GGA GCG AAG CTC TGC AAC ATA TTC TCC ACC AGC TTC ACC GTT   480

Glu Thr Ala Thr Arg Glu Tyr Lys Lys Ser Phe Lys Gln Thr Trp Gly Asn Asn Met Gly   180
GAG ACT GCC ACG AGG GAA TAC AAG AAA AGT TTC AAG CAG ACT TGG GGA AAC AAC ATG GGA   540

Lys Ala Ser Asp Val Gln Ile Lys Asp Phe Asn Gly Thr Asp Tyr Thr Arg Ile Thr Phe   200
AAA GCC TCC GAT GTG CAG ATC AAG GAC TTC AAT GGC ACT GAC TAC ACA CGC ATC ACA TTC   600

Ser Pro Asp Leu Ala Lys Phe Lys Met Asp Arg Leu Asp Glu Asp Ile Val Ala Leu Met   220
AGT CCC GAT CTG GCC AAG TTC AAG ATG GAC CGT CTC GAT GAA GAT ATT GTG GCT CTA ATG   660

Ser Arg Arg Ala Tyr Asp Val Ala Ala Ser Ser Lys Gly Val Ser Val Phe Leu Asn Gly   240
TCG CGT CGC GCC TAC GAT GTG GCC GCC TCA TCC AAG GGA GTA TCC GTC TTC TTA AAC GGC   720

Asn Lys Leu Gly Val Arg Asn Phe Lys Asp Tyr Ile Asp Leu His Ile Lys Asn Thr Asp   260
AAC AAA CTG GGT GTA CGC AAC TTC AAG GAC TAT ATT GAT TTG CAC ATC AAG AAC ACG GAC   780

Asp Asp Ser Gly Pro Pro Ile Lys Ile Val His Glu Val Ala Asn Glu Arg Trp Glu Val   280
GAC GAT TCT GGC CCA CCG ATC AAG ATA GTC CAC GAG GTT GCC AAC GAG CGT TGG GAG GTG   840

Ala Cys Cys Pro Ser Asp Arg Gly Phe Gln Gln Val Ser Phe Val Asn Ser Ile Ala Thr   300
GCT TGC TGT CCT TCA GAT CGA GGC TTC CAA CAG GTC TCG TTT GTC AAC TCG ATA GCT ACC   900

Tyr Lys Gly Gly Arg His Val Asp His Val Val Asp Asn Leu Ile Lys Gln Leu Leu Glu   320
TAC AAG GGC GGT CGG CAT GTG GAC CAT GTG GTA GAC AAT CTC ATT AAG CAG CTG CTC GAG   960

Val Leu Lys Lys Lys Asn Lys Gly Gly Ile Asn Ile Lys Pro Phe Gln Val Arg Asn His   340
GTT CTG AAG AAA AAG AAC AAA GGT GGC ATC AAC ATC AAG CCA TTC CAG GTC CGA AAC CAT  1020

Leu Trp Val Phe Val Asn Cys Leu Ile Glu Asn Pro Thr Phe Asp Ser Gln Thr Lys Glu   360
CTG TGG GTT TTT GTC AAC TGC TTG ATA GAG AAC CCC ACT TTC GAC TCG CAG ACC AAG GAG  1080

Asn Met Thr Leu Gln Gln Lys Gly Phe Gly Ser Lys Cys Thr Leu Ser Glu Lys Phe Ile   380
AAC ATG ACT CTG CAA CAA AAG GGC TTC GGT TCC AAG TGT ACC CTC TCG GAA AAG TTC ATC  1140
```

Asn Asn Met Ser Lys Ser Gly Ile Val Glu Ser Val Leu Ala Trp Ala Lys Phe Lys Ala 400
AAC AAC ATG TCC AAG TCT GGC ATC GTG GAG TCT GTG CTG GCA TGG GCC AAG TTC AAG GCC 1200

Gln Asn Asp Ile Ala Lys Thr Gly Gly Arg Lys Ser Ser Lys Ile Lys Gly Ile Pro Lys 420
CAA AAT GAC ATT GCC AAG ACG GGC GGT CGC AAG TCA AGC AAG ATC AAG GGC ATT CCC AAG 1260

Leu Glu Asp Ala Asn Glu Ala Gly Gly Lys Asn Ser Ile Lys Cys Thr Leu Ile Leu Thr 440
CTG GAG GAC GCT AAC GAG GCG GGT GGA AAG AAC TCG ATT AAA TGC ACC CTC ATC CTC ACC 1320

Glu Gly Asp Ser Ala Lys Ser Leu Ala Val Ser Gly Leu Gly Val Ile Gly Arg Asp Leu 460
GAG GGA GAC TCA GCC AAG TCA TTG GCC GTA TCC GGT TTG GGC GTG ATC GGA CGA GAT CTC 1380

Tyr Gly Val Phe Pro Leu Arg Gly Lys Leu Leu Asn Val Arg Glu Ala Asn Phe Lys Gln 480
TAC GGC GTG TTC CCG CTT AGG GGT AAA CTT CTA AAT GTG CGG GAA GCT AAT TTC AAG CAG 1440

Leu Ser Glu Asn Ala Glu Ile Asn Asn Leu Cys Lys Ile Ile Gly Leu Gln Tyr Lys Lys 500
CTT TCG GAG AAT GCC GAA ATC AAC AAT TTA TGC AAG ATA ATT GGC TTG CAA TAC AAA AAG 1500

Lys Tyr Leu Thr Glu Asp Asp Leu Lys Thr Leu Arg Tyr Gly Lys Val Met Ile Met Thr 520
AAG TAC CTC ACT GAG GAC GAT CTA AAA ACT CTG CGC TAT GGC AAA GTG ATG ATC ATG ACA 1560

Asp Gln Asp Gln Asp Gly Ser His Ile Lys Gly Leu Leu Ile Asn Phe Ile His Thr Asn 540
GAT CAG GAT CAG GAT GGC TCC CAC ATC AAG GGT CTC TTG ATC AAC TTT ATC CAC ACC AAT 1620

Trp Pro Glu Leu Leu Arg Leu Pro Phe Leu Glu Glu Phe Ile Thr Pro Ile Val Lys Ala 560
TGG CCA GAG CTT CTG CGT CTG CCC TTC CTC GAG GAG TTC ATT ACA CCA ATT GTG AAG GCA 1680

Thr Lys Lys Asn Glu Glu Leu Ser Phe Tyr Ser Leu Pro Glu Phe Glu Glu Trp Lys Asn 580
ACC AAG AAA AAC GAG GAG CTG TCA TTC TAC TCG CTA CCC GAA TTC GAG GAG TGG AAA AAT 1740

Asp Thr Ala Asn His His Thr Tyr Asn Ile Lys Tyr Tyr Lys Gly Leu Gly Thr Ser Thr 600
GAT ACA GCT AAT CAC CAT ACG TAC AAT ATA AAG TAC TAT AAG GGT TTG GGT ACT TCG ACC 1800

Ser Lys Glu Ala Lys Glu Tyr Phe Gln Asp Met Asp Arg His Arg Ile Leu Phe Lys Tyr 620
TCC AAG GAG GCG AAA GAG TAT TTT CAA GAC ATG GAT CGC CAT CGC ATC TTA TTT AAG TAC 1860

Asp Gly Ser Val Asp Asp Glu Ser Ile Val Met Ala Phe Ser Lys Lys His Ile Glu Ser 640
GAT GGA TCG GTG GAT GAT GAG AGC ATT GTT ATG GCC TTC TCC AAG AAG CAC ATC GAG TCG 1920

Arg Lys Val Trp Leu Thr Asn His Met Asp Glu Val Lys Arg Arg Lys Glu Leu Gly Leu 660
CGA AAG GTA TGG CTT ACC AAC CAC ATG GAC GAG GTG AAG CGT CGC AAG GAG CTC GGT TTG 1980

Pro Glu Arg Tyr Leu Tyr Thr Lys Gly Thr Lys Ser Ile Thr Tyr Ala Asp Phe Ile Asn 680
CCA GAG CGA TAT CTC TAC ACC AAG GGC ACC AAG AGC ATC ACC TAT GCA GAC TTT ATC AAT 2040

Leu Glu Leu Val Leu Phe Ser Asn Ala Asp Asn Glu Arg Ser Ile Pro Ser Leu Val Asp 700
CTG GAG TTG GTG CTG TTC TCG AAT GCT GAC AAT GAG CGC TCC ATT CCT AGT CTG GTG GAT 2100

Gly Leu Lys Pro Gly Gln Arg Lys Val Met Phe Thr Cys Phe Lys Arg Asn Asp Lys Arg 720
GGC TTG AAG CCG GGT CAG CGG AAG GTG ATG TTC ACT TGC TTC AAG AGG AAT GAC AAG CGT 2160

Glu Val Lys Val Ala Gln Leu Ser Gly Ser Val Ala Glu Met Ser Ala Tyr His His Gly 740
GAG GTG AAG GTG GCC CAG CTA TCC GGA TCA GTG GCG GAG ATG TCA GCC TAT CAC CAC GGA 2220

Glu Val Ser Leu Gln Met Thr Ile Val Asn Leu Ala Gln Asn Phe Val Gly Ala Asn Asn 760
GAG GTA TCC CTA CAG ATG ACA ATC GTG AAT CTG GCT CAG AAT TTT GTG GGT GCC AAC AAC 2280

Ile Asn Leu Leu Glu Pro Arg Gly Gln Phe Gly Thr Arg Leu Ser Gly Gly Lys Asp Cys 780
ATT AAT CTG CTT GAG CCA CGC GGT CAA TTT GGT ACC CGC TTG TCG GGA GGC AAA GAT TGC 2340

Ala Ser Ala Arg Tyr Ile Phe Thr Ile Met Ser Pro Leu Thr Arg Leu Ile Tyr His Pro 800
GCC AGC GCT CGT TAC ATT TTC ACT ATA ATG TCT CCC TTG ACG CGA CTC ATC TAC CAT CCT 2400

```
Leu Asp Asp Pro Leu Leu Asp Tyr Gln Val Asp Asp Gly Gln Lys Ile Glu Pro Leu Trp   820
TTG GAC GAT CCA CTT TTG GAT TAC CAG GTG GAC GAT GGC CAA AAG ATC GAG CCA CTA TGG  2460

Tyr Leu Pro Ile Ile Pro Met Val Leu Val Asn Gly Ala Glu Gly Ile Gly Thr Gly Trp   840
TAT CTG CCC ATC ATA CCG ATG GTA TTG GTA AAC GGA GCC GAG GGC ATA GGA ACT GGA TGG  2520

Ser Thr Lys Ile Ser Asn Tyr Asn Pro Arg Glu Ile Met Lys Asn Leu Arg Lys Met Ile   860
TCC ACG AAG ATA TCC AAC TAC AAT CCT CGT GAG ATT ATG AAA AAT CTA AGG AAG ATG ATA  2580

Asn Gly Gln Glu Pro Ser Val Met His Pro Trp Tyr Lys Asn Phe Leu Gly Arg Met Glu   880
AAC GGA CAA GAG CCA AGT GTG ATG CAT CCG TGG TAC AAG AAC TTT TTA GGA CGC ATG GAG  2640

Tyr Val Ser Asp Gly Arg Tyr Ile Gln Thr Gly Asn Ile Gln Ile Leu Ser Gly Asn Arg   900
TAT GTT TCG GAT GGT CGT TAT ATT CAG ACT GGT AAC ATT CAA ATT TTG TCC GGA AAC CGT  2700

Leu Glu Ile Ser Glu Leu Pro Val Gly Val Trp Thr Gln Asn Tyr Lys Glu Asn Val Leu   920
TTA GAA ATC AGT GAA CTC CCT GTG GGC GTA TGG ACG CAA AAC TAC AAG GAA AAT GTC CTG  2760

Glu Pro Leu Ser Asn Gly Thr Glu Lys Val Lys Gly Ile Ile Ser Glu Tyr Arg Glu Tyr   940
GAG CCT TTA TCA AAC GGC ACC GAA AAG GTT AAG GGT ATT ATT TCC GAG TAC AGG GAG TAT  2820

His Thr Asp Thr Thr Val Arg Phe Val Ile Ser Phe Ala Pro Gly Glu Phe Glu Arg Ile   960
CAT ACA GAC ACC ACC GTT CGC TTT GTG ATC AGT TTC GCA CCT GGA GAA TTT GAG CGC ATT  2880

His Ala Glu Glu Gly Gly Phe Tyr Arg Val Phe Lys Leu Thr Thr Thr Leu Ser Thr Asn   980
CAT GCA GAA GAA GGT GGT TTC TAC CGA GTG TTC AAA CTT ACC ACG ACG CTG TCC ACC AAC  2940

Gln Met His Ala Phe Asp Gln Asn Asn Cys Leu Arg Arg Phe Pro Thr Ala Ile Asp Ile  1000
CAG ATG CAT GCG TTC GAC CAG AAC AAC TGT CTG CGA CGC TTC CCC ACC GCG ATC GAT ATC  3000

Leu Lys Glu Tyr Tyr Lys Leu Arg Arg Glu Tyr Tyr Ala Arg Arg Arg Asp Phe Leu Val  1020
CTT AAA GAG TAT TAC AAA CTG CGA CGA GAG TAC TAC GCC CGC CGG AGG GAC TTT TTG GTC  3060

Gly Gln Leu Thr Ala Gln Ala Asp Arg Leu Ser Asp Gln Ala Arg Phe Ile Leu Glu Lys  1040
GGC CAG CTC ACG GCG CAG GCT GAT CGC CTC AGT GAT CAG GCG CGC TTT ATT CTC GAA AAG  3120

Cys Glu Lys Lys Leu Val Val Glu Asn Lys Gln Arg Lys Ala Met Cys Asp Glu Leu Leu  1060
TGT GAG AAG AAG CTG GTG GTG GAG AAC AAG CAG CGC AAG GCC ATG TGT GAT GAG TTG TTG  3180

Lys Arg Gly Tyr Arg Pro Asp Pro Val Lys Glu Trp Gln Arg Arg Ile Lys Met Glu Asp  1080
AAG CGT GGA TAT CGT CCC GAT CCC GTC AAG GAG TGG CAG CGC CGC ATT AAA ATG GAG GAT  3240

Ala Glu Gln Ala Asp Glu Glu Asp Glu Glu Glu Glu Glu Ala Ala Pro Ser Val Ser Ser  1100
GCC GAG CAA GCT GAC GAA GAG GAC GAA GAA GAA GAG GAG GCA GCT CCC AGT GTC AGC TCA  3300

Lys Ala Lys Lys Glu Lys Glu Val Asp Pro Glu Lys Ala Phe Lys Lys Leu Thr Asp Val  1120
AAG GCT AAG AAG GAA AAG GAA GTC GAT CCG GAA AAG GCC TTT AAG AAA CTA ACT GAT GTC  3360

Lys Lys Phe Asp Tyr Leu Leu Gly Met Ser Met Trp Met Leu Thr Glu Glu Lys Lys Asn  1140
AAA AAG TTT GAC TAC CTT CTG GGT ATG TCC ATG TGG ATG TTG ACA GAG GAG AAG AAA AAC  3420

Glu Leu Leu Lys Gln Arg Asp Thr Lys Leu Ser Glu Leu Glu Ser Leu Arg Lys Lys Thr  1160
GAG CTA CTC AAA CAG CGC GAC ACT AAA CTG TCC GAG TTG GAA AGT CTG CGC AAG AAG ACG  3480

Pro Glu Met Leu Trp Leu Asp Asp Leu Asp Ala Leu Glu Ser Lys Leu Asn Glu Val Glu  1180
CCG GAG ATG CTT TGG CTG GAT GAT TTG GAT GCC CTC GAA TCG AAG TTA AAT GAA GTA GAG  3540

Glu Lys Glu Arg Ala Glu Glu Gln Gly Ile Asn Leu Lys Thr Ala Lys Ala Leu Lys Gly  1200
GAG AAG GAG CGC GCA GAA GAG CAG GGG ATC AAT CTC AAG ACC GCG AAG GCT TTG AAG GGC  3600

Gln Lys Ser Ala Ser Ala Lys Gly Arg Lys Val Lys Ser Met Gly Gly Gly Ala Gly Ala  1220
CAG AAG TCG GCC TCA GCT AAG GGC AGA AAG GTT AAG TCA ATG GGA GGC GGA GCG GGT GCC  3660
```

```
Gly Asp Val Phe Pro Asp Pro Asp Gly Glu Pro Val Glu Phe Lys Ile Thr Glu Glu Ile  1240
GGA GAC GTA TTC CCA GAT CCC GAT GGA GAA CCC GTC GAG TTT AAG ATC ACC GAA GAA ATC  3720

Ile Lys Lys Met Ala Ala Ala Ala Lys Val Ala Gln Ala Ala Lys Glu Pro Lys Lys Pro  1260
ATT AAG AAA ATG GCA GCG GCT GCC AAG GTA GCT CAG GCG GCT AAG GAG CCG AAG AAA CCC  3780

Lys Glu Pro Lys Glu Pro Lys Val Lys Lys Glu Pro Lys Gly Lys Gln Ile Lys Ala Glu  1280
AAA GAG CCC AAG GAG CCA AAA GTG AAG AAG GAA CCG AAG GGC AAG CAG ATT AAA GCC GAA  3840

Pro Asp Ala Ser Gly Asp Glu Val Asp Glu Phe Asp Ala Met Val Glu Gly Gly Ser Lys  1300
CCT GAT GCC AGC GGC GAT GAG GTG GAC GAA TTT GAT GCA ATG GTC GAG GGC GGC TCT AAG  3900

Thr Ser Pro Lys Ala Lys Lys Ala Val Val Lys Lys Glu Pro Gly Glu Lys Lys Pro Arg  1320
ACC TCA CCC AAG GCC AAG AAG GCG GTT GTC AAG AAG GAG CCG GGA GAA AAA AAG CCG CGC  3960

Gln Lys Lys Glu Asn Gly Gly Gly Leu Lys Gln Ser Lys Ile Asp Phe Ser Lys Ala Lys  1340
CAA AAG AAG GAG AAC GGC GGC GGG CTC AAG CAG AGC AAG ATT GAC TTC AGC AAG GCC AAG  4020

Ala Lys Lys Ser Asp Asp Asp Val Glu Glu Val Thr Pro Arg Ala Glu Arg Pro Gly Arg  1360
GCG AAA AAG AGC GAC GAT GAC GTG GAG GAA GTA ACT CCT CGT GCA GAG CGT CCT GGA CGT  4080

Arg Gln Ala Ser Lys Lys Ile Asp Tyr Ser Ser Leu Phe Ser Asp Glu Glu Glu Asp Gly  1380
CGA CAG GCC AGT AAG AAG ATA GAC TAC AGC TCG CTC TTC TCG GAT GAG GAG GAA GAC GGT  4140

Gly Asn Val Gly Ser Asp Asp Asp Gly Asn Ala Ser Asp Asp Asp Ser Pro Lys Arg Pro  1400
GGC AAC GTT GGC TCT GAT GAT GAT GGC AAT GCC AGC GAT GAT GAT TCG CCC AAG CGT CCA  4200

Ala Lys Arg Gly Arg Glu Asp Glu Ser Ser Gly Gly Ala Lys Lys Lys Ala Pro Pro Lys  1420
GCT AAG CGT GGA CGA GAG GAT GAG AGC AGT GGA GGA GCC AAA AAG AAA GCT CCG CCC AAG  4260

Lys Arg Arg Ala Val Ile Glu Ser Asp Asp Asp Asp Ile Glu Ile Asp Glu Asp Asp Asp  1440
AAG AGG CGA GCT GTA ATT GAA AGC GAT GAT GAT GAT ATC GAG ATT GAT GAA GAT GAT GAT  4320

Asp Asp Ser Asp Phe Asn Cys  1447
GAT GAT TCT GAT TTT AAT TGT  4341
```

Homo sapiens Top2

```
Met Glu Val Ser Pro Leu Gln Pro Val Asn Glu Asn Met Gln Val Asn Lys Ile Lys Lys    20
ATG GAA GTG TCA CCA TTG CAG CCT GTA AAT GAA AAT ATG CAA GTC AAC AAA ATA AAG AAA    60

Asn Glu Asp Ala Lys Lys Arg Leu Ser Val Glu Arg Ile Tyr Gln Lys Lys Thr Gln Leu    40
AAT GAA GAT GCT AAG AAA AGA CTG TCT GTT GAA AGA ATC TAT CAA AAG AAA ACA CAA TTG   120

Glu His Ile Leu Leu Arg Pro Asp Thr Tyr Ile Gly Ser Val Glu Leu Val Thr Gln Gln    60
GAA CAT ATT TTG CTC CGC CCA GAC ACC TAC ATT GGT TCT GTG GAA TTA GTG ACC CAG CAA   180

Met Trp Val Tyr Asp Glu Asp Val Gly Ile Asn Tyr Arg Glu Val Thr Phe Val Pro Gly    80
ATG TGG GTT TAC GAT GAA GAT GTT GGC ATT AAC TAT AGG GAA GTC ACT TTT GTT CCT GGT   240

Leu Tyr Lys Ile Phe Asp Glu Ile Leu Val Asn Ala Ala Asp Asn Lys Gln Arg Asp Pro   100
TTG TAC AAA ATC TTT GAT GAG ATT CTA GTT AAT GCT GCG GAC AAC AAA CAA AGG GAC CCA   300

Lys Met Ser Cys Ile Arg Val Thr Met Ile Arg Lys Gln Leu Ile Ser Ile Trp Asn Asn   120
AAA ATG TCT TGT ATT AGA GTC ACA ATG ATC CGG AAA CAA TTA ATT AGT ATA TGG AAT AAT   360

Gly Lys Gly Ile Pro Val Val Glu His Lys Val Glu Lys Met Tyr Val Pro Ala Leu Ile   140
GGA AAA GGT ATT CCT GTT GTT GAA CAC AAA GTT GAA AAG ATG TAT GTC CCA GCT CTC ATA   420

Phe Gly Gln Leu Leu Thr Ser Ser Asn Tyr Asp Asp Asp Glu Lys Lys Val Thr Gly Gly   160
TTT GGA CAG CTC CTA ACT TCT AGT AAC TAT GAT GAT GAT GAA AAG AAA GTG ACA GGT GGT   480
```

```
Arg Asn Gly Tyr Gly Ala Lys Leu Cys Asn Ile Phe Ser Thr Lys Phe Thr Val Glu Thr   180
CGA AAT GGC TAT GGA GCC AAA TTG TGT AAC ATA TTC AGT ACC AAA TTT ACT GTG GAA ACA   540

Ala Ser Arg Glu Tyr Lys Lys Met Phe Lys Gln Thr Trp Met Asp Asn Met Gly Arg Ala   200
GCC AGT AGA GAA TAC AAG AAA ATG TTC AAA CAG ACA TGG ATG GAT AAT ATG GGA AGA GCT   600

Gly Glu Met Glu Leu Lys Pro Phe Asn Gly Glu Asp Tyr Thr Cys Ile Thr Phe Gln Pro   220
GGT GAG ATG GAA CTC AAG CCC TTC AAT GGA GAA GAT TAT ACA TGT ATC ACC TTT CAG CCT   660

Asp Leu Ser Lys Phe Lys Met Gln Ser Leu Asp Lys Asp Ile Val Ala Leu Met Val Arg   240
GAT TTG TCT AAG TTT AAA ATG CAA AGC CTG GAC AAA GAT ATT GTT GCA CTA ATG GTC AGA   720

Arg Ala Tyr Asp Ile Ala Gly Ser Thr Lys Asp Val Lys Val Phe Leu Asn Gly Asn Lys   260
AGA GCA TAT GAT ATT GCT GGA TCC ACC AAA GAT GTC AAA GTC TTT CTT AAT GGA AAT AAA   780

Leu Pro Val Lys Gly Phe Arg Ser Tyr Val Asp Met Tyr Leu Lys Asp Lys Leu Asp Glu   280
CTG CCA GTA AAA GGA TTT CGT AGT TAT GTG GAC ATG TAT TTG AAG GAC AAG TTG GAT GAA   840

Thr Gly Asn Ser Leu Lys Val Ile His Glu Gln Val Asn His Arg Trp Glu Val Cys Leu   300
ACT GGT AAC TCC TTG AAA GTA ATA CAT GAA CAA GTA AAC CAC AGG TGG GAA GTG TGT TTA   900

Thr Met Ser Glu Lys Gly Phe Gln Gln Ile Ser Phe Val Asn Ser Ile Ala Thr Ser Lys   320
ACT ATG AGT GAA AAA GGC TTT CAG CAA ATT AGC TTT GTC AAC AGC ATT GCT ACA TCC AAG   960

Gly Gly Arg His Val Asp Tyr Val Ala Asp Gln Ile Val Thr Lys Leu Val Asp Val Val   340
GGT GGC AGA CAT GTT GAT TAT GTA GCT GAT CAG ATT GTG ACT AAA CTT GTT GAT GTT GTG  1020

Lys Lys Lys Asn Lys Gly Gly Val Ala Val Lys Ala His Gln Val Lys Asn His Met Trp   360
AAG AAG AAG AAC AAG GGT GGT GTT GCA GTA AAA GCA CAT CAG GTG AAA AAT CAC ATG TGG  1080

Ile Phe Val Asn Ala Leu Ile Glu Asn Pro Thr Phe Asp Ser Gln Thr Lys Glu Asn Met   380
ATT TTT GTA AAT GCC TTA ATT GAA AAC CCA ACC TTT GAC TCT CAG ACA AAA GAA AAC ATG  1140

Thr Leu Gln Pro Lys Ser Phe Gly Ser Thr Cys Gln Leu Ser Glu Lys Phe Ile Lys Ala   400
ACT TTA CAA CCC AAG AGC TTT GGA TCA ACA TGC CAA TTG AGT GAA AAA TTT ATC AAA GCT  1200

Ala Ile Gly Cys Gly Ile Val Glu Ser Ile Leu Asn Trp Val Lys Phe Lys Ala Gln Val   420
GCC ATT GGC TGT GGT ATT GTA GAA AGC ATA CTA AAC TGG GTG AAG TTT AAG GCC CAA GTC  1260

Gln Leu Asn Lys Lys Cys Ser Ala Val Lys His Asn Arg Ile Lys Gly Ile Pro Lys Leu   440
CAG TTA AAC AAG AAG TGT TCA GCT GTA AAA CAT AAT AGA ATC AAG GGA ATT CCC AAA CTC  1320

Asp Asp Ala Asn Asp Ala Gly Gly Arg Asn Ser Thr Glu Cys Thr Leu Ile Leu Thr Glu   460
GAT GAT GCC AAT GAT GCA GGG GGC CGA AAC TCC ACT GAG TGT ACG CTT ATC CTG ACT GAG  1380

Gly Asp Ser Ala Lys Thr Leu Ala Val Ser Gly Leu Gly Val Val Gly Arg Asp Lys Tyr   480
GGA GAT TCA GCC AAA ACT TTG GCT GTT TCA GGC CTT GGT GTG GTT GGG AGA GAC AAA TAT  1440

Gly Val Phe Pro Leu Arg Gly Lys Ile Leu Asn Val Arg Glu Ala Ser His Lys Gln Ile   500
GGG GTT TTC CCT CTT AGA GGA AAA ATA CTC AAT GTT CGA GAA GCT TCT CAT AAG CAG ATC  1500

Met Glu Asn Ala Glu Ile Asn Asn Ile Ile Lys Ile Val Gly Leu Gln Tyr Lys Lys Asn   520
ATG GAA AAT GCT GAG ATT AAC AAT ATC ATC AAG ATT GTG GGT CTT CAG TAC AAG AAA AAC  1560

Tyr Glu Asp Glu Asp Ser Leu Lys Thr Leu Arg Tyr Gly Lys Ile Met Ile Met Thr Asp   540
TAT GAA GAT GAA GAT TCA TTG AAG ACG CTT CGT TAT GGG AAG ATA ATG ATT ATG ACA GAT  1620

Gln Asp Gln Asp Gly Ser His Ile Lys Gly Leu Leu Ile Asn Phe Ile His His Asn Trp   560
CAG GAC CAA GAT GGT TCC CAC ATC AAA GGC TTG CTG ATT AAT TTT ATC CAT CAC AAC TGG  1680

Pro Ser Leu Leu Arg His Arg Phe Leu Glu Glu Phe Ile Thr Pro Ile Val Lys Val Ser   580
CCC TCT CTT CTG CGA CAT CGT TTT CTG GAG GAA TTT ATC ACT CCC ATT GTA AAG GTA TCT  1740
```

```
Lys Asn Lys Gln Glu Met Ala Phe Tyr Ser Leu Pro Glu Phe Glu Glu Trp Lys Ser Ser   600
AAA AAC AAG CAA GAA ATG GCA TTT TAC AGC CTT CCT GAA TTT GAA GAG TGG AAG AGT TCT  1800

Thr Pro Asn His Lys Lys Trp Lys Val Lys Tyr Tyr Lys Gly Leu Gly Thr Ser Thr Ser   620
ACT CCA AAT CAT AAA AAA TGG AAA GTC AAA TAT TAC AAA GGT TTG GGC ACC AGC ACA TCA  1860

Lys Glu Ala Lys Glu Tyr Phe Ala Asp Met Lys Arg His Arg Ile Gln Phe Lys Tyr Ser   640
AAG GAA GCT AAA GAA TAC TTT GCA GAT ATG AAA AGA CAT CGT ATC CAG TTC AAA TAT TCT  1920

Gly Pro Glu Asp Asp Ala Ala Ile Ser Leu Ala Phe Ser Lys Lys Gln Ile Asp Asp Arg   660
GGT CCT GAA GAT GAT GCT GCT ATC AGC CTG GCC TTT AGC AAA AAA CAG ATA GAT GAT CGA  1980

Lys Glu Trp Leu Thr Asn Phe Met Glu Asp Arg Arg Gln Arg Lys Leu Leu Gly Leu Pro   680
AAG GAA TGG TTA ACT AAT TTC ATG GAG GAT AGA AGA CAA CGA AAG TTA CTT GGG CTT CCT  2040

Glu Asp Tyr Leu Tyr Gly Gln Thr Thr Thr Tyr Leu Thr Tyr Asn Asp Phe Ile Asn Lys   700
GAG GAT TAC TTG TAT GGA CAA ACT ACC ACA TAT CTG ACA TAT AAT GAC TTC ATC AAC AAG  2100

Glu Leu Ile Leu Phe Ser Asn Ser Asp Asn Glu Arg Ser Ile Pro Ser Met Val Asp Gly   720
GAA CTT ATC TTG TTC TCA AAT TCT GAT AAC GAG AGA TCT ATC CCT TCT ATG GTG GAT GGT  2160

Leu Lys Pro Gly Gln Arg Lys Val Leu Phe Thr Cys Phe Lys Arg Asn Asp Lys Arg Glu   740
TTG AAA CCA GGT CAG AGA AAG GTT TTG TTT ACT TGC TTC AAA CGG AAT GAC AAG CGA GAA  2220

Val Lys Val Ala Gln Leu Ala Gly Ser Val Ala Glu Met Ser Ser Tyr His His Gly Glu   760
GTA AAG GTT GCC CAA TTA GCT GGA TCA GTG GCT GAA ATG TCT TCT TAT CAT CAT GGT GAG  2280

Met Ser Leu Met Met Thr Ile Ile Asn Leu Ala Gln Asn Phe Val Gly Ser Asn Asn Leu   780
ATG TCA CTA ATG ATG ACC ATT ATC AAT TTG GCT CAG AAT TTT GTG GGT AGC AAT AAT CTA  2340

Asn Leu Leu Gln Pro Ile Gly Gln Phe Gly Thr Arg Leu His Gly Gly Lys Asp Ser Ala   800
AAC CTC TTG CAG CCC ATT GGT CAG TTT GGT ACC AGG CTA CAT GGT GGC AAG GAT TCT GCT  2400

Ser Pro Arg Tyr Ile Phe Thr Met Leu Ser Ser Leu Ala Arg Leu Leu Phe Pro Pro Lys   820
AGT CCA CGA TAC ATC TTT ACA ATG CTC AGC TCT TTG GCT CGA TTG TTA TTT CCA CCA AAA  2460

Asp Asp His Thr Leu Lys Phe Leu Tyr Asp Asp Asn Gln Arg Val Glu Pro Glu Trp Tyr   840
GAT GAT CAC ACG TTG AAG TTT TTA TAT GAT GAC AAC CAG CGT GTT GAG CCT GAA TGG TAC  2520

Ile Pro Ile Ile Pro Met Val Leu Ile Asn Gly Ala Glu Gly Ile Gly Thr Gly Trp Ser   860
ATT CCT ATT ATT CCC ATG GTG CTG ATA AAT GGT GCT GAA GGA ATC GGT ACT GGG TGG TCC  2580

Cys Lys Ile Pro Asn Phe Asp Val Arg Glu Ile Val Asn Asn Ile Arg Arg Leu Met Asp   880
TGC AAA ATC CCC AAC TTT GAT GTG CGT GAA ATT GTA AAT AAC ATC AGG CGT TTG ATG GAT  2640

Gly Glu Glu Pro Leu Pro Met Leu Pro Ser Tyr Lys Asn Phe Lys Gly Thr Ile Glu Glu   900
GGA GAA GAA CCT TTG CCA ATG CTT CCA AGT TAC AAG AAC TTC AAG GGT ACT ATT GAA GAA  2700

Leu Ala Pro Asn Gln Tyr Val Ile Ser Gly Glu Val Ala Ile Leu Asn Ser Thr Thr Ile   920
CTG GCT CCA AAT CAA TAT GTG ATT AGT GGT GAA GTA GCT ATT CTT AAT TCT ACA ACC ATT  2760

Glu Ile Ser Glu Leu Pro Val Arg Thr Trp Thr Gln Thr Tyr Lys Glu Gln Val Leu Glu   940
GAA ATC TCA GAG CTT CCC GTC AGA ACA TGG ACC CAG ACA TAC AAA GAA CAA GTT CTA GAA  2820

Pro Met Leu Asn Gly Thr Glu Lys Thr Pro Pro Leu Ile Thr Asp Tyr Arg Glu Tyr His   960
CCC ATG TTG AAT GGC ACC GAG AAG ACA CCT CCT CTC ATA ACA GAC TAT AGG GAA TAC CAT  2880

Thr Asp Thr Thr Val Lys Phe Val Val Lys Met Thr Glu Glu Lys Leu Ala Glu Ala Glu   980
ACA GAT ACC ACT GTG AAA TTT GTT GTG AAG ATG ACT GAA GAA AAA CTG GCA GAG GCA GAG  2940

Arg Val Gly Leu His Lys Val Phe Lys Leu Gln Thr Ser Leu Thr Cys Asn Ser Met Val  1000
AGA GTT GGA CTA CAC AAA GTC TTC AAA CTC CAA ACT AGT CTC ACA TGC AAC TCT ATG GTG  3000
```

Leu Phe Asp His Val Gly Cys Leu Lys Lys Tyr Asp Thr Val Leu Asp Ile Leu Arg Asp 1020
CTT TTT GAC CAC GTA GGC TGT TTA AAG AAA TAT GAC ACG GTG TTG GAT ATT CTA AGA GAC 3060

Leu Phe Glu Leu Arg Leu Lys Tyr Tyr Gly Leu Arg Lys Glu Trp Leu Leu Gly Met Leu 1040
CTT TTT GAA CTC AGA CTT AAA TAT TAT GGA TTA AGA AAA GAA TGG CTC CTA GGA ATG CTT 3120

Gly Ala Glu Ser Ala Lys Leu Asn Asn Gln Ala Arg Phe Ile Leu Glu Lys Ile Asp Gly 1060
GGT GCT GAA TCT GCT AAA CTG AAT AAT CAG GCT CGC TTT ATC TTA GAG AAA ATA GAT GGC 3180

Lys Ile Ile Ile Glu Asn Lys Pro Lys Lys Glu Leu Ile Lys Val Leu Ile Gln Arg Gly 1080
AAA ATA ATC ATT GAA AAT AAG CCT AAG AAA GAA TTA ATT AAA GTT CTG ATT CAG AGG GGA 3240

Tyr Asp Ser Asp Pro Val Lys Ala Trp Lys Glu Ala Gln Gln Lys Val Pro Asp Glu Glu 1100
TAT GAT TCG GAT CCT GTG AAG GCC TGG AAA GAA GCC CAG CAA AAG GTT CCA GAT GAA GAA 3300

Glu Asn Glu Glu Ser Asp Asn Glu Lys Glu Thr Glu Lys Ser Asp Ser Val Thr Asp Ser 1120
GAA AAT GAA GAG AGT GAC AAC GAA AAG GAA ACT GAA AAG AGT GAC TCC GTA ACA GAT TCT 3360

Gly Pro Thr Phe Asn Tyr Leu Leu Asp Met Pro Leu Trp Tyr Leu Thr Lys Glu Lys Lys 1140
GGA CCA ACC TTC AAC TAT CTT CTT GAT ATG CCC CTT TGG TAT TTA ACC AAG GAA AAG AAA 3420

Asp Glu Leu Cys Arg Leu Arg Asn Glu Lys Glu Gln Glu Leu Asp Thr Leu Lys Arg Lys 1160
GAT GAA CTC TGC AGG CTA AGA AAT GAA AAA GAA CAA GAG CTG GAC ACA TTA AAA AGA AAG 3480

Ser Pro Ser Asp Leu Trp Lys Glu Asp Leu Ala Thr Phe Ile Glu Glu Leu Glu Ala Val 1180
AGT CCA TCA GAT TTG TGG AAA GAA GAC TTG GCT ACA TTT ATT GAA GAA TTG GAG GCT GTT 3540

Glu Ala Lys Glu Lys Gln Asp Glu Gln Val Gly Leu Pro Gly Lys Gly Gly Lys Ala Lys 1200
GAA GCC AAG GAA AAA CAA GAT GAA CAA GTC GGA CTT CCT GGG AAA GGG GGG AAG GCC AAG 3600

Gly Lys Lys Thr Gln Met Ala Glu Val Leu Pro Ser Pro Arg Gly Gln Arg Val Ile Pro 1220
GGG AAA AAA ACA CAA ATG GCT GAA GTT TTG CCT TCT CCG CGT GGT CAA AGA GTC ATT CCA 3660

Arg Ile Thr Ile Glu Met Lys Ala Glu Ala Glu Lys Lys Asn Lys Lys Lys Ile Lys Asn 1240
CGA ATA ACC ATA GAA ATG AAA GCA GAG GCA GAA AAG AAA AAT AAA AAG AAA ATT AAG AAT 3720

Glu Asn Thr Glu Gly Ser Pro Gln Glu Asp Gly Val Glu Leu Glu Gly Leu Lys Gln Arg 1260
GAA AAT ACT GAA GGA AGC CCT CAA GAA GAT GGT GTG GAA CTA GAA GGC CTA AAA CAA AGA 3780

Leu Glu Lys Lys Gln Lys Arg Glu Pro Gly Thr Lys Thr Lys Lys Gln Thr Thr Leu Ala 1280
TTA GAA AAG AAA CAG AAA AGA GAA CCA GGT ACA AAG ACA AAG AAA CAA ACT ACA TTG GCA 3840

Phe Lys Pro Ile Lys Lys Gly Lys Lys Arg Asn Pro Trp Pro Asp Ser Glu Ser Asp Arg 1300
TTT AAG CCA ATC AAA AAA GGA AAG AAG AGA AAT CCC TGG CCT GAT TCA GAA TCA GAT AGG 3900

Ser Ser Asp Glu Ser Asn Phe Asp Val Pro Pro Arg Glu Thr Glu Pro Arg Arg Ala Ala 1320
AGC AGT GAC GAA AGT AAT TTT GAT GTC CCT CCA CGA GAA ACA GAG CCA CGG AGA GCA GCA 3960

Thr Lys Thr Lys Phe Thr Met Asp Leu Asp Ser Asp Glu Asp Phe Ser Asp Phe Asp Glu 1340
ACA AAA ACA AAA TTC ACA ATG GAT TTG GAT TCA GAT GAA GAT TTC TCA GAT TTT GAT GAA 4020

Lys Thr Asp Asp Glu Asp Phe Val Pro Ser Asp Ala Ser Pro Pro Lys Thr Lys Thr Ser 1360
AAA ACT GAT GAT GAA GAT TTT GTC CCA TCA GAT GCT AGT CCA CCT AAG ACC AAA ACT TCC 4080

Pro Lys Leu Ser Asn Lys Glu Leu Lys Pro Gln Lys Ser Val Val Ser Asp Leu Glu Ala 1380
CCA AAA CTT AGT AAC AAA GAA CTG AAA CCA CAG AAA AGT GTC GTG TCA GAC CTT GAA GCT 4140

Asp Asp Val Lys Gly Ser Val Pro Leu Ser Ser Ser Pro Pro Ala Thr His Phe Pro Asp 1400
GAT GAT GTT AAG GGC AGT GTA CCA CTG TCT TCA AGC CCT CCT GCT ACA CAT TTC CCA GAT 4200

Glu Thr Glu Ile Thr Asn Pro Val Pro Lys Lys Asn Val Thr Val Lys Lys Thr Ala Ala 1420
GAA ACT GAA ATT ACA AAC CCA GTT CCT AAA AAG AAT GTG ACA GTG AAG AAG ACA GCA GCA 4260

```
Lys Ser Gln Ser Ser Thr Ser Thr Thr Gly Ala Lys Lys Arg Ala Ala Pro Lys Gly Thr  1440
AAA AGT CAG TCT TCC ACC TCC ACT ACC GGT GCC AAA AAA AGG GCT GCC CCA AAA GGA ACT  4320

Lys Arg Asp Pro Ala Leu Asn Ser Gly Val Ser Gln Lys Pro Asp Pro Ala Lys Thr Lys  1460
AAA AGG GAT CCA GCT TTG AAT TCT GGT GTC TCT CAA AAG CCT GAT CCT GCC AAA ACC AAG  4380

Asn Arg Arg Lys Arg Lys Pro Ser Thr Ser Asp Asp Ser Asp Ser Asn Phe Glu Lys Ile  1480
AAT CGC CGC AAA AGG AAG CCA TCC ACT TCT GAT GAT TCT GAC TCT AAT TTT GAG AAA ATT  4440

Val Ser Lys Ala Val Thr Ser Lys Lys Ser Lys Gly Glu Ser Asp Asp Phe His Met Asp  1500
GTT TCG AAA GCA GTC ACA AGC AAG AAA TCC AAG GGG GAG AGT GAT GAC TTC CAT ATG GAC  4500

Phe Asp Ser Ala Val Ala Pro Arg Ala Lys Ser Val Arg Ala Lys Lys Pro Ile Lys Tyr  1520
TTT GAC TCA GCT GTG GCT CCT CGG GCA AAA TCT GTA CGG GCA AAG AAA CCT ATA AAG TAC  4560

Leu Glu Glu Ser Asp Glu Asp Asp Leu Phe  1530
CTG GAA GAG TCA GAT GAA GAT GAT CTG TTT  4590
```

Subject Index